AF411429

Artificial Intelligence in Medical Imaging: The Beginning of a New Era

Artificial Intelligence in Medical Imaging: The Beginning of a New Era

Editor

Cosimo Nardi

Basel • Beijing • Wuhan • Barcelona • Belgrade • Novi Sad • Cluj • Manchester

Editor
Cosimo Nardi
University of Florence
Florence
Italy

Editorial Office
MDPI
St. Alban-Anlage 66
4052 Basel, Switzerland

This is a reprint of articles from the Special Issue published online in the open access journal *Applied Sciences* (ISSN 2076-3417) (available at: https://www.mdpi.com/journal/applsci/special_issues/AI_Medical_Imaging_New_Era).

For citation purposes, cite each article independently as indicated on the article page online and as indicated below:

Lastname, A.A.; Lastname, B.B. Article Title. *Journal Name* **Year**, *Volume Number*, Page Range.

ISBN 978-3-0365-9484-2 (Hbk)
ISBN 978-3-0365-9485-9 (PDF)
doi.org/10.3390/books978-3-0365-9485-9

Contents

About the Editor

Cosimo Nardi

Cosimo Nardi is an assistant professor in the Department of Biomedical, Experimental and Clinical Sciences at the University of Florence, Italy. He is a medical doctor specialized in radiology who mainly deals with head and neck disease, radiation doses, cone beam computed tomography, and artificial intelligence related to images. He is author of more than 70 peer-reviewed scientific papers, most of which are dedicated to maxillofacial disease. He acted as principal investigator for University of Florence regarding the project DRAGON "Rapid and secure AI imaging based diagnosis, stratification, follow-up and preparedness for coronavirus pandemics" as part of the Horizon 2020 European funding Programme. He is a council member of both the PhD's College of Clinical Sciences of the University of Florence and Head and Neck section of the Italian Society of Radiology. He is also the director of the Master on MRI applications and techniques at the University of Florence.

Preface

Many applications have started using artificial intelligence. But what is artificial intelligence? How can you use it in the medical field? This Special Issue will provide you with the answers you were looking for, emphasizing the strength of this innovative tool and focusing on its application in the medical field.

Cosimo Nardi
Editor

applied
sciences

Editorial

Special Issue on Artificial Intelligence in Medical Imaging: The Beginning of a New Era

Cosimo Nardi

Department of Experimental and Clinical Biomedical Sciences "Mario Serio", University of Florence, Viale Morgagni 50, 50134 Florence, Italy; cosimo.nardi@unifi.it

Citation: Nardi, C. Special Issue on Artificial Intelligence in Medical Imaging: The Beginning of a New Era. *Appl. Sci.* **2023**, *13*, 11562. https://doi.org/10.3390/app132011562

Received: 10 October 2023
Accepted: 17 October 2023
Published: 23 October 2023

Artificial intelligence (AI) can be considered the real revolution of the 21st century. This approach is increasingly used in everyday life and is also expanding in the medical field. Clinical practice has always focused on the evaluation of radiological parameters obtained from a few data [1]. AI introduces a change, as it requires a large dataset to work, but in turn avoids user-based evaluation. This aspect can be especially helpful in diagnostics, in which it is becoming very popular [2,3]. AI is finding wide application in the tumor field and for skin cancer [4], being able to detect variations in the image that would not be visible only with human eyes. Its application is increasingly widespread, even in neuroimaging analysis [5].

AI is based on the use of models that need to be implemented. Choosing the correct model is an important aspect as some models may work better than others, even according to the right training and testing sets [6,7]. Another important aspect, especially in diagnostics, is image segmentation, because mis-segmentation could introduce errors that can lead to misinterpretations. For this reason, AI itself is under study to create systems that implement automatic segmentation without errors [8]. Spatial resolution is one of the main problems when it comes to studying images. That is why new systems have been created to display images in super resolution. The problem is that this requires large storage memories and expensive calculations that have led to the study of a new "light" system to create images with super resolution (and less heavy) to be used in medicine [9]. The problem of memory and computational costs is quite recurrent considering AI; in fact, new methods are under study to select the most characteristic features and eliminate redundant ones. The problem is that in medicine it is not easy to eliminate some features even if redundant as they can diversify one disease from another [10]. Another approach may lie in the creation of small subnets that contain specific characteristics indicative of particular diseases [11].

The large number of features extracted is a problem, but so is the size of the dataset. In fact, in medicine, very often the dataset is not enough to train a model. That is why data augmentation was introduced to generate larger datasets from available data [12].

However, even today the biggest problem related to AI in medicine is the concept of the "black box". Doctors do not trust this approach at all because they do not know what happens within the system that provides output to a given input. This is why explainable AI is spreading lately, a discipline that tries to explain what happens within the systems of AI [13].

At the same time, other innovative systems are spreading such as augmented reality, which is an immersive technology that together with AI could be used to improve the performance of surgeons during interventions [14,15]. That is why different approaches to superimpose the images created in specific parts of the body and their accurate representations are being studied [16]. In addition, AI can also be used in a new emerging discipline: radiomics. This discipline allows us to extrapolate information by evaluating the voxels themselves, their arrangement and the relationships that exist. Along with AI, radiomics can create a system that takes an incoming image and evaluating changes in characteristics can provide a diagnosis or prognosis [17]. AI is becoming increasingly used as new approaches develop. For example, active learning ensures achieving great performance by using as few high-quality sample annotations as possible [18].

Finally, AI can be used in fields related to medicine introducing social benefits. One example is the possibility of studying nonvoluntary facial microexpressions. This can be used in the field of safety, psychology and medicine [19].

Much remains to be accomplished to replace humans in the medical field, although the introduction of AI has begun to bring many benefits, especially as a support system.

Conflicts of Interest: The author declares no conflict of interest.

References

1. Pietragalla, M.; Bicci, E.; Calistri, L.; Lorini, C.; Bonomo, P.; Borghesi, A.; Lo Casto, A.; Mungai, F.; Bonasera, L.; Maggiore, G.; et al. Magnetic Resonance with Diffusion and Dynamic Perfusion-Weighted Imaging in the Assessment of Early Chemoradiotherapy Response of Naso-Oropharyngeal Carcinoma. *Appl. Sci.* **2023**, *13*, 2799. [CrossRef]
2. Gilberg, L.; Teodorescu, B.; Maerkisch, L.; Baumgart, A.; Ramaesh, R.; Gomes Ataide, E.J.; Koç, A.M. Deep Learning Enhances Radiologists' Detection of Potential Spinal Malignancies in CT Scans. *Appl. Sci.* **2023**, *13*, 8140. [CrossRef]
3. Huynh, H.N.; Tran, A.T.; Tran, T.N. Region-of-Interest Optimization for Deep-Learning-Based Breast Cancer Detection in Mammograms. *Appl. Sci.* **2023**, *13*, 6894. [CrossRef]
4. Mukadam, S.B.; Patil, H.Y. Skin Cancer Classification Framework Using Enhanced Super Resolution Generative Adversarial Network and Custom Convolutional Neural Network. *Appl. Sci.* **2023**, *13*, 1210. [CrossRef]
5. Hu, M.; Nardi, C.; Zhang, H.; Ang, K.-K. Applications of Deep Learning to Neurodevelopment in Pediatric Imaging: Achievements and Challenges. *Appl. Sci.* **2023**, *13*, 2302. [CrossRef]
6. Hassen Mohammed, H.; Elharrouss, O.; Ottakath, N.; Al-Maadeed, S.; Chowdhury, M.E.H.; Bouridane, A.; Zughaier, S.M. Ultrasound Intima-Media Complex (IMC) Segmentation Using Deep Learning Models. *Appl. Sci.* **2023**, *13*, 4821. [CrossRef]
7. Dang Nguyen, N.A.; Huynh, H.N.; Tran, T.N. Improvement of the Performance of Scattering Suppression and Absorbing Structure Depth Estimation on Transillumination Image by Deep Learning. *Appl. Sci.* **2023**, *13*, 10047. [CrossRef]
8. Benedetti, P.; Femminella, M.; Reali, G. Mixed-Sized Biomedical Image Segmentation Based on U-Net Architectures. *Appl. Sci.* **2023**, *13*, 329. [CrossRef]
9. Wang, B.; Yan, B.; Jeon, G.; Yang, X.; Liu, C.; Zhang, Z. Lightweight Dual Mutual-Feedback Network for Artificial Intelligence in Medical Image Super-Resolution. *Appl. Sci.* **2022**, *12*, 12794. [CrossRef]
10. Nadimi-Shahraki, M.H.; Asghari Varzaneh, Z.; Zamani, H.; Mirjalili, S. Binary Starling Murmuration Optimizer Algorithm to Select Effective Features from Medical Data. *Appl. Sci.* **2023**, *13*, 564. [CrossRef]
11. Ali, R.; Hardie, R.C.; Narayanan, B.N.; Kebede, T.M. IMNets: Deep Learning Using an Incremental Modular Network Synthesis Approach for Medical Imaging Applications. *Appl. Sci.* **2022**, *12*, 5500. [CrossRef]
12. Yu, X.; Wang, S.; Hu, J. Guided Random Mask: Adaptively Regularizing Deep Neural Networks for Medical Image Analysis by Potential Lesions. *Appl. Sci.* **2022**, *12*, 9099. [CrossRef]
13. Alghamdi, H.S. Towards Explainable Deep Neural Networks for the Automatic Detection of Diabetic Retinopathy. *Appl. Sci.* **2022**, *12*, 9435. [CrossRef]
14. Franzò, M.; Pica, A.; Pascucci, S.; Marinozzi, F.; Bini, F. Hybrid System Mixed Reality and Marker-Less Motion Tracking for Sports Rehabilitation of Martial Arts Athletes. *Appl. Sci.* **2023**, *13*, 2587. [CrossRef]
15. Barcali, E.; Iadanza, E.; Manetti, L.; Francia, P.; Nardi, C.; Bocchi, L. Augmented Reality in Surgery: A Scoping Review. *Appl. Sci.* **2022**, *12*, 6890. [CrossRef]
16. Simoni, A.; Barcali, E.; Lorenzetto, C.; Tiribilli, E.; Rastrelli, V.; Manetti, L.; Nardi, C.; Iadanza, E.; Bocchi, L. Innovative Tool for Automatic Detection of Arterial Stenosis on Cone Beam Computed Tomography. *Appl. Sci.* **2023**, *13*, 805. [CrossRef]
17. Calamandrei, L.; Mariotti, L.; Bicci, E.; Calistri, L.; Barcali, E.; Orlandi, M.; Landini, N.; Mungai, F.; Bonasera, L.; Bonomo, P.; et al. Morphological, Functional and Texture Analysis Magnetic Resonance Imaging Features in the Assessment of Radiotherapy-Induced Xerostomia in Oropharyngeal Cancer. *Appl. Sci.* **2023**, *13*, 810. [CrossRef]
18. Wu, M.; Li, C.; Yao, Z. Deep Active Learning for Computer Vision Tasks: Methodologies, Applications, and Challenges. *Appl. Sci.* **2022**, *12*, 8103. [CrossRef]
19. Talluri, K.K.; Fiedler, M.-A.; Al-Hamadi, A. Deep 3D Convolutional Neural Network for Facial Micro-Expression Analysis from Video Images. *Appl. Sci.* **2022**, *12*, 11078. [CrossRef]

Article

IMNets: Deep Learning Using an Incremental Modular Network Synthesis Approach for Medical Imaging Applications

Redha Ali [1,*], Russell C. Hardie [1], Barath Narayanan Narayanan [1,2] and Temesguen M. Kebede [1]

[1] Department of Electrical and Computer Engineering, University of Dayton, 300 College Park, Dayton, OH 45469, USA; rhardie@udayton.edu (R.C.H.); narayananb1@udayton.edu (B.N.N.); tmessay1@udayton.edu (T.M.K.)

[2] Sensors and Software Systems Division, University of Dayton Research Institute, 1700 South Patterson Blvd., Dayton, OH 45409, USA

* Correspondence: almahdir1@udayton.edu

Abstract: Deep learning approaches play a crucial role in computer-aided diagnosis systems to support clinical decision-making. However, developing such automated solutions is challenging due to the limited availability of annotated medical data. In this study, we proposed a novel and computationally efficient deep learning approach to leverage small data for learning generalizable and domain invariant representations in different medical imaging applications such as malaria, diabetic retinopathy, and tuberculosis. We refer to our approach as Incremental Modular Network Synthesis (IMNS), and the resulting CNNs as Incremental Modular Networks (IMNets). Our IMNS approach is to use small network modules that we call SubNets which are capable of generating salient features for a particular problem. Then, we build up ever larger and more powerful networks by combining these SubNets in different configurations. At each stage, only one new SubNet module undergoes learning updates. This reduces the computational resource requirements for training and aids in network optimization. We compare IMNets against classic and state-of-the-art deep learning architectures such as AlexNet, ResNet-50, Inception v3, DenseNet-201, and NasNet for the various experiments conducted in this study. Our proposed IMNS design leads to high average classification accuracies of 97.0%, 97.9%, and 88.6% for malaria, diabetic retinopathy, and tuberculosis, respectively. Our modular design for deep learning achieves the state-of-the-art performance in the scenarios tested. The IMNets produced here have a relatively low computational complexity compared to traditional deep learning architectures. The largest IMNet tested here has 0.95 M of the learnable parameters and 0.08 G of the floating-point multiply–add (MAdd) operations. The simpler IMNets train faster, have lower memory requirements, and process images faster than the benchmark methods tested.

Keywords: medical imaging; deep learning; malaria detection; diabetic retinopathy; tuberculosis detection; modular networks

Citation: Ali, R.; Hardie, R.C.; Narayanan, B.N.; Kebede, T.M. IMNets: Deep Learning Using an Incremental Modular Network Synthesis Approach for Medical Imaging Applications. *Appl. Sci.* **2022**, *12*, 5500. https://doi.org/10.3390/app12115500

Academic Editor: Jan Egger

Received: 23 April 2022
Accepted: 27 May 2022
Published: 29 May 2022

Publisher's Note: MDPI stays neutral with regard to jurisdictional claims in published maps and institutional affiliations.

1. Introduction

1.1. Background

Recently, deep learning with convolutional neural networks (CNNs) has proven to be highly effective for computer-aided detection (CAD) in medical image analysis. The trend in CNN architectures recently has been towards ever deeper and wider networks with dense connectivity. For example, ViT-G/14 [1] and ViT-MoE-15B [2] were the top two CNN architectures in the ImageNet Large-Scale Visual Recognition Challenge (ILSVRC) competition in 2021 [3]. The ViT-G/14 and ViT-MoE-15B architectures contain 1.843 G and 14.70 G parameters, respectively. Furthermore, ViT-G/14 requires 965.3 G floating point operations (FLOPs) per single image, which is a very computationally costly and power-hungry solution. Perhaps even more significantly, larger networks require more training data to be able to generalize to new data [4]. In many medical image analysis applications, access to properly-labeled truth imagery is limited, especially for rare diseases [4]. Data collection and truthing in medical imaging can be cost-intensive, time-consuming, and requires expert analysis. Transfer learning can help to reduce the amount of application-specific data required for training. However, large amounts of data may still be needed to obtain the desired reproducibility and generalizability, even with transfer learning [5,6]. Data augmentation is another approach to dealing with limited training data. However, data augmentation can be very challenging in some medical imaging modalities such as chest radiographs [7].

Modular CNN architectures are a promising approach for complex problem-solving that may be able to help address the challenges described above. Some modular methods are inspired by the structure and function of the human brain. Recent findings in neuroscience reveal a high level of modularity and hierarchy of neural structure in the human brain [8]. In the early 1980s, neuroscientific research categorized the central nervous system (CNS) in the human brain as a massively parallel and self-organizing modular system [9–11]. The CNS consists of distinctive regions. Each region develops as a functional module. The modules are densely connected and interact with one another to accomplish complex perception and cognitive tasks in an efficient manner [9]. Traditional CNN architectures often use repeating structures such as layers or groups of layers. However, the networks are generally trained as one monolithic entity with all learnable parameters being updated simultaneously.

1.2. Applications

Malaria is a deadly disease that is considered endemic in many countries around the world [12]. In the year 2020, the World Health Organization (WHO) reported an estimated 229 million cases of malaria worldwide, which caused an estimated 409,000 deaths [13]. Malaria occurs in humans via protozoa within the blood cells of the genus Plasmodium. These parasites are transmitted by the bite of a female Anopheles mosquito [14]. The mosquito bite injects the Plasmodium into the affected person's blood, and then the Plasmodium parasites pass quickly to the liver to mature and replicate [15]. The most common imaging modality for detecting parasites in a thin blood smear sample is microscopical imaging [16]. While microscopy is relatively low-cost and widely accessible, diagnosis efficiency depends on the experience of parasitologists [17]. False-positive or false-negative diagnoses can lead to inappropriate or unnecessary prescriptions that can cause side effects in patients. Due to the global shortage of parasitologists in impoverished urban areas accurately processing the large number of specimens encountered is not always possible [18,19]. Thus, CAD systems can be highly beneficial in this application.

Another disease for which CAD systems can help by providing accurate early detection is diabetic retinopathy (DR). This condition is a typical development of diabetes, affecting the retina's small blood vessels, leading to vision deterioration [20]. The research described in [21] studies the offloading footwear to prevent and lower mortality rates in high-risk diabetic feet. A recent study has reported that DR affects the vision of 2.6 million people in the world [20,22]. Several retinal imaging systems can be utilized to detect the indication of diabetic retinopathy, including color fundus photography, fluorescein angiography, B-scan ultrasonography, and optical coherence tomography [23]. The retina images that we use in our study have been captured using fundus photography under a variety of imaging conditions. Early-stage diagnosis of DR grading is integral to prevent the occurrence of blindness. Hence, CAD systems could help save millions of people from potentially preventable vision loss and blindness by improving early detection.

The third and final application we consider here is pulmonary tuberculosis (TB). This disease is a significant public health issue causing more than 9 million expected new cases and roughly 1.4 million deaths every year [24]. The detection of TB on chest radiographs (CRs) is essential for diagnosing TB. Chest radiography imaging (e.g., X-ray or computed tomography (CT) imaging) is easy to perform with fast diagnosis and has a high sensitivity for diagnosing TB infection. However, CRs are the fastest and most affordable form of imaging and require significantly less radiation, data memory, and processing time than CT scans [25]. The WHO recommends using CRs to screen and triage people for TB [26]. Note that CRs are among the first procedures of examination related to suspects' lung disease. They are low-cost and widely accessible for health care providers. The use of CAD systems for TB detection can help radiologist workflow so they may be able to process more cases with greater accuracy.

1.3. Related Works

Various approaches have been proposed in the literature for medical imaging CAD systems, including those for malaria, DR and TB [27–43]. The work described in [28] aims to improve malaria parasite detection using tiny red blood smear patches; they utilize several existing deep convolutional neural networks in place of handcrafted feature extraction. The study claims that using preprocessing techniques such as standardization, normalization, and stain normalization does not improve the overall performance model. An effective multi-magnification deep residual neural network (MM-ResNet) has been trained on microscopic phone image datasets of malaria blood smears obtained from the AI research group at Makerere University. The MM-ResNet-50 end-to-end framework takes three different images of size as inputs. It concatenates each ResNet-50 at the second to the last layer, followed by a final fully connected layer [29]. VGG-16 and VGG-19 have been trained on the National Institutes of Health (NIH) malaria dataset using hyperparameter tuning techniques described in [30], the CNN used to automate the screening of malaria in low-resource countries achieves an accuracy of 0.9600. A survey article on image analysis of microscopic blood slides uses many machine learning techniques for malaria detection [31]. Patient information was considered, such as nationality, age, gender, body region, and symptomatology of a patient as a part of features engineering for malaria detection. Furthermore, they examined six machine learning algorithms, including support vector machine (SVM), random forest (RF), multilayer perceptron (MLP), AdaBoost, gradient boosting (GB), and CatBoost to classify infected and non-infected cells [32]. A fast CNN architecture present in [33] is used to classify thin blood smeary images. This paper studied the performance of transfer learning approaches for various pre-trained CNN architectures, including AlexNet, ResNet-50, VGG-16, and DenseNet-201. Furthermore, they studied the performance of a traditional machine learning algorithm using a bag-of-features model with SVM.

Many deep learning methods, originally proposed for the ILSVRC [44], have been adapted to the medical image application. Among these are meta-algorithms for DR detection, which combine five CNN architectures into one predictive model [34]. Zhang et al. [35] fine-tuned ResNet-50 that pre-trained on the ImageNet dataset. The work described in [36] developed a real-time smartphone app to detect and classify DR by using a pre-trained Inception v3 model with a transfer learning technique. A hybrid machine learning technique is introduced in [37] to detect and grade DR severity level. The study compares simple transfer learning-based approaches using seven pre-trained networks. Another fine-tuned, pre-trained approach for DR detection is presented in [38] using a cosine annealing strategy to decay the learning rate. The transfer learning method for TB described in [39] was used to neutrophil cluster detection. An automatic TB screening system presented in [40] is based on transfer learning from lower convolutional layers of pre-trained networks. The method in [42] uses a simple segmentation approach to classify the images' foreground and background. The segmented objects are then fed to a trained CNN to classify the objects into bacilli and non-bacilli. A total of four state-of-the-art 3D CNN models are used to detect the spatial location of lesions and classify the candidates into miliary, infiltrative, caseous, tuberculoma, and cavitary types in [43]. A multi-strategy fast non-dominated solution ranking algorithm with high robustness is described in [45].

Of particular relevance to our work is the Net2Net method introduced by Chen et al. [46] The method is modular in that it allows two neural networks to mimic the behavior of a more complex network. The Net2Net is an effective technique to transfer the prior knowledge from a trained neural network (teacher network) to a new deeper, or wider network (student network). The Net2Net approach implemented in [46] combines two neural networks to form a larger network. It does so by either increasing the width or the depth of the network. The method replicates the teacher network weights to expand the student network size either in width or depth. After replicating, the new addition is initialized to be an identity network. This method can guarantee that the student model can perform just as well as the teacher network at the start of training. The student model obtains good accuracy much faster than training the larger network from scratch. While Net2Net is a practical and innovative approach that works very well for knowledge transfer, the Net2Net method has a few limitations. For example, the current implementation of Net2Net in [46] uses only two networks. Furthermore, there are restrictions on the networks in terms of kernel sizes, activation functions, and initialization, so as to achieve the stated network properties.

Another related modular technique that is designed to work with small amounts of training data is presented in [47]. The module uses the entire CNN network as modules. It combines pre-trained modules with untrained modules, allowing the new network to learn discriminative features. The pre-trained models VGG-16 and ResNet-50 were used. The module fine-tunes the VGG-16 model on the Stanford Cars dataset by replacing the last three layers with two consecutive fully connected layers, softmax, and loss function. Then, the module merges the fixed VGG16 features with a ResNet-50. The output of both models was then fed to two fully connected layers, softmax, and loss function.

1.4. Contributions

In this paper, we propose a novel and a computationally efficient deep learning approach for medical image analysis using CNNs. We refer to our approach as Incremental Modular Network Synthesis (IMNS), and the resulting CNNs as Incremental Modular Networks (IMNets). Our IMNS approach is to use small network modules that we call SubNets that are capable of generating salient features for a particular problem. Compared with other modular methods in the literature, our IMNS approach has some distinct features. First, we begin with small compact SubNet modules to keep the computational complexity low. Second, we build networks using both series and parallel arrangements in a sequential incremental manner. This provides freedom of building nearly any custom network without restriction. The essential feature of our approach is that we start by training one small SubNet and lock in those network parameters. We add depth or width to that initial network and train only the new SubNet at a time. We do this incrementally until we achieve the desired network performance. Our approach guarantees the freedom of choosing any configuration for the initial network, including the number of layers, the kernel size, series network incremental or parallel network incremental. To the best of our knowledge this kind of modular network synthesis approach has not been previously employed in medical image CAD applications.

1.5. Paper Organization

The remainder of the paper is organized as follows. A description of the datasets used is presented in Section 2. In Section 3, we describe the proposed IMNS method and resulting IMNets. Section 4 presents the experimental results. Finally, we offer discussion and conclusions in Section 5.

2. Materials

In this paper, we utilized three different datasets to study the performance of our method. First, we utilized a publicly available dataset provided by the NIH [48] for malaria detection. Second, we used the publicly available Asia Pacific Tele-Ophthalmology Society (APTOS) 2019 blindness detection challenge dataset [49] for DR detection. Lastly, for TB detection, we make use of a publicly available Shenzhen chest radiograph dataset [50].

2.1. Malaria Dataset

Malaria dataset provided segmented cell samples that have been obtained from the thin blood smear slide images from the Malaria Screener research activity [48]. According to the NIH, all images were manually labeled by a proficient slide reader at the Mahidol-Oxford Tropical Medicine Research Unit in Bangkok, Thailand [48]. The dataset comprises 27,558 cell images with equal representation of parasitized and uninfected cells. We randomly divided the dataset into 80% for training and 20% for testing representations regarding each class. Moreover, we split the training dataset into 90% and 10% for training and validation sets. Table 1 shows the hold-out validation distribution of the malaria dataset and the number of training, validation, and testing samples. Figure 1 shows the raw sample, which tends to have different illumination conditions. Therefore, we pre-processed all images by applying the color constancy technique [51] to ensure the perceived color of each image remained the same under different illumination conditions. Results of the color constancy outputs for the input images in Figure 1 are shown in Figure 2.

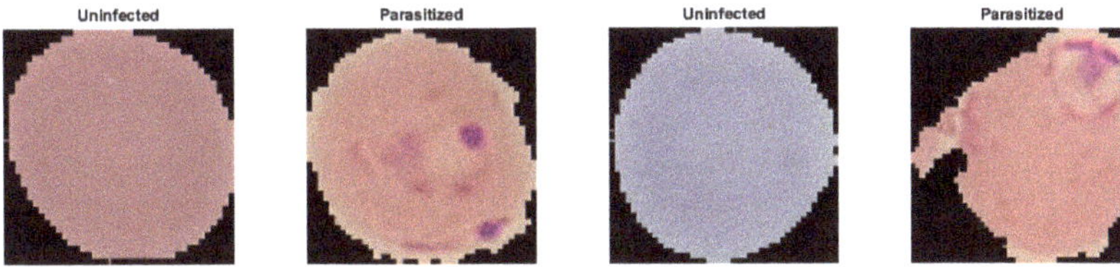

Figure 1. Raw parasitized and uninfected sample images for malaria detection labeled by expert slide readers.

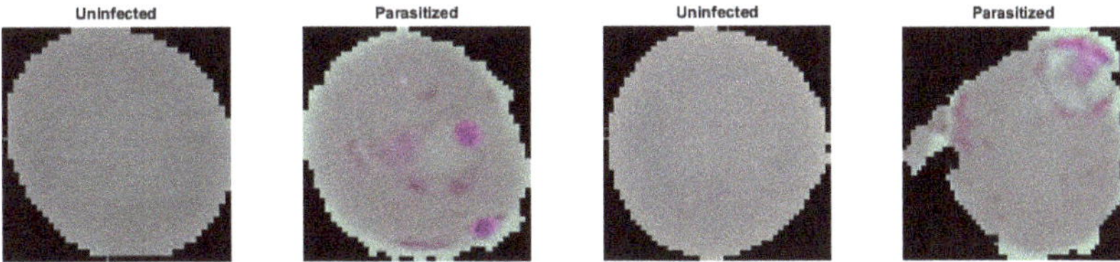

Figure 2. Malaria detection images from Figure 1 after color constancy processing.

Table 1. The hold-out validation distribution of the data source for each application and the number of training, validation, and testing cases.

Applications Datasets	Malaria	Diabetic Retinopathy	Tuberculosis
Images size	112×112	299×299	299×299
No. of training set	19842	2637	477
No. of validation set	2204	293	53
No. of testing set	5512	732	132

2.2. Diabetic Retinopathy Dataset

The technicians in the Aravind Eye Hospital in India have collected retinal images from patients who live in rural areas aiming to detect and prevent diabetic retinopathy [49]. Trained doctors then reviewed these images to provide the diagnosis. This APTOS 2019 dataset consists of 5590 retinal image samples. The dataset has been split up into training and testing cases by the challenge host organization. The training dataset is comprised of 3662 samples. The testing dataset contains 1928 samples, but the labels for the testing dataset are not publicly available yet. The dataset contains five classes, including No DR and the other four stages of DR (Mild DR, Moderate DR, Proliferative DR, and Severe DR). In this study, we grouped the dataset into two possible disease categories, normal and DR classes. The four types of DR diseases have been grouped together in the DR class. Moreover, since the testing dataset labels are not available, we solely used the training dataset provided as part of the APTOS 2019 challenge. The training dataset was randomly split into 80% for training and 20% for testing. Then, the training dataset is divided into 90% and 10% for training and validation sets. Table 1 shows the number of training, validation, and testing samples for DR dataset. Figure 3 shows random samples of labeled images from the APTOS 2019 DR dataset after we grouped them into two classes.

We have studied and visualized the dataset, and we found that the images contain artifacts, varying sizes, different optic nerve angles, and were captured under different lighting conditions so that some are underexposed or overexposed. To handle this variability, we propose applying pre-processing techniques to seek to normalize the data for these factors. The eye image pre-processing technique consists of four steps:

1. We find the mask of the orange portion of the eye and separate it from the black background.
2. We locate the optic nerve that appears as a bright disk in the images. This is achieved by applying a Gaussian low-pass filter with a spatial standard deviation approximately equal to the radius of the optic nerve disk. The brightest pixel after the blurring operation generally is located near the center of the optic nerve.
3. We compare the location of the optic nerve center to the center of the eye mask to determine the orientation of the eye. We then rotate the image so that optic nerve is consistently on the right of center in the resulting image.
4. Finally, we crop, zero pad, and interpolate to obtain the same size images. We do so in such a way as to not change the aspect ratio of image, as this would contaminate the geometric integrity of the data.

This simple pre-processing technique renders the retinal images in the database more uniform and allows the CAD system to achieve improved performance. Examples of the retinal images from Figure 3 after implementing the pre-processing steps described above are shown in Figure 4.

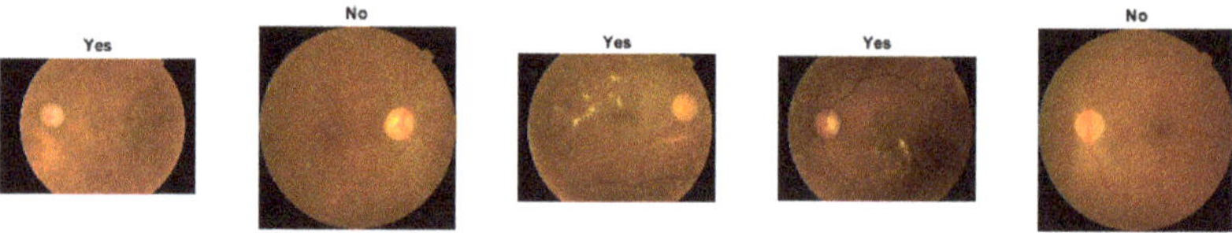

Figure 3. Raw retinal images of a healthy retina (normal class) and DR damage blood vessels in the retina (DR class).

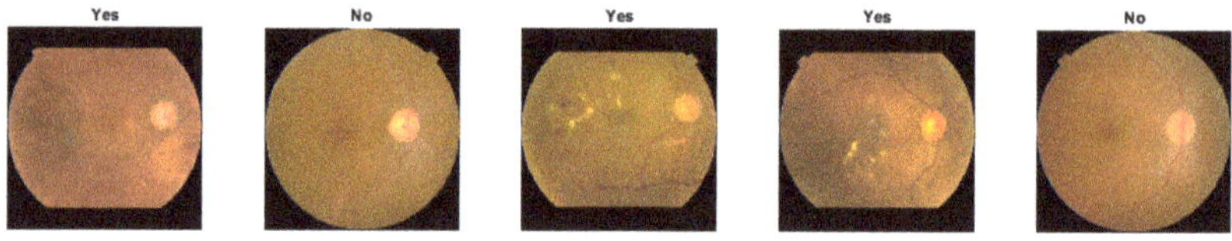

Figure 4. Retinal images from Figure 3 after applying the proposed pre-processing steps to normalize the images in the database.

2.3. Tuberculosis Dataset

We utilized the Shenzhen dataset [50] for TB detection that holds 326 normal CR cases and 326 CR with active pulmonary tuberculosis. The chest radiograph images in the Shenzhen dataset have been collected by Shenzhen No. 3 Hospital in Shenzhen,

Guangdong province, China. In our experimental study using these data, we perform a hold-out validation. We randomly divide the dataset into groups of 72% for training, 8% for validation, and 20% for testing. Table 1 shows the hold-out validation distribution of the TB dataset.

The CR images are in JPEG format with a resolution of 3000×3000 pixels. Some example labeled CR images from the Shenzhen dataset are shown in Figure 5. Figure 5 shows that some CR samples in the dataset have an inverse intensity polarity. We find that it is critical to network performance to make all of the CR images have the same polarity. Therefore, all cases are reviewed manually and inverted as needed. For this research, we converted all the images to a size of 299×299. The example images Figure 5 after the corrective inversion processing are shown in Figure 6.

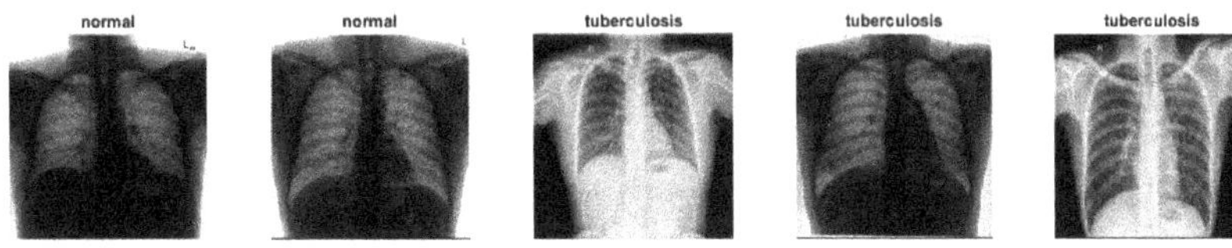

Figure 5. Chest radiograph samples from the Shenzhen dataset labeled by radiologists as normal and tuberculosis. Starting from the left, the first, second, and fourth chest radiograph images have inverse polarity.

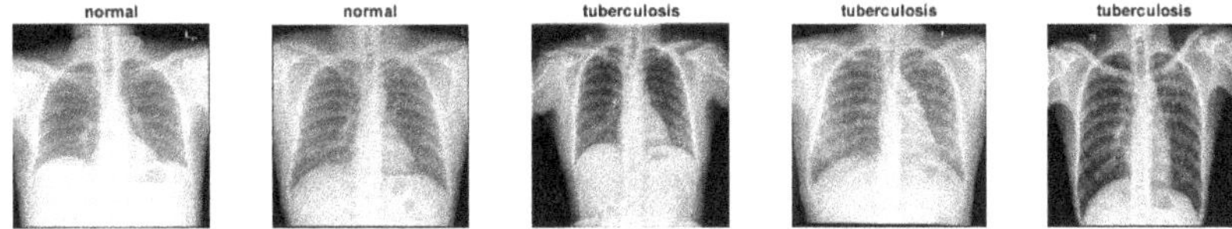

Figure 6. Preprocessed chest radiograph images from the Shenzhen dataset to provide polarity uniformity.

3. Methods

In this section, describe the details of the proposed IMNS method. We begin with an overview. Next, we explain the details of each SubNet and how they work together. Then, we present the specific IMNet architecture used in our experimental study. Finally, we end this section with a discussion of our network training process.

3.1. Overview

The inspiration for the IMNS approach comes from children's building blocks. We propose that CNN architectures can be assembled using modular components in a manner that is akin to building a structure with a child's building blocks. Each module requires only an incremental additional training process. This allows for a potentially massive network without the computational cost of training the final network at one time, which could be prohibitive. The proposed IMNS uses a unique hybrid learning strategy that successfully combines multiple SubNet to produce complementary information.

In our approach, each SubNets module is added incrementally onto existing architecture in either a series or parallel fashion. These two scenarios are illustrated in Figure 7. Note that in Figure 7a, a new SubNet is added in series to the feature computation layers of the current IMNet. The classification layers are moved to the end of the network as shown. Note also that the learnable parameters of the current IMNet are locked-in, and only the learnable parameters of the new SubNet are updated. For large networks, this dramatically reduces the computational demands of the back-propagation updates. At some stages of the IMNS process, the user may wish to expand the network in parallel. This is shown in Figure 7b. As before the classification layers are moved to the end, and only the new

Subnet is updated in the back-propagation learning algorithm. One new operation that is needed here is the concatenation layer that takes the feature maps generated from the current IMNet and concatenates them with the feature maps generated by the new SubNet. We concatenate these feature maps in the channel dimension.

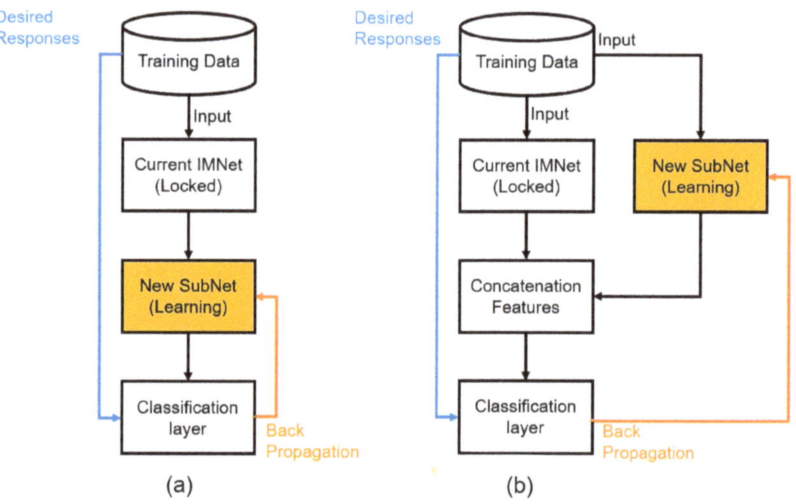

Figure 7. Illustration of the IMNS workflow for building IMNets. (**a**) Addition of a series SubNet, (**b**) addition of a parallel SubNet.

3.2. SubNet Architecture

The individual SubNet architectures considered here are shown in Figure 8. The feature generating SubNets are comprised of a selected number of the layer groups shown in Figure 8a. The classification layers are shown in Figure 8b. Figure 8a shows the convolutional layer structure where each convolutional layer followed by a batch normalization layer, rectified linear units (ReLU), and max pooling of window size 2×2 with a stride of 2 to downsample the feature maps. Note that the number and size of convolution filters present in each layer may differ. The classification block consists of one fully connected layer, softmax function, and cross-entropy loss function as shown in Figure 8b.

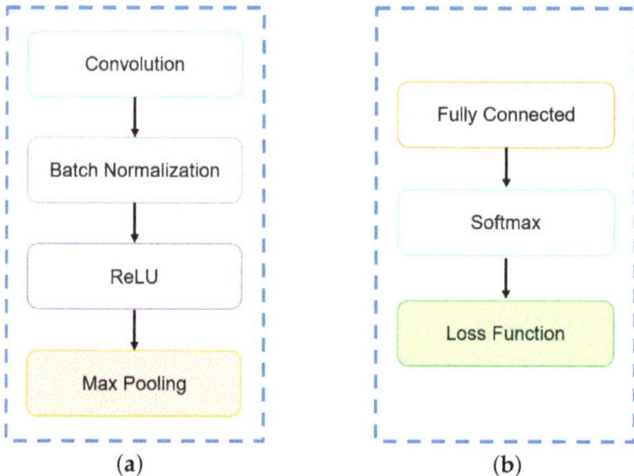

Figure 8. (**a**) Convolutional layer structure. (**b**) Classification layer structure.

Let us formally define the output of a SubNet made up of $L-1$ layer groups such as those shown in Figure 8a followed by an L'th classification layer as shown in Figure 8b. To begin, let us define one minibatch of input data as

$$\mathbf{X} = \{\mathbf{X}_1, \mathbf{X}_2, \ldots, \mathbf{X}_N\}, \tag{1}$$

where $\mathbf{X}_n \in \mathbb{R}^{H \times W \times D}$ is the n'th exemplar from the minibatch. These inputs represent potentially multi-channel images with H rows, W columns, and a channel depth of D. Consider the case of classification with M distinct classes. Let the truth for each exemplar be denoted as $\mathbf{y}_n = [y_{n,1}, y_{n,2}, \ldots, y_{n,M}]^T \in \mathbb{R}^M$, for $n = 1, 2, \ldots, N$.

Let us define the n'th exemplar, $\mathbf{X}_n$, as the input to Layer group 1 of the network. Let this be represented in lexicographical notation as the $HWD \times 1$ vector $\mathbf{x}_n^0$. Note that this is formed by reshaping the $3D$ data-cube in $\mathbf{X}_n$ into a column vector. The output of each convolutional layer group shown in Figure 8a can be expressed as

$$\mathbf{x}_n^l = g(\mathbf{W}^l \mathbf{x}_n^{l-1} + \mathbf{b}^l), \tag{2}$$

for layer group $l = 1, 2, \ldots, L-1$ and exemplar $n = 1, 2, \ldots, N$ within the minibatch. The weights of all of the convolution kernels for layer group l are represented in the weight matrix $\mathbf{W}^l$. The dimensions of $\mathbf{W}^1$ are $HWN_f^1 \times HWD$ where N_f^1 is the number of filters in layer group $l = 1$. The dimensions are reduced in subsequent layer groups due to the max pooling layers employed. Bias terms are represented in the vector $\mathbf{b}^l$. Note that $\mathbf{x}_n^l$ is the output $3D$ feature map cube of the current layer l in lexicographical form as a vector.

The ReLU and max pooling layers illustrated in Figure 8a are jointly represented with the nested function

$$g(\mathbf{x}) = \text{MaxPool}(\text{Max}(\mathbf{0}, \mathbf{x})). \tag{3}$$

The maximum of each element and 0 provides the ReLU operation. The ReLU activation function $g(\cdot)$ is used here to overcome the vanishing gradient problem associated with some other activation functions and allows the network to learn faster and perform better. The MaxPool$(\cdot)$ operator uses 2×2 spatial sub-sampling kernel to reduce the size of the feature maps by a factor of 2 in each spatial dimension of each channel.

After the convolution layers groups, we implement the classification layer group as shown in Figure 8b. The fully connected layer is similar to that in Equation (2), except here the output size is equal to the number of classes, M, and the weight matrix connects every input and output. It does not employ convolution kernels. Furthermore, there is no ReLU or max pooling. The fully connected layer function may be represented as

$$\mathbf{x}_n^L = \mathbf{W}^L \mathbf{x}_n^{L-1} + \mathbf{b}^L, \tag{4}$$

where $\mathbf{x}_n^L = [x_{n,1}^L, x_{n,2}^L, \ldots x_{n,M}^L]^T$ is the output. The vector $\mathbf{x}^{L-1}$ is the final feature map from the $L-1$ convolution layer groups. The biases for the fully connected layer are contained in $\mathbf{b}^L$.

After the fully connected layer, we have the so-called soft-max operation that normalizes the output and is given by

$$\hat{\mathbf{y}}_n = [\hat{y}_{n,1}, \hat{y}_{n,2}, \ldots, \hat{y}_{n,M}]^T = \text{Softmax}(\mathbf{x}_n^L), \tag{5}$$

where

$$\hat{y}_{n,m} = \frac{e^{x_{n,m}^L}}{\sum_{j=1}^M e^{x_{n,j}^L}}. \tag{6}$$

Note that the outputs of the softmax operation, $\hat{y}_{n,m}$, are in the range $[0, 1]$ and

$$\sum_{m=1}^M \hat{y}_{n,m} = 1. \tag{7}$$

All of the mathematical details mentioned above can be compactly summarized as follows

$$\hat{\mathbf{y}} = f(\mathbf{X}, \phi), \tag{8}$$

where $\hat{\mathbf{y}} = [\hat{y}_1, \hat{y}_2, \ldots, \hat{y}_N]^T$ is the predicated labels and all of the learnable parameters are given by

$$\phi = \{\mathbf{W}^l, \mathbf{b}^l | l \in \{1, 2, \ldots, L\}\}. \tag{9}$$

Note that the function $f(\cdot)$ is the overall SubNet predictor module and ϕ denotes the learnable parameters of the network. The learnable parameters are updated after each minibatch based on the empirical risk computed over that minibatch. The empirical cross-entropy error function used here is given by

$$R_{emp}(\mathbf{X}, \phi) = -\frac{1}{M} \sum_{n=1}^{N} \sum_{m=1}^{M} y_{n,m} \times \ln(\hat{y}_{n,m}). \tag{10}$$

Note that $R_{emp}(\cdot)$ depends on two arguments, the minibatch data $\mathbf{X}$ and the learnable parameters in ϕ. The variable N is the number of examples in the minibatch, and M is the number of classes. The variable $y_{n,m}$ is the truth labels and $\hat{y}_{n,m}$ is the predicated labels of our model. Once the loss is computed for one minibatch, back-propagation is used to update the learnable parameters in ϕ for the SubNet using the adaptive moment estimation (Adam) optimizer [52].

3.3. Series and Parallel Combinations

Consider the series combination of two SubNets: $A + B$. Let SubNet A have L_A convolution layers that follow Equation (2), and SubNet B has L_B. The combined network would have a total of $L = L_A + L_B + 1$ layers, where the final layer is the one fully connected layer as shown in Equation (4). The parameters for SubNet A are

$$\phi_A = \{\mathbf{W}^l, \mathbf{b}^l | l \in \{1, 2, \ldots, L_A\}\}. \tag{11}$$

These are fixed after the training for SubNet A. The parameters for the SubNet B convolution layers, plus the fully connected layer are given by

$$\phi_{B+} = \{\mathbf{W}^l, \mathbf{b}^l | l \in \{L_A + 1, L_A + 2, \ldots, L_A + L_B + 1\}\}. \tag{12}$$

The parameters in ϕ_{B+} are updated during the training of $A + B$. This output of the series layers goes to the softmax layer as before using Equation (5). This scenario is illustrated in Figure 7a.

Next, consider two parallel SubNets: $A \parallel B$. Again, let SubNet A have L_A convolution layers that follow Equation (2), and SubNet B has L_B. Let us define the convolution layer parameters for each SubNet as

$$\phi_A = \{\mathbf{W}_A^l, \mathbf{b}_A^l | l \in \{1, 2, \ldots, L_A\}\} \tag{13}$$

and

$$\phi_B = \{\mathbf{W}_B^l, \mathbf{b}_B^l | l \in \{1, 2, \ldots, L_B\}\}. \tag{14}$$

The output of the SubNet A convolution layers is given by

$$\mathbf{x}_{A,n}^l = g(\mathbf{W}_A^l \mathbf{x}_{A,n}^{l-1} + \mathbf{b}_A^l), \tag{15}$$

where $l = 1, 2, \ldots, L_A$. The output of the the SubNet B convolution layers is given by

$$\mathbf{x}_{B,n}^l = g(\mathbf{W}_B^l \mathbf{x}_{B,n}^{l-1} + \mathbf{b}_B^l), \tag{16}$$

where $l = 1, 2, \ldots, L_B$. Note that the inputs to the two parallel SubNets are the same so that we have $\mathbf{x}^0_{A,n} = \mathbf{x}^0_{B,n} = \mathbf{X}_n$. Let the fully connected output layer be designated as Layer $L = \mathrm{Max}(L_A, L_B) + 1$. The output of this fully connected layer with the final feature maps concatenated is given by

$$\mathbf{x}^L_n = \mathbf{W}_{A||B} \begin{bmatrix} \mathbf{x}^{L_A}_{A,n} \\ \mathbf{x}^{L_B}_{B,n} \end{bmatrix} + \mathbf{b}_{A||B}. \tag{17}$$

This output goes to the softmax layer as before using Equation (5). The parameters in ϕ_A are fixed and the parameters in ϕ_B along with $\mathbf{W}_{A||B}$ and $\mathbf{b}_{A||B}$ are updated. This scenario is illustrated in Figure 7b.

3.4. Proposed IMNet Architecture

In general, the IMNS can be used to create a limitless number of final architectures by combining the proposed SubNets, or other SubNet architectures. Here, we propose one specific example that we believe effectively balances performance and computational complexity for the medical imaging applications mentioned in Section 1. The proposed IMNet architecture is illustrated in Figure 9. Figure 9 shows five SubNets, A, B, C, D, and E, that are incrementally added to produce the final network. We use relatively small and compact SubNet modules to maintain a small computational cost. The details for each SubNet are provided in Table 2. Figure 7a shows the workflow for adding the SubNet in series, and Figure 7b shows the addition of a parallel SubNet.

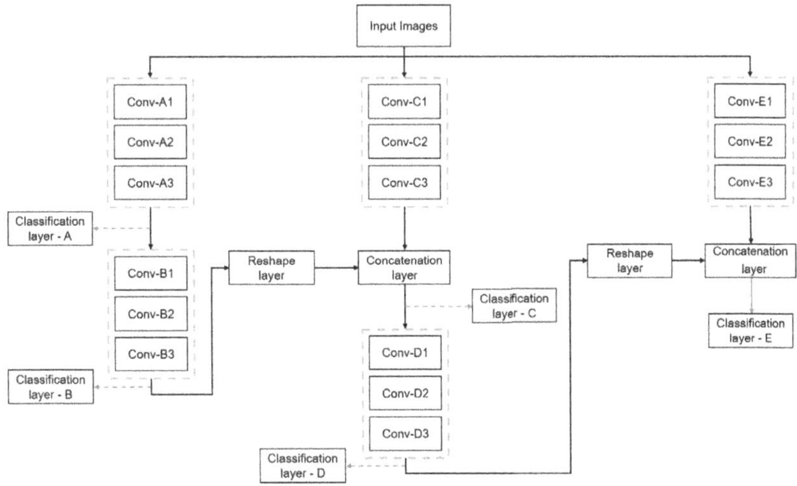

Figure 9. IMNet architecture used here in the experimental results. SubNets A, B, C, D and E are added incrementally in order to produce the full network shown. Details of the SubNets are provided in Table 2.

First, we use all of the available minibatches to train the SubNet A and minimize the loss to obtain the optimum parameters using Equation (10). After training for SubNet A is complete, we lock in the learnable parameters for this module and refer to it as IMNet A. This network is used to generate the feature maps that will be the input to the new SubNet B. The first convolutional layer of the SubNet B receives all the feature maps from $L - 1$ layer of SubNet A as input. In this case, the SubNet B is connected in a series configuration that can be denoted as $A + B$ and we refer to this as IMNet A, B.

Next, we lock the current IMNet A, B, and add a new SubNet C in parallel. This combination of SubNets may be expressed as $(A + B)||C$ and we refer to this as IMNet A-C for notational convenience. We use the equations mentioned above to generate feature maps using IMNet A, B and concatenate these with the feature maps generated by the new SubNet C. We use a reshape layer to match the feature map of IMNet A, B with output feature maps of SubNet C and concatenate the feature maps in the depth dimension. SubNet D is then added in series to produce the configuration $(A + B) || C + D$, denoted here as IMNet A-D. Finally, we lock IMNet A-D and add the last new SubNet E in a parallel configuration. This IMNS sequence can be represented as $[(A + B) || C + D] || E$ and we denote this as IMNet A-E. We selected this configuration because we find that alternating between series and parallel SubNets is generally effective, as these two additions tend to complement each other.

Table 2. SubNet architectures used in the IMNet in Figure 9.

Model	Layers	Filter Size	Total Parameters	MAdd
SubNet A	Conv-A2 Conv-A2 Conv-A3	$3 \times 3 \times 8$ $3 \times 3 \times 16$ $3 \times 3 \times 32$	0.018M	9.94M
SubNet B	Conv-B1 Conv-B2 Conv-B3	$3 \times 3 \times 64$ $3 \times 3 \times 128$ $3 \times 3 \times 256$	0.390M	11.94M
SubNet C	Conv-C1 Conv-C2 Conv-C3	$1 \times 1 \times 8$ $1 \times 1 \times 16$ $1 \times 1 \times 32$	0.021M	1.10M
SubNet D	Conv-D1 Conv-D2 Conv-D3	$3 \times 3 \times 64$ $3 \times 3 \times 128$ $3 \times 3 \times 256$	0.390M	43.65M
SubNet E	Conv-E1 Conv-E2 Conv-E3	$1 \times 3 \times 64$ $3 \times 1 \times 128$ $1 \times 3 \times 256$	0.165M	13.82M

3.5. Network Training

We study the performance of our proposed approach by utilizing the following data separation: 72% of the samples of each class are assigned to the training set, 8% to the validation set, and the remaining 20% to the test set. All of the image processing and classification stages are implemented using MATLAB deep learning platform [53] version r2020b. The hardware used is a Windows PC equipped with with Intel Xeon CPU E5-1630 v4 @ 3.70 GHz and 32 GB of RAM. Network training and testing are accelerated using an NVIDIA TITAN RTX GPU. We trained the network and tuned our hyperparameters for the proposed IMNet architecture solely on the training and validation datasets. All of the IMNets are trained from scratch with randomly initialized weights. We choose the Adam optimization technique [52] to accelerate the convergence time and find the global minimum cost function for all networks. We chose an initial learning rate of 0.001 with different mini-batch sizes for each application and a validation frequency of 50. Note that the validation frequency details how many iterations pass before re-validating during training. In our configurations we validate every 50 iterations. Note that the learning rate is kept adaptive to accelerate the learning process and prevent over-fitting. The learning rate is scheduled to decrease by a factor of 0.1 after one half of epochs are completed. We also use a training policy called "ValidationPatience" and set this parameter to 50. This value specifies the number of times that the validation loss can be larger than the smallest value achieved before the training process halts. Furthermore, in order to prevent overfitting and to improve model generalization, we apply a simple and effective regularization technique known as $L2$ regularization [54] with a value of 0.0001.

3.6. Statistical Analysis

It is important to assess the efficacy of classification algorithms to aid in method comparisons, method selection, understanding system limitations, and to identify opportunities for future improvement. The metrics we use as performance and efficiency metrics are balanced accuracy ($BACC$), specificity ($SPEC$), sensitivity ($SENS$), ROC curves, AUC, and testing time. These metrics defined in [55] provide an objective quantitative picture of the efficacy of the systems tested. We used the two-sided t-test to compare model performances. A $p < 0.05$ was considered statistically significant. All statistical analyses were performed with the statistical package of MATLAB version r2020b. In addition, to test the reproducibility of the model, we repeated such an experiment 10 times and reported mean and standard deviation (SD).

4. Experiment Results

In this section, we present the results obtained using our proposed approaches. In order to demonstrate the efficacy of our proposed algorithm, we compare our IMNS model results against those from well established and state-of-the-art CNN models including AlexNet [56], ResNet-50 [57],Inception v3 [58], DenseNet-201 [59], and NasNet [60]. For these large benchmark networks, we use transfer learning. The weights are imported from MATLAB deep learning toolbox [61] version r2020b. The pre-trained weights are imported from pre-trained networks. The pre-trained networks have been trained on a subset of the ImageNet database [62], which is used in the ILSVRC [44]. Approximately 1.4 million images have been used to train these networks to classify images into 1000 object classes. Fine-tuning a pre-trained network is more efficient than training a network from scratch. This is important with networks of these sizes. For IMNets, we use the training methodology described in Section 3.5. Furthermore, note that our results use publicly available datasets, as described in Section 2, to allow for independently reproducible results. We present the results for our IMNet in several forms to show the evolution in performance using IMNS starting with IMNet *A* and going to IMNet *A-E*, as shown in Figure 9. To quantitatively evaluate the results, we employ the performance metrics defined in Section 3.6.

4.1. Quantitative Results Summary

We applied the IMNS method to each of the datasets described in Section 2. In particular, we consider the detection of malaria, DR, and TB. The results for these three experiments are, respectively, summarized in Tables 3–5.

Table 3 shows the performance metrics for the IMNS method with various IMNets for malaria detection using blood smear slide images. Note that here IMNet *A-E* had a significantly higher BACC (97.0 ± 0.36) than AlexNet, ResNet-50, DenseNet-201, and NasNet (96.2 ± 0.22 [$p < 0.05$], 96.5 ± 0.51 [$p < 0.05$], 96.2 ± 0.43 [$p < 0.05$], and 96.7 ± 0.12 [$p < 0.05$], respectively). In addition, our proposed IMNet *A-E* outperforms the Inception v3 in this experiment (96.8 ± 0.39 [$p < 0.05$]). Furthermore, note that IMNet *A-D* took only 11.71 seconds to process 5512 samples ($9\times$ faster than Inception v3). The highest AUC in this experiment is achieved with IMNet *A-D*. Note also that in this application the addition of SubNet *E* lowers all of the metrics. This may suggest that the IMNS process can be halted as further improvement is not expected with additional modules.

Table 3. Malaria dataset results showing hold-out validation performance on the test set using our IMNS method and benchmark methods.

Model	BACC (%)	SPEC (%)	SENS (%)	AUC	Testing Time (s)
AlexNet	96.8 ± 0.39	96.0 ± 1.50	94.1 ± 1.05	0.985 ± 0.002	81.01
ResNet-50	96.5 ± 0.51	97.8 ± 0.34	95.3 ± 1.07	0.992 ± 0.003	88.08
DenseNet-201	96.2 ± 0.43	97.2 ± 0.75	95.2 ± 1.06	0.992 ± 0.002	157.43
Inception v3	96.8 ± 0.39	97.6 ± 0.74	96.0 ± 1.15	0.993 ± 0.001	104.20
NasNet	96.7 ± 0.12	97.6 ± 0.65	95.8 ± 0.69	0.993 ± 0.001	92.35
IMNet *A*	96.8 ± 0.39	96.0 ± 1.50	94.1 ± 1.05	0.985 ± 0.002	**11.18**
IMNet *A, B*	96.1 ± 0.50	97.2 ± 0.21	95.1 ± 0.89	0.991 ± 0.003	11.23
IMNet *A-C*	96.4 ± 0.30	97.1 ± 0.27	95.7 ± 0.55	0.993 ± 0.001	11.58
IMNet *A-D*	$\mathbf{97.0 \pm 0.36}$	$\mathbf{97.9 \pm 0.39}$	$\mathbf{96.1 \pm 0.63}$	$\mathbf{0.995 \pm 0.001}$	11.71
IMNet *A-E*	96.7 ± 0.19	97.5 ± 0.60	95.8 ± 0.55	0.994 ± 0.001	12.26

Table 4. Diabetic retinopathy dataset results showing hold-out validation performance on the test set using our IMNS method and benchmark methods.

Model	BACC (%)	SPEC (%)	SENS (%)	AUC	Testing Time (s)
AlexNet	97.2 ± 0.52	96.9 ± 0.82	97.4 ± 0.76	0.994 ± 0.003	48.83
ResNet-50	97.9 ± 0.73	97.3 ± 1.08	98.5 ± 0.94	$\mathbf{0.997 \pm 0.001}$	41.71
DenseNet-201	$\mathbf{98.0 \pm 0.39}$	$\mathbf{98.0 \pm 0.27}$	97.9 ± 0.82	0.996 ± 0.002	64.49
Inception v3	97.8 ± 0.41	97.2 ± 0.81	$\mathbf{98.5 \pm 0.74}$	0.995 ± 0.002	43.12
NasNet	97.0 ± 0.53	96.7 ± 0.85	97.3 ± 0.81	0.994 ± 0.001	55.41
IMNet *A*	92.2 ± 3.90	93.8 ± 3.26	90.6 ± 8.85	0.980 ± 0.009	**6.85**
IMNet *A, B*	96.1 ± 0.80	94.8 ± 2.18	97.4 ± 1.28	0.991 ± 0.003	7.25
IMNet *A-C*	97.0 ± 0.50	96.6 ± 0.82	97.4 ± 0.91	0.995 ± 0.001	7.49
IMNet *A-D*	97.7 ± 0.39	97.7 ± 1.01	97.7 ± 0.98	0.996 ± 0.001	7.51
IMNet *A-E*	97.9 ± 0.23	98.0 ± 0.65	97.7 ± 0.35	0.996 ± 0.001	7.63

Table 5. Tuberculosis dataset results showing hold-out validation performance on the test set using our IMNS method and benchmark methods.

Model	BACC (%)	SPEC (%)	SENS (%)	AUC	Testing Time (s)
AlexNet	86.1 ± 2.91	85.9 ± 3.53	86.3 ± 6.17	0.927 ± 0.017	0.775
ResNet-50	87.7 ± 2.46	85.5 ± 4.11	90.0 ± 3.26	0.926 ± 0.016	2.51
DenseNet-201	87.6 ± 2.30	84.7 ± 2.71	$\mathbf{90.4 \pm 3.89}$	0.931 ± 0.019	2.74
Inception v3	85.5 ± 3.18	81.1 ± 5.13	89.8 ± 4.66	0.910 ± 0.028	1.28
NasNet	84.2 ± 2.31	80.3 ± 5.62	88.1 ± 4.81	0.900 ± 0.029	1.18
IMNet *A*	80.2 ± 4.68	82.5 ± 12.3	78.0 ± 14.9	0.899 ± 0.044	**0.234**
IMNet *A, B*	82.8 ± 4.50	82.6 ± 9.37	83.0 ± 10.1	0.918 ± 0.041	0.249
IMNet *A-C*	85.9 ± 5.27	81.6 ± 12.1	90.3 ± 6.53	0.937 ± 0.034	0.258
IMNet *A-D*	87.8 ± 4.06	$\mathbf{87.3 \pm 5.86}$	88.4 ± 4.66	0.944 ± 0.025	0.285
IMNet *A-E*	$\mathbf{88.6 \pm 2.25}$	85.3 ± 3.36	89.0 ± 5.50	$\mathbf{0.953 \pm 0.018}$	0.301

The results summary for DR detection in retinal images are shown in Table 4. Here the IMNet *A-E* achieved a higher BACC (97.92 ± 0.23) which is significantly better than AlexNet and NasNet (97.20 ± 0.52 [$p < 0.05$], and 97.05 ± 0.53 [$p < 0.05$]). The BACC of IMNet *A-E* is competitive with ResNet-50, DenseNet-201 and Inception v3 (97.9 ± 0.73 [$p = 0.86$], 98.0 ± 0.39 [$p = 0.42$], and 97.8 ± 0.41 [$p = 0.66$], respectively). The DenseNet-201 gives the best BACC here and the ResNet-50 model does have a slightly higher AUC than IMNet *A-E*. However, IMNet *A-E* processes 732 images in 7.63 s, as compared with 41.71 seconds for ResNet-50. As can be seen by the different IMNet results in Table 4, the BACC score rises with the addition of each SubNet during the IMNS process in this experiment.

The results summary for TB detection in chest radiographs is presented in Table 5. The highest BACC of 88.6 ± 2.25 is achieved with IMNet *A-E*, which is significantly higher than AlexNet, Inception v3, and NasNet (86.1 ± 2.91 [p<0.05], 85.5 ± 3.18 [$p < 0.05$], and 84.2 ± 2.31 [$p < 0.05$], respectively). The IMNet *A-E* produces a higher BACC than ResNet-50 and DenseNet-201 (87.7 ± 2.46 [$p = 0.30$], and 87.6 ± 2.30 [$p = 0.21$]). The highest AUC of 0.953 ± 0.018 is achieved with IMNet *A-E* which is significantly higher than the best of benchmark methods, DenseNet-201, (0.931 ± 0.019 [$p < 0.05$]). Note that these IMNets outperform the large scale models in this application with far less computational cost and computational time. Results in Table 5 also indicate a modest but consistent boost in the performance as we add more SubNets during the IMNS process.

Moreover, we compare our IMNets against different current state-of-the-art methods. For malaria application, our IMNet *A-D* has a comparative AUC score of 0.995 compared with the current state-of-the-art methods with lower computational complexity, including Rajaraman et al. (0.993) [63], Rahman et al. (0.993) [28], and Rajaraman et al. (0.991) [48]. For DR application, IMNet *A-D* and IMNet *A-E* produced a comparable AUC of 0.995 and relatively lower computational complexity with the following proposed methods, including Gulshan et al. (0.991) [64], Chetoui et al. (0.986) [38], and Sahlsten et al. (0.987) [65]. Finally, IMNet *A-E* has a comparative AUC score of 0.953 compared with the following state-of-the-art methods, including Meraj et al. (0.920) [66], Sathitratanacheewin et al. (0.850) [67], and Hwang et al. (0.926) [40].

Figures 10–12 show ROC curves for malaria, DR, and TB, respectively. The ROC curves provide further insight because they illustrate classifier performance for a range of operating points. For clarity, we only show ROC curves for the top five models in each application. For malaria detection, the IMNet *A-D* obtained the best result in terms of AUC and an area of (0.995 ± 0.001). However, IMNet *A-E* obtained a competitive AUC score for both DR and TB with areas of (0.996 ± 0.001) and (0.949 ± 0.019), respectively.

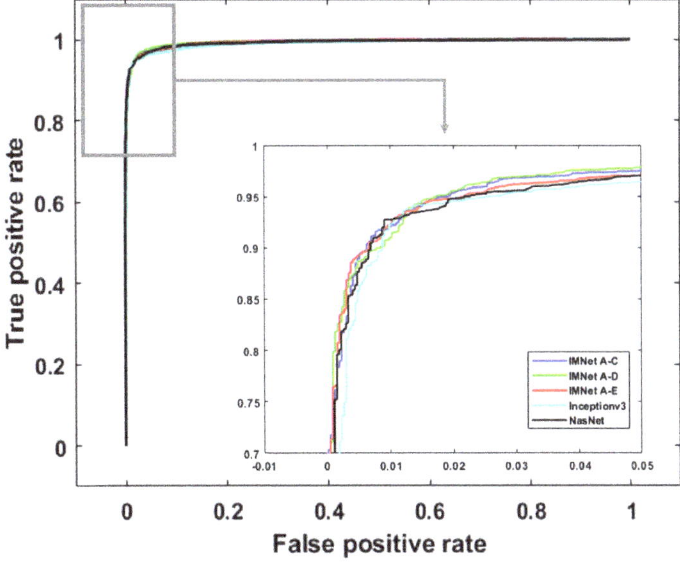

Figure 10. Malaria dataset ROC curve for the five best performing networks.

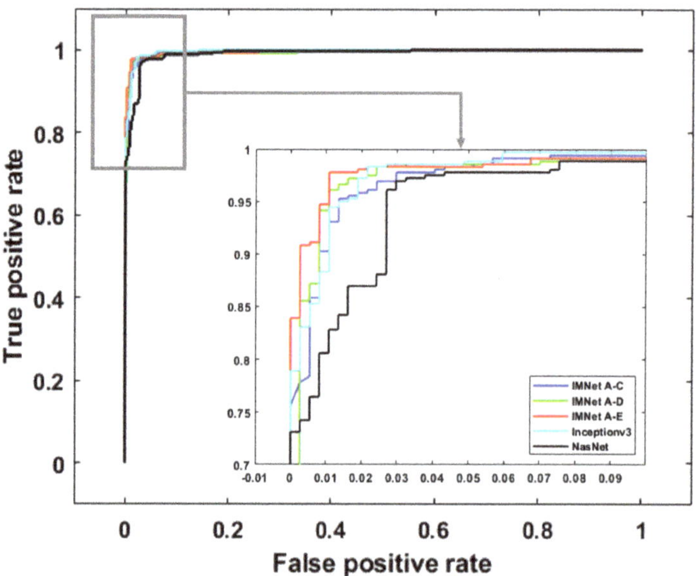

Figure 11. Diabetic retinopathy dataset ROC curve for the five best performing networks.

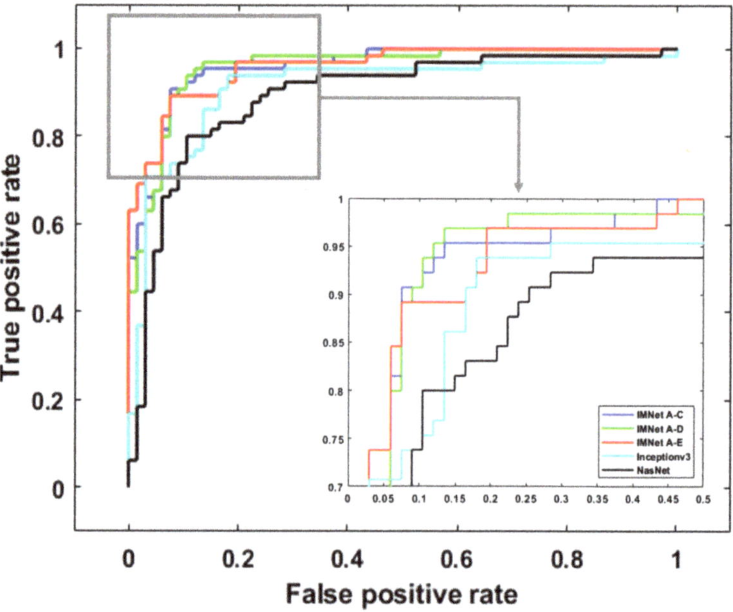

Figure 12. Tuberculosis dataset ROC curve for the five best performing networks.

4.2. Computational Complexity Comparison

In this section, we compare the computational complexity of AlexNet, ResNet-50, Inception v3, DenseNet-201, NasNet, and IMNets by counting the number of multiplications and additions required to process a single image. Furthermore, we compare between all mentioned models the total number of learnable parameters within each CNN model. We calculate the number of learnable parameters for each layer, and then sum up the learnable parameters in each layer to obtain the total amount of learnable parameters in the entire network. Figures 13 and 14 and Table 6 show the results of our computational complexity study. In Figure 13, we show balanced accuracy on the malaria dataset versus the number of learnable parameters. On the other hand, in Figure 13 we show balanced accuracy versus the number of floating-point multiply–add (MAdd) operations for the same dataset. Note that the composite MAdd operations are determined for the input images size of 112×112 reported in Section 2.1. The diameter of each circle is proportional to the the total number of learnable parameters for Figure 13, and the circle size is the MAdd for Figure 14. Note that the IMNets have fewer learnable parameters, and fewer MAdd operations, as shown in Table 6.

The numerical values for the total number of learnable parameters and MAdd counts are listed in Table 2 for the malaria dataset networks. Note that IMNet *A-E* (the largest IMNet tested here) has fewer parameters than AlexNet by a factor of approximately 64, and by a factor of approximately 6 compared with NasNet. In terms of the MAdd count, IMNet *A-E* has fewer than AlexNet by a factor of approximately 9, and fewer than NasNet by a factor of approximately 61.

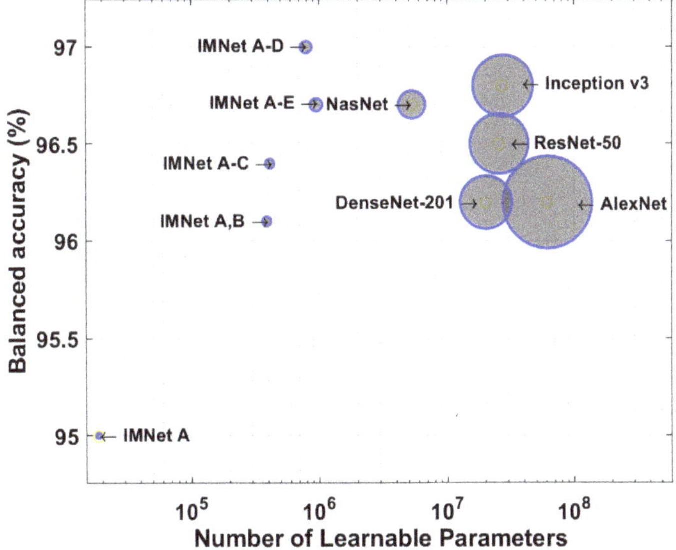

Figure 13. Balanced accuracy on the malaria dataset versus the number of learnable parameters. The computational cost is measured based on the number of MAdd operations to process a single example. The diameter of each circle is proportional to the total number of learnable parameters of the network.

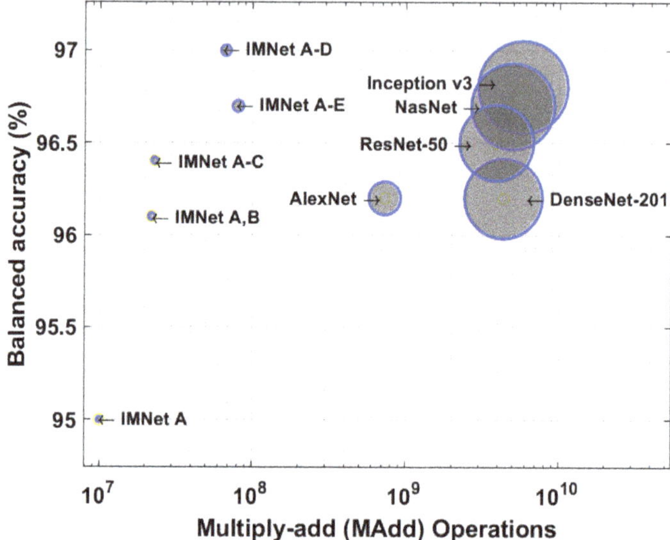

Figure 14. Balanced accuracy on the malaria dataset versus the number of floating-point multiply–add (MAdd) operations. The computational cost is measured based on the number of MAdd operations to process a single example. The diameter of each circle is proportional to the MAdd of the network.

Table 6. Resource usage for IMNets in comparison to benchmark models for the malaria dataset networks.

Model	Total Parameters	MAdd
AlexNet	61.10M	0.72G
ResNet-50	25.56M	3.87G
Inception v3	27.16M	5.72G
DenseNet-201	20.01M	4.29G
NasNet	5.290M	4.93G
IMNet A	0.018M	0.0099G
IMNet A, B	0.390M	0.0218G
IMNet A-C	0.412M	0.0229G
IMNet A-D	0.790M	0.0666G
IMNet A-E	0.955M	0.0804G

4.3. Visual Explanations

Figures 15–17 show the class activation mapping (CAM) [68] outcomes for malaria, DR, and TB, respectively. The examples are for different IMNets on test samples that had been identified as a true positives by the medical professionals. The CAM outputs can give us more confidence in our models' predictions as they highlight the discriminative regions used by a model to identify a positive class in the dataset. Our goal is to investigate and understand which image region has contributed more to the final model prediction. The idea of the CAM is the following: the probabilities predicted by the network are mapped back to the final convolutional layer to highlight the discriminative regions that are specific to that class [68]. CAM is the output of the activation map after the last convolutional layer for a particular class. CAM is the global average pooling layer applied following the last convolutional layer based on the spatial location in order to generate the weights [68]. Therefore, it allows distinguishing the areas within an image that differentiates the class [68].

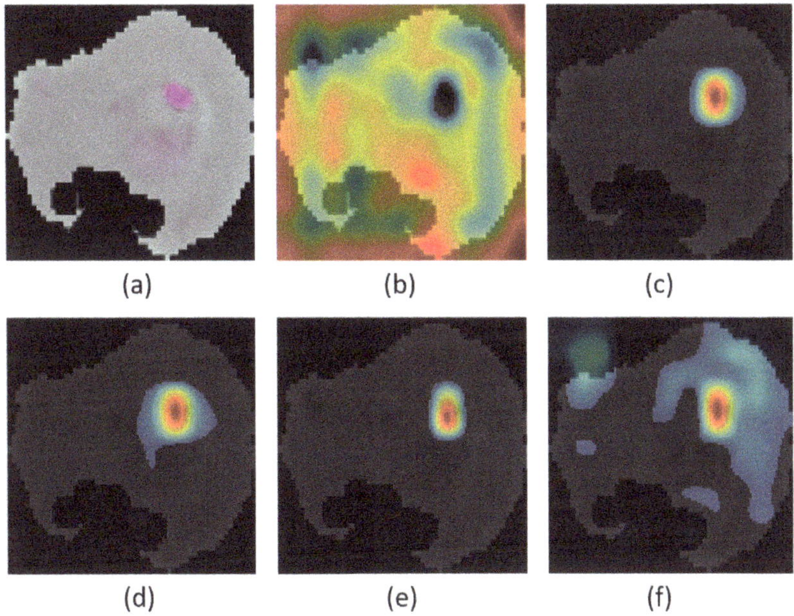

Figure 15. CAM visualization on malaria dataset for a test sample using various IMNets: (**a**) Original sample, (**b**) IMNet *A*, (**c**) IMNet *A, B*, (**d**) IMNet *A-C*, (**e**) IMNet *A-D*, and (**f**) IMNet *A-E*.

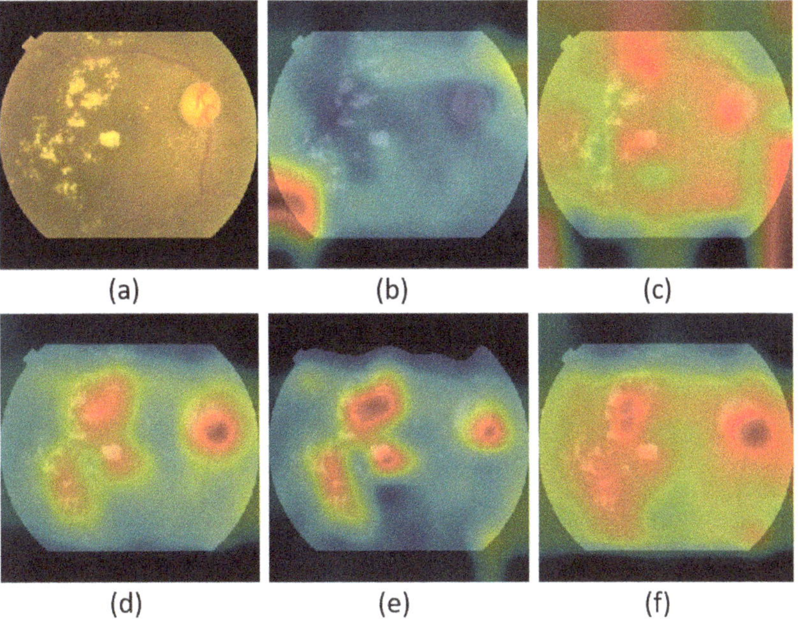

Figure 16. CAM visualization on DR dataset for a test sample using various IMNets: (**a**) Original sample, (**b**) IMNet *A*, (**c**) IMNet *A, B*, (**d**) IMNet *A-C*, (**e**) IMNet *A-D*, and (**f**) IMNet *A-E*.

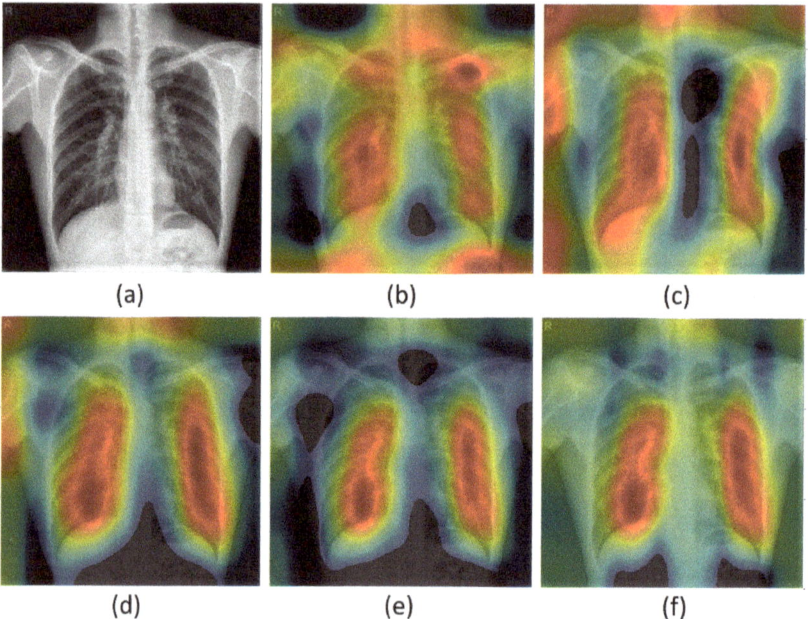

Figure 17. CAM visualization on TB dataset for a test sample using various IMNets: (**a**) Original sample, (**b**) IMNet *A*, (**c**) IMNet *A, B*, (**d**) IMNet *A-C*, (**e**) IMNet *A-D*, and (**f**) IMNet *A-E*.

Consider the malaria detection CAM results in Figure 15. The original sample image is shown in Figure 15a. The nucleic acids carry three components: parasites, white blood cells, and platelets highlighted in a bluish-purple color [69], as shown in the original sample image. The other images in Figure 15 are CAM results overlaid on the original image for IMNet *A* through IMNet *A-E*. Note that the red regions in the CAM images correspond to the spatial regions of most significance to the classifier. In the case of the CAM result for IMNet *A*, shown in Figure 15b, the attention is distributed and not well focused on the clinically significant portion of the thin smear image. On the other hand, as the IMNS process continues and modules are added, the CAM results do show that attention becomes more focused over the stain on the thin smear example to identify the presence of parasites. The CAM results showing the most focus on the nucleus is IMNet *A-D*, and this is the best performing IMNet as shown in Table 3.

The CAM results for DR are shown in Figure 16. The input retinal image is shown in Figure 16a. Note that the key aspect of detecting or diagnosing DR is the presence of retinal lesions. There are two main types of lesion defects, white lesions and red lesions. The hard and soft exudates are collectively referred to as white lesions. The red lesions are microaneurysms and hemorrhages [70]. The original image contains hard and soft exudates. We can tell that IMNet *A-C*, IMNet *A-D*, and IMNet *A-E*, focused on these hard and soft exudates that appear as white spots on the original image. Interestingly, these networks also appear to be focusing attention on the optic disk, which is the bright disk in the upper right side of the retinal image. This may be because its color and size resemble that of the large white lesions.

Finally, the CAM for pulmonary tuberculosis is shown in Figure 17. The original CR image with tuberculosis is shown in Figure 17a. Note that there are multiple light areas in the mid-zone lung with fibrotic shadows of primary pulmonary TB. The CAM results for our IMNet models show that attention is focused on these regions. As a result, our model performs well and generally provides an accurate interpretation. Although this example looks good, in many instances, the IMNet *A* and IMNet *A, B* CAM results show focus on these clinically significant regions and insignificant regions such as shoulders and background as well. This is consistent with the relatively average classifier performance for that network provided in Table 5. However, all of the other IMNets perform well and tend to produce what we believe are clinically appropriate CAM results.

5. Conclusions and Discussion

In this research, we have proposed IMNS as a new method for designing and training deep learning models. The resulting networks are referred to as IMNets. We have demonstrated the efficacy of the proposed method in detecting three diseases using three different imaging modalities. The best performing IMNets in our study achieved a balanced accuracy of 97.0%, 97.9%, and 88.6% and AUC of 0.995, 0.996, and 0.949 for the detection of malaria, DR, TB, respectively. Our modular approach starts with a single SubNet and we add one additional SubNet at a time, either in series or in parallel with the previous network. Only the new SubNet weights are updated at each stage of IMNS. This approach keeps the computational complexity low and allows the network to train well with a relatively small training set.

The performance of IMNets rivals, and in some cases exceeds, that of much larger state-of-the-art networks where transfer learning is employed. We attribute this to the relatively small training sets available and the limitations of transfer learning. Since the pre-trained networks are trained for a different application, significant adaption may be required for a new task. Large networks can be very powerful where there are sufficient data to properly train them. However, the large networks, with a high number of learnable parameters, can become a liability when only small training sets are available. In other words, for large pre-trained models to be helpful, both extensive data from the same domain and large computational resources are required. It remains the case that large truthed datasets for medical imaging applications are often difficult to come by. This behoves us to explore more compact networks and training strategies such as the proposed IMNS.

Monolithic deep learning with transfer learning may suffer from overfitting issues, due to limited training data in many medical image analysis applications. In addition, the computational cost grows with deeper and wider monolithic networks. The building-block IMNS approach addresses these issues by employing relatively small SubNets and training only one SubNet at a time. As we can see in the results section, our IMNS provides results that rival or exceed many popular large-scale models in the experiments presented here. Moreover, our IMNets trained faster, had lower memory requirements, and processed test images more quickly than the benchmark methods tested.

From a learning perspective, we believe IMNS has several benefits over monolithic deep learning. As with other modular approaches, complex problems are addressed using several small SubNets, rather than one large monolithic network. We believe this helps to mitigate the complex optimization difficulties and vanishing gradient problems that monolithic CNN approaches face. Furthermore, our results suggest that the IMNS allows for the effective transfer of prior knowledge from the fixed portion of the IMNet to a new SubNet. In future work, we plan to extend the architecture of IMNets in two ways. First, we will investigate the impact of combining these SubNets in different configurations. Moreover, we will also examine different SubNet architectures.

Author Contributions: Conceptualization, R.A. and R.C.H.; methodology, R.A. and R.C.H.; software, R.A. and R.C.H.; validation, R.A., R.C.H. and B.N.N.; writing—original draft preparation, R.A.; writing—review and editing, R.A., R.C.H., B.N.N. and T.M.K.; supervision, R.C.H., B.N.N. and T.M.K. All authors have read and agreed to the published version of the manuscript.

Funding: This research received no external funding.

Institutional Review Board Statement: Not applicable.

Informed Consent Statement: Not applicable.

Data Availability Statement: The segmented cells from the thin blood smear slide images for the parasitized and uninfected classes are available at https://lhncbc.nlm.nih.gov/, accessed on 22 May 2022 The publicly available datasets for DR detection can be found at https://www.kaggle.com/competitions/aptos2019-blindness-detection/data, accessed on 22 May 2022 The Shenzhen dataset for TB detection is available at https://lhncbc.nlm.nih.gov/, accessed on 22 May 2022.

Acknowledgments: In no particular order, we thank Vijayan Asari, John S. Loomis and Youssef N. Raffoul for their helpful discussions and suggestions.

Conflicts of Interest: The authors declare that there is no conflict of interest.

References

1. Zhai, X.; Kolesnikov, A.; Houlsby, N.; Beyer, L. Scaling vision transformers. *arXiv* **2021**, arXiv:2106.04560.
2. Riquelme, C.; Puigcerver, J.; Mustafa, B.; Neumann, M.; Jenatton, R.; Pinto, A.S.; Keysers, D.; Houlsby, N. Scaling Vision with Sparse Mixture of Experts. *arXiv* **2021**, arXiv:2106.05974.
3. Image Classification on ImageNe. Available online: https://paperswithcode.com/sota/image-classification-on-imagenet (accessed on 6 July 2021).
4. D'souza, R.N.; Huang, P.Y.; Yeh, F.C. Structural analysis and optimization of convolutional neural networks with a small sample size. *Sci. Rep.* **2020**, *10*, 1–13. [CrossRef] [PubMed]
5. Arsenovic, M.; Karanovic, M.; Sladojevic, S.; Anderla, A.; Stefanovic, D. Solving current limitations of deep learning based approaches for plant disease detection. *Symmetry* **2019**, *11*, 939. [CrossRef]
6. Cremer, C.Z. Deep limitations? Examining expert disagreement over deep learning. *Prog. Artif. Intell.* **2021**, *26*, 1–16. [CrossRef]
7. Lv, X.; Zhang, X. Generating chinese classical landscape paintings based on cycle-consistent adversarial networks. In Proceedings of the 2019 6th International Conference on systems and Informatics (ICSAI), Shanghai, China, 2–4 November 2019; pp. 1265–1269.
8. Chen, K. Deep and Modular Neural Networks. In *Handbook of Computational Intelligence*; Springer: Berlin/Heidelberg, Germany, 2015.
9. Albright, T.D.; Jessell, T.M.; Kandel, E.R.; Posner, M.I. Neural science: A century of progress and the mysteries that remain. *Neuron* **2000**, *25*, S1–S55. [CrossRef]
10. Fodor, J.A. *The Modularity of Mind*; MIT Press: Cambridge, MA, USA, 1983.
11. Edelman, G.M. *Neural Darwinism: The Theory of Neural Group Selection*; Basic Books: New York, NY, USA, 1987.
12. O'Connell, K.A.; Gatakaa, H.; Poyer, S.; Njogu, J.; Evance, I.; Munroe, E.; Solomon, T.; Goodman, C.; Hanson, K.; Zinsou, C.; et al. Got ACTs? Availability, price, market share and provider knowledge of anti-malarial medicines in public and private sector outlets in six malaria-endemic countries. *Malar. J.* **2011**, *10*, 326. [CrossRef]
13. WHO. *World Malaria Report 2020: 20 Years of Global Progress and Challenges*; WHO: Geneva, Switzerland, 2020.
14. Mace, K.E.; Arguin, P.M.; Tan, K.R. Malaria surveillance—United States, 2015. *MMWR Surveill. Summ.* **2018**, *67*, 1. [CrossRef]
15. Posfai, D.; Sylvester, K.; Reddy, A.; Ganley, J.G.; Wirth, J.; Cullen, Q.E.; Dave, T.; Kato, N.; Dave, S.S.; Derbyshire, E.R. Plasmodium parasite exploits host aquaporin-3 during liver stage malaria infection. *PLoS Pathog.* **2018**, *14*, e1007057. [CrossRef]
16. Dey, N.; Ashour, A.S.; Borra, S. *Classification in BioApps: Automation of Decision Making*; Springer: Berlin/Heidelberg, Germany, 2017; Volume 26.
17. WHO. *Malaria Microscopy: Quality Assurance Manual, Version 2*; WHO: Geneva, Switzerland, 2016.
18. Yang, F.; Poostchi, M.; Yu, H.; Zhou, Z.; Silamut, K.; Yu, J.; Maude, R.J.; Jaeger, S.; Antani, S. Deep learning for smartphone-based malaria parasite detection in thick blood smears. *IEEE J. Biomed. Health Inform.* **2019**, *24*, 1427–1438. [CrossRef]
19. Dong, Y.; Jiang, Z.; Shen, H.; Pan, W.D.; Williams, L.A.; Reddy, V.V.; Benjamin, W.H.; Bryan, A.W. Evaluations of deep convolutional neural networks for automatic identification of malaria infected cells. In Proceedings of the 2017 IEEE EMBS International Conference on Biomedical & Health Informatics (BHI), Chicago, IL, USA, 16–19 February 2017; pp. 101–104.
20. Zheng, Y.; He, M.; Congdon, N. The worldwide epidemic of diabetic retinopathy. *Indian J. Ophthalmol.* **2012**, *60*, 428. [PubMed]
21. Zhang, X.; Wang, H.; Du, C.; Fan, X.; Cui, L.; Chen, H.; Deng, F.; Tong, Q.; He, M.; Yang, M.; et al. Custom-Molded Offloading Footwear Effectively Prevents Recurrence and Amputation, and Lowers Mortality Rates in High-Risk Diabetic Foot Patients: A Multicenter, Prospective Observational Study. *Diabetes Metab. Syndr. Obes. Targets Ther.* **2022**, *15*, 103. [CrossRef] [PubMed]
22. Flaxman, S.R.; Bourne, R.R.; Resnikoff, S.; Ackland, P.; Braithwaite, T.; Cicinelli, M.V.; Das, A.; Jonas, J.B.; Keeffe, J.; Kempen, J.H.; et al. Global causes of blindness and distance vision impairment 1990–2020: A systematic review and meta-analysis. *Lancet Glob. Health* **2017**, *5*, e1221–e1234. [CrossRef]
23. Salz, D.A.; Witkin, A.J. Imaging in diabetic retinopathy. *Middle East Afr. J. Ophthalmol.* **2015**, *22*, 145. [PubMed]
24. Harding, E. WHO global progress report on tuberculosis elimination. *Lancet Respir. Med.* **2020**, *8*, 19. [CrossRef]
25. Narayanan, B.N.; Hardie, R.C.; Krishnaraja, V.; Karam, C.; Davuluru, V.S.P. Transfer-to-transfer learning approach for computer aided detection of COVID-19 in chest radiographs. *AI* **2020**, *1*, 539–557. [CrossRef]
26. Organization, W.H. *World Malaria Report 2015*; World Health Organization: Geneva, Switzerland, 2016.
27. Ali, R.; Hardie, R.C.; Ragb, H.K. Ensemble lung segmentation system using deep neural networks. In Proceedings of the 2020 IEEE Applied Imagery Pattern Recognition Workshop (AIPR), Washington, DC, USA, 13–15 October 2020; pp. 1–5.
28. Rahman, A.; Zunair, H.; Rahman, M.S.; Yuki, J.Q.; Biswas, S.; Alam, M.A.; Alam, N.B.; Mahdy, M. Improving malaria parasite detection from red blood cell using deep convolutional neural networks. *arXiv* **2019**, arXiv:1907.10418.

29. Pattanaik, P.; Mittal, M.; Khan, M.Z.; Panda, S. Malaria detection using deep residual networks with mobile microscopy. *J. King Saud-Univ.-Comput. Inf. Sci.* **2022**, *34*, 1700–1705. [CrossRef]

30. Zhao, O.S.; Kolluri, N.; Anand, A.; Chu, N.; Bhavaraju, R.; Ojha, A.; Tiku, S.; Nguyen, D.; Chen, R.; Morales, A.; et al. Convolutional neural networks to automate the screening of malaria in low-resource countries. *PeerJ* **2020**, *8*, e9674. [CrossRef]

31. Poostchi, M.; Silamut, K.; Maude, R.J.; Jaeger, S.; Thoma, G. Image analysis and machine learning for detecting malaria. *Transl. Res.* **2018**, *194*, 36–55. [CrossRef]

32. Lee, Y.W.; Choi, J.W.; Shin, E.H. Machine learning model for predicting malaria using clinical information. *Comput. Biol. Med.* **2020**, *129*, 104151. [CrossRef] [PubMed]

33. Narayanan, B.N.; Ali, R.; Hardie, R.C. Performance analysis of machine learning and deep learning architectures for malaria detection on cell images. In *Applications of Machine Learning*; International Society for Optics and Photonics: Bellingham, WA, USA, 2019; Volume 11139, p. 111390W.

34. Qummar, S.; Khan, F.G.; Shah, S.; Khan, A.; Shamshirband, S.; Rehman, Z.U.; Khan, I.A.; Jadoon, W. A deep learning ensemble approach for diabetic retinopathy detection. *IEEE Access* **2019**, *7*, 150530–150539. [CrossRef]

35. Zhang, S.; Wu, H.; Murthy, V.; Wang, X.; Cao, L.; Schwartz, J.; Hernandez, J.; Rodriguez, G.; Liu, B.J. The application of deep learning for diabetic retinopathy prescreening in research eye-PACS. In *Medical Imaging 2018: Imaging Informatics for Healthcare, Research, and Applications*; International Society for Optics and Photonics: Bellingham, WA, USA, 2018; Volume 10579, p. 1057913.

36. Majumder, S.; Elloumi, Y.; Akil, M.; Kachouri, R.; Kehtarnavaz, N. A deep learning-based smartphone app for real-time detection of five stages of diabetic retinopathy. In *Real-Time Image Processing and Deep Learning 2020*; International Society for Optics and Photonics: Bellingham, WA, USA, 2020; Volume 11401, p. 1140106.

37. Narayanan, B.N.; Hardie, R.C.; De Silva, M.S.; Kueterman, N.K. Hybrid machine learning architecture for automated detection and grading of retinal images for diabetic retinopathy. *J. Med Imaging* **2020**, *7*, 034501. [CrossRef] [PubMed]

38. Chetoui, M.; Akhloufi, M.A. Explainable end-to-end deep learning for diabetic retinopathy detection across multiple datasets. *J. Med. Imaging* **2020**, *7*, 044503. [CrossRef]

39. Niazi, M.K.K.; Beamer, G.; Gurcan, M.N. An application of transfer learning to neutrophil cluster detection for tuberculosis: efficient implementation with nonmetric multidimensional scaling and sampling. In *Medical Imaging 2018: Digital Pathology*; International Society for Optics and Photonics: Bellingham, WA, USA, 2018; Volume 10581, p. 1058108.

40. Hwang, S.; Kim, H.E.; Jeong, J.; Kim, H.J. A novel approach for tuberculosis screening based on deep convolutional neural networks. In *Medical imaging 2016: Computer-Aided Diagnosis*; SPIE: Bellingham, WA, USA, 2016; Volume 9785, pp. 750–757.

41. Wu, E.Q.; Zhou, M.; Hu, D.; Zhu, L.; Tang, Z.; Qiu, X.Y.; Deng, P.Y.; Zhu, L.M.; Ren, H. Self-Paced Dynamic Infinite Mixture Model for Fatigue Evaluation of Pilots' Brains. *IEEE Trans. Cybern.* **2020**. [CrossRef]

42. Panicker, R.O.; Kalmady, K.S.; Rajan, J.; Sabu, M. Automatic detection of tuberculosis bacilli from microscopic sputum smear images using deep learning methods. *Biocybern. Biomed. Eng.* **2018**, *38*, 691–699. [CrossRef]

43. Li, X.; Zhou, Y.; Du, P.; Lang, G.; Xu, M.; Wu, W. A deep learning system that generates quantitative CT reports for diagnosing pulmonary tuberculosis. *Appl. Intell.* **2021**, *51*, 4082–4093. [CrossRef]

44. Russakovsky, O.; Deng, J.; Su, H.; Krause, J.; Satheesh, S.; Ma, S.; Huang, Z.; Karpathy, A.; Khosla, A.; Bernstein, M.; et al. Imagenet large scale visual recognition challenge. *Int. J. Comput. Vis.* **2015**, *115*, 211–252. [CrossRef]

45. Deng, W.; Zhang, X.; Zhou, Y.; Liu, Y.; Zhou, X.; Chen, H.; Zhao, H. An enhanced fast non-dominated solution sorting genetic algorithm for multi-objective problems. *Inf. Sci.* **2022**, *585*, 441–453. [CrossRef]

46. Chen, T.; Goodfellow, I.; Shlens, J. Net2Net: Accelerating Learning via Knowledge Transfer. *arXiv* **2016**, arXiv:1511.05641.

47. Anderson, A.; Shaffer, K.; Yankov, A.; Corley, C.D.; Hodas, N.O. Beyond Fine Tuning: A Modular Approach to Learning on Small Data. *arXiv* **2016**, arXiv:1611.01714.

48. Rajaraman, S.; Antani, S.K.; Poostchi, M.; Silamut, K.; Hossain, M.A.; Maude, R.J.; Jaeger, S.; Thoma, G.R. Pre-trained convolutional neural networks as feature extractors toward improved malaria parasite detection in thin blood smear images. *PeerJ* **2018**, *6*, e4568. [CrossRef] [PubMed]

49. APTOS 2019 Blindness Detection. Available online: https://www.kaggle.com/c/aptos2019-blindness-detection/overview (accessed on 14 December 2020).

50. Jaeger, S.; Candemir, S.; Antani, S.; Wáng, Y.X.J.; Lu, P.X.; Thoma, G. Two public chest X-ray datasets for computer-aided screening of pulmonary diseases. *Quant. Imaging Med. Surg.* **2014**, *4*, 475. [PubMed]

51. Finlayson, G.D.; Trezzi, E. Shades of gray and colour constancy. In Proceedings of the Color and Imaging Conference. Society for Imaging Science and Technology, Scottsdale, AZ, USA, 9–12 November 2004; Volume 2004, pp. 37–41.

52. Kingma, D.P.; Ba, J. Adam: A Method for Stochastic Optimization. *arXiv* **2015**, arXiv:1412.6980.

53. Kim, P. Matlab deep learning. *With Machine Learning, Neural Networks and Artificial Intelligence*; Springer: Berkeley, CA, USA, 2017; Volume 130.

54. Ng, A.Y. Feature selection, L 1 vs. L 2 regularization, and rotational invariance. In Proceedings of the 21st International Conference on Machine Learning, Banff, AB, Canada, 4–8 July 2004; p. 78.

55. Brinker, T.J.; Hekler, A.; Utikal, J.S.; Grabe, N.; Schadendorf, D.; Klode, J.; Berking, C.; Steeb, T.; Enk, A.H.; Von Kalle, C. Skin cancer classification using convolutional neural networks: Systematic review. *J. Med. Internet Res.* **2018**, *20*, e11936. [CrossRef]

56. Krizhevsky, A.; Sutskever, I.; Hinton, G.E. Imagenet classification with deep convolutional neural networks. *Adv. Neural Inf. Process. Syst.* **2012**, *25*, 1097–1105. [CrossRef]

57. He, K.; Zhang, X.; Ren, S.; Sun, J. Deep residual learning for image recognition. In Proceedings of the IEEE Conference on Computer Vision and Pattern Recognition, Las Vegas, NV, USA, 27–30 June 2016; pp. 770–778.
58. Szegedy, C.; Vanhoucke, V.; Ioffe, S.; Shlens, J.; Wojna, Z. Rethinking the inception architecture for computer vision. In Proceedings of the IEEE Conference on Computer Vision and Pattern Recognition, New York, NY, USA, 17–22 June 2016; pp. 2818–2826.
59. Huang, G.; Liu, Z.; Van Der Maaten, L.; Weinberger, K.Q. Densely connected convolutional networks. In Proceedings of the IEEE Conference on Computer Vision and Pattern Recognition, Honolulu, HI, USA, 21–26 July 2017; pp. 4700–4708.
60. Zoph, B.; Vasudevan, V.; Shlens, J.; Le, Q.V. Learning transferable architectures for scalable image recognition. In Proceedings of the IEEE Conference on Computer Vision and Pattern Recognition, Salt Lake City, UT, USA, 18–23 June 2018; pp. 8697–8710.
61. MATLAB Deep Learning Toolbox Documentation. Available online: https://www.mathworks.com/help/deeplearning/ (accessed on 6 July 2021).
62. Deng, J.; Dong, W.; Socher, R.; Li, L.J.; Li, K.; Fei-Fei, L. Imagenet: A large-scale hierarchical image database. In Proceedings of the 2009 IEEE Conference on Computer Vision and Pattern Recognition, Miami, FL, USA, 20–25 June 2009; pp. 248–255.
63. Rajaraman, S.; Jaeger, S.; Antani, S.K. Performance evaluation of deep neural ensembles toward malaria parasite detection in thin-blood smear images. *PeerJ* **2019**, *7*, e6977. [CrossRef]
64. Gulshan, V.; Peng, L.; Coram, M.; Stumpe, M.C.; Wu, D.; Narayanaswamy, A.; Venugopalan, S.; Widner, K.; Madams, T.; Cuadros, J.; et al. Development and validation of a deep learning algorithm for detection of diabetic retinopathy in retinal fundus photographs. *JAMA* **2016**, *316*, 2402–2410. [CrossRef]
65. Sahlsten, J.; Jaskari, J.; Kivinen, J.; Turunen, L.; Jaanio, E.; Hietala, K.; Kaski, K. Deep learning fundus image analysis for diabetic retinopathy and macular edema grading. *Sci. Rep.* **2019**, *9*, 1–11. [CrossRef]
66. Meraj, S.S.; Yaakob, R.; Azman, A.; Rum, S.; Shahrel, A.; Nazri, A.; Zakaria, N.F. Detection of pulmonary tuberculosis manifestation in chest X-rays using different convolutional neural network (CNN) models. *Int. J. Eng. Adv. Technol. (IJEAT)* **2019**, *9*, 2270–2275. [CrossRef]
67. Sathitratanacheewin, S.; Sunanta, P.; Pongpirul, K. Deep learning for automated classification of tuberculosis-related chest X-Ray: dataset distribution shift limits diagnostic performance generalizability. *Heliyon* **2020**, *6*, e04614. [CrossRef] [PubMed]
68. Zhou, B.; Khosla, A.; Lapedriza, A.; Oliva, A.; Torralba, A. Learning deep features for discriminative localization. In Proceedings of the IEEE Conference on Computer Vision and Pattern Recognition, Las Vegas, NV, USA, 27–30 June 2016; pp. 2921–2929.
69. Prasad, K.; Winter, J.; Bhat, U.M.; Acharya, R.V.; Prabhu, G.K. Image analysis approach for development of a decision support system for detection of malaria parasites in thin blood smear images. *J. Digit. Imaging* **2012**, *25*, 542–549. [CrossRef] [PubMed]
70. Borsos, B.; Nagy, L.; Iclanzan, D.; Szilágyi, L. Automatic detection of hard and soft exudates from retinal fundus images. *Acta Unitiversitatis-Sapientiae-Inform.* **2019**, *11*, 65–79. [CrossRef]

MDPI

Article

Augmented Reality in Surgery: A Scoping Review

Eleonora Barcali [1,2], Ernesto Iadanza [3,*], Leonardo Manetti [4], Piergiorgio Francia [1], Cosimo Nardi [2] and Leonardo Bocchi [1]

1   Department of Information Engineering, University of Florence, 50139 Florence, Italy; eleonora.barcali@unifi.it (E.B.); piergiorgiofrancia5@gmail.com (P.F.); leonardo.bocchi@unifi.it (L.B.)
2   Department of Biomedical Experimental and Clinical Sciences "Mario Serio", University of Florence, 50139 Florence, Italy; cosimo.nardi@unifi.it
3   Department of Medical Biotechnologies, University of Siena, 53100 Siena, Italy
4   Epica Imaginalis, 50019 Sesto Fiorentino, Italy; l.manetti@imaginalis.it
*   Correspondence: ernesto.iadanza@unisi.it

Abstract: Augmented reality (AR) is an innovative system that enhances the real world by superimposing virtual objects on reality. The aim of this study was to analyze the application of AR in medicine and which of its technical solutions are the most used. We carried out a scoping review of the articles published between 2019 and February 2022. The initial search yielded a total of 2649 articles. After applying filters, removing duplicates and screening, we included 34 articles in our analysis. The analysis of the articles highlighted that AR has been traditionally and mainly used in orthopedics in addition to maxillofacial surgery and oncology. Regarding the display application in AR, the Microsoft HoloLens Optical Viewer is the most used method. Moreover, for the tracking and registration phases, the marker-based method with a rigid registration remains the most used system. Overall, the results of this study suggested that AR is an innovative technology with numerous advantages, finding applications in several new surgery domains. Considering the available data, it is not possible to clearly identify all the fields of application and the best technologies regarding AR.

Keywords: augmented reality; image guided surgery; surgery

Citation: Barcali, E.; Iadanza, E.; Manetti, L.; Francia, P.; Nardi, C.; Bocchi, L. Augmented Reality in Surgery: A Scoping Review. *Appl. Sci.* **2022**, *12*, 6890. https://doi.org/10.3390/app12146890

Academic Editors: João M. F. Rodrigues and Dimitris Mourtzis

Received: 5 May 2022
Accepted: 4 July 2022
Published: 7 July 2022

Publisher's Note: MDPI stays neutral with regard to jurisdictional claims in published maps and institutional affiliations.

1. Introduction

Imaging is known to play an increasingly important role in many surgery domains [1]. Its origin can be dated back to 1895 when W. C. Roentgen discovered the existence of X-rays [2]. While in the course of the twentieth century, X-rays have found increasing application, in more recent years, other techniques have been developed and acquiring data from the internal structures of the human body has become more and more useful [1,3–5]. All this facilitated an increasing use of images to guide surgeons during interventions, leading to the affirmation of image-guided surgery (IGS) [6]. In this sense, the need for reducing surgery evasiveness, by supporting physicians in the diagnosis and preoperative phases as well as during surgeries themselves, led to the use of different solutions such as the 3D visualization of anatomical parts and the application of augmented reality (AR) in surgery [1,3,4]. Augmented reality consists in merging the real word with virtual objects (VOs) generated by computer graphic systems, creating a world for the user that is augmented with VOs. The first application of AR in medicine dates back to 1968 when Sutherland created the first head-mounted display [7]. The term AR is often used in conjunction with virtual reality (VR). The difference between them is that VR creates a digital artificial environment by stimulating the senses of the user and simulating the external world through computer graphic systems [8], while AR overlays computer-generated images onto the real world, increasing the user perception and showing something that would otherwise not be perceptible as reported by Park et al. in [1] and Desselle et al. in [9].

The application of AR in IGS can be an increasingly important opportunity for the treatment of patients. In particular, AR allows one to see 3D images projected directly onto patients thanks to the use of special displays. All this can facilitate the perception of the reality examined and lighten the task of the operators themselves compared to the traditional approach consisting in 2D preoperative images displayed on 2D monitors [1,5].

In this way, doctors can directly see 3D images projected onto patients using special displays, described in the next paragraph, instead of using 2D preoperative images displayed on 2D monitors that require the doctor to mentally transform them into 3D objects as well as remove the sight from the patient [1,5].

The purpose of this review is providing an overview of AR by describing which medical applications it can be used in and which aspects characterize this technology to provide doctors with information on this emerging tool. We would like it to be a starting point for more in-depth research and applications in the clinical field. In order to better understand AR application, this review started by describing some key technological aspects such as: tracking, registration and displays.

2. Theoretical Background

This section describes the main aspects leading to the visualization of the VOs superimposed on the real world. The workflow of augmented-reality-enabled systems is shown in Figure 1. This Figure 1 shows that once the virtual model has been rendered, tracking and recording are the two basic steps. In this sense, tracking and registration provide the correct spatial positioning of the VOs with respect to the real world [10]. This result is possible because, with monitoring, the spatial characteristics of an object are detected and measured. Specifically, with regard to AR, tracking indicates the operations necessary to determine the device's six degrees of freedom, 3D location and orientation within the environment, necessary to calculate the real time user's point of view. Tracking can be performed outdoors and indoors. We focused on the latter. Two methods of indoor tracking are then distinguishable: outside-in and inside-out. In the outside-in method, the sensors are placed in a stationary place in the environment and sense the device location, often resorting to marker-based systems [11]. In the inside-out method, the camera or the sensors are placed on the actual device whose spatial features are to be tracked in the environment. In this case, the device aims to determine how its position changes in relation to the environment, as for the head-mounted displays (HMDs). The inside-out tracking can be marker-based or marker-less. The marker-based vision technique, making use of optical sensors, measures the device pose starting from the recognition of some fiducial markers placed in the environment. This method can also hyperlink physical objects to web-based content using graphic tags or automatic identification technologies such as radio-frequency-identification (RFId) systems [12]. The marker-less method, conversely, does not require fiducial markers. It bases its measures on the recognition of distinct characteristics, present in the environment, that in turn are used to localize the position of the device in combination with computer vision and image-processing techniques. Registration involves the matching and alignment of tracked spatial features obtained from the real world (RW) with the corresponding points of the VOs to reach an optimal overlapping between them [1]. The accuracy of this process allows an accurate representation of the virtual reality over the real world and determines the natural appearance of an augmented image [13]. The registration phase is connected to the tracking one. Based on the ways these two are accomplished, the process is defined as manual, fully automatic or semiautomatic. The manual one refers to manual registration and manual tracking. It consists in finding landmarks both on the model and the patient and consequently manually orienting and resizing of the obtained preoperative 3D model displayed on the operative monitor to make it match real images. The fully automatic process is the most complex one, especially with soft tissues. Since real world objects change their shapes with time, the same deformation needs to be applied to the VOs to address the fact that any deformation during surgery, due to events such as respiration, can result in an inaccurate real-time registration, subsequently causing an imprecise overlapping between 3D VOs and ROs. Finally, the semiautomatic process associates the automatic tracking with the manual registration. The identification of landmark structures, both on the obtained 3D model and on the real structures, occurs automatically, while its overlay on the model, and its orienting and resizing, occurs manually. This aspect is what differentiates the automatic process from the semiautomatic one. The latter provides the overlay of the AR images on real life statically and manually, while the former makes the 3D virtual models dynamically match the actual structures [1,14–16]. For the visualization of the VOs onto the real world, several AR display technologies exist, usually classified in head, body and world devices, depending on the place where they are located [7,17]. World devices are located in a fixed place. This category includes desktop displays used as AR displays, and projector-based displays. The former are equipped with a webcam, a virtual mirror showing the scene framed by the camera and a virtual showcase, allowing the user to see the scene, alongside additional information. Projector-based displays cast virtual objects directly onto the corresponding real-world objects' surfaces. With body devices, we usually refer to handheld Android-based platforms, such as tablets or mobile phones. These devices use the camera for capturing the actual scenes in real time, while some sensors (e.g., gyroscopes and accelerometers and magnetometer) can determine their rotation. These devices usually resort to fiducial image targets for the tracking-registration phase [18]. Finally, the HMDs are near eye displays, wearable devices consisting in sort of glasses that have the advantage of leaving the hands free to perform other tasks. HMDs are mainly of two types: video see-through and optical see-through. The first ones refer to special lenses that let the user see the external real world through a camera whose frames are in turn combined with VOs. In this way, the external environment is recorded in real time and the final images overlaying the VOs are produced directly over the user's lenses. Differently, the optical see-through devices consist of an optical combiner or holographic waveguides, the lenses, that enable the overlay of images transmitted by a projector over the same lenses through which a normal visualization of the real world is allowed. In this way the user visualizes directly the reality augmented with the VOs overlaid onto it [7,19]. Figure 2 shows an example of HMD.

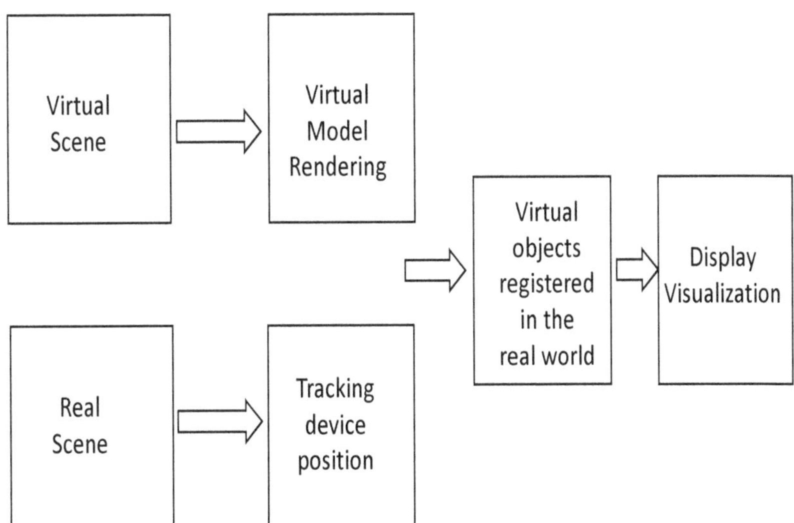

Figure 1. Workflow of augmented-reality-enabled systems.

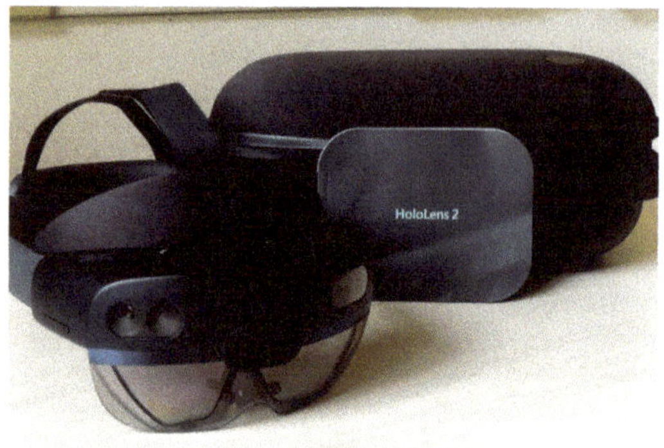

Figure 2. Example of HMD, HoloLens 2 (Microsoft, WA, USA) .

The different techniques are summarized in Figure 3. The aim of this study was to describe the state of the art relating to the use of AR in the surgery domain. The description and analyses of the various procedures used to create the virtual images represented a further objective. This scoping review aims to provide a summary of the surgical fields in which this new technology finds its best application providing doctors with an overview of the key aspects behind viewing accurate virtual images superimposed on the real world. The research highlighted that the marker-based tracking and the rigid registration are currently the most used systems to acquire data, as reported in the following paragraphs.

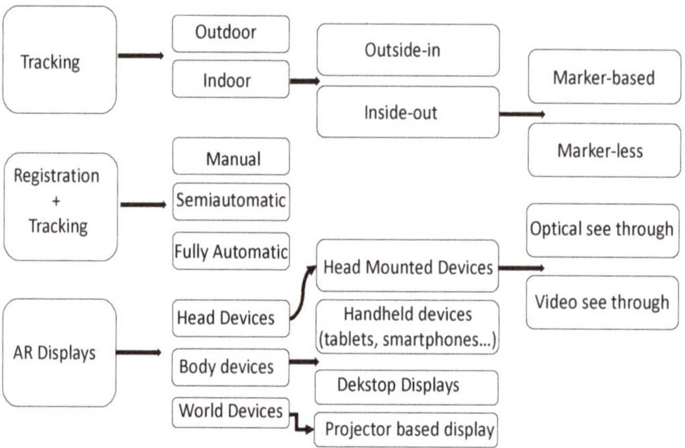

Figure 3. Summary of the techniques.

3. Materials and Methods

We followed the PRISMA Guidelines for scoping reviews [20]. The results are shown in Figure 4. The histogram in Figure 5 shows the trend of the number of publications from 1982 to 2021 present on Scopus searching English articles for "augmented reality in surgery". Between 2020 and 2021, the number of publications increased by 40%. In 2022, at the time of writing, 50 articles have already been published and indexed on Scopus.

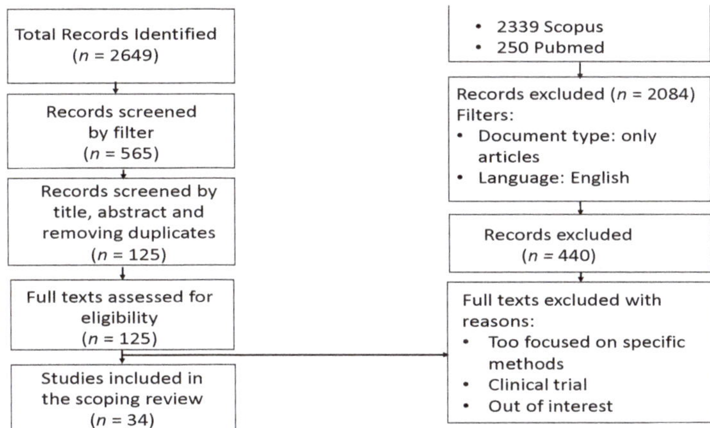

Figure 4. Criteria for the inclusion of articles.

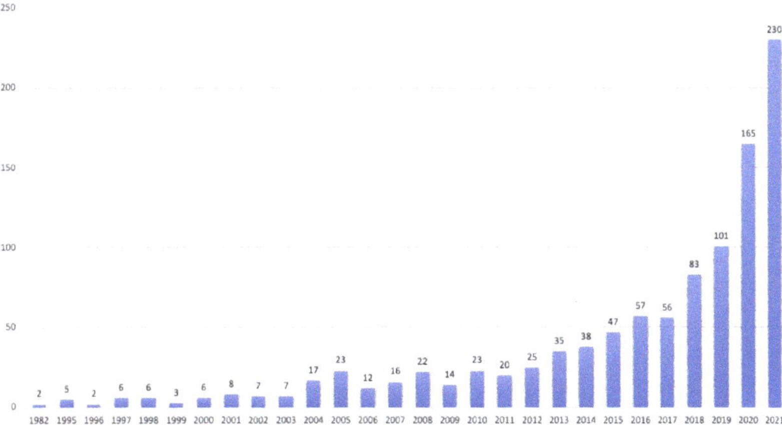

Figure 5. Trend of Publications on Augmented Reality in Surgery over the Years.

3.1. Inclusion Criteria

The studies included in the review need to be related to the main topic: augmented reality. We limited the selection by imposing restrictions on the document type (articles only) and on the language (English only). The query was limited to a relatively short period of time, (2019–February 2022) ensuring the attention was focused on the innovations introduced in the latest years. The queries we used during our searches were: "TITLE-ABS-KEY ("augmented reality" AND surgery) AND (LIMIT-TO (DOCTYPE, "ar")) AND (LIMIT-TO (LANGUAGE, "English")) AND (LIMIT-TO (PUBYEAR, 2022) OR LIMIT-TO (PUBYEAR, 2021) OR LIMIT-TO (PUBYEAR, 2020) OR LIMIT-TO (PUBYEAR, 2019))" for Scopus and Record on Pubmed.

3.2. Selection of Sources Criteria

The inclusion criteria were applied to filter the found articles. Additional documents were then added based on citations from excluded articles, deemed interesting for this review but not caught by the query because of the limitations that we decided to set. The team established two reviewers, E.B. and P.F. In both searches; they screened independently all the articles, starting from the abstracts and the titles, choosing the ones deemed pertinent according to their own judgement. The articles chosen by both reviewers were directly integrated in the list of articles to be downloaded. The studies that were chosen from only one of the two reviewers were integrated in the list only after the agreement of a third reviewer, L.B., who took the final decision whether to include or discard the article from the final review. Starting from this list, full texts of these studies were downloaded and the process of choice was repeated based on the content of the studies found, thus obtaining the final list of articles to be included.

4. Results

The initial search yielded a total of 2649 articles. After applying filters, removing duplicates and screening the studies based on abstracts and titles, 125 studies remained, from which those included in the study were chosen. The final summary refers to a total of 34 articles. The reason for not including some articles is related to their content, in some cases deemed too specific, concerning clinical trials or topics outside the field of interest. The list of AR applications in the different surgery domains as reported in the selected articles is shown in Table 1. We decided to create Table 1, containing an overview of the AR applications in different areas and methods present in the chosen articles. The Table 1 is organized as follows: the first column shows the author (or authors) of the article, the second the application to which the article refers, the third the technology used for processing, the fourth the display used to view the virtual object merged with reality, the fifth the registration method used in the article, the sixth the error made in terms of approximations and the seventh the data set that was used in the article.

Evaluating all the selected articles, both in the filtered research and those added manually, we decided to summarize the main aspects of the AR applications in three schemes reported in Tables 2–4. The aspects we decided to analyze and report as percentage of application in the analyzed studies are the ones described in the Section 2. For what concerns the application of AR in different fields, the scheme in Table 2 shows that this technique has been traditionally mainly used in orthopedics. Lately, the innovation has been represented by its increasingly widespread application in maxillofacial surgery, in addition to oncology. However, the numerous areas in which AR is used confirm the important role that this technology may have in the future in the health field. The scheme in Table 3 shows how the projection over the patient is, at the moment, the least used method, while the Optical Viewer by Microsoft, HoloLens, is the most used one. The first model (HoloLens 1) together with the second one (HoloLens 2) amounts to 38% of the scheme in Table 3. For what concerns tracking and registration, reported in the scheme in Table 4, the marker-based method paired with rigid registration remains the most used system. Once we analyzed all the articles listed in Table 2, we decided to delve into the applications more recurrent in our research and which, in our opinion, seemed to have the most interesting clinical implications. The applications we decided to investigate are reported below.

Table 1. Most recent Augmented reality application for each field and method that resulted from the research.

Author	Application	Technology	Display	Registration	Error	Data Set
Schwam [21]	Lateral skull surgery	BrainLab Curve™, Surgical Theate and Zeiss OPMI PENTERO 900	Microscope-based HUD	Marker-less, rigid	Not reported	40 patients.
Coelho [22]	Antenatal Treatment of Myelomeningocele. Preoperative and post operative simulation to make it happen	Unity Engine, Google ARCore libraries, ray casting target object rendering	Application for smartphone and tablets	object placed based on the rendering,	Not reported	1 pregnant woman at 27 weeks of gestation.
Gouveia [23]	Left breast cancer: Oncology	Contrast-enhanced MRI Horos R software v2.4.0	HoloLens AR Headset	Marker-based, rigid,	Not reported	57 menopausal woman.
Chen [24]	Knee surgery arthroscopy	CT scanner, optical tracking system,	Glasses-free 3D display	Marker-based, rigid	Mean: 0.32 mm. Reduced error targets of 2.10 mm and 2.70 mm	Experiments: preclinical on knee phantom and in-vitro swine knee.
Golse [25]	Liver resection	3D segmentation. CT	Standard mobile external monitor	Real time Marker-less, non-rigid	7.9 mm root mean square error for internal landmark registration	In vivo: 5 patients, ex vivo: native tumor-free.
Gsaxner [26]	Head and neck Carcinoma: Training	CT, PET-CT and MRI scans. Instant calibration	HoloLens AR Headset	Marker-less, rigid	Between a few mm of up to 2 cm.	11 health care professionals.
Molina [27]	Spinal navigation	iCT scans. Gertzbein-Robbins (GS) scale	AR-HMD Xvision (Augmedics)	Marker-based, rigid	Linear deviation: 2.07 mm. Angular deviation: 2.41°.	78-yr-old female.
Ackermann [28]	Osteotomy cuts and reorientation of the acetabular fragment: navigation system	CT scan	Microsoft HoloLens	Marker-based, rigid	Osteotomy starting points: 10.8 mm. Osteotomy directions: 5.4. Reorientation errors: $x = 6.7°$, $y = 7.0°$, $z = 0.9°$. LCE angle postoperative error: 4.5°	2 fresh-frozen human cadaverous hips.
Liu [29]	Medical training and telemonitored surgery	2 color digital cameras.	Microsoft HoloLens	Marker-based, rigid	The overall tracking one: less than 2.5 mm. The overall guidance one: less than 2.75 mm	Ex vivo arm phantom, in vivo rabbit model.

Table 1. *Cont.*

Author	Application	Technology	Display	Registration	Error	Data Set
Collins [30]	Uterus: Laparoscopy	MR or CT and monocular laparoscopes	Monitor	Marker-less, rigid	Distribution increase towards the cervix (2 mm for 15 views up to 8 mm for 2 views)	Phantom and videos recorded during laparoscopic surgery.
Arpaia [31]	Neurosurgery	Equipment of the brain computer interface.	Epson Moverio BT-350 glasses.	Not reported	Not reported	10 runs on the same patient.
Shrestha [32]	Bowel	CT scans and endoscope camera intraoperatively.	Monitor	Marker-based, rigid	The overlay error accuracy was 0.24777px. Performance was 44fps	People with three different ages: 15–25, 26–35, 35–60.
Wei [33]	Plastic surgery	Google Face API	Android display	Rigid, Marker-based	Not reported	4 benchmarks data set.
Lee [34]	thyroid surgery	CT. Semiautomatic registration	AR screen. Master surgical robot screen.	Marker-based, rigid	Mean ± SD = 1.9 ± 1.5 mm	9 patients.
Hussain [35]	Ear surgery	Without tracking system, CT. Microscope 2D real time video	DDM. Bronchoscopy monitor	Marker-less, rigid	Surgical instrument tip position one: 0.3 ± 0.22 mm	6 artificial human temporal bone specimens.
Feufel [36]	Ultrasound guided needle placement	Reflective markers Ultrasound transducer	Microsoft HoloLens	Marker-based, rigid	Mean error of 7.4 mm	20 participants.
Carl [37]	Aneurysm surgery: indocyanine green (ICG) hagiography	CT, 3D rotational (DynaCT) or Time-of-flight magnetic resonance angiography. Automatic registration	Operating microscope HUD	Marker-based, rigid	Target registration one: 0.71 ± 0.21 mm	20 patients with 22 aneurysm.
Chan [38]	Transoral robotic surgery	CT	3D Surgeon's console	Marker-based, rigid	Not reported	2 cadavers.
Ferraguti [39]	Percutaneous Nephrolithotomy	Ct or MRI, 3 electrodes. Real time registration.	Microsoft HoloLens	Marker-based, rigid	Translation and orientation norm between 2 transformation matrices: 15.80 mm and 4.12°	11 samples.
Auloge [40]	Percutaneous vertebroplasty	Cone-beam CT	Monitor	Marker-based, rigid	Not reported	2 groups of 10 patients.
Libaw [41]	Inhaled induction of general anesthesia, pediatric	iPhone 7.	AR headset	Not reported	Not reported	3 patients: 8 an 10 years old.

Table 1. *Cont.*

Author	Application	Technology	Display	Registration	Error	Data Set
Pietruski [42]	Fibula free flap harvest	7 markers. Actual, virtual registration. Sagittal surgical saw (GB129R) with a tracking adapter	HMD: Moverio BT-200 Smart Glasses, Epson	Marker-based, rigid	Not reported	756 osteotomies simulated.
Jiang [43]	Vascular localization system	CTA scan. No ionic contrast agent. Registration real time.	Microsoft HoloLens	Marker-based, rigid	Minimum 1.35 mm. Maximum 3.18 mm	7 operators.
Samei [44]	Laparoscopic radical prostatectomy	MRI. 3 transformations.	From Da Vinci to pc	Marker-based, rigid	Not reported	Agar prostate phantom ex vivo. 12 patients in vivo.
Rose [45]	Otolaryngology - head and neck surgery	CT, MeshLab and Unity.	Microsoft HoloLens	Marker-based, rigid	In measurement of accuracy: 2.47 ± 0.46 mm (1.99, 3.30)	A phantom.
Carl [46]	Transsphenoidal Surgery	C-arm radiographic fluoroscopy. Registration using iCT.	Operating microscopes HUD	Marker-based, rigid	Target registration error of 0.83 ± 0.44 mm	288 cases of transsphenoidal surgery.
Sharma [47]	Jaw surgery	Ct scan. Virtual scenes. Stereo views.	monitor	Marker-less, rigid	Alignment error 0.59 ± 0.62 mm	20 different samples after jaw surgery.
Abdel [48]	Foot sarcoma: Oncology	NDI Polaris. Smartphone AR application: FINO	Samsung galaxy	Marker-based, rigid	Not reported	A 39-year-old male patient.
Melero [49]	Rehabilitation: upper limbs	Myo armband. EMG data. Microsoft Kinect sensor	Monitor	Marker-less, rigid	Not reported	3 subjects, with 10 trials for each subject.
Tu [50]	Orthopedics	C++ application on pc. C♯ application in Unity. Connection via TCP/IP	HoloLens 2	Marker-based, rigid	Distance error: 1.61 ± 0.44 mm. 3D angle error: $1.46 \pm 0.46°$	Phantom and cadaver experiment.
Cofano [51]	Spine Surgery	Ct, TeamViewer software and holosurgery	HoloLens 2	Marker-less, rigid	not reported	2 patients.
Heinrich [52]	Training	Not specified	HoloLens 1	Marker-based, rigid	Error rates ($p = 0.047$)	10 surgical trainees.

Table 2. Augmented Reality Applications.

Application	Percentage of Application
Telemonitoring	4%
Maxillofacial	23%
Liver Surgery	4%
Pediatric	4%
Orthopedics	27%
Oncology	19%
Training	8%
Puncture Surgery	7%
Bowel Surgery	4%

Table 3. Percentage of distribution of the displays of Augmented Reality used in medical applications evaluated in our study.

Type of Display	Percentage of Application
Smartphone	14%
Video see through Device	14%
Generic Head Mounted Display	17%
Unspecified Display	14%
Projected Directly over the Patient	3%
HoloLens 2	10%
HoloLens 1	28%

Table 4. Augmented Reality tracking and registration methods.

Tracking and Registration Methods	Percentage of Application
Marker based and Non-rigid Registration	4%
Markerless rigid Registration	20%
Markerless Non-rigid Registration	8%
Markeerbased and rigid Registration	68%

4.1. Oncology

AR application is frequent in oncology, being used for osteosarcoma [53], mandibular [54], kidney and prostate cancer [55], meningioma [56], urological cancer, intracranial [57], neuro-oncological [58], and cancer of the liver [14]. Indeed, AR application ensures an accurate visualization of the tumor, identifying its edges and position during surgeries. The capability to visualize the real anatomical structures, such as convolutions, grooves, blood vessels and nervous tracts, allows control during their resection, and permits surgeons to try to eradicate the tumor while removing as little of the surrounding healthy tissue as possible [59–62]. Adequate planning also provides bone information that, together with information about the tumor, can lead to its successful removal [54]. Furthermore, the application of the AR to the innovative twin digital simulation technique can also be a medical support tool. In particular, this solution may allow oncologists to monitor and control the patient in addition to predicting the outcome of cancer through the development of appropriate simulation models and the creation of appropriate data sets [63].

4.2. Orthopedics

The application of AR in orthopedics [64–66] is relatively recent, dating back to the beginning of the 2000s [65]. The purpose of applying AR to orthopedic computer systems for computer-assisted surgery (CAS) is to increase the accuracy during surgeries, improving the possible outcomes and at the same time decreasing procedure-related complications. AR application can also contribute to the reduction of both surgery time and radiographic doses for both patients and surgery teams. AR avoids the use of X-rays to see through the patients, reducing their exposure time [67].

4.3. Spinal Surgeries

AR is often used in spinal surgeries [68,69]. In this application, the accuracy is fundamental since an imprecision in the placement of an instrument can lead to spinal cord, nerve root or vascular injuries [70]. Open methods and direct visualization supporting the placement of the instrumentation, such as pedicle screw, have historically characterized this type of surgery [70]. The use of AR in spine surgery dates back to 1997 when Peuchot and his team developed a system for visualizing a vertebra during surgery [71]. For the past 20 years, Minimally Invasive Surgery (MIS) has been under investigation. Many articles have targeted study of Minimally Invasive Surgery (MIS) over the past 20 years. This has led to the introduction of new approaches such as the inoperative navigation that introduces several advantages to visualizing anatomy and precisely guiding surgeries. Furthermore, MIS ensures a higher level of accuracy, while minimizing possible damage to contiguous structures, providing access to deeper ones and improving dynamics and logistics in the operating room. The union of AR and MIS allows the surgeon to see more accurately inside the patient, possibly visualizing the preoperative planned drilling trajectory over the display, ensuring advantages in terms of accuracy, reduction of radiation exposure, blood loss and hospital stay. The drawbacks are mainly related to high costs and to the steepness of the learning curve, still too high [72].

4.4. Neurosurgery

The use of AR is quite frequent also in neurosurgery. Its application in this area has already been tackled in oncology, but it finds its maximum utility in neuronavigation [73]. It can help surgeons in reducing the consequences of the treatment, improving the quality of the surgery and reducing the operation time [74–77]. The first neuronavigation system (NNS) dates back to 1986. The advantage offered by AR associated to NSS consists in the mapping of the preoperative images directly onto the patient's visible surface, thus showing its anatomy on it [73,78].

4.5. Surgical Training and Medical Education

AR is assuming a fundamental and emerging role also for what concerns surgical training and medical education [79–83]. Its introduction results in providing students with a better anatomic conceptualization and allows surgical simulations to improve their performances [84]. AR ensures the possibility to practice surgeries without risks for the patient, saving the need of a supervisor and consequently reducing costs for the structures [85]. It also provides an increasing acquisition of skills such as speed, ability to multitask, accuracy, hand–eye coordination and bimanual operation. The evolution of this system has led to the use of telemonitoring, where experienced surgeons can train students remotely, and also to take part in consultations among experts located in different countries [86].

5. Discussion

Augmented reality is an innovative technology that presents several advantages, with new applications still in development. Knowing about this technology is every day becoming more important and can provide information to medical doctors and encourage new applications and deeper research. The reason for its increasing success is connected to the possibility it offers to visualize and interact with digital objects without having to lose view of the real world to watch the monitor displaying the medical imaging of the area of interest [1]. Moreover, research has shown its capacity to reduce the exposure to ionizing radiation. This aspect is important because it is well known that ionizing radiations can have harmful effects with possible effects on biomolecules such as DNA, lipids, proteins, and cancer risks [87–89]. One study [71] calculated the average of the staff radiation exposure using AR that, compared to the literature values, decreased to less than 0.01%. Moreover, the absorbed dose of the patient exposition resulted in a decrease of its value up to 32% compared to the quantities due to conventional techniques [71]. All these aspects may allow the diffusion of AR and the possibility of assuming it as a systemic tool in medicine. The analysis of the studies considered showed that AR finds applications in many surgery domains and especially in the field of maxillofacial surgery, orthopedics and oncology. In particular, oncology is one of the areas of application particularly indicated. In this sense, AR finds a lot of applications in different kinds of cancer with the aim of facilitating and reducing the consequences of the treatment as well as improving outcomes [14,53–58]. Even with regard to orthopedics, the use of AR can be particularly recommended and is aimed at promoting the quality of surgical interventions, and therefore improving the outcomes as well as reducing the risk of complications [64–67]. In this sense, spinal surgeries represent an important area of application of AR where it can represent an important resource available to surgeons [68,69]. Regarding the available display technology, the results obtained show that the Optical Viewer by Microsoft, HoloLens, is the most used [36,39,90]. The marker-based method paired with the rigid registration was the most used solution in the context of AR tracking and registration methods [42–46]. In this regard, it is clear that the goal is to be able to reduce or eliminate the problems associated with tissue deformations. Unfortunately, the limited number of data available did not allow for more in-depth analyses on this issue. AR is a technology that is every day becoming more popular. Here we provided an idea of what it is, which technologies it is formed from and in which applications it is more popularly used. Unfortunately, some limitations still affect the application of AR in the surgical field. From our study, we noticed that the output is too much related to the accuracy of the registration and tracking systems that need to be as reliable as possible. Errors during those mentioned phases could lead to a misalignment of the VOs with the real world [91,92]. Mainly for the HMDs, the different field of view between human vision and visors represents an obstacle too [93,94]. Finally, one of the biggest issues that affects this technology is the vergence–accommodation conflict. In nature, the point where the eyes verge and focus is the same, while AR displays are featured by a fixed focal distance; consequently, the points of vergence and focus may be different. This causes discomfort, fatigue and different eye depth perception [95–97]. Some limitations characterize this study since the purpose of the review consisted in providing a contemporary view, but the results may exclude longitudinal trends. A potential problem in this study may also be the possible underrepresentation of documents about AR in surgery. Not all the studies published in the years analyzed may have been identified, despite trying to be as comprehensive as possible (according to the filters chosen). For our search, we used only those terms indicated in the Section 3, but others could have been chosen. Moreover, it is possible that some papers were excluded as they did not include those specific words, but their synonyms. Furthermore, our search was attempted using two multidisciplinary databases, Pubmed and Scopus, but others could have yielded additional studies. We decided to use only English terms and include only English articles. We did not reach out to experts on the topic for a consultation about additional studies that we may not have included.

6. Conclusions

AR is a technology that is increasingly being applied in surgery. This is due to the numerous advantages it offers although it is still an evolving technology. Since AR allows an accurate visualization of the anatomical structures and a good control of the activities performed during surgical resections, the fields in which it is most commonly used are orthopedics and oncology. For what concerns the displays, Microsoft HoloLens Viewer is the most used method. Likewise, the marker-based system combined with rigid registration is the most common solution for tracking and registration. The need for high accuracy of registration and tracking systems, as well as VOs misalignment problems and the possible vergence–accommodation conflict are important limitations. The latter can hinder the use of AR in surgery. The results of this study, as well as presenting the technological solutions used, show that AR can be applied in different fields of surgery. All of this can favor the realization of further studies aimed at overcoming the current limitations on AR in the clinical setting as well as promoting its application. Considering the significant role that AR can play within the treatment of a large numbers of patients, further studies are needed to better define all possible fields of application of AR and the best technological solutions to be used.

Author Contributions: E.B., L.M., E.I. and L.B. designed the study. E.B., P.F. and L.B. performed the bibliographic research and organized the results. E.I., P.F., C.N. and L.M. aided in interpreting the results and wrote the final version of the manuscript with the support of all authors. All authors have read and agreed to the published version of the manuscript.

Funding: This work was supported by Fondazione Cassa di Risparmio di Firenze, Florence, Italy (grant number 2020.1515). The authors thank Ian Webster PGCE for revising the English content.

Institutional Review Board Statement: Not applicable.

Informed Consent Statement: Not applicable.

Data Availability Statement: Not applicable.

Conflicts of Interest: The authors declare no conflict of interest.

References

1. Park, B.J.; Hunt, S.J.; Martin, C., III; Nadolski, G.J.; Wood, B.; Gade, T.P. Augmented and Mixed Reality: Technologies for Enhancing the Future of IR. *J. Vasc. Interv. Radiol.* **2020**, *31*, 1074–1082. [CrossRef] [PubMed]
2. Villarraga-Gómez, H.; Herazo, E.L.; Smith, S.T. X-ray computed tomography: From medical imaging to dimensional metrology. *Precis. Eng.* **2019**, *60*, 544–569. [CrossRef]
3. Cutolo, F. Augmented Reality in Image-Guided Surgery. In *Encyclopedia of Computer Graphics and Games*; Lee, N., Ed.; Springer International Publishing: Cham, Switzterland, 2017; pp. 1–11._78-1. [CrossRef]
4. Allison, B.; Ye, X.; Janan, F. *MIXR: A Standard Architecture for Medical Image Analysis in Augmented and Mixed Reality*; IEEE Computer Society: Washington, DC, USA , 2020; pp. 252–257. [CrossRef]
5. Marmulla, R.; Hoppe, H.; Mühling, J.; Eggers, G. An augmented reality system for image-guided surgery: This article is derived from a previous article published in the journal International Congress Series. *Int. J. Oral Maxillofac. Surg.* **2005**, *34*, 594–596. [CrossRef] [PubMed]
6. Peters, T.M. Image-guidance for surgical procedures. *Phys. Med. Biol.* **2006**, *51*, R505–R540. [CrossRef] [PubMed]
7. Eckert, M.; Volmerg, J.S.; Friedrich, C.M. Augmented Reality in Medicine: Systematic and Bibliographic Review. *JMIR Publ.* **2019**, *7*, e10967 . [CrossRef] [PubMed]
8. Kim, Y.; Kim, H.; Kim, Y.O. Virtual Reality and Augmented Reality in Plastic Surgery: A Review. *Arch. Plast. Surg.* **2017**, *44*, 179–187. [CrossRef] [PubMed]
9. Desselle, M.R.; Brown, R.A.; James, A.R.; Midwinter, M.J.; Powell, S.K.; Woodruff, M.A. Augmented and Virtual Reality in Surgery. *Comput. Sci. Eng.* **2020**, *22*, 18–26. [CrossRef]
10. Pérez-Pachón, L.; Poyade, M.; Lowe, T.; Gröning, F. Image Overlay Surgery Based on Augmented Reality: A Systematic Review. In *Biomedical Visualisation. Advances in Experimental Medicine and Biology*; Springer International Publishing: Cham, Switzterland , 2020; Volume 1260, pp. 175–195._10. [CrossRef]
11. Zafari, F.; Gkelias, A.; Leung, K.K. A Survey of Indoor Localization Systems and Technologies. *IEEE Commun. Surv. Tutor.* **2019**, *21*, 2568–2599. [CrossRef]
12. Cheng, J.; Chen, K.; Chen, W. Comparison of marker-based AR and markerless AR: A case study on indoor decoration system. In Proceedings of the Lean and Computing in Construction Congress (LC3): Proceedings of the Joint Conference on Computing in Construction (JC3), Heraklion, Greece, 4–7 July 2017; pp. 483–490. [CrossRef]
13. Thangarajah, A.; Wu, J.; Madon, B.; Chowdhury, A.K. Vision-based registration for augmented reality-a short survey. In Proceedings of the 2015 IEEE International Conference on Signal and Image Processing Applications (ICSIPA), Kuala Lumpur, Malaysia, 19–21 October 2015; pp. 463–468. [CrossRef]
14. Quero, G.; Lapergola, A.; Soler, L.; Shahbaz, M.; Hostettler, A.; Collins, T.; Marescaux, J.; Mutter, D.; Diana, M.; Pessaux, P. Virtual and Augmented Reality in Oncologic Liver Surgery. *Surg. Oncol. Clin. N. Am.* **2019**, *28*, 31–44. [CrossRef]
15. Tuceryan, M.; Greer, D.S.; Whitaker, R.T.; Breen, D.E.; Crampton, C.; Rose, E.; Ahlers, H.K. Calibration requirements and procedures for a monitor-based augmented reality system. *IEEE Trans. Vis. Comput. Graph.* **1995**, *1*, 255–273. [CrossRef]
16. Maybody, M.; Stevenson, C.; Solomon, S.B. Overview of Navigation Systems in Image-Guided Interventions. *Tech. Vasc. Interv. Radiol.* **2013**, *16*, 136–143. [CrossRef]
17. Zhanat, M.; Vslor, H.A. Augmented Reality for Robotics: A Review. *Robotics* **2020**, *9*, 21. [CrossRef]

18. Mourtzis, D.; Angelopoulos, J.; Panopoulos, N. Intelligent Predictive Maintenance and Remote Monitoring Framework for Industrial Equipment Based on Mixed Reality. *Front. Mech. Eng.* **2020**, *6*, 578379. [CrossRef]
19. Bruce, T. H. A Survey of Visual, Mixed, and Augmented Reality Gaming. *Assoc. Comput. Mach.* **2012**, *10*, 1. doi: 10.1145/238 1876.2381879 . [CrossRef]
20. Tricco, A.C.; Lillie, E.; Zarin, W.; O'Brien, K.K.; Colquhoun, H.; Levac, D.; Moher, D.; Peters, M.D.; Horsley, T.; Weeks, L.; et al. PRISMA Extension for Scoping Reviews (PRISMA-ScR): Checklist and Explanation. *Ann. Intern. Med.* **2018**, *169*, 467–473. [CrossRef]
21. Schwam, Z.G.; Kaul, V.F.; Bu, D.D.; Iloreta, A.M.C.; Bederson, J.B.; Perez, E.; Cosetti, M.K.; Wanna, G.B. The utility of augmented reality in lateral skull base surgery: A preliminary report. *Am. J. Otolaryngol.* **2021**, *42*, 102942. [CrossRef]
22. Coelho, G.; Trigo, L.; Faig, F.; Vieira, E.V.; da Silva, H.P.G.; Acácio, G.; Zagatto, G.; Teles, S.; Gasparetto, T.P.D.; Freitas, L.F.; et al. The Potential Applications of Augmented Reality in Fetoscopic Surgery for Antenatal Treatment of Myelomeningocele. *World Neurosurg.* **2022**, *159*, 27–32. [CrossRef]
23. Gouveia, P.F.; Costa, J.; Morgado, P.; Kates, R.; Pinto, D.; Mavioso, C.; Anacleto, J.; Martinho, M.; Lopes, D.S.; Ferreira, A.R.; et al. Breast cancer surgery with augmented reality. *Breast* **2021**, *56*, 14–17. [CrossRef]
24. Chen, F.; Cui, X.; Han, B.; Liu, J.; Zhang, X.; Liao, H. Augmented reality navigation for minimally invasive knee surgery using enhanced arthroscopy. *Comput. Methods Programs Biomed.* **2021**, *201*, 105952. [CrossRef]
25. Golse, N.; Petit, A.; Lewin, M.; Vibert, E.; Cotin, S. Augmented Reality during Open Liver Surgery Using a Markerless Non-rigid Registration System. *J. Gastrointest. Surg.* **2021**, *25*, 662–671. [CrossRef]
26. Gsaxner, C.; Pepe, A.; Li, J.; Ibrahimpasic, U.; Wallner, J.; Schmalstieg, D.; Egger, J. Augmented Reality for Head and Neck Carcinoma Imaging: Description and Feasibility of an Instant Calibration, Markerless Approach. *Comput. Methods Programs Biomed.* **2020**, *200*, 105854. [CrossRef] [PubMed]
27. Molina, C.; Sciubba, D.; Greenberg, J.; Khan, M.; Withamm, T. Clinical Accuracy, Technical Precision, and Workflow of the First in Human Use of an Augmented-Reality Head-Mounted Display Stereotactic Navigation System for Spine Surgery. *Oper. Neurosurg.* **2021**, *20*, 300–309. [CrossRef] [PubMed]
28. Ackermann, J.; Florentin, L.; Armando, H.; Jess, S.; Mazda, F.; Stefan, R.; Patrick, Z.; Furnstahl, P. Augmented Reality Based Surgical Navigation of Complex Pelvic Osteotomies—A Feasibility Study on Cadavers. *Appl. Sci.* **2021**, *11*, 1228. [CrossRef]
29. Peng, L.; Chenmeng, L.; Changlin, X.; Zeshu, Z.; Junqi, M.; Jian, G.; Pengfei, S.; Ian, V.; Pawlik, T.M.; Chengbiao, D.; et al. A Wearable Augmented Reality Navigation System for Surgical Telementoring Based on Microsoft HoloLens. *Ann. Biomed. Eng.* **2021**, *49*, 287–298. [CrossRef]
30. Collins, T.; Pizarro, D.; Gasparini, S.; Bourdel, N.; Chauvet, P.; Canis, M.; Calvet, L.; Bartoli, A. Augmented Reality Guided Laparoscopic Surgery of the Uterus. *IEEE Trans. Med. Imaging* **2021**, *40*, 371–380. [CrossRef] [PubMed]
31. Arpaia, P.; Benedetto, E.D.; Duraccio, L. Design, implementation, and metrological characterization of a wearable, integrated AR-BCI hands-free system for health 4.0 monitoring. *Measurement* **2021**, *177*, 109280. [CrossRef]
32. Shrestha, G.; Alsadoon, A.; Prasad, P.W.C.; Al-Dala'in, T.; Alrubaie, A. A novel enhanced energy function using augmented reality for a bowel: Modified region and weighted factor. *Multimed. Tools Appl.* **2021**, *80*, 17893–17922. [CrossRef]
33. Wei, W.; Ho, E.; McCay, K.; Damasevicius, R.; Maskeliunas, R.; Esposito, A. Assessing Facial Symmetry and Attractiveness using Augmented Reality. *Pattern Anal. Appl.* **2021**, 1–17. [CrossRef]
34. Lee, D.; Yu, H.W.; Kim, S.; Yoon, J.; Lee, K.; Chai, Y.J.; Choi, J.Y.; Kong, H.J.; Lee, K.E.; Cho, H.S.; et al. Vision-based tracking system for augmented reality to localize recurrent laryngeal nerve during robotic thyroid surgery. *Sci. Rep.* **2020**, *10*, 8437. [CrossRef]
35. Hussain, R.; Lalande, A.; Marroquin, R.; Guigou, C.; Bozorg Grayeli, A. Video-based augmented reality combining CT-scan and instrument position data to microscope view in middle ear surgery. *Sci. Rep.* **2020**, *10*, 6767. [CrossRef]
36. Rüger, C.; Feufel, M.; Moosburner, S.; Özbek, C.; Pratschke, J.; Sauer, I. Ultrasound in augmented reality: A mixed-methods evaluation of head-mounted displays in image-guided interventions. *Int. J. Comput. Assist. Radiol. Surg.* **2020**, *15*, 1895–1905. [CrossRef]
37. Carl, B.; Bopp, M.H.A.; Benescu, A.; Saß, B.; Nimsky, C. Indocyanine green angiography visualized by augmented reality in aneurysm surgery. *World Neurosurg.* **2020**, *142*, e307–e315. [CrossRef]
38. Chan, J.Y.K.; Holsinger, F.C.; Liu, S.; Sorger, J.M.; Azizian, M.; Tsang, R.K.Y. Augmented reality for image guidance in transoral robotic surgery. *J. Robot. Surg.* **2019**, *14*, 579–583. [CrossRef]
39. Ferraguti, F.; Minelli, M.; Farsoni, S.; Bazzani, S.; Bonfè, M.; Vandanjon, A.; Puliatti, S.; Bianchi, G.; Secchi, C. Augmented Reality and Robotic-Assistance for Percutaneous Nephrolithotomy. *IEEE Robot. Autom. Lett.* **2020**, *5*, 4556–4563. [CrossRef]
40. Auloge, P.; Cazzato, R.; Ramamurthy, N.; De Marini, P.; Rousseau, C.; Garnon, J.; Charles, Y.; Steib, J.P.; Gangi, A. Augmented reality and artificial intelligence-based navigation during percutaneous vertebroplasty: a pilot randomised clinical trial. *Eur. Spine J.* **2020**, *29*, 1580–1589. [CrossRef]
41. Libaw, J.; Sinskey, J. Use of Augmented Reasameility During Inhaled Induction of General Anesthesia in 3 Pediatric Patients: A Case Report. *A&A Pract.* **2020**, *14*, e01219. [CrossRef]
42. Pietruski, P.; Majak, M.; Swiatek-Najwer, E.; Żuk, M.; Popek, M.; Jaworowski, J.; Mazurek, M. Supporting Fibula Free Flap Harvest With Augmented Reality: A Proof-of-Concept Study. *Laryngoscope* **2019**, *130*, 1173–1179. [CrossRef]

43. Jiang, T.; Yu, D.; Wang, Y.; Zan, T.; Wang, S.; Li, Q. HoloLens-Based Vascular Localization System: Precision Evaluation Study With a Three-Dimensional Printed Model. *J. Med. Internet Res.* **2020**, *22*, e16852. [CrossRef]

44. Samei, G.; Tsang, K.; Kesch, C.; Lobo, J.; Hor, S.; Mohareri, O.; Chang, S.; Goldenberg, S.L.; Black, P.C.; Salcudean, S. A partial augmented reality system with live ultrasound and registered preoperative MRI for guiding robot-assisted radical prostatectomy. *Med. Image Anal.* **2020**, *60*, 101588. [CrossRef]

45. Rose, A.; Kim, H.; Fuchs, H.; Frahm, J.M. Development of augmented-reality applications in otolaryngology-head and neck surgery: Augmented Reality Applications. *Laryngoscope* **2019**, *129*, S1–S11. [CrossRef]

46. Carl, B.; Bopp, M.; Voellger, B.; Saß, B.; Nimsky, C. Augmented reality in transsphenoidal surgery. *World Neurosurg.* **2019**, *125*, e873–e883. [CrossRef] [PubMed]

47. Sharma, A.; Alsadoon, A.; Prasad, P.W.C.; Al-Dala'in, T.; Haddad, S. A novel augmented reality visualization in jaw surgery: enhanced ICP based modified rotation invariant and modified correntropy. *Multimed. Tools Appl.* **2021**, *80*, 1–25. [CrossRef]

48. Abdel Al, S.; Abou Chaar, M.K.; Mustafa, A.; Al-Hussaini, M.; Barakat, F.; Asha, W. Innovative Surgical Planning in Resecting Soft Tissue Sarcoma of the Foot Using Augmented Reality With a Smartphone. *J. Foot Ankle Surg.* **2020**, *59*, 1092–1097. [CrossRef]

49. Melero, M.; Hou, A.; Cheng, E.; Tayade, A.; Lee, S.C.; Unberath, M.; Navab, N. Upbeat: Augmented Reality-Guided Dancing for Prosthetic Rehabilitation of Upper Limb Amputees. *J. Healthc. Eng.* **2019**, *2019*, 1–9. [CrossRef] [PubMed]

50. Tu, P.; Yao, G.; Lungu, A.; Li, D.; Wang, H.; Chen, X. Augmented Reality Based Navigation for Distal Interlocking of Intramedullary Nails Utilizing Microsoft HoloLens 2. *Comput. Biol. Med.* **2021**, *133*, 104402. [CrossRef] [PubMed]

51. Cofano, F.; Di Perna, G.; Bozzaro, M.; Longo, A.; Marengo, N.; Zenga, F.; Zullo, N.; Cavalieri, M.; Damiani, L.; Boges, D.; et al. Augmented Reality in Medical Practice: From Spine Surgery to Remote Assistance. *Front. Surg.* **2021**, *8*, 657901. [CrossRef] [PubMed]

52. Heinrich, F.; Huettl, F.; Schmidt, G.; Paschold, M.; Kneist, W.; Huber, T.; Hansen, C. HoloPointer: A virtual augmented reality pointer for laparoscopic surgery training. *Int. J. CARS* **2021**, *16*, 161–168. [CrossRef] [PubMed]

53. Brookes, M.J.; Chan, C.D.; Baljer, B.; Wimalagunaratna, S.; Crowley, T.P.; Ragbir, M.; Irwin, A.; Gamie, Z.; Beckingsale, T.; Ghosh, K.M.; et al. Surgical Advances in Osteosarcoma. *Cancers* **2021**, *13*, 388. [CrossRef]

54. Kraeima, J.; Glas, H.; Merema, B.; Vissink, A.; Spijkervet, F.; Witjes, M. Three-dimensional virtual surgical planning in the oncologic treatment of the mandible. *Oral Dis.* **2021**, *27*, 14–20. [CrossRef]

55. Wake, N.; Nussbaum, J.E.; Elias, M.I.; Nikas, C.V.; Bjurlin, M.A. 3D Printing, Augmented Reality, and Virtual Reality for the Assessment and Management of Kidney and Prostate Cancer: A Systematic Review. *Urology* **2020**, *143*, 20–32. [CrossRef]

56. Alexandre, L.; Torstein, M.; Karl, S.; Marco, C. Augmented reality in intracranial meningioma surgery: A case report and systematic review. *J. Neurosurg. Sci.* **2020**, *64*, 369–376. [CrossRef]

57. Lee, C.; Wong, G.K.C. Virtual reality and augmented reality in the management of intracranial tumors: A review. *J. Clin. Neurosci.* **2019**, *62*, 14–20. [CrossRef]

58. Gerard, I.J.; Kersten-Oertel, M.; Petrecca, K.; Sirhan, D.; Hall, J.A.; Collins, D.L. Brain shift in neuronavigation of brain tumors: A review. *Med. Image Anal.* **2017**, *35*, 403–420. [CrossRef]

59. Inoue, D.; Cho, B.; Mori, M.; Kikkawa, Y.; Amano, T.; Nakamizo, A.; Yoshimoto, K.; Mizoguchi, M.; Tomikawa, M.; Hong, J.; et al. Preliminary study on the clinical application of augmented reality neuronavigation. *J. Neurol. Surg. Part A Cent. Eur. Neurosurg.* **2013**, *74*, 71–76. [CrossRef]

60. Besharati, T.L.; Mehran, M. Augmented reality-guided neurosurgery: accuracy and intraoperative application of an image projection technique. *J. Neurosurg.* **2015**, *123*, 206–211. [CrossRef]

61. Cabrilo, I.; Sarrafzadeh, A.; Bijlenga, P.; Landis, B.N.; Schaller, K. Augmented reality-assisted skull base surgery. *Neurochirurgie* **2014**, *60*, 304–306. [CrossRef]

62. Contreras López, W.; Navarro, P.; Crispin, S. Intraoperative clinical application of augmented reality in neurosurgery: A systematic review. *Clin. Neurol. Neurosurg.* **2019**, *177*, 6–11. [CrossRef]

63. Mourtzis, D.; Angelopoulos, J.; Panopoulos, N.; Kardamakis, D. A Smart IoT Platform for Oncology Patient Diagnosis based on AI: Towards the Human Digital Twin. *Procedia CIRP* **2021**, *104*, 1686–1691. [CrossRef]

64. Casari, F.A.; Navab, N.; Hruby, L.A.; Philipp, K.; Ricardo, N.; Romero, T.; de Lourdes dos Santos Nunes, F.; Queiroz, M.C.; Furnstahl, P.; Mazda, F. Augmented Reality in Orthopedic Surgery Is Emerging from Proof of Concept Towards Clinical Studies: a Literature Review Explaining the Technology and Current State of the Art. *Curr. Rev. Musculoskelet. Med.* **2021**, *14*, 192–203. [CrossRef]

65. Bagwe, S.; Singh, K.; Kashyap, A.; Arora, S.; Maini, L. Evolution of augmented reality applications in Orthopaedics: A systematic review. *J. Arthrosc. Jt. Surg.* **2021**, *8*, 84–90. [CrossRef]

66. Negrillo-Cardenas, J.; Jimenez-Perez, J.R.; Feito, F.R. The role of virtual and augmented reality in orthopedic trauma surgery: From diagnosis to rehabilitation. *Comput. Methods Programs Biomed.* **2020**, *191*, 105407. [CrossRef] [PubMed]

67. Jud, L.; Fotouhi, J.; Andronic, O.; Aichmair, A.; Osgood, G.; Navab, N.; Farshad, M. Applicability of augmented reality in orthopedic surgery—A systematic review. *BMC Musculoskelet. Disord.* **2020**, *21*, 103. [CrossRef] [PubMed]

68. Molina, C.A.; Phillips, F.M.; Poelstra, K.A.; Colman, M.; Khoo, L.T. 151. A cadaveric precision and accuracy analysis of augmented reality mediated percutaneous pedicle implant insertion. *Spine J.* **2020**, *20*, S74. [CrossRef]

69. Burstrom, G.; Persson, O.; Edstrom, E.; Elmi-Terander, A. Augmented reality navigation in spine surgery: A systematic review. *Acta Neurochir.* **2021**, *163*, 843–852. [CrossRef]

70. Frank, Y.; Georgios, M.; Kosuke, S.; Jeremy, S. Current innovation in virtual and augmented reality in spine surgery. *Ann. Transl. Med.* **2021**, *9*, 94. [CrossRef]
71. Sakai, D.; Joyce, K.; Sugimoto, M.; Horikita, N.; Hiyama, A.; Sato, M.; Devitt, A.; Watanabe, M. Augmented, virtual and mixed reality in spinal surgery: A real-world experience. *J. Vasc. Intervetional Radiol.* **2020**, *3*, 28. [CrossRef]
72. Vadalà, G.; Salvatore, S.D.; Ambrosio, L.; Russo, F.; Papalia, R.; Denaro, V. Robotic Spine Surgery and Augmented Reality Systems: A State of the Art. *Neurospine* **2020**, *17*, 88–100. [CrossRef]
73. Liu, T.; Yonghang, T.; Chengming, Z.; Lei, W.; Jun, Z.; Junjun, P.; Shi, J. Augmented reality in neurosurgical navigation: A survey. *Int. J. Med Robot. Comput. Assist. Surg. MRCAS* **2020**, *16*, e2160. [CrossRef]
74. Deng, W.; Li, F.; Wang, M.; Song, Z. Easy-to-Use Augmented Reality Neuronavigation Using a Wireless Tablet PC. *Stereotact. Funct. Neurosurg.* **2014**, *92*, 17–24. [CrossRef]
75. Gumprecht, H.K.; Widenka, D.C.; Lumenta, C.B. BrainLab VectorVision Neuronavigation System: Technology and clinical experiences in 131 cases. *Neurosurgery* **1999**, *44*, 97–104. [CrossRef]
76. Grunert, P.; Darabi, K.; Espinosa, J.; Filippi, R. Computer-aided navigation in neurosurgery. *Neurosurg. Rev.* **2003**, *26*, 73–99. [CrossRef]
77. Cleary, K.; Peters, T.M. Image-Guided Interventions: Technology Review and Clinical Applications. *Annu. Rev. Biomed. Eng.* **2010**, *12*, 119–142. [CrossRef]
78. Incekara, F.; Smits, M.; Dirven, C.; Vincent, A. Clinical Feasibility of a Wearable Mixed-Reality Device in Neurosurgery. *World Neurosurg.* **2018**, *118*, e422–e427. [CrossRef]
79. Moro, C.; Phelps, C.; Redmond, P.; Stromberga, Z. HoloLens and mobile augmented reality in medical and health science education: A randomised controlled trial. *Br. J. Educ. Technol.* **2020**, *52*, 680–694. [CrossRef]
80. Kumar, N.; Pandey, S.; Rahman, E. A Novel Three-Dimensional Interactive Virtual Face to Facilitate Facial Anatomy Teaching Using Microsoft HoloLens. *Aesthetic Plast. Surg.* **2021**, *45*, 1005–1011. [CrossRef]
81. Moro, C.; Phelps, C.; Jones, D.; Stromberga, Z. Using Holograms to Enhance Learning in Health Sciences and Medicine. *Med. Sci. Educ.* **2020**, *30*, 1351–1352. [CrossRef]
82. Parsons, D.; Mac Callum, K. Current Perspectives on Augmented Reality in Medical Education: Applications, Affordances and Limitations. *Adv. Med Educ. Pract.* **2021**, *12*, 77–91. [CrossRef]
83. Williams, M.A.; McVeigh, J.; Handa, A.I.; Regent, L. Augmented reality in surgical training: A systematic review. *Postgrad. Med. J.* **2020**, *96*, 537–542. [CrossRef]
84. Cao, C.; Cerfolio, R.J. Virtual or Augmented Reality to Enhance Surgical Education and Surgical Planning. *Thorac. Surg. Clin.* **2019**, *29*, 329–337. [CrossRef]
85. Yeung, A.W.K.; Tosevska, A.; Klager, E.; Eibensteiner, F.; Laxar, D.; Jivko, S.; Marija, G.; Sebastian, Z.; Stefan, K.; Rik, C.; et al. Virtual and Augmented Reality Applications in Medicine: Analysis of the Scientific Literature. *J. Med Internet Res.* **2021**, *23*, e25499. [CrossRef]
86. McKnight, R.R.; Pean, C.A.; Buck, J.S.; Hwang, J.S.; Hsu, J.R.; Pierrie, S.N. Virtual Reality and Augmented Reality—Translating Surgical Training into Surgical Technique. *Curr. Rev. Musculoskelet. Med.* **2020**, *13*, 663–674. [CrossRef]
87. Fazel, R.; Krumholz, H.M.; Wang, Y.; Ross, J.S.; Chen, J.; Ting, H.H.; Shah, N.D.; Nasir, K.; Einstein, A.J.; Nallamothu, B.K. Exposure to low-dose ionizing radiation from medical imaging procedures. *N. Engl. J. Med.* **2009**, *361*, 849–857. [CrossRef]
88. Hong, J.Y.; Han, K.; Jung, J.H.; Kim, J.S. Association of exposure to diagnostic low-dose ionizing radiation with risk of cancer among youths in South Korea. *JAMA Netw. Open* **2019**, *2*, e1910584. [CrossRef]
89. Reisz, J.; Bansal, N.; Qian, J.; Zhao, W.; Furdui, C. Effects of ionizing radiation on biological molecules–mechanisms of damage and emerging methods of detection. *Antioxidants Redox Signal.* **2014**, *21*, 260–292. [CrossRef]
90. Peng, H.; Ding, C. Minimum redundancy and maximum relevance feature selection and recent advances in cancer classification. *Feature Sel. Data Min.* **2005**, *3*, 185–205. [CrossRef]
91. Singh, V.K.; Ali, A.; Nair, P.S. A Report on Registration Problems in Augmented Reality. *Int. J. Eng. Res. Technol.* **2014**, *3*, 819–821.
92. Chen, Y.; Wang, Q.; Chen, H.; Song, X.; Tang, H.; Tian, M. An overview of augmented reality technology. *J. Phys. Conf. Ser.* **2019**, *1237*, 022082. [CrossRef]
93. Lee, Y.H.; Zhan, T.; Wu, S.T. Prospects and challenges in augmented reality displays. *Virtual Real. Intell. Hardw.* **2019**, *1*, 10–20. [CrossRef]
94. Ren, D.; Goldschwendt, T.; Chang, Y.; Höllerer, T. Evaluating wide-field-of-view augmented reality with mixed reality simulation. In Proceedings of the 2016 IEEE Virtual Reality (VR), Greenville, SC, USA, 19–23 March 2016; pp. 93–102. [CrossRef]
95. Zhou, Y.; Zhang, J.; Fang, F. Vergence-accommodation conflict in optical see-through display: Review and prospect. *Results Opt.* **2021**, *5*, 100160. [CrossRef]
96. Erkelens, I.M.; MacKenzie, K.J. 19-2: Vergence-Accommodation Conflicts in Augmented Reality: Impacts on Perceived Image Quality. *SID Symp. Dig. Tech. Pap.* **2020**, *51*, 265–268. [CrossRef]
97. Kim, J.; Kane, D.; Banks, M.S. The rate of change of vergence–Accommodation conflict affects visual discomfort. *Vis. Res.* **2014**, *105*, 159–165. [CrossRef] [PubMed]

*applied
sciences*

Review

Deep Active Learning for Computer Vision Tasks: Methodologies, Applications, and Challenges

Mingfei Wu, Chen Li * and Zehuan Yao

College of Computer, National University of Defense Technology, Changsha 410073, China
* Correspondence: lichen14@nudt.edu.cn

Abstract: Active learning is a label-efficient machine learning method that actively selects the most valuable unlabeled samples to annotate. Active learning focuses on achieving the best possible performance while using as few, high-quality sample annotations as possible. Recently, active learning achieved promotion combined with deep learning-based methods, which are named deep active learning methods in this paper. Deep active learning plays a crucial role in computer vision tasks, especially in label-insensitive scenarios, such as hard-to-label tasks (medical images analysis) and time-consuming tasks (autonomous driving). However, deep active learning still has some challenges, such as unstable performance and dirty data, which are future research trends. Compared with other reviews on deep active learning, our work introduced the deep active learning from computer vision-related methodologies and corresponding applications. The expected audience of this vision-friendly survey are researchers who are working in computer vision but willing to utilize deep active learning methods to solve vision problems. Specifically, this review systematically focuses on the details of methods, applications, and challenges in vision tasks, and we also introduce the classic theories, strategies, and scenarios of active learning in brief.

Keywords: deep learning; active learning; computer vision; artificial intelligence

Citation: Wu, M.; Li, C.; Yao, Z. Deep Active Learning for Computer Vision Tasks: Methodologies, Applications, and Challenges. *Appl. Sci.* **2022**, *12*, 8103. https://doi.org/10.3390/app12168103

Academic Editor: Cosimo Nardi

Received: 11 July 2022
Accepted: 9 August 2022
Published: 12 August 2022

Publisher's Note: MDPI stays neutral with regard to jurisdictional claims in published maps and institutional affiliations.

1. Introduction

With the rapid development of deep learning, the performance of computer vision tasks has achieved breakthroughs benefiting from large-scale annotated datasets, such as ImageNet [1], Cityscapes [2], and AbdomenCT-1K [3]. These datasets provide direct supervision for model training. Meanwhile, there are useless, uninformative examples, which serve as risks to overwhelm the training. Apart from the noise inside the labeled data, there are always scenarios where unlabeled data is abundant. However, manual labeling is high cost, such as medical image analysis, autonomous driving, anomaly detection, and related issues in computer vision tasks. Specifically, taking the Cityscapes [2] dataset as an example, the pixel-wise annotation will cost more than 90 min per image in the urban street segmentation task. Similarly, in the medical image tumor segmentation task, it is more challenging for medical experts to detect the mm-size objects from 3D volume data, which is more time-consuming and medical knowledge–demanding.

Under the above conditions, maximizing the model's performance with a limited annotation budget becomes the primary concern. In order to figure out this problem, active learning becomes a promising solution to improve data efficiency and relieve the high annotation burden. As a subfield of machine learning, the core idea of active learning [4] is to find the most valuable samples through some heuristic strategies, so the model can achieve or even exceed the expected performance with as few labeled samples as possible. The intuition of active learning is that not all samples in a dataset have the same significance for model training. For example, some samples contain more noise that hinders training. Besides this, some samples are too similar to be worth labeling. Therefore, the goal of active learning is to select as few valuable or ambiguous samples as possible via the designed strategy and promote the performance with the selected samples interactively. The above iterative training process between optimization and annotator is the primary active learning mechanism, and human annotations exist in each training interaction. Consequently, the essence of active learning is the *Human-in-the-Loop* (HITL) computing systems, where human expertise is joint in the computer-based systems [5]. Humans (such as doctors in clinical diagnosis) are part of the intelligent system and participate in the model training process by providing judgments or domain-knowledge that influence the final output of the system [6]. More details are introduced in Algorithm 1 and other surveys [5,6].

Algorithm 1: The pool-based active learning workflow

    Input   : A labeled data pool L, an unlabeled data pool $\mathcal{U}$, annotators A.
    Output: A well-trained model $\mathcal{M}$ with least labeling cost.
1 $\mathcal{M}$ initialization;
2 **repeat**
3 | Train the model $\mathcal{M}$ with L;
4 | Obtain the representation $\mathcal{R}$ of all samples $x \in \mathcal{U}$, $\mathcal{R} = \mathcal{M}(x)$;
5 | Query the top-K informative samples $\mathcal{K}$ via selection strategies, according to the representation $\mathcal{R}$;
6 | Annotate the samples $\mathcal{K}$ and obtain the labels $Y^{\mathcal{K}} = A(\mathcal{K})$;
7 | Update $L = L \cup \{\mathcal{K}, Y^{\mathcal{K}}\}$, update $\mathcal{U} = \mathcal{U}/\mathcal{K}$.
8 **until** *End conditions*;

Settles's active learning literature survey [4] systematically summarized classic active learning methods in 2004. More basic definitions and formulations can be found in this survey [4]. Active learning has been rapidly growing and booming with various novel methodologies, applications and challenges in recent years. Settles's survey provided the basic theory for active learning, especially the traditional AL. Differently, our work focuses on deep learning–based active learning theories in computer vision tasks, which is named deep active learning in this paper.

Apart from this, we summarize the latest surveys [6–9] about deep learning in Table 1. Previous surveys systematically introduced the deep active learning in many fields, such as CV and NLP. After studied the existing surveys about deep active learning and their references, we found that there was not any review designed for CV researcher. Hence, we decided to re-organize existing works and introduce latest research from a CV-related perspective in this manuscript. The biggest difference between this manuscript and above-mentioned works is that the expected audience of this review are researchers who are working in computer vision but willing to utilize deep active learning methods to solve CV problems. Active learning is still new to them. As such, we organized this manuscript from the perspective of a CV researcher, and introduced the deep active learning from CV-related methodologies and corresponding applications. This CV-friendly survey is our new contribution to the community.

The remainder of this review is as follows: First, we argue that it is necessary to introduce the basic concepts and methodologies of traditional active learning for the newcomers. Then, Section 2 introduces the three basic candidate selection strategies in active learning and give responding examples, and then we detail the pool-based strategy. Section 3 introduces the common querying scenarios in active learning. Then, we focus on the methodologies integrated deep learning and active learning. Section 4 details the recent methodologies for deep learning-based active learning. Section 5 details the applications of deep active learning, especially in computer vision tasks. Finally, Section 6 summarizes the existing challenges of deep active learning in computer vision tasks, which are the future works in this field. Section 7 concludes the survey.

Table 1. The latest surveys about deep active learning.

Title	Main Contents	Publication
Samuel Budd et al. A survey on active learning and human-in-the-loop deep learning for medical image analysis [6]	▷ Investigate the active learning in the medical image analysis. ▷ Propose the considerations in the deep learning–based active learning, including noisy oracles, weakly supervised learning, multi-task learning, annotation interface, and variable learning costs. ▷ Discuss the future prospective and unanswered questions in the medical image analysis.	Medical Image Analysis. 2021, 71, 102062
Punit Kumar et al. Active Learning Query Strategies for Classification, Regression, and Clustering: A Survey [7]	▷ Summarize the active learning query strategies for three tasks, including classification, regression, and clustering. ▷ Classify the query strategies under classification into: informative-based, representative-based, informative-and-representative-based, and others. ▷ Summarize the empirical evaluation of active learning query strategies. ▷ Present the implementation, application, and challenges of the active learning in brief.	Journal of Computer Science and Technology. 2020, 35, 913–945.
Pengzhen Ren et al. A Survey of Deep Active Learning [8]	▷ Classify the existing works in the deep active learning. ▷ Summarize the deep active learning applications, including vision and NLP. ▷ Especially, in the visual data processing tasks, it discusses image classification and recognition, object detection and semantic segmentation, and video processing.	ACM Computing Surveys. 54.9 (2021): 1–40.
Xueying Zhan et al. A Comparative Survey of Deep Active Learning [9]	▷ Categorize deep active learning sampling methods and querying strategies. ▷ Compare deep active learning algorithms across common used datasets. ▷ Conduct experiments to explore influence factors of deep active learning ▷ Release a deep active learning toolkit, named DeepAL+.	arXiv:2203. 13450, 2022.

2. Candidate Selection Strategies in Active Learning

In the classic active learning framework, one of the two most important components is how to develop a criterion for evaluating the "worthiness" of unlabeled samples. After evaluation, the selection strategies decide whether one sample is worthy of being labeled by the annotator according to its worthiness. The selected samples are regarded as candidates. Finally, an appropriate selection strategy reduces the labeling cost and thus has important implications in active learning. Due to this knowledge being beyond the main concern of this review, Table 2 introduces four basic selection strategies in brief, and more details can be found in existing active learning surveys [6–9].

Table 2. Summary of candidate selection strategies in active learning.

Strategies	Methodologies	Typical Works
Random selection	$\triangleright$ **Random sampling** is to use random numbers to select samples from the unlabeled dataset for labeling. $\triangleright \mathbb{X} = \underset{x \in U}{Random}(x)$	N.A.
Uncertainty-based selection	$\triangleright$ **Least confidence** is to select the sample with the smallest probability of the top1 predicted class. In practice, the opposite of the maximum predicted probability is often taken as the uncertainty score of the sample. $\triangleright \mathbb{X} = \underset{x \in U}{\arg\max}[1 - P(\hat{y} \mid x)] = \underset{x \in U}{\arg\min} P(\hat{y} \mid x)$	Li et al. [10] Agrawal et al. [11,12]
	$\triangleright$ **Margin sampling** is to calculate the difference between the probabilities of the top1 and the top2 predicted class. Then the samples with the smallest difference are defined as hard-to-classify samples for labeling. $\triangleright \mathbb{X} = \underset{x \in U}{\arg\min}(P(\hat{y}_1 \mid x) - P(\hat{y}_2 \mid x))$	Ajay J et al. [13] Zhou et al. [14]
	$\triangleright$ **Multi-class level uncertainty** is to select the two samples that are the farthest from the classification hyperplane of multi-class and take their distance difference as the score. $\triangleright \mathbb{X} = \underset{x \in U}{\arg\min}\left\{D(x, \hat{y}_1) - \max_{y \neq \hat{y}_1} D(x, y)\right\}$	Gu et al. [15] Yang et al. [16]
	$\triangleright$ **Maximize entropy** is to utilize the methodology that larger entropy denotes higher uncertainty. The sample with the largest entropy is selected as candidate. $\triangleright \mathbb{X} = \underset{x \in U}{\arg\max} E_x = \underset{x \in U}{\arg\min}\left\{\sum_{i=1}^{Y} P(y_i \mid x) \times \log P(y_i \mid x)\right\}$	Yu et al. [17] Ozdemir et al. [18]
Diversity-based selection	$\triangleright$ **Angle-based** measurement is to to measure diversity by calculating the undirected angles between the induced hyper-planes. $\triangleright \left\|\cos(\angle(h_i, h_j))\right\| = \dfrac{\left\|\langle\phi(x_i),\phi(x_j)\rangle\right\|}{\|\phi(x_i)\|\|\phi(x_j)\|} = \left\|k(x_i, x_j)\right\| / \sqrt{k(x_i, x_i)k(x_j, x_j)}$	Brinker et al. [19]
	$\triangleright$ **Redundancy-based** measurement is to measure the diversity as the redundancy between unlabeled points via symmetric KL divergence [20] between the two vectors of probability values. $\triangleright R(x_i, x_j) = \sum_{y}^{\|Y\|} \left[P(y \mid x_i) - P(y \mid x_j)\right] \log\left[P(y \mid x_i) / P(y \mid x_j)\right]$	Shayok et al. [21] Zhou et al. [22]
Committee-based selection [23]	$\triangleright$ **Vote entropy–based** measurement is to select the hard sample voted by the Committee. The models in the committee distinguish samples into different classes. The predicted results toward one sample with the largest entropy is classified as hard sample and needs to vote for labeling. $\triangleright \mathbb{X} = \underset{x \in U}{\arg\min}\left\{-\frac{1}{N}\sum_{i=1}^{Y}\left[Vote(y_i) \log \frac{Vote(y_i)}{N}\right]\right\}$	Yan et al. [24] Dagan et al. [25]
	$\triangleright$ **Average KL divergence–based** measurement is to measure the deviation of those unlabeled samples via calculating the average KL divergence of the committee $\mathcal{C}$. $\mathbb{X} = \underset{x \in U}{\arg\max} \frac{1}{N}\sum_{i=1}^{N} KL(P_{\mathcal{M}_i}\|P_{\mathcal{C}}) = \underset{x \in U}{\arg\max} \frac{1}{N}\sum_{i=1}^{N}\sum_{j}^{\|Y\|} P_{\mathcal{M}_j}(y_j \mid x) \log \frac{P_{\mathcal{M}_i}(y_j|x)}{P_{\mathcal{C}}(y_j|x)}$	Dagan et al. [25]

Note: $\mathbb{X}$ represents the selected sample, U is the unlabeled data pool, $\hat{y}_1$ is the predicted class with max probability, $P(\hat{y}_1 \mid x)$ is the conditional probability where the input is x, and the predicted class is $\hat{y}_1$. $\hat{y}_2$ denotes the top2 predicted class. $D(x, y)$ represents the distance from the sample x to the classification hyperplane of class y. $\phi(x_i)$ is the normalization function, $k(x_i, x_j) = \langle\phi(x_i), \phi(x_j)\rangle$ is the kernel function. $|Y|$ is the total number of classes, $P(y \mid x)$ is the conditional probability where the input is x_i, and the predicted class is y. $Vote(y_i)$ denotes the number of models that voted for the current class y_i, and $\sum_{i=1}^{|Y|} Vote(y_i) = N$. $\mathcal{M}_i$ represents a specific model in the committee. $KL(P_1\|P_2)$ is the KL divergence between two distributions P_1 and P_2.

2.1. Random Selection Strategies

Random sampling uses random numbers to select samples from the unlabeled dataset for labeling. There is no interaction with the model's prediction in the above process, which is the most naive selection strategy. Consequently, it is often used as the basic comparison experiment in active learning.

2.2. Uncertainty-Based Selection Strategies

The uncertainty-based selecting method is the simplest and most common strategy, which assumes that the samples closest to the classification hyperplane have richer information than others. It selects the most uncertain samples according to the predicted value of the samples by the current model.

Typical uncertainty-based selection strategies includes Least confidence, Margin sampling, Multi-class level uncertainty, and Maximize entropy. This survey briefly summarized the above strategies in Table 2. More details can be found in existing surveys [6–9]. Consequently, the machine learning model can quickly improve its performance by learning the labels of the samples with substantial uncertainty.

2.3. Diversity-Based Selection Strategy

The above uncertainty-based selection strategy can effectively sample a single candidate for annotation, but it is often ineffective when there are multiple candidates. At this time, the selection strategy based on the diversity of sample feature distribution comes into being. Diversity tends to reflect the prediction consistency among the samples, i.e., higher diversity values denote more inconsistency between the candidate sample and the entire unlabeled pool [26]. Typical diversity-based measurement strategies includes angles-based and redundancy-based perspectives. We also summarized above strategies in Table 2 and provided the comparison between uncertainty-based and diversity-based selections in Figure 1.

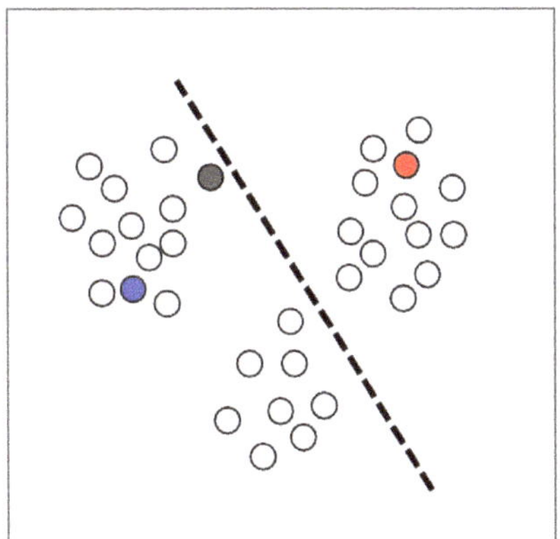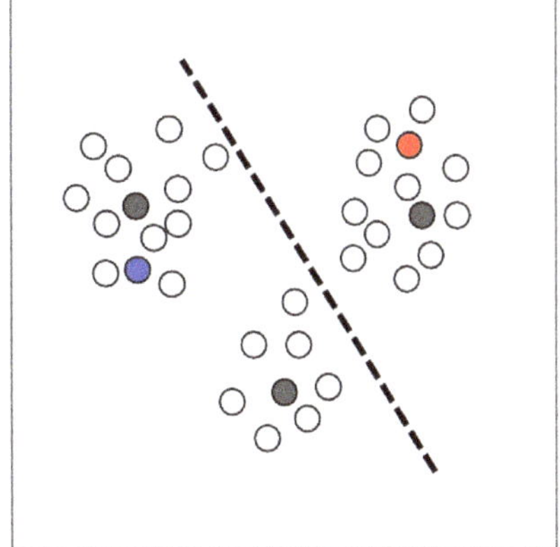

(**a**) Uncertainty-based selection strategies   (**b**) Diversity-based selection strategies

Figure 1. Illustrations of different candidate selection strategies in active learning. The dashed line represents the classification hyper-plane of the existing model. The hollow circles represent unlabeled data, the colored circles represent labeled data, and the gray circles represent selected candidates. The gray circle in subfigure (**a**) represents the least confident sample selected by the uncertainty-based strategy. In subfigure (**b**), the three gray circles represent the most representative samples selected by the diversity-based strategy.

2.4. Committee-Based Selection Strategy

Committee-based selection strategy [23] is based on version space reduction, and its core idea is to preferentially select unlabeled samples that can reduce the version space to the greatest extent. The committee-based selection strategy's motivation is that the most informative selections are the samples where the committee predicts the most inconsistent. Typical committee-based selection strategies include vote entropy and average KL divergence, which are listed in Table 2. There are four basic steps in the committee-based selection strategy:

1. Multiple models $\{\mathcal{M}_1, \ldots, \mathcal{M}_N\}$ are used to construct a committee $\mathcal{C}$ for voting, i.e., $\mathcal{C} = \{\mathcal{M}_1, \ldots, \mathcal{M}_N\}$.
2. The models in the committee $\mathcal{C}$ are then trained on the labeled dataset L and get different parameters.
3. All models in the committee make predictions separately on unlabeled samples from $\mathcal{U}$. The samples with the richest information are voted.
4. The samples which obtain the most disagreements are selected as candidates for labeling.

3. Common Querying Scenarios in Active Learning

According to the application scenarios, active learning methods can be divided into three types: query synthesis scenario, stream-based scenario, and pool-based scenario. We briefly summarize the above querying scenarios in Table 3 and introduce the pool-based scenario in detail.

Table 3. Summary of common querying scenarios in active learning.

Scenarios	Concepts	Limitations	Publications
Membership query synthesis	▷ The membership query synthesis is to generate new unlabeled instances for querying by itself instead of selecting samples from the real-world distribution [27].	▷ It may encounter troubles when the generated data is too arbitrary for the annotator to recognize or does not contains any semantic information.	[28,29]
Stream-based sampling	▷ The stream-based scenario [30] is to sample from the natural distribution instead of the synthesized one. ▷ In this scenario, the selection process is similar to a pipeline. The unlabeled sample is firstly input into the model one by one. ▷ Then, the active learning strategy needs to decide whether to pass it to the annotator for labeling or reject it directly.	▷ It is necessary for the model to immediately decide based on a single input rather than the comprehensive consideration of this batch. ▷ The active learning system may suffer from the absence of knowledge of unseen areas.	[31–33]
Pool-based sampling	▷ The pool-based sampling scenario is to selects the most valuable samples from an unlabeled data pool for labeling according to the informativeness [34]. ▷ The unlabeled data pool is sampled from the natural distribution instead of synthesized samples.	▷ It is computationally expensive because every iteration requires the informativeness evaluation for the whole pool.	[35–37]

Among the above mainstreams, pool-based active learning is more compatible with batch-based modern deep learning training mechanisms. Compared with the stream-based selective sampling, the pool-based scenario is able to consider every sample based on this batch comprehensively. Consequently, it has become the most common scenario in computer vision tasks. Moreover, the most related works introduced in this review also belong to the pool-based active learning scenario.

Figure 2 is a classic pool-based active learning framework. A single batch of unlabeled samples is input to the model from the unlabeled data pool during the training process. Then, the query strategy selects the most valuable samples for labeling according to the informativeness. After that, these labeled samples are added to the labeled dataset to continue training the model.

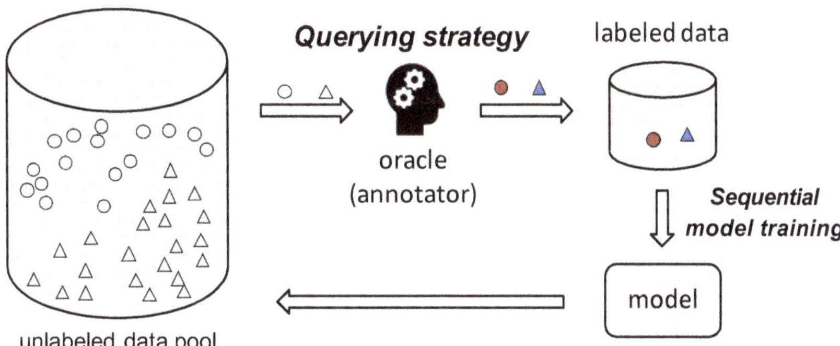

Figure 2. The classic pool-based active learning workflow.

Sequentially, we formally define the pool-based active learning method in Algorithm 1. The End conditions include that the performance of the model meets requirements, or the budgets for annotation run out, or the selected samples are hard for annotators to label.

As shown in Figure 2, there are the labeled pool $L = \{(x_1^l, y_1^l), \cdots, (x_{N_l}^l, y_{N_l}^l)\}$ and the unlabeled pool $\mathcal{U} = \{x_1^u, \cdots, x_{N_u}^u\}$ at the beginning. N_l and N_u are the number of labeled and unlabeled samples, respectively. Then, the data from the labeled pool is fed into the machine learning model for supervised training. After that, the well-trained model is utilized to extract the representation $\mathcal{R}^u = \{(r_1^u), \cdots, (r_{N_u}^u)\}$ of the unlabeled pool data. Based on $\mathcal{R}^u$, the informativeness is calculated according to the query strategy. The top-K samples $\mathcal{K} = \{x_1^k, \cdots, x_{N_k}^k\}$ are selected out to the oracle (human annotators) and obtain the labels $Y^{\mathcal{K}} = \{y_1^k, \cdots, y_{N_k}^k\}$. Finally, the labeled pool L will be updated to $L = \{(x_1^l, y_1^l), \cdots, (x_{N_l}^l, y_{N_l}^l), (x_1^k, y_1^k), \cdots, (x_{N_k}^k, y_{N_k}^k)\}$. With the updated L, the machine learning model will promote the performance in return. Meanwhile, the size of unlabeled pool $\mathcal{U}$ is reduced to $N_u - N_k$. The above loop will be repeated until one of the end conditions is met. Since the selected samples $\mathcal{K}$ from the unlabeled pool $\mathcal{U}$ are the most significant ones for training, the size of L to learn a model can often be much smaller than the size required in classic supervised learning without active learning.

4. Deep Active Learning Methods

Recent developments are dedicated to multi-label active learning, hybrid active learning, and deep learning–based active learning. In the upcoming sections, we will detail deep learning–based active learning.

4.1. Deep Active Learning for CNNs

As we introduced in Section 2.2, uncertainty is one of the most used metrics to select candidate in active learning. We summarize the uncertainty estimation methods in Table 4. At the same time, Bayesian methods are known for their ability to capture underlying model uncertainty. The classic Bayesian active learning framework consists of an unlabeled data pool $\mathcal{U}$, the labeled data pool L, and a Bayesian model $\mathcal{M}$ trained on the current L. The output of the Bayesian model $\mathcal{M}$ is $p(y|x, \mathcal{M}, L)$, where the input data is x and the prediction $y \in \{1, ..., c\}$ in the classification tasks. In Bayesian deep learning, the model $\mathcal{M}$ is replaced by a Convolutional Neural Network (CNN) with prior probability distributions. Gal et al. [38] proposed a Bernoulli-based approximate variational inference method. After that, they [39] proposed to capture model uncertainty by using the Monte Carlo dropout regularization as a variational Bayesian approximation. Consequently, it is natural to utilize Bayesian methods to actively select valuable candidates.

Gal et al. [40] introduced the Bayesian Convolutional Neural Networks into the active learning framework. They demonstrated that such combination improved performance over existing active learning methods on the image classification dataset MNIST [41] (achieving 5% test error) and skin cancer diagnosis dataset ISIC 2016 [42] (achieving 0.71/0.75 AUC). Bayesian Active Learning by Disagreement (BALD) [43] was proposed to be the basic selection strategy, where Shannon entropy [44] was utilized to measure the "information content". The discrepancy of Shannon entropy denoted the difference between the information entropy of a certain sample and the average information entropy of the dataset. The larger the difference, the more information the sample contained relative to the average. Finally, the BALD strategy pushed the samples with the largest Shannon entropy. The above process is formulated as follows.

$$\mathbb{X} = \arg\max_{\mathbf{x} \in \mathcal{U}} \mathbb{I}(x) = \arg\max_{\mathbf{x} \in \mathcal{U}} SE(\mathbf{y} \mid \mathbf{x}, L) - \mathbb{E}_{\theta \sim P(\theta|L)} [SE(\mathbf{y} \mid \mathbf{x}, \mathcal{M})], \tag{1}$$

where $\mathbb{I}(\cdot)$ is the mutual information. Higher mutual information means the model's predictions are more uncertain. $SE(\mathbf{y} \mid \mathbf{x}, L)$ and $SE(\mathbf{y} \mid \mathbf{x}, \mathcal{M})$ are represented by the Shannon Entropy of the prediction $P_{\mathcal{M}}(\mathbf{y} \mid \mathbf{x})$ and the mean distribution $P_{\mathcal{M}}(\mathbf{y} \mid L)$, respectively.

Ensemble learning is a well-known technique in machine learning that improves performance by integrating different models and combining their results [45]. Ref. [46] explored the difference between ensemble-based methods against Monte Carlo dropout methods on image classification tasks MNIST [41] (achieving 90% test set accuracy with roughly 12,200 labeled images), CIFAR-10 [47] (achieving 91.5% accuracy) and diabetic retinopathy (DR) detection (https://www.kaggle.com/competitions/diabetic-retinopathy-detection/rules, accessed on 9 July 2022) (achieving 0.983 AUC). They conducted extensive experiments from 11 aspects and found that the former outperformed the latter and was a more reliable measure of uncertainty.

Table 4. Summary of uncertainty estimation methods in deep active learning.

Type	Methodology	Equation
MC dropout	▷ In practice, the MC dropout usually trains the CNN with the labeled data pool L with dropout. ▷ After training, it generates a new dropout mask for the model parameters $\mathcal{M}_t$ and performing T forward inference. ▷ The output is the average of T results.	$p(y \mid x, L) = \frac{1}{T} \sum_{t=1}^{T} p(y \mid x, \mathcal{M}_t)$
Deep Ensembles	▷ The ensemble-based approaches design N neural networks $\{\mathcal{M}_1 \cdots \mathcal{M}_N\}$ at first. ▷ These networks share same architecture but initialized from different weights. ▷ Then networks are trained with the labeled data pool L. ▷ The average of the outputs of the N networks is the final output.	$p(y \mid x, L) = \frac{1}{N} \sum_{i=1}^{N} p(y \mid x, \mathcal{M}_i)$

Bayesian-based methods addressed the problem of uncertainty-based candidate selection strategies, but there was another obstacle that needed to be solved. The biggest difference of CNN-based deep learning (DL) methods and traditional active learning (AL) methods is that AL methods query candidates one by one while DL methods load a batch size of samples at the same time. Ozan Sener and Silvio Savarese [48] conducted experiments on vision tasks with traditional active learning methods and found that previous AL works did not perform well for CNN-based vision tasks due to the batch settings during model training. They attributed this ineffectiveness to batch sampling. In order to solve it, they defined the active learning as a *Core-Set* selection problem, where the algorithm aims to train on a small mount of labeled samples instead of the whole dataset such that the trained model is able to get competitive performance over the model trained on the whole dataset. They defined the core-set selection problem as the following optimization:

$$\min_{L^1:|L^1|\leq b} \left| \frac{1}{n} \sum_{i\in[n]} loss(\mathbf{x}_i, y_i; \mathcal{M}_{L^0\cup L^1}) - \frac{1}{|L^0+L^1|} \sum_{j\in L^0\cup L^1} loss(\mathbf{x}_j, y_j; \mathcal{M}_{L^0\cup L^1}) \right|, \quad (2)$$

where L^0 represents the randomly selected samples at the beginning, L^1 represents newly selected samples from the entire dataset under budget b. n is the number of samples in the entire dataset. $\mathcal{M}_{L^0\cup L^1}$ denotes the trained model under the subset L^0 and L^1. *loss* is the loss function, where the authors suggested the cross-entropy loss for effective training. In order to prove their hypothesis, they conducted experiments on active learning for fully supervised models and weakly supervised models. Specifically, they used dataset CIFAR10/100 [47] for image classification and dataset SVHN [49] for digit classification.

Based on the *Core-Set* strategy, the combination of batch-based active learning and deep learning has been a researcher topic in the community. the goal of batch-based deep active learning is to select the most informative batch or mini-batch $\mathcal{B}^* = \{x_1^*, x_2^*, \ldots, x_n^*\}$ from the loaded batches $\mathcal{B}$, where $\mathcal{B}$ belongs to unlabeled pool $\mathcal{U}$, and n is the batch size. We formulate above process as follows.

$$\mathcal{B}^* = \arg\max_{\mathcal{B}\subseteq U} \Phi(\mathcal{B}, \mathcal{M}(L)), \quad (3)$$

where $\Phi(\cdot)$ is the score function to measure the informativeness of the batch $\mathcal{B}$, L is the labeled data pool, $\mathcal{M}$ is the trained model.

After that, most of the related research was devoted to the innovation of the scoring function $\Phi(\cdot)$. David Janz et al. [50] adopted the idea of Bayesian Active Learning by Disagreement (BALD) [43] into scoring function. Specifically, they utilized the mutual information $\mathbb{I}(\cdot)$ as score function and selected samples with the maximum gain of information, where $\Phi_{BALD}(\mathcal{B}, \mathcal{M}(L)) = \sum_{i=1}^{n} \mathbb{I}(y_i; \mathcal{M} \mid x_i, L)$. However, BALD just considered the mutual information between one single sample x_i and model parameters $\mathcal{M}(L)$, and did not take the relationship between samples in an batch. As an extension of BALD, BatchBALD [51] promoted the BALD by estimating the mutual information $\mathbb{I}(\cdot)$ between all samples in an batch and the model, which was formulated as $\Phi_{BatchBALD}(\mathcal{B}, \mathcal{M}(L)) = \mathbb{I}(y_{1:n}; \mathcal{M} \mid x_{1:n}, L)$.

Yoo et al. [52] proposed a novel loss prediction module into the target model. This module consisted of global average pooling (GAP), full connected layer (FC) and ReLU, capturing multi-level features. Then the features was concentrated and calculated the loss prediction. All unlabeled samples were evaluated by this module via mini-batch. The batch with the top-K predicted losses selected as candidates and then labeled to update training set. the proposed module was evaluated in Image Classification task CIFAR-10 [47] (achieving accuracy of 0.9101), Object Detection task PASCAL VOC [53] (achieving 0.7338 mAP), and Human Pose Estimation task MPII [54] (achieving 0.8046 PCKh@0.5).

4.2. Generative Adversarial Active Learning

According to the analysis of [55], the *Core-Set* strategy [48] is very inefficient in high-dimensional representation learning due to its inherent distance-based computation. This obstacle is well addressed by leveraging GAN or VAE for representation learning from high-dimensional data.

Generative Adversarial Networks (GAN) is a novel representation learning method proposed by Goodfellow [56]. Its core idea is "Generative" and "Adversarial". The GAN network structure contains two models. One is the generator $\mathbb{G}$ and the other is the discriminator $\mathbb{D}$. The generator generally uses a deconvolutional neural network or a fully connected neural network to synthesize new data (e.g., images). At the same time, the discriminator is a CNN-based binary classifier to distinguish whether the input is from the natural distribution or synthesized one from the generator.

The discriminator is trained first to make a good judgment so that the real and generated samples can be better distinguished. Then the parameters of the generator can be updated more accurately. The goal of discriminator is that $P(\mathbb{D}(x)|x \in real) = 1$ while $P(\mathbb{D}(x)|x \in fake) = 0$. Then the generator and discriminator in GANs are trained against each other in a two-player game. The weights and biases of the discriminator and generator are trained through back-propagation until they reach a dynamic equilibrium state with unlabeled samples. In order to discriminate samples and classify them, the discriminator usually utilizes the cross-entropy loss to measure similarity, which is formulated as follows [56]:

$$\mathcal{L}_{\mathbb{D}} = \frac{1}{N} \sum_{i=1}^{N} \left[\log \mathbb{D}\left(x^{(i)}\right) + \log\left(1 - \mathbb{D}\left(\mathbb{G}\left(z^{(i)}\right)\right)\right) \right] \tag{4}$$

$$\mathcal{L}_{\mathbb{G}} = \frac{1}{N} \sum_{i=1}^{N} \log\left(1 - \mathbb{D}\left(\mathbb{G}\left(z^{(i)}\right)\right)\right) \tag{5}$$

$z \sim p_z(z)$ is the generated distribution and $x \sim p_{\text{data}}(x)$ is the real distribution. N denotes the batch size. As a consequence, the objective function of GAN is shown as follows [56]:

$$\mathcal{L} = \arg \min_{\mathbb{G}} \max_{\mathbb{D}} V(\mathbb{G}, \mathbb{D}) = \mathbb{E}_{x \sim p_{\text{data}}(x)}[\log \mathbb{D}(x)] + \mathbb{E}_{z \sim p_z(z)}[\log(1 - \mathbb{D}(\mathbb{G}(z)))], \tag{6}$$

where $V(\mathbb{G}, \mathbb{D})$ is the difference between the generated distribution p_z and the actual distribution p_{data}. max is to train the discriminator to discriminate the sample to the greatest extent. min is to train the generator to minimize the difference between the generated sample and the actual sample. When the algorithm converges, the samples generated by the generator can confuse the discriminator and cannot distinguish right from wrong. In other words, the generator should try its best to generate more realistic results to deceive the discriminator. The discriminator should try its best to distinguish the truth from the false and not be deceived by the generator. The network reaches the ideal state when the generated result is actual (discrimination probability is 0.5).

Zhu et al. [57] firstly proposed a novel query synthesis-based active learning method GAAL fused with GAN. GAAL was trained on datasets MNIST [41] (achieving accuracy 70.44%) and CIFAR-10 [47] while tested on the dataset USPS [58]. The workflow of GAAL can be concluded as Algorithm 2.

Algorithm 2: The synthesis-based active learning method workflow

Input : A labeled data pool L, an unlabeled data pool $\mathcal{U}$, annotators A,
   a generator $\mathbb{G}$ and a discriminator $\mathbb{D}$.
Output: A well-trained model $\mathcal{M}$ with the least labeling cost.
1   $\mathcal{M}$ initialization;
2   **repeat**
3   Train the model $\mathcal{M}$ with the labeled dataset L is fed into the for supervised training;
4   Train the generator $\mathbb{G}$ and discriminator $\mathbb{D}$ with all unlabeled dataset $\mathcal{U}$ via Equation (6);
5   Synthesize instances with the generator $\mathbb{G}$ for querying;
6   Obtain the representation $\mathcal{R}$ of all synthesized instances;
7   Query the top-K synthesized instances $\mathcal{K}$ via uncertain strategies, according to the representation $\mathcal{R}$;
8   Label $\mathcal{K}$ from the annotators as the ground truth $Y^{\mathcal{K}}$;
9   Update $L = L \cup \{\mathcal{K}, Y^{\mathcal{K}}\}$, update $\mathcal{U} = \mathcal{U} \cup \mathcal{K}$.
10   **until** *End conditions*;

GAAL inspired GAN-based generative adversarial methods in active learning. Consequently, the latter works were devoted to studying pool-based Generative Adversarial Active Learning. Tran et al. [59] proposed a framework of Bayesian Generative Active Learning (BGAL) to solve multi-classification tasks when the amount of labeled data is small. The proposed BGAL was validated on MNIST [41] (achieving accuracy 99.68%), CIFAR-10/100 [47] (achieving accuracy 91.13%), and SVHN [49] (achieving accuracy 69.69%). Mayer et al. [60] proposed a pool-based active learning strategy(ASAL). Compared to traditional pool-based strategies for exhaustive uncertainty search from unlabeled pools, ASAL utilized GAN to generate the most representative samples from unlabeled pools, resulting in more efficient active learning techniques. ASAL was validated on the datasets MNIST [41] (reducing 300 labeled samples), CIFAR-10 [47] (reducing 500 labeled samples), CelebA [61] (reducing 750 labeled samples), SVHN [49], and LSUN Scenes [62]. Sinha et al. [63] proposed a pool-based semi-supervised active learning algorithm VAAL. VAAL obtained great improvement in experimental results on classification and segmentation. VAAL achieved great improvement in experimental results, including CIFAR10/100 [47] (achieving accuracy of 90.16%/63.14%), Caltech-256 [64] (achieving 1.01% improvement on Coreset), ImageNet [1], Cityscapes [2] (achieving mIoU 62.95), and BDD100K [65] (achieving mIoU 44.95).

The above methods were devoted to directly generating the representative samples by solving some optimization problems, and then improving the efficiency of screening samples for active learning. More details are summarized in Table 5. Apart from that, Huijser et al. [66] firstly proposed to use a GAN to generate a batch of samples along the decision boundary of the current classifier. Next, they determined the location where the category changed from the generated samples through visualization and added it to the set of samples to be labeled. Finally, the method's effectiveness was verified by a large number of image classification experiments. The method can reduce the burden of data annotation by requiring the annotator to label decision boundaries instead of samples.

Table 5. Summary of generative methods in deep active learning.

Methods	Innovation	Architecture	Comments
GAAL [57]	▷ The first novel query synthesis-based active learning method GAAL fused with GAN. ▷ GAAL combined query synthesis with the uncertainty sampling principle and adaptively synthesized training instances for querying to increase learning speed. ▷ The DCGAN was implemented to replace the unlabeled pool in previous work.	▷ Generator: CNN ▷ Discriminator: CNN ▷ Predictor: SVM ▷ Score-function: Uncertainty	▷ GAAL was the first work integrated active learning and generative methods. ▷ GAAL provided rich representation training samples for active learning via GAN. ▷ GAAL was limited by the generated abnormal instances if the GAN was not optimized correctly. ▷ GAAL is limited by the binary classification setting.
BGAL [59]	▷ BGAL integrated deep active learning and data augmentation methods to generate informative samples and expand the labeled data set to improve the accuracy of model classification. ▷ BGAL also integrated ACGAN [67] and VAE-GAN [68] into a novel generative model named VAE-ACGAN, where the VAE decoder was the generator of the GAN. ▷ VAE-ACGAN generated new synthetic instances on the query samples. ▷ The learner and the VAE-ACGAN were jointly trained in this work.	▷ Generator: VAE ▷ Discriminator: Bayesian CNN ▷ Predictor: Resnet18 ▷ Score-function: MC-dropout	▷ BGAL extended the GAAL by combined more robust data augmentation techniques. ▷ The combination of data augmentation and active learning obtained consistent improvement on classification than single methods. ▷ The computation efficiency need to be improved due to the computational cost is high.
ASAL [60]	▷ ASAL consists of uncertainty sampling, adversarial sample generation, and sample matching. ▷ In order to approximate the underlying data distribution from the unlabeled data pool, ASAL utilized a GAN to generate adversarial samples. ▷ ASAL designed an efficient matching algorithm, where an uncertainty score was calculated to measure the similarity between the unlabeled samples and the generated samples. ▷ ASAL selected the most similar samples from the pool and performs annotation.	▷ Generator: CNN with matching ▷ Discriminator: CNN ▷ Predictor: CNN ▷ Score-function: Entropy	▷ ASAL was the first pool-based generative active learning method. ▷ The main contribution of ASAL was to select the most similar sample from pool instead of directly annotating it via a matching algorithm. ▷ ASAL utilized the entropy for uncertainty estimation and was applied in the multi-label classification.
VAAL [63]	▷ VAAL utilized adversarial learning to promote active learning. ▷ A variational autoencoder (VAE) was used to extract image features, and then a discriminator decided whether the image was labeled or unlabeled. ▷ The VAE hoped to trick the discriminator into judging all samples as labeled data, but the discriminator hoped to accurately distinguish unlabeled samples in the data pool. ▷ The annotator labeled the unlabeled samples selected based on this method.	▷ Generator: VAE ▷ Discriminator: MLP ▷ Predictor: VGG16 ▷ Score-function: Confidence	▷ VAAL provided a computational efficient sampling method with the best accuracy and time cost.

4.3. Semi-Supervised Active Learning

Vision tasks based on supervised learning require a large amount of labeled data for model training. These labeled data not only provide direct supervision signals but also limit the generalization ability of the model [69]. At the same time, the acquisition of these data is challenging due to the cost and time of annotation [26]. To alleviate this limitation, methods based on semi-supervised learning have become another mainstream. It studies model training under the premise of limited labeled data. It expects higher performance to balance the dilemma between performance and cost, which is a perfect fit for active learning, thus bringing a lot of research and practical value.

Semi-supervised learning (SSL) can train the model with a small amount of labeling cost. Unlike active learning, SSL methods usually select samples with confident prediction results instead of uncertain samples and then label them directly by the model instead of annotators. However, it is still impossible to guarantee that these high-confidence prediction results do not contain noise or erroneous results due to model prediction accuracy errors. Thus, these predictions could not directly participate in model training because the pseudo-labels may make the model abnormal. In contrast, active learning selects the samples with the most uncertain prediction results to be labeled by annotators, which can be used as the ground truth without noise and thus ensure the quality of labels. Therefore, the combination of semi-supervised learning methods and active learning methods can complement each other to a certain extent.

McCallumzy et al. [70] firstly proposed the above idea that combined Query-by-Committee active learning and Expectation-Maximization (EM) algorithms, using the naive Bayes method as a classifier and conducting experiments on text classification tasks. Subsequently, Muslea et al. [71] promoted the above work and proposed the joint testing method (Co-Testing), where two classifiers were trained in different perspectives. After that, samples were jointly selected for annotation. Finally, new labeled data were joined into the expectation-maximization (Co-EM) algorithm. Sequentially, Zhou et al. [72] combined the above semi-supervised learning and active learning methods and then validated that both of them are beneficial to the image retrieval task.

In addition, the self-training algorithm is one of the primary methods in SSL, and its core steps are shown in Algorithm 3. Firstly, the self-training algorithm initializes the model with a small number of labeled samples to ensure the initial performance of the model. Then, the algorithm selects appropriate samples and calculates their corresponding predicted labels based on the predicted representations. Finally, the labeled dataset is updated with new pseudo-labeled samples and the model is trained again in the next iteration. The main challenge of the semi-supervised learning algorithm is that SSL is easy to introduce a large number of noise samples during the training process, so the model cannot learn the correct information. Some researchers mitigate noisy samples by constructing Co-Training [73] and Tri-Training [74] algorithms of multiple classifiers.

Algorithm 3: The workflow of basic self-training algorithm

---

    input : A labeled data pool L, an unlabeled data pool $\mathcal{U}$, self-training threshold θ.
    output: A well-trained model $\mathcal{M}$ with the least labeling cost.
1 $\mathcal{M}$ initialization;
2 **repeat**
3 | Train the model $\mathcal{M}$ with the labeled dataset L;
4 | Obtain the representation $\mathcal{R}$ of all unlabeled samples from $\mathcal{U}$;
5 | Evaluate the confidence C of each sample according to the representation $\mathcal{R}$;
6 | Select the samples $\mathcal{K}$ that meet the threshold and their corresponding model
        | prediction labels $Y^{\mathcal{K}}$;
7 | Update $L = L \cup \{\mathcal{K}, Y^{\mathcal{K}}\}$, update $\mathcal{U} = \mathcal{U}/\mathcal{K}$.
8 **until** *End conditions*;

Apart from this, the authors of Refs. [75–77] integrated semi-supervised learning and active learning skillfully. They combined uncertainty-based selection strategies and self-training methods and made full use of their respective advantages while making up for their shortcomings. Consequently, their works achieved remarkable results.

However, the above semi-supervised active learning methods have not dealt with the noisy samples effectively, so it still had a significant impact on the model.

Song et al. [78] combined active Learning and semi-supervised Learning with inconsistent prediction and utilized data augmentations. These works achieved remarkable performance in CIFAR-10/100 [47] (improving 1.47%/1.16% accuracy) and SVHN [49] (improving 0.43% accuracy) classification tasks. However, they still suffered from data augmentation because there were too little data augmentations to estimate inconsistency. Sequentially, Guo et al. [79] proposed REVIVAL method and obtained more semantic distribution information via learning the continuous local distribution of unlabeled samples from feature space.

Despite the progress, most active learning algorithms suffer from data waste problems because they ignore that most of the data in unlabeled pool is not actively annotated, which can further enhance the performance via semi-supervised learning (SSL).

4.4. Active Contrastive Learning

Semi-supervised learning still needs some labeled data to carry out training, but self-supervised learning extracts representation by mining data instead of annotation. Contrastive learning is one of the most successful paradigms for self-supervised learning. The key idea of contrastive learning is to learn its relationship by comparing the similarity of different samples in the dataset. Thus, how to define the positive pairs (similar samples) and negative pairs (dissimilar samples) is the main issue in contrastive learning. The workflow of basic contrastive active learning is reported in the Algorithm 4. For arbitrary data x, the goal of contrastive learning is to learn an encoder $f(\cdot)$ such that:

$$\text{score}(f(x), f(x^+)) >> \text{score}(f(x), f(x^-)), \tag{7}$$

where $f(x)$ denotes the global features. $f(x^+)$ denotes the local features from positive samples. $f(x^-)$ denotes the local features from negative samples. The $\text{score}(\vec{X_1}, \vec{X_2})$ is the function to measure the similarity of vectors $\vec{X_1}$ and $\vec{X_2}$. Euclidean distance and cosine similarity are two classical score functions, which are formulated as follows.

$$Euclidean\left(\vec{X_1}, \vec{X_2}\right) = \left\| F\left(\vec{X_1}\right) - F\left(\vec{X_2}\right) \right\|_2 \tag{8}$$

$$Cosine\left(\vec{X_1}, \vec{X_2}\right) = \frac{\vec{X_1}^T \vec{X_2}}{|\vec{X_1}||\vec{X_2}|} \tag{9}$$

After that, it optimizes an objective that pulls the positive pairs together while pushing the negative pairs away in the representation space. The loss function InfoNCE [80] is usually used in the related research of contrastive learning, which is formulated as follows:

$$L_{NCE} = -\mathbf{E}_X \left[\log \frac{\text{score}(f(x_i), f(x_i^+))}{\text{score}(f(x_i), f(x_i^+)) + \sum_{j \neq i} \text{score}\left(f(x_i), f\left(x_j^-\right)\right)} \right], \tag{10}$$

where the corresponding sample x has one positive and N-1 negative pairs. By minimizing the InfoNCE loss, contrastive learning is to make the features of $f(x)$ more similar to the features of positive samples $f(x^+)$, and less similar to the features of N-1 negative samples $f(x_j^-)(j \in 1, \ldots, N-1)$. Finally, it can maximize a lower bound on the mutual information between $f(x)$ and $f(x^+)$ [81].

McAllester et al. [82] analyzed the theoretical shortcomings of contrastive learning, where they argued that the learned representations of contrastive learning were high relative to the size of negative samples. For example, MoCo [83–85] and SimCLR [86,87] obtained success due to the various data argumentation with large memory bank and extremely large batch size, respectively. However, Saunshi et al. [88] validated that a larger size of negative samples does not always learn better representations, leading to better performance. They found that the larger batch size would likely generate more redundant samples, thus affecting the efficiency of contrastive learning.

In order to address the above problems, active learning was introduced into contrastive learning by assisting the selection of negative samples [89]. When they carried out the cross-modal contrastive representation learning, uncertainty and diversity were used to sample the negative samples, thus reducing the redundancy actively.

Furthermore, previous active learning works assume that the labeled and unlabeled data pools follow the same class distribution. When applying these works to mismatched class distribution tasks, it suffered from performance degradation sharply. Du et al. [90] focused on this problem. They firstly defined the score function:

$$score = \theta(x_i, x_j) = \exp\left(\frac{f(x_i)^\top f(x_j)}{\tau \cdot \|f(x_i)\| \|f(x_j)\|}\right) \tag{11}$$

Then, they used contrastive learning to select semantic and distinctive features and then selected the most informative unlabeled samples $v^\star$ with minimax selection scheme for querying.

$$v^\star = \underset{v_i \in \mathcal{V}-L}{\arg\min} \ \underset{v_j \in \mathcal{N}(i)}{\max} \ d(x_i, x_j), \tag{12}$$

where $d(x_i, x_j)$ calculates the Euclidean distance between embeddings of two nodes x_i and x_j, $\mathcal{N}(i)$ represents the neighbor set of node i, $\mathcal{V}$ and L denotes the node set and labeled set, respectively. Finally, they generalized contrastive learning to active learning with the following modified loss function:

$$L_{CAL}(x_i, x_i') = -\log \frac{\theta(x_i, x_i') + \lambda \cdot \sum_{x_p \in \mathcal{P}(i)} \theta(x_i, x_p)}{\theta(x_i, x_i') + \lambda \cdot \sum_{x_p \in \mathcal{P}(i)} \theta(x_i, x_p) + \sum_{j \neq i} \left[\theta(x_i, x_j') + \theta(x_i, x_j')\right]}, \tag{13}$$

where $\mathcal{P}(i)$ denotes the set of positive embedding whose label is the same with node v_i.

Algorithm 4: The workflow of basic contrastive active learning algorithm

---

    input : A labeled data pool L, an unlabeled data pool $\mathcal{U}$.
    output: A well-trained model $\mathcal{M}$ with the least labeling cost.
1 $\mathcal{M}$ initialization;
2 **repeat**
3   construct the positive and negative sample pairs;
4   Train the model $\mathcal{M}$ with the labeled dataset L by minimizing the contrastive loss objective;
5   Calculate the distance between each sample in the batch;
6   Select the candidate $\mathcal{K}$ according the distance of embeddings or other certain estimations;
7   Query $\mathcal{K}$'s label $Y^{\mathcal{K}}$;
8   Update the positive embeddings set $\mathcal{P}(\cdot)$ with samples embeddings having the same label as $Y^{\mathcal{K}}$;
9   Update $L = L \cup \{\mathcal{K}, Y^{\mathcal{K}}\}$, update $\mathcal{U} = \mathcal{U}/\mathcal{K}$.
10 **until** *End conditions*;

Zhu et al. [91] integrated graph neural networks (GNNs)-based active learning with contrastive learning. They denoised the selected candidates by considering the neighborhood propagation scheme in GNNs. Krishnan et al. [92] proposed the featuresim score, which selected balanced, diverse, and informative samples (samples in-between clusters and from edge of clusters) from each class. Gao et al. [93] applied active learning and contrastive learning to the fine-grained off-road semantic segmentation task. They used different semantic attributes for weak supervision and defined the image patches that share the same label as positive pairs while the rest were negative pairs. Besides this, they proposed a risk evaluation method to evaluate high-risk predictions and selected for additional annotation.

4.5. Other Deep Active Learning

Unsupervised domain adaptation (UDA) is a type of cross-domain transfer learning, where the source samples are annotated, and the labels of target samples are absent during training [94]. The goal of UDA is to minimize the discrepancy in distribution between the source domain and the target domain and to learn a robust generalized representation without target annotations [95]. At present, only a few works [96,97] have utilized active learning methods to solve domain adaptation challenges and to improve the performance of the source domain model in the target domain. Recently, Ning et al. [98] first proposed a novel framework that combines active learning and unsupervised domain adaptation to assist cross-domain semantic segmentation tasks. Specifically, they clustered multiple anchors from the source domain to adopt the multi-center distribution. After that, they queried from the unlabeled target samples to the most complimentary samples from the source domain as candidates. The active learning method modeled the data distribution in both the source and target domain and, thus, captured more comprehensive information for domain adaptation.

5. Applications

Recently, computer vision is achieving a breakthrough with deep learning and booms wide applications, which require large amounts of labeled data. Meanwhile, it is impossible to abandon labels entirely or give up unlabeled data in practical applications. Under this condition, active learning can provide a more reasonable expedient, i.e., to annotate valuable data instead of all data.

According to the detailed introductions of deep active learning in the previous sections, we can conclude that the deep active learning methods can play a significant role in the label-intensive vision tasks, helping to reduce labeling costs. In other words, active learning applications are for such conditions, i.e., how to save the workload of labeling and make the model achieve satisfactory performance under a large amount of unlabeled data. Here, we only introduce some extensive-studied applications related to active learning, especially deep active learning.

5.1. Deep Learning-Based Autonomous Driving

In supervised deep learning, a large amount of labeled data needs to be collected for training [99,100], especially in the scorching field of autonomous driving. In this field, the perception of the environment of unmanned vehicles is particularly important [101,102]. The perception of the model directly affects the quality of decision making and plays a vital role in the safety of unmanned vehicles [103,104]. However, there are many complexities environments in autonomous driving scenes. In order to ensure the performance of the model, most companies need to collect the images, point clouds, and radar data from the actual scene for training. Such massive amounts of data are challenging to collect and label. Nevertheless, active learning is able to select the most valuable samples (or via the uncertainty estimation of the current model prediction) and then manually label them. Finally, we can carry out continued model training, thereby improving the model's performance as much as possible, improving stability and security. In this section, we introduce and compare the applications on deep active learning–based autonomous driving. The overview is summarized in the Table 6.

Table 6. Summary of applications on deep learning-based autonomous driving.

Applications	Comments	Implementation	Evaluation
Autonomous navigation [105]	▷ Proposed a framework for learning autonomous policies for navigation tasks from demonstrations.	▷ Network: 3 × (Conv + Pool) + FC. ▷ Score-function Entropy: $E(X) = -\sum_{i=1}^{N} P(x_i) \log P(x_i)$	▷ Reach the flag: error rate = 2.48%. ▷ Follow the line: error rate = 4.06%. ▷ Reach the correct object: error rate = 0.86%. ▷ Eat all disks: error rate = 1.70%
Weather and light classification [106]	▷ Released the first public dataset for weather and light level classification focused on autonomous driving.	▷ Target network: Resnet18 ▷ Loss-prediction module [52]: 4 × (GAP+FC+ReLU)+Concat+FC. ▷ Selection strategy: High loss samples.	▷ Weather1 [107]: accuracy = 98.80% ▷ Weather2 [108]: F1 score = 0.872 ▷ Proposed dataset [106]: F1 score = 0.772
3D object detection [109]	▷ The first work that introduced active learning into 3D object detection in autonomous driving.	▷ 3D Detector: VoxelNet ▷ Score-function: Diversity: $\min_{\mathcal{B} \subseteq \mathcal{U}} \max_{\mathbf{x}_i \in \mathcal{B}} \min_{\mathbf{x}_j \in \mathcal{B} \cup L}$ $d_{temp+spat+feat}(\mathbf{x}_i, \mathbf{x}_j)$ s.t.$cost(\mathcal{B}) \leq budget$	▷ nuScenes [110]: mAP = 45.02.
Lane detection [111]	▷ The first work that introduced active learning into lane detection in autonomous driving.	▷ Student model: ResNet-122 (for PLN [112]) ResNet-18 (for UFLD [113]) ▷ Teacher model: SENet-154 (for PLN [112]) ResNet-101 (for UFLD [113]) ▷ Score-function: Combined the uncertainty and diversity metrics.	▷ CULane [114] and LLA-MAS [115]. (F1 score not reported)
Crowd counting [116]	▷ The first work that used predictive uncertainty for sample selection pertaining to crowd counting task.	▷ Local feature block: VGG16 ▷ Non-local feature block: Transformer ▷ Score-function: Informativeness difference : $\text{Diff}(X_i) = (Mean_i(X_i, \mathcal{M}_1) - Mean_i(X_i, \mathcal{M}_2))^2$	▷ UCF-QNRF [117]: MAE = 86; MSE = 146. ▷ UCF CC [118]: MAE = 210; MSE = 305.4. ▷ ShanghaiTech-A [119]: MAE = 61.5; MSE = 103.4. ▷ ShanghaiTech-B [119]: MAE = 7.5; MSE = 11.9. ▷ NWPU [120]: MAE = 78; MSE = 448.
Crowd counting [121]	▷ Proposed a partition-based sample selection with weights (PSSW) strategy to actively select and annotate both diverse and dissimilar samples for network training.	▷ Backbone: VGG16 pretrained by imagenet ▷ Score-function: Diverse in density and dissimilar to previous selections.	▷ ShanghaiTech-A [119]: MAE = 80.4; MSE = 138.8. ▷ ShanghaiTech-B [119]: MAE = 12.7; MSE = 20.4. ▷ UCF CC [118]: MAE = 318.7; MSE = 421.6. ▷ Mall [122]: MAE = 3.8; MSE = 5.4. ▷ TRANCOS [123]: MAE = 7.5. ▷ DCC [124]: MAE = 4.5.

Hussein et al. [105] introduced active learning into the autonomous navigation application. In order to address the challenge of generalizing a model over unseen data, they utilized the entropy to measure the confidence of prediction and then labeled the low-confident samples for iterative training. Dhananjaya et al. [106] focused on the harsh weather and low light conditions during driving. They proposed a related dataset containing 60k images from videos, which consisted of various weather conditions (clear, rain, and snow), light levels (bright, moderate, and low), and street types (asphalt, grass, and cobblestone). Under the proposed dataset, previous deep learning–based autonomous driving algorithms suffered from accuracy degradation. The authors introduced an active learning framework to reduce the redundancy from adjacent frames in the video and find the optimal subset for training. Peng et al. [111] designed a novel metric combined with uncertainty and diversity to measure the informativeness of samples. The uncertainty was utilized to estimate the valuable knowledge and noise, while the diversity was used to reduce data redundancy. Liang et al. [109] took the advantage of the multimodal information provided in LiDAR point clouds, and proposed a diversity-based acquisition function that enforces spatial and temporal diversity in the selected samples. Besides this, they investigated the cold-start problem of active learning and demonstrated that the proposed diversity-based methods was able to select better initial batch at early batches, resulting in better performance. Ranjan et al. [116] focused on the domain adaptation of crowd counting. Based on the Query-By-Committee sampling strategy, they constructed the committee with two CTN networks and estimated the density and uncertainty of predictions from committee. Afterwards, they selected the informative samples from the target domain for active learning. Zhao et al. [121] selected the most informative samples via diverse in density and dissimilar to previous selections. The diversity was evaluated by separating the unlabeled set into different density partitions. The dissimilarity was evaluated by considering local crowd density and global crowd count.

5.2. Intelligent Medical Assisted Diagnosis

In the medical field, the development of deep learning has brought revolutionary development to many aspects, including diagnosis [125,126]. However, the above data-driven methods inevitably require a large amount of labeled data [127,128]. However, labeling medical images is time-consuming and labor-intensive, which also requires specific professional knowledge [129,130]. Therefore, it is efficient to use active learning to select samples that are difficult to predict by the model for selective labeling. There is much research studying active learning in the medical field. We summarized the most typical works in the Table 7.

Table 7. Summary of applications on deep learning-based intelligent medical assisted diagnosis.

Applications	Comments	Implementation	Evaluation
Medical image detection and classification [22,26]	▷ Combined active learning, incremental fine-tuning, and transfer learning.	▷ Network: AlexNet pretrained by imagenet ▷ Selection strategy: Entropy: $e_i^j = -\sum_{k=1}^{\|Y\|} p_i^{j,k} \log p_i^{j,k}$ Diversity: $d_i(j,l) = \sum_{k=1}^{\|Y\|} \left(p_i^{j,k} - p_i^{l,k} \right) \log \frac{p_i^{j,k}}{p_i^{l,k}}$	▷ polyp detection: ↓ 86% labels. ▷ pulmonary embolism detection: ↓ 80% labels. ▷ colonoscopy frame classification: ↓ 82% labels. ▷ scene classification: ↓ 35% labels.
COVID-19 Lung Ultrasound Multi-symptom Classification [131]	▷ The first work that introduced active learning into ultrasound classification for COVID-19-assisted diagnosis.	▷ Backbone: ResNet50 pretrained by imagenet ▷ Score-function: Least confidence: $LC(x) = \max_{1 \le i \le l} p(l_i \mid x)$t Multi-label entropy: $MLE(x) = \sum_{i=1}^{l} (p(l_i \mid x) \log p(l_i \mid x) + p\left(\overline{l_i} \mid x\right) \log p\left(\overline{l_i} \mid x\right))$	▷ COVID19-LUSMS v1: ↓ 80% labels.
Brain tumor Classification [132]	▷ Sampling candidates by discarding subsets of training samples with the highest and lowest uncertainty scores.	▷ Network: AlexNet pretrained by imagenet ▷ Score-function: Combined entropy and Kullback–Leibler(KL) divergence: $E(X) = -\sum_{i=1}^{N} P(x_i) \log P(x_i)$ $D(p\|q) = \sum_{i=1}^{n} P(x_i) \log \frac{P(x_i)}{Q(x_i)}$	▷ MICCAI BRATS [133–135]: ↓ 40% labels.
Diabetic retinopathy classification [136]	▷ The first work that introduced active learning into lane detection in autonomous driving.	▷ Bayesian convolutional neural network (BCNN): Monte-Carlo drop-out ▷ Teacher model: SENet-154 (for PLN [112]) ResNet-101 (for UFLD [113]) ▷ Score-function: entropy.	▷ APTOS 2019 [137]: AUC = 0.99 (multi-class classification) Accuracy = 92% (multi-class classification) Accuracy = 85% (BCNN in Active Learning)
Histopathology image analysis [138]	▷ The first work that proposed an AL framework (PathAL) to dynamically identify important samples to annotate and to distinguish noisy from hard samples in the training set.	▷ Backbone: EfficientNet-B0 [139] ▷ Noisy sample detector: O2U-Net [140] Curriculum Sample Classification: CurriculumNet [141] ▷ Score-function: Distinguished noisy samples from hard ones, and selected the most informative samples to be annotated.	▷ PANDA [142]: quadratic weighted kappa = 89.5.
Gastric adenocarcinoma and colorectal cancer [143]	▷ The first work that explored the identification of the most informative region of patches and proposed a patch location system to select patches.	▷ Backbone: ResNet-18 ▷ Loss-prediction module [52]: 4 × (GAP+FC+ReLU)+Concat+FC. ▷ Score-function: $\Delta\mathbf{p}_*^{\mathcal{G}} = \arg\max_{\Delta\mathbf{p}^{\mathcal{G}}} \mathcal{H}\left(\Delta\mathbf{p}^{\mathcal{G}} + \mathbf{p}^{\mathcal{G}}\right).$	▷ TCGA [144,145]: AUC = 0.933. accuracy = 92.7%.

Zhou et al. [22,26] introduced transfer learning and data enhancement into active learning. By measuring the uncertainty and diversity, the proposed AIFT framework achieved SOTA performance in the biomedical image analysis. Liu et al. [131] introduced active learning into ultrasound classification for COVID19-assisted diagnosis. In order to actively reduce the labeling efforts, the proposed method combined least confidence and entropy selection strategies. Hao et al. [132] combined entropy and Kullback–Leibler(KL) divergence for uncertainty-based sampling. Apart from the active learning, they also carried out transfer learning from imagenet-pretrained AlexNet [146] to MRI (MICCAI BRATS 2019 dataset) [133–135]. The proposed transfer learning framework reduced the annotation cost while maintaining the stability and robustness of the model performance for brain tumor classification.

Ahsan et al. [136] integrated Bayesian-based CNN and uncertainty-based active learning method, where active learning was applied to the pool-based sampling and query by committee scenarios. Wang et al. [147] formulated the active learning as a Markov decision process and introduced a deep reinforcement learning algorithm for the selection of the most informative samples. The proposed method was validated in four kinds of lung disease detection with CT images (chestCT (https://tianchi.aliyun.com/competition/entrance/231724/introduction, accessed on 11 July 2022)) and diabetic retinopathy in digital color fundus photograph (Retinopathy (https://www.kaggle.com/competitions/diabetic-retinopathy-detectio, accessed on 11 July 2022)). Smit et al. [148] pretrained the active learning framework with contrastive learning and utilized the cosine similarity to classify unseen images. The proposed method was validated in the eight common chest observations in X-ray images (CheXpert [149]). Shen et al. [143] first explored the identification of the most informative region of patches and proposed a patch location system to select patches. The proposed method was validated in three gastric adenocarcinoma and colorectal cancer datasets from The Cancer Genome Atlas (TCGA [144]). After that, they continued to explore the whole-slide histopathology image annotation with active learning. In this work [150], they incorporated spatial distribution representation and histopathology tissue informativeness for uncertainty sampling. Li et al. [138] adopted the semi-supervised idea that selected confident samples from the unlabeled set and automatically utilized their corresponding predictions as pseudo-labels for training. They proposed the PathAL framework, where annotators and co-training label the other "informative" sample with the above pseudo-labels.

6. Challenges

Although the motivation of deep active learning is to reduce the amount of annotation in practical applications and provide an efficient learning solution for deep learning, the current active learning methods still have some challenges in practical application, which can be summarized into the following four aspects.

6.1. Inefficient Serial Human-in-the-Loop Collaboration

The essence of active learning is still a process of continuous interaction between computers and annotators, which will undoubtedly cause inconvenience in interaction. The process of most active learning methods is still to select a batch of candidates and send them to annotators for labeling and expect annotators to label them as soon as possible and return the labeled samples back, and finally, the model continues to train and then select candidates again. This is a serial process, which means that when annotators are labeling, the model cannot be trained or perform any other operations. It is necessary to wait for the end of manual labeling before the next round of iterative training can be performed.

For example, we assume that there is an active learning labeling system in the medical scene. For the computer, the strategy first selects some samples and sends them to the doctor for labeling, and then is in the idle period waiting for labeling. For the doctor, after receiving the samples, it is time-consuming to label and then return it to the model training, and wait for the subsequent feedback from the model. In this way, the doctor and the model wait for each other's operations, reducing efficiency and convenience. Consequently, an efficient parallel strategy for active learning is expected to be proposed in the future.

6.2. Dirty Data and Noisy Oracle

Most of the existing deep learning research assumes that the data is independent and identically distributed and uses publicly available datasets. These datasets contain little to no dirty data (noise, imbalance). However, in industrial practice, data sources are far from the ideal dataset with more dirty data. For example, there are categories with fewer samples or fewer categories with more samples (sample category distribution imbalance). The uncertainty selection strategy is widely used in active learning, but it is hard to evaluate the uncertainty of noisy samples. At the same time, the oracle's annotation is considered ground truth, but it may also contain errors [151]. Consequently, it is unconfident when these noisy samples or labels are used for active learning. Such samples may not improve the model's performance but even worsen the performance.

6.3. Difficult to Cross-Domain Transfer

No matter what selection strategy is used in the existing active learning, it is based on the current data distribution of the source domain. Industrial practice requires a more general and generalizable active learning strategy, so that they can transfer between different domains and tasks with considerable performance.

As a sub-field of transfer learning, cross-domain adaptation has been extensively studied in the recent years [152,153]. Prabhu et al. [154] demonstrated that existing model uncertainty-based or diversity-based active learning methods based solely on are ineffective for domain adaptation. Xie et al. [155] introduced an energy-based strategy to select the most representative and informative target data to assist the adaptation. However, we are disappointed to find that most active learning strategies are domain-designed, and there is no guarantee that the active learning strategy can achieve competitive performance when cross-domain transfer. For example, there is already an active learning method designed for cat and dog classification tasks based on the uncertainty selection strategy, and it has achieved better performance. Now, if we transfer it into a new task for husky and labrador classification, the performance may degrade. If the new task is organ or tumor classification in medical images, redesigning a new active learning method is more recommended than using the previous method, but it wastes time and cost. Fu et al. [156] proposed the transferable query selection (TQS) strategy to select the most informative samples under domain shift. The TQS consists of transferable committee, transferable uncertainty, and transferable domainness. Besides these, rare works have studied the unsupervised domain adaptation with active learning. Consequently, an active learning strategy with robust cross-domain transferring ability is expected to be proposed in the future to solve this challenge.

6.4. Unstable Performance

The biggest challenge that hinders the practical application of active learning methods is the unstable performance. As introduced in previous sections, active learning is to select candidates according to some strategy. These selected samples are significant for the sequential training and evaluation, especially at the beginning.

As we expected, deep active learning usually outperforms random sampling, especially when high-redundant data distribution. However, we have to admit that current active learning may still perform worse than random sampling in the early stage when the data distribution is diverse and has low redundancy. Random sampling can collect more representative samples than active learning under this condition, and the model receptive field is more comprehensive, thus obtaining better initialization. This phenomenon is named the cold-start problem in active learning, which is shown in Figure 3. When the application scenario of active learning has the data distribution mentioned above, it must afford the cost of additional selection samples than random sampling in the early stage of training. If the performance is worse than random sampling, this part of the cost has already been invested and cannot be recovered.

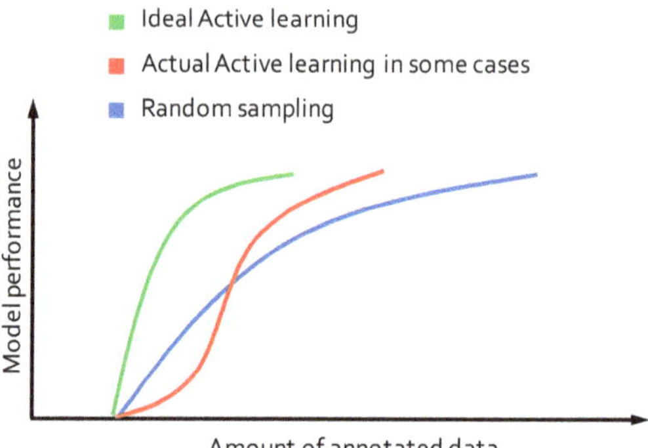

Figure 3. Illustration of the cold-start in active learning. The green curve denotes the ideal active learning process. The red curve denotes the actual active learning process. The blue curve denotes the training process of the random selection strategy without any active learning.

Therefore, the industry has stricter requirements for active learning in practical applications, and it is almost necessary to work if the designed strategy is directly applied. If not, those selected samples are still marked, and time and money are lost. Such harsh requirements and unstable performance lead people to prefer to save this cost and turn to directly adopting random sampling, but design a better model or use a better optimization strategy to achieve more stable performance.

Zhou et al. [22] explored the cold-start problem and found the reasons were the scarcity of labeled dataset and the instability of the model at the beginning. They addressed this problem by cooperating with the random sampling method. They obtained better performance in early stages and improvement during sequential steps. Another solution is pretrained active learning, which means that, before carrying out active learning, we initialized the model with pretrained weights and gave the stability to the model. Typical self-supervised pretraining methods such as MoCo [83] or Genesis [157] utilize the unlabeled data pool and have the potential to address the cold start problems in active learning.

7. Conclusions

This paper reviewed the fundamental theories of active learning, including the candidate selection strategies and querying scenarios. Besides this, we conducted a comprehensive analysis of deep learning–based active learning, including generative adversarial active learning, semi-supervised active learning, active contrastive learning, and unsupervised active domain adaptation. Meanwhile, active learning applications in computer vision tasks were detailed, such as deep learning-based autonomous driving and intelligent medical assisted diagnosis. Lastly, we summarized some challenges in current deep active learning methods for future research.

Author Contributions: Writing—original draft preparation, M.W.; writing—review and editing, C.L.; project administration, funding acquisition, Z.Y. All authors have read and agreed to the published version of the manuscript.

Funding: This research was funded by the National Key Research and Development Program of China (No. 2018YFB0204301) and Natural Science Foundation of Hunan Province of China (No. 2022JJ30666).

Institutional Review Board Statement: Not applicable.

Informed Consent Statement: Not applicable.

Data Availability Statement: Not applicable.

Conflicts of Interest: The authors declare no conflict of interest.

References

1. Deng, J.; Dong, W.; Socher, R.; Li, L.J.; Li, K.; Li, F.-F. Imagenet: A large-scale hierarchical image database. In Proceedings of the IEEE Conference on Computer Vision and Pattern Recognition Miami, FL, USA, 20–25 June 2009; pp. 248–255.
2. Cordts, M.; Omran, M.; Ramos, S.; Rehfeld, T.; Enzweiler, M.; Benenson, R.; Franke, U.; Roth, S.; Schiele, B. The cityscapes dataset for semantic urban scene understanding. In Proceedings of the IEEE Conference on Computer Vision and Pattern Recognition Las Vegas, NV, USA, 27–30 June 2016; pp. 3213–3223.
3. Ma, J.; Zhang, Y.; Gu, S.; Zhu, C.; Ge, C.; Zhang, Y.; An, X.; Wang, C.; Wang, Q.; Liu, X.; et al. AbdomenCT-1K: Is Abdominal Organ Segmentation A Solved Problem. *IEEE Trans. Pattern Anal. Mach. Intell. (TPAMI)* **2021**. [CrossRef] [PubMed]
4. Settles, B. *Active Learning Literature Survey*; Computer Sciences Technical Report 1648; University of Wisconsin–Madison: Madison, WI, USA, 2004.
5. Netzer, E.; Geva, A.B. Human-in-the-loop active learning via brain computer interface. *Ann. Math. Artif. Intell.* **2020**, *88*, 1191–1205. [CrossRef]
6. Budd, S.; Robinson, E.C.; Kainz, B. A survey on active learning and human-in-the-loop deep learning for medical image analysis. *Med. Image Anal.* **2021**, *71*, 102062. [CrossRef] [PubMed]
7. Kumar, P.; Gupta, A. Active learning query strategies for classification, regression, and clustering: A survey. *J. Comput. Sci. Technol.* **2020**, *35*, 913–945. [CrossRef]
8. Ren, P.; Xiao, Y.; Chang, X.; Huang, P.Y.; Chen, X.; Wang, X. A survey of deep active learning. *ACM Comput. Surv. (CSUR)* **2021**, *54*, 1–40. [CrossRef]
9. Zhan, X.; Wang, Q.; Huang, K.H.; Xiong, H.; Dou, D.; Chan, A.B. A comparative survey of deep active learning. *arXiv* **2022**, arXiv:2203.13450.
10. Li, M.; Sethi, I.K. Confidence-based active learning. *IEEE Trans. Pattern Anal. Mach. Intell.* **2006**, *28*, 1251–1261.
11. Agrawal, A.; Tripathi, S.; Vardhan, M. Multicore based least confidence query sampling strategy to speed up active learning approach for named entity recognition. *Computing* **2021**, 1–19. [CrossRef]
12. Agrawal, A.; Tripathi, S.; Vardhan, M. Active learning approach using a modified least confidence sampling strategy for named entity recognition. *Prog. Artif. Intell.* **2021**, *10*, 113–128. [CrossRef]
13. Joshi, A.J.; Porikli, F.; Papanikolopoulos, N. Multi-class active learning for image classification. In Proceedings of the 2009 IEEE Conference on Computer Vision and Pattern Recognition, Miami, FL, USA 20–25 June 2009; pp. 2372–2379. [CrossRef]
14. Zhou, J.; Sun, S. Improved margin sampling for active learning. In Proceedings of the Chinese Conference on Pattern Recognition, Changsha, China, 17–19 November 2014; Springer: Berlin/Heidelberg, Germany, 2014; pp. 120–129.
15. Gu, Y.; Jin, Z.; Chiu, S.C. Active learning combining uncertainty and diversity for multi-class image classification. *IET Comput. Vis.* **2015**, *9*, 400–407. [CrossRef]
16. Yang, Y.; Ma, Z.; Nie, F.; Chang, X.; Hauptmann, A.G. Multi-class active learning by uncertainty sampling with diversity maximization. *Int. J. Comput. Vis.* **2015**, *113*, 113–127. [CrossRef]
17. Yu, D.; Varadarajan, B.; Deng, L.; Acero, A. Active learning and semi-supervised learning for speech recognition: A unified framework using the global entropy reduction maximization criterion. *Comput. Speech Lang.* **2010**, *24*, 433–444. [CrossRef]
18. Ozdemir, F.; Peng, Z.; Tanner, C.; Fuernstahl, P.; Goksel, O. Active learning for segmentation by optimizing content information for maximal entropy. In *Deep Learning in Medical Image Analysis and Multimodal Learning for Clinical Decision Support*; Springer: Cham, Switzerland, 2018; pp. 183–191.
19. Brinker, K. Incorporating diversity in active learning with support vector machines. In Proceedings of the 20th International Conference on Machine Learning, Washington, DC, USA, 21 August 2003; pp. 59–66.
20. Kukar, M. Transductive reliability estimation for medical diagnosis. *Artif. Intell. Med.* **2003**, *29*, 81–106. [CrossRef]
21. Chakraborty, S.; Balasubramanian, V.; Sun, Q.; Panchanathan, S.; Ye, J. Active batch selection via convex relaxations with guaranteed solution bounds. *IEEE Trans. Pattern Anal. Mach. Intell. (TPAMI)* **2015**, *37*, 1945–1958. [CrossRef]
22. Zhou, Z.; Shin, J.Y.; Gurudu, S.R.; Gotway, M.B.; Liang, J. Active, continual fine tuning of convolutional neural networks for reducing annotation efforts. *Med. Image Anal.* **2021**, *71*, 101997. [CrossRef]
23. Seung, H.S.; Opper, M.; Sompolinsky, H. Query by Committee. In Proceedings of the Fifth Annual Workshop on Computational Learning Theory, Pittsburgh, PA, USA, 1 July 1992; pp. 287–294.
24. Yan, Y.; Rosales, R.; Fung, G.; Dy, J. Active Learning from Crowds. In Proceedings of the 28th International Conference on Machine Learning, Bellevue, WA, USA, 28 June 2011; Getoor, L., Scheffer, T., Eds.; ACM: New York, NY, USA, 2011; pp. 1161–1168.
25. Dagan, I.; Engelson, S.P. Committee-based sampling for training probabilistic classifiers. In *Machine Learning Proceedings 1995*; Elsevier: Amsterdam, The Netherlands, 1995; pp. 150–157.
26. Zhou, Z.; Shin, J.; Zhang, L.; Gurudu, S.; Gotway, M.; Liang, J. Fine-tuning convolutional neural networks for biomedical image analysis: Actively and incrementally. In Proceedings of the IEEE Conference on Computer Vision and Pattern Recognition, Honolulu, HI, USA, 21–26 July 2017; pp. 7340–7351.
27. Angluin, D. Queries and Concept Learning. *Mach. Learn.* **1988**, *2*, 319–342. [CrossRef]
28. Schumann, R.; Rehbein, I. Active learning via membership query synthesis for semi-supervised sentence classification. In Proceedings of the 23rd Conference on Computational Natural Language Learning, Hong Kong, China, 3–4 November 2019; pp. 472–481.

29. Alabdulmohsin, I.; Gao, X.; Zhang, X. Efficient active learning of halfspaces via query synthesis. In Proceedings of the Twenty-Ninth AAAI Conference on Artificial Intelligence, Austin, TX, USA, 25–30 January 2015.

30. Atlas, L.; Cohn, D.; Ladner, R. Training Connectionist Networks with Queries and Selective Sampling. In *Advances in Neural Information Processing Systems*; Touretzky, D., Ed.; Morgan-Kaufmann: Burlington, MA, USA, 1989; Volume 2.

31. Balasubramanian, V.; Chakraborty, S.; Panchanathan, S. Generalized query by transduction for online active learning. In Proceedings of the IEEE 12th International Conference on Computer Vision (ICCV) Workshops, Kyoto, Japan, 27 September–4 October 2009; pp. 1378–1385.

32. Ho, S.S.; Wechsler, H. Query by transduction. *IEEE Trans. Pattern Anal. Mach. Intell. (TPAMI)* **2008**, *30*, 1557–1571.

33. Monteleoni, C.; Kaariainen, M. Practical online active learning for classification. In Proceedings of the 2007 IEEE Conference on Computer Vision and Pattern Recognition (CVPR), Minneapolis, MN, USA, 17–22 June 2007; pp. 1–8.

34. Lewis, D.D.; Gale, W.A. A Sequential Algorithm for Training Text Classifiers. In Proceedings of the 17th Annual International ACM SIGIR Conference on Research and Development in Information Retrieval, Dublin, Ireland, 3–6 July 1994; pp. 3–12.

35. Wu, D. Pool-based sequential active learning for regression. *IEEE Trans. Neural Netw. Learn. Syst.* **2018**, *30*, 1348–1359. [CrossRef]

36. Zhan, X.; Liu, H.; Li, Q.; Chan, A.B. A Comparative Survey: Benchmarking for Pool-based Active Learning. In Proceedings of the 30th International Joint Conference on Artificial Intelligence (IJCAI 2021), Virtual, 19–27 August 2021; pp. 4679–4686.

37. Sugiyama, M.; Nakajima, S. Pool-based active learning in approximate linear regression. *Mach. Learn.* **2009**, *75*, 249–274. [CrossRef]

38. Gal, Y.; Ghahramani, Z. Dropout as a bayesian approximation: Representing model uncertainty in deep learning. In Proceedings of the International Conference on Machine Learning, New York, NY, USA, 19–24 June 2016; pp. 1050–1059.

39. Gal, Y.; Ghahramani, Z. Bayesian convolutional neural networks with Bernoulli approximate variational inference. *arXiv* **2015**, arXiv:1506.02158.

40. Gal, Y.; Islam, R.; Ghahramani, Z. Deep bayesian active learning with image data. In Proceedings of the International Conference on Machine Learning (ICML), Sydney, NSW, Australia, 6–11 August 2017; pp. 1183–1192.

41. LeCun, Y.; Bottou, L.; Bengio, Y.; Haffner, P. Gradient-based learning applied to document recognition. *Proc. IEEE* **1998**, *86*, 2278–2324. [CrossRef]

42. Codella, N.C.; Gutman, D.; Celebi, M.E.; Helba, B.; Marchetti, M.A.; Dusza, S.W.; Kalloo, A.; Liopyris, K.; Mishra, N.; Kittler, H.; et al. Skin lesion analysis toward melanoma detection: A challenge at the 2017 international symposium on biomedical imaging (isbi), hosted by the international skin imaging collaboration (isic). In Proceedings of the 2018 IEEE 15th International Symposium on Biomedical Imaging (ISBI 2018), Washington, DC, USA, 4–7 April 2018; pp. 168–172.

43. Houlsby, N.; Huszár, F.; Ghahramani, Z.; Lengyel, M. Bayesian active learning for classification and preference learning. *arXiv* **2011**, arXiv:1112.5745.

44. Shannon, C.E. A mathematical theory of communication. *ACM SIGMOBILE Mob. Comput. Commun. Rev.* **2001**, *5*, 3–55. [CrossRef]

45. Hansen, L.K.; Salamon, P. Neural network ensembles. *IEEE Trans. Pattern Anal. Mach. Intell.* **1990**, *12*, 993–1001. [CrossRef]

46. Beluch, W.H.; Genewein, T.; Nürnberger, A.; Köhler, J.M. The power of ensembles for active learning in image classification. In Proceedings of the IEEE Conference on Computer Vision and Pattern Recognition, Salt Lake City, UT, USA, 18–23 June 2018; pp. 9368–9377.

47. Krizhevsky, A. Learning Multiple Layers of Features from Tiny Images. 2009. Available online: https://www.cs.toronto.edu/~kriz/learning-features-2009-TR.pdf (accessed on 10 July 2022).

48. Sener, O.; Savarese, S. Active Learning for Convolutional Neural Networks: A Core-Set Approach. In Proceedings of the International Conference on Learning Representations (ICLR), Vancouver, BC, Canada, 30 April–3 May 2018.

49. Netzer, Y.; Wang, T.; Coates, A.; Bissacco, A.; Wu, B.; Ng, A.Y. Reading digits in natural images with unsupervised feature learning. In Proceedings of the NIPS Workshop on Deep Learning and Unsupervised Feature Learning, Granada, Spain, 16 December 2011; p. 5.

50. Janz, D.; van der Westhuizen, J.; Hernández-Lobato, J.M. Actively learning what makes a discrete sequence valid. *arXiv* **2017**, arXiv:1708.04465.

51. Kirsch, A.; Van Amersfoort, J.; Gal, Y. Batchbald: Efficient and diverse batch acquisition for deep bayesian active learning. In Proceedings of the NIPS'19: Proceedings of the 33rd International Conference on Neural Information Processing Systems, Vancouver, BC, Canada, 8–14 December 2019; Volume 32.

52. Yoo, D.; Kweon, I.S. Learning loss for active learning. In Proceedings of the IEEE/CVF Conference on Computer Vision and Pattern Recognition, Long Beach, CA, USA, 15–20 June 2019; pp. 93–102.

53. Everingham, M.; Van Gool, L.; Williams, C.K.; Winn, J.; Zisserman, A. The pascal visual object classes (voc) challenge. *Int. J. Comput. Vis. (IJCV)* **2010**, *88*, 303–338. [CrossRef]

54. Andriluka, M.; Pishchulin, L.; Gehler, P.; Schiele, B. 2d human pose estimation: New benchmark and state of the art analysis. In Proceedings of the IEEE Conference on Computer Vision and Pattern Recognition (CVPR), Columbus, OH, USA, 23–28 June 2014; pp. 3686–3693.

55. François, D. High-dimensional data analysis. From Optimal Metric to Feature Selection. Ph.D. Thesis, Université Catholique de Louvain, Ottignies-Louvain-la-Neuve, Belgium, 2008; pp. 54–55.

56. Goodfellow, I.; Pouget-Abadie, J.; Mirza, M.; Xu, B.; Warde-Farley, D.; Ozair, S.; Courville, A.; Bengio, Y. Generative Adversarial Nets. In Proceedings of the Advances in Neural Information Processing Systems 27 (NIPS 2014), Montreal, QC, USA, 8–13 December 2014; Volume 27.

57. Zhu, J.; Bento, J. Generative Adversarial Active Learning. *arXiv* **2017**, arXiv:1702.07956.

58. LeCun, Y.; Boser, B.; Denker, J.S.; Henderson, D.; Howard, R.E.; Hubbard, W.; Jackel, L.D. Backpropagation Applied to Handwritten Zip Code Recognition. *Neural Comput.* **1989**, *1*, 541–551. [CrossRef]

59. Tran, T.; Do, T.T.; Reid, I.; Carneiro, G. Bayesian generative active deep learning. In Proceedings of the International Conference on Machine Learning (ICML), Long Beach, CA, USA, 9–15 June 2019; pp. 6295–6304.

60. Mayer, C.; Timofte, R. Adversarial Sampling for Active Learning. In Proceedings of the 2020 IEEE Winter Conference on Applications of Computer Vision (WACV), Snowmass, CO, USA, 1–5 March 2020; pp. 3060–3068.

61. Liu, Z.; Luo, P.; Wang, X.; Tang, X. Deep Learning Face Attributes in the Wild. In Proceedings of the 2015 IEEE International Conference on Computer Vision (ICCV), Santiago, Chile, 7–13 December 2015; pp. 3730–3738. [CrossRef]

62. Yu, F.; Zhang, Y.; Song, S.; Seff, A.; Xiao, J. LSUN: Construction of a Large-scale Image Dataset using Deep Learning with Humans in the Loop. *arXiv* **2015**, arXiv:1506.03365.

63. Sinha, S.; Ebrahimi, S.; Darrell, T. Variational Adversarial Active Learning. In Proceedings of the 2019 IEEE/CVF International Conference on Computer Vision (ICCV), Seoul, Korea, 27 October 2019; pp. 5971–5980. [CrossRef]

64. Griffin, G.; Holub, A.; Perona, P. Caltech-256 Object Category Dataset. Available online: https://data.caltech.edu/records/20087 (accessed on 10 July 2022).

65. Yu, F.; Chen, H.; Wang, X.; Xian, W.; Chen, Y.; Liu, F.; Madhavan, V.; Darrell, T. Bdd100k: A diverse driving dataset for heterogeneous multitask learning. In Proceedings of the IEEE/CVF Conference on Computer Vision and Pattern Recognition (CVPR), Seattle, WA, USA, 13–19 June 2020; pp. 2636–2645.

66. Huijser, M.; Gemert, J.C.v. Active Decision Boundary Annotation with Deep Generative Models. In Proceedings of the 2017 IEEE International Conference on Computer Vision (ICCV), Venice, Italy, 22–29 October 2017; pp. 5296–5305.

67. Odena, A.; Olah, C.; Shlens, J. Conditional image synthesis with auxiliary classifier gans. In Proceedings of the International Conference on Machine Learning (ICML), Sydney, NSW, Australia, 6–11 August 2017; pp. 2642–2651.

68. Larsen, A.B.L.; Sønderby, S.K.; Larochelle, H.; Winther, O. Autoencoding beyond pixels using a learned similarity metric. In Proceedings of the International Conference on Machine Learning (ICML), New York, NY, USA, 19–24 June 2016; pp. 1558–1566.

69. Li, C.; Chen, W.; Luo, X.; He, Y.; Tan, Y. Adaptive Pseudo Labeling for Source-Free Domain Adaptation in Medical Image Segmentation. In Proceedings of the ICASSP 2022—2022 IEEE International Conference on Acoustics, Speech and Signal Processing (ICASSP), Singapore, 22–27 May 2022; pp. 1091–1095.

70. McCallum, A.; Nigam, K. Employing EM and Pool-Based Active Learning for Text Classification. In Proceedings of the Fifteenth International Conference on Machine Learning (ICML), Madison, WI, USA, 24–27 July 1998; pp. 350–358.

71. Muslea, I.; Minton, S.; Knoblock, C.A. Active+ semi-supervised learning= robust multi-view learning. In Proceedings of the Fifteenth International Conference on Machine Learning (ICML), Sydney, NSW, Australia, 8–12 July 2002; Volume 2, pp. 435–442.

72. Zhou, Z.H.; Chen, K.J.; Jiang, Y. Exploiting unlabeled data in content-based image retrieval. In Proceedings of the European Conference on Machine Learning (ECML), Pisa, Italy, 20–24 September 2004; pp. 525–536.

73. Blum, A.; Mitchell, T. Combining Labeled and Unlabeled Data with Co-Training. In Proceedings of the Eleventh Annual Conference on Computational Learning Theory (COLT), Madison, WI, USA, 24–26 July 1998; pp. 92–100.

74. Zhou, Z.H.; Li, M. Tri-training: Exploiting unlabeled data using three classifiers. *IEEE Trans. Knowl. Data Eng.* **2005**, *17*, 1529–1541. [CrossRef]

75. Han, W.; Coutinho, E.; Ruan, H.; Li, H.; Schuller, B.; Yu, X.; Zhu, X. Semi-supervised active learning for sound classification in hybrid learning environments. *PLoS ONE* **2016**, *11*, e0162075. [CrossRef] [PubMed]

76. Tomanek, K.; Hahn, U. Semi-supervised active learning for sequence labeling. In Proceedings of the 47th Annual Meeting of the Association of Computational Linguistics (ACL), Singapore, 2–7 August 2009; pp. 1039–1047.

77. Tur, G.; Hakkani-Tür, D.; Schapire, R.E. Combining active and semi-supervised learning for spoken language understanding. *Speech Commun.* **2005**, *45*, 171–186. [CrossRef]

78. Song, S.; Berthelot, D.; Rostamizadeh, A. Combining mixmatch and active learning for better accuracy with fewer labels. *arXiv* **2019**, arXiv:1912.00594.

79. Guo, J.; Shi, H.; Kang, Y.; Kuang, K.; Tang, S.; Jiang, Z.; Sun, C.; Wu, F.; Zhuang, Y. Semi-supervised active learning for semi-supervised models: Exploit adversarial examples with graph-based virtual labels. In Proceedings of the IEEE/CVF International Conference on Computer Vision (ICCV), Montreal, QC, Canada, 11–17 October 2021; pp. 2896–2905.

80. Van den Oord, A.; Li, Y.; Vinyals, O. Representation learning with contrastive predictive coding. *arXiv* **2018**, arXiv:1807.03748.

81. Poole, B.; Ozair, S.; Van Den Oord, A.; Alemi, A.; Tucker, G. On variational bounds of mutual information. In Proceedings of the International Conference on Machine Learning (ICML), Long Beach, CA, USA, 9–15 June 2019; pp. 5171–5180.

82. McAllester, D.; Stratos, K. Formal limitations on the measurement of mutual information. In Proceedings of the International Conference on Artificial Intelligence and Statistics, (PMLR), Palermo, Italy, 3–5 June 2020; pp. 875–884.

83. He, K.; Fan, H.; Wu, Y.; Xie, S.; Girshick, R. Momentum contrast for unsupervised visual representation learning. In Proceedings of the IEEE/CVF Conference on Computer Vision and Pattern Recognition (CVPR), Seattle, WA, USA, 14–19 June 2020; pp. 9729–9738.

84. Chen, X.; Fan, H.; Girshick, R.; He, K. Improved baselines with momentum contrastive learning. *arXiv* **2020**, arXiv:2003.04297.
85. Chen, X.; Xie, S.; He, K. An Empirical Study of Training Self-Supervised Vision Transformers. In Proceedings of the IEEE/CVF International Conference on Computer Vision (ICCV), Montreal, QC, Canada, 11–17 October 2021; pp. 9640–9649.
86. Chen, T.; Kornblith, S.; Norouzi, M.; Hinton, G. A simple framework for contrastive learning of visual representations. In Proceedings of the International Conference on Machine Learning (ICML), PMLR, Vienna, Austria, 12–18 July 2020; pp. 1597–1607.
87. Chen, T.; Kornblith, S.; Swersky, K.; Norouzi, M.; Hinton, G.E. Big self-supervised models are strong semi-supervised learners. *Adv. Neural Inf. Process. Syst. (NIPS)* **2020**, *33*, 22243–22255.
88. Saunshi, N.; Plevrakis, O.; Arora, S.; Khodak, M.; Khandeparkar, H. A theoretical analysis of contrastive unsupervised representation learning. In Proceedings of the International Conference on Machine Learning (ICML), PMLR, Long Beach, CA, USA, 10–15 June 2019; pp. 5628–5637.
89. Ma, S.; Zeng, Z.; McDuff, D.; Song, Y. Active Contrastive Learning of Audio-Visual Video Representations. In Proceedings of the International Conference on Learning Representations (ICLR), New Orleans, LA, USA, 3–7 May 2021.
90. Du, P.; Zhao, S.; Chen, H.; Chai, S.; Chen, H.; Li, C. Contrastive coding for active learning under class distribution mismatch. In Proceedings of the IEEE/CVF International Conference on Computer Vision (ICCV), Montreal, QC, Canada, 11–17 October 2021; pp. 8927–8936.
91. Zhu, Y.; Xu, W.; Liu, Q.; Wu, S. When contrastive learning meets active learning: A novel graph active learning paradigm with self-supervision. *arXiv* **2020**, arXiv:2010.16091.
92. Krishnan, R.; Ahuja, N.; Sinha, A.; Subedar, M.; Tickoo, O.; Iyer, R. Improving robustness and efficiency in active learning with contrastive loss. *arXiv* **2021**, arXiv:2109.06873.
93. Gao, B.; Zhao, X.; Zhao, H. An Active and Contrastive Learning Framework for Fine-Grained Off-Road Semantic Segmentation. *arXiv* **2022**, arXiv:2202.09002.
94. Li, C.; Luo, X.; Chen, W.; He, Y.; Wu, M.; Tan, Y. AttENT: Domain-Adaptive Medical Image Segmentation via Attention-Aware Translation and Adversarial Entropy Minimization. In Proceedings of the 2021 IEEE International Conference on Bioinformatics and Biomedicine (BIBM), Houston, TX, USA, 9–12 December 2021; pp. 952–959.
95. Li, C.; Chen, W.; Wu, M.; Luo, X.; He, Y.; Tan, Y. Tri-Directional Tasks Complementary Learning for Unsupervised Domain Adaptation of Cross-modality Medical Image Semantic Segmentation. In Proceedings of the 2021 IEEE International Conference on Bioinformatics and Biomedicine (BIBM), Houston, TX, USA, 9–12 December 2021; pp. 1406–1411.
96. Chattopadhyay, R.; Fan, W.; Davidson, I.; Panchanathan, S.; Ye, J. Joint transfer and batch-mode active learning. In Proceedings of the International Conference on Machine Learning (ICML), PMLR, Atlanta, GA, USA, 16–21 June 2013; pp. 253–261.
97. Huang, S.J.; Zhao, J.W.; Liu, Z.Y. Cost-effective training of deep cnns with active model adaptation. In Proceedings of the 24th ACM SIGKDD International Conference on Knowledge Discovery & Data Mining (KDD), London, UK, 19–23 August 2018; pp. 1580–1588.
98. Ning, M.; Lu, D.; Wei, D.; Bian, C.; Yuan, C.; Yu, S.; Ma, K.; Zheng, Y. Multi-anchor active domain adaptation for semantic segmentation. In Proceedings of the IEEE/CVF International Conference on Computer Vision (ICCV), Montreal, QC, Canada, 11–17 October 2021; pp. 9112–9122.
99. He, Y.; Zhang, L.; Chen, W.; Luo, X.; Jia, X.; Li, C. CenterRepp: Predict Central Representative Point Set's Distribution For Detection. In Proceedings of the 2020 25th International Conference on Pattern Recognition (ICPR), Milan, Italy, 10–15 January 2021; pp. 8960–8967.
100. Jia, X.; Chen, W.; Li, C.; Liang, Z.; Wu, M.; Tan, Y.; Huang, L. Multi-scale cost volumes cascade network for stereo matching. In Proceedings of the 2021 IEEE International Conference on Robotics and Automation (ICRA), New Orleans, LA, USA, 3–7 May 2021; pp. 8657–8663.
101. He, Y.; Chen, W.; Li, C.; Luo, X.; Huang, L. Fast and Accurate Lane Detection via Graph Structure and Disentangled Representation Learning. *Sensors* **2021**, *21*, 4657. [CrossRef]
102. Chen, W.; Luo, X.; Liang, Z.; Li, C.; Wu, M.; Gao, Y.; Jia, X. A Unified Framework for Depth Prediction from a Single Image and Binocular Stereo Matching. *Remote Sens.* **2020**, *12*, 588. [CrossRef]
103. Jia, X.; Chen, W.; Liang, Z.; Luo, X.; Wu, M.; Li, C.; He, Y.; Tan, Y.; Huang, L. A joint 2D-3D complementary network for stereo matching. *Sensors* **2021**, *21*, 1430. [CrossRef]
104. He, Y.; Chen, W.; Liang, Z.; Chen, D.; Tan, Y.; Luo, X.; Li, C.; Guo, Y. Fast and Accurate Lane Detection via Frequency Domain Learning. In Proceedings of the 29th ACM International Conference on Multimedia (MM), Virtual, 20–24 October 2021; pp. 890–898.
105. Hussein, A.; Gaber, M.M.; Elyan, E. Deep active learning for autonomous navigation. In Proceedings of the International Conference on Engineering Applications of Neural Networks, Aberdeen, UK, 2–5 September 2016; Springer: Cham, Switzerland, 2016; pp. 3–17.
106. Dhananjaya, M.M.; Kumar, V.R.; Yogamani, S. Weather and light level classification for autonomous driving: Dataset, baseline and active learning. In Proceedings of the 2021 IEEE International Intelligent Transportation Systems Conference (ITSC), Indianapolis, IN, USA, 19–22 September 2021; pp. 2816–2821.
107. Ajayi, G. Multi-Class Weather Dataset for Image Classification. 2018. Available online: https://data.mendeley.com/datasets/4drtyfjtfy/1 (accessed on 11 July 2022).

108. Zhao, B.; Li, X.; Lu, X.; Wang, Z. A CNN–RNN architecture for multi-label weather recognition. *Neurocomputing* **2018**, *322*, 47–57. [CrossRef]
109. Liang, Z.; Xu, X.; Deng, S.; Cai, L.; Jiang, T.; Jia, K. Exploring Diversity-based Active Learning for 3D Object Detection in Autonomous Driving. *arXiv* **2022**, arXiv:2205.07708.
110. Caesar, H.; Bankiti, V.; Lang, A.H.; Vora, S.; Liong, V.E.; Xu, Q.; Krishnan, A.; Pan, Y.; Baldan, G.; Beijbom, O. nuscenes: A multimodal dataset for autonomous driving. In Proceedings of the IEEE/CVF Conference on Computer Vision and Pattern Recognition (CVPR), Seattle, WA, USA, 14–19 June 2020; pp. 11621–11631.
111. Peng, F.; Wang, C.; Liu, J.; Yang, Z. Active Learning for Lane Detection: A Knowledge Distillation Approach. In Proceedings of the IEEE/CVF International Conference on Computer Vision (ICCV), Montreal, QC, Canada, 11–17 October 2021; pp. 15152–15161.
112. Chen, Z.; Liu, Q.; Lian, C. Pointlanenet: Efficient end-to-end cnns for accurate real-time lane detection. In Proceedings of the 2019 IEEE Intelligent Vehicles Symposium (IV), Paris, France, 9–12 June 2019; pp. 2563–2568.
113. Qin, Z.; Wang, H.; Li, X. Ultra fast structure-aware deep lane detection. In *Proceedings of the European Conference on Computer Vision (ECCV)*; Springer: Cham, Switzerland, 2020; pp. 276–291.
114. Pan, X.; Shi, J.; Luo, P.; Wang, X.; Tang, X. Spatial as deep: Spatial cnn for traffic scene understanding. In Proceedings of the AAAI Conference on Artificial Intelligence (AAAI), New Orleans, LA, USA, 2–7 February 2018; Volume 32.
115. Behrendt, K.; Soussan, R. Unsupervised labeled lane markers using maps. In Proceedings of the IEEE/CVF International Conference on Computer Vision Workshops, Seoul, Korea, 27–28 October 2019.
116. Ranjan, V.; Wang, B.; Shah, M.; Hoai, M. Uncertainty estimation and sample selection for crowd counting. In Proceedings of the Asian Conference on Computer Vision (ACCV), Kyoto, Japan, 30 November–4 December 2020.
117. Idrees, H.; Tayyab, M.; Athrey, K.; Zhang, D.; Al-Maadeed, S.; Rajpoot, N.; Shah, M. Composition loss for counting, density map estimation and localization in dense crowds. In Proceedings of the European Conference on Computer Vision (ECCV), Munich, Germany, 8–14 September 2018; pp. 532–546.
118. Idrees, H.; Saleemi, I.; Seibert, C.; Shah, M. Multi-source multi-scale counting in extremely dense crowd images. In Proceedings of the IEEE Conference on Computer Vision and Pattern Recognition (CVPR), Portland, OR, USA, 25–27 June 2013; pp. 2547–2554.
119. Zhang, Y.; Zhou, D.; Chen, S.; Gao, S.; Ma, Y. Single-image crowd counting via multi-column convolutional neural network. In Proceedings of the IEEE Conference on Computer Vision and Pattern Recognition (CVPR), Las Vegas, NV, USA, 26 June–1 July 2016; pp. 589–597.
120. Wang, Q.; Gao, J.; Lin, W.; Li, X. NWPU-crowd: A large-scale benchmark for crowd counting and localization. *IEEE Trans. Pattern Anal. Mach. Intell. (TPAMI)* **2020**, *43*, 2141–2149. [CrossRef]
121. Zhao, Z.; Shi, M.; Zhao, X.; Li, L. Active crowd counting with limited supervision. In *Proceedings of the European Conference on Computer Vision (ECCV)*; Springer: Cham, Switzerland, 2020; pp. 565–581.
122. Chen, K.; Loy, C.C.; Gong, S.; Xiang, T. Feature Mining for Localised Crowd Counting. In Proceedings of the British Machine Vision Conference (BMVC), Guildford, UK, 3–7 September 2012; pp. 21.1–21.11. [CrossRef]
123. Guerrero-Gómez-Olmedo, R.; Torre-Jiménez, B.; López-Sastre, R.; Maldonado-Bascón, S.; Onoro-Rubio, D. Extremely overlapping vehicle counting. In Proceedings of the Iberian Conference on Pattern Recognition and Image Analysis (IbPRIA), Santiago, Spain, 17–19 June 2015; pp. 423–431.
124. Marsden, M.; McGuinness, K.; Little, S.; Keogh, C.E.; O'Connor, N.E. People, penguins and petri dishes: Adapting object counting models to new visual domains and object types without forgetting. In Proceedings of the IEEE Conference on Computer Vision and Pattern Recognition (CVPR), Salt Lake City, UT, USA, 18–22 June 2018; pp. 8070–8079.
125. Li, C.; Chen, W.; Luo, X.; Wu, M.; Jia, X.; Tan, Y.; Wang, Z. Application of U-Shaped Convolutional Neural Network Based on Attention Mechanism in Liver CT Image Segmentation. In Proceedings of the International Conference on Medical Imaging and Computer-Aided Diagnosis, Oxford, UK, 20–21 January 2020; Springer: Singapore, 2020; pp. 198–206.
126. Ze-Huan, Y.; Wei, C.; Chen, L.; Hao-Yi, Y.; Yu-Lin, H.; Yu-Song, T.; Fei, L. Automatic Diagnosis of Vaginal Microecological Pathological Images Based on Deep Learning. *Prog. Biochem. Biophys.* **2021**, *48*, 1348–1357.
127. Li, C.; Chen, W.; Tan, Y. Point-sampling method based on 3D U-net architecture to reduce the influence of false positive and solve boundary blur problem in 3D CT image segmentation. *Appl. Sci.* **2020**, *10*, 6838. [CrossRef]
128. Li, C.; Tan, Y.; Chen, W.; Luo, X.; He, Y.; Gao, Y.; Li, F. ANU-Net: Attention-based Nested U-Net to exploit full resolution features for medical image segmentation. *Comput. Graph.* **2020**, *90*, 11–20. [CrossRef]
129. Li, C.; Tan, Y.; Chen, W.; Luo, X.; Gao, Y.; Jia, X.; Wang, Z. Attention unet++: A nested attention-aware u-net for liver ct image segmentation. In Proceedings of the 2020 IEEE International Conference on Image Processing (ICIP), Abu Dhabi, United Arab Emirates, 25–28 October 2020; pp. 345–349.
130. Li, C.; Chen, W.; Tan, Y. Render u-net: A unique perspective on render to explore accurate medical image segmentation. *Appl. Sci.* **2020**, *10*, 6439. [CrossRef]
131. Liu, L.; Lei, W.; Wan, X.; Liu, L.; Luo, Y.; Feng, C. Semi-supervised active learning for COVID-19 lung ultrasound multi-symptom classification. In Proceedings of the 2020 IEEE 32nd International Conference on Tools with Artificial Intelligence (ICTAI), Virutal, 9–11 November 2020; pp. 1268–1273.
132. Hao, R.; Namdar, K.; Liu, L.; Khalvati, F. A transfer learning–based active learning framework for brain tumor classification. *Front. Artif. Intell.* **2021**, *4*, 635766. [CrossRef]

133. Menze, B.H.; Jakab, A.; Bauer, S.; Kalpathy-Cramer, J.; Farahani, K.; Kirby, J.; Burren, Y.; Porz, N.; Slotboom, J.; Wiest, R.; et al. The multimodal brain tumor image segmentation benchmark (BRATS). *IEEE Trans. Med. Imaging (TMI)* **2014**, *34*, 1993–2024. [CrossRef]
134. Bakas, S.; Akbari, H.; Sotiras, A.; Bilello, M.; Rozycki, M.; Kirby, J.S.; Freymann, J.B.; Farahani, K.; Davatzikos, C. Advancing the cancer genome atlas glioma MRI collections with expert segmentation labels and radiomic features. *Sci. Data* **2017**, *4*, 170117. [CrossRef]
135. Bakas, S.; Reyes, M.; Jakab, A.; Bauer, S.; Rempfler, M.; Crimi, A.; Shinohara, R.T.; Berger, C.; Ha, S.M.; Rozycki, M.; et al. Identifying the best machine learning algorithms for brain tumor segmentation, progression assessment, and overall survival prediction in the BRATS challenge. *arXiv* **2018**, arXiv:1811.02629.
136. Ahsan, M.A.; Qayyum, A.; Qadir, J.; Razi, A. An Active Learning Method for Diabetic Retinopathy Classification with Uncertainty Quantification. *arXiv* **2020**, arXiv:2012.13325.
137. Lam, C.; Yi, D.; Guo, M.; Lindsey, T. Automated detection of diabetic retinopathy using deep learning. *AMIA Summits Transl. Sci. Proc.* **2018**, *2018*, 147.
138. Li, W.; Li, J.; Wang, Z.; Polson, J.; Sisk, A.E.; Sajed, D.P.; Speier, W.; Arnold, C.W. PathAL: An Active Learning Framework for Histopathology Image Analysis. *IEEE Trans. Med. Imaging* **2021**, *41*, 1176–1187. [CrossRef]
139. Tan, M.; Le, Q. Efficientnet: Rethinking model scaling for convolutional neural networks. In Proceedings of the International Conference on Machine Learning (ICML), PMLR, Long Beach, CA, USA, 10–15 June 2019; pp. 6105–6114.
140. Huang, J.; Qu, L.; Jia, R.; Zhao, B. O2u-net: A simple noisy label detection approach for deep neural networks. In Proceedings of the IEEE/CVF International Conference on Computer Vision (ICCV), Seoul, Korea, 27 October–2 November 2019; pp. 3326–3334.
141. Guo, S.; Huang, W.; Zhang, H.; Zhuang, C.; Dong, D.; Scott, M.R.; Huang, D. Curriculumnet: Weakly supervised learning from large-scale web images. In Proceedings of the European Conference on Computer Vision (ECCV), Munich, Germany, 8–14 September 2018; pp. 135–150.
142. Bulten, W.; Kartasalo, K.; Chen, P.H.C.; Ström, P.; Pinckaers, H.; Nagpal, K.; Cai, Y.; Steiner, D.F.; van Boven, H.; Vink, R.; et al. Artificial intelligence for diagnosis and Gleason grading of prostate cancer: The PANDA challenge. *Nat. Med.* **2022**, *28*, 154–163. [CrossRef]
143. Shen, Y.; Ke, J. Representative Region Based Active Learning For Histological Classification Of Colorectal Cancer. In Proceedings of the 2021 IEEE 18th International Symposium on Biomedical Imaging (ISBI), Nice, France, 13–16 April 2021; pp. 1730–1733.
144. Comprehensive molecular profiling of lung adenocarcinoma. *Nature* **2014**, *511*, 543–550. [CrossRef]
145. Kather, J.N.; Krisam, J.; Charoentong, P.; Luedde, T.; Herpel, E.; Weis, C.A.; Gaiser, T.; Marx, A.; Valous, N.A.; Ferber, D.; et al. Predicting survival from colorectal cancer histology slides using deep learning: A retrospective multicenter study. *PLoS Med.* **2019**, *16*, e1002730. [CrossRef]
146. Russakovsky, O.; Deng, J.; Su, H.; Krause, J.; Satheesh, S.; Ma, S.; Huang, Z.; Karpathy, A.; Khosla, A.; Bernstein, M.; et al. Imagenet large scale visual recognition challenge. *Int. J. Comput. Vis. (IJCV)* **2015**, *115*, 211–252. [CrossRef]
147. Wang, J.; Yan, Y.; Zhang, Y.; Cao, G.; Yang, M.; Ng, M.K. Deep reinforcement active learning for medical image classification. In Proceedings of the International Conference on Medical Image Computing and Computer-Assisted Intervention (MICCAI), Lima, Peru, 4–8 October 2020; pp. 33–42.
148. Smit, A.; Vrabac, D.; He, Y.; Ng, A.Y.; Beam, A.L.; Rajpurkar, P. MedSelect: Selective Labeling for Medical Image Classification Combining Meta-Learning with Deep Reinforcement Learning. *arXiv* **2021**, arXiv:2103.14339.
149. Irvin, J.; Rajpurkar, P.; Ko, M.; Yu, Y.; Ciurea-Ilcus, S.; Chute, C.; Marklund, H.; Haghgoo, B.; Ball, R.; Shpanskaya, K.; et al. Chexpert: A large chest radiograph dataset with uncertainty labels and expert comparison. In Proceedings of the AAAI Conference on Artificial Intelligence (AAAI), Honolulu, HI, USA, 27 January–1 February 2019; Volume 33, pp. 590–597.
150. Shen, Y.; Ke, J. Su-Sampling Based Active Learning For Large-Scale Histopathology Image. In Proceedings of the 2021 IEEE International Conference on Image Processing (ICIP), Anchorage, AK, USA, 19–22 September 2021; pp. 116–120.
151. Younesian, T.; Zhao, Z.; Ghiassi, A.; Birke, R.; Chen, L.Y. QActor: Active Learning on Noisy Labels. In Proceedings of the Asian Conference on Machine Learning, PMLR, Virtual, 17–19 November 2021; pp. 548–563.
152. Guan, H.; Liu, M. Domain adaptation for medical image analysis: A survey. *IEEE Trans. Biomed. Eng.* **2021**, *69*, 1173–1185. [CrossRef] [PubMed]
153. Choudhary, A.; Tong, L.; Zhu, Y.; Wang, M.D. Advancing medical imaging informatics by deep learning-based domain adaptation. *Yearb. Med. Inform.* **2020**, *29*, 129–138. [CrossRef] [PubMed]
154. Prabhu, V.; Chandrasekaran, A.; Saenko, K.; Hoffman, J. Active domain adaptation via clustering uncertainty-weighted embeddings. In Proceedings of the IEEE/CVF International Conference on Computer Vision (ICCV), Montreal, QC, Canada, 11–17 October 2021; pp. 8505–8514.
155. Xie, B.; Yuan, L.; Li, S.; Liu, C.H.; Cheng, X.; Wang, G. Active learning for domain adaptation: An energy-based approach. In Proceedings of the AAAI Conference on Artificial Intelligence (AAAI), Virtual, 24–27 February 2022; Volume 36, pp. 8708–8716.
156. Fu, B.; Cao, Z.; Wang, J.; Long, M. Transferable query selection for active domain adaptation. In Proceedings of the IEEE/CVF Conference on Computer Vision and Pattern Recognition (CVPR), Virtual, 19–25 June 2021; pp. 7272–7281.
157. Zhou, Z.; Sodha, V.; Pang, J.; Gotway, M.B.; Liang, J. Models genesis. *Med. Image Anal.* **2021**, *67*, 101840. [CrossRef] [PubMed]

applied
sciences

Article

Guided Random Mask: Adaptively Regularizing Deep Neural Networks for Medical Image Analysis by Potential Lesions

Xiaorui Yu [1], Shuqi Wang [1,2,*] and Junjie Hu [3,*]

[1] National Engineering Research Center for Biomaterials, Sichuan University, Chengdu 610065, China; yuxiaoruiscu@163.com
[2] Sichuan Provincial Clinical Research Center for Respiratory Diseases, West China Hospital, Chengdu 610065, China
[3] College of Computer Science, Sichuan University, Chengdu 610065, China
* Correspondence: shuqi@scu.edu.cn (S.W.); hujunjie@scu.edu.cn (J.H.)

Abstract: Data augmentation is a critical regularization method that contributes to numerous state-of-the-art results achieved by deep neural networks (DNNs). The visual interpretation method demonstrates that the DNNs behave like object detectors, focusing on the discriminative regions in the input image. Many studies have also discovered that the DNNs correctly identify the lesions in the input, which has been confirmed in the current work. However, for medical images containing complicated lesions, we observe the DNNs focus on the most prominent abnormalities, neglecting sub-clinical characteristics that may also help diagnosis. We speculate this bias may hamper the generalization ability of DNNs, potentially causing false predicted results. Based on this consideration, a simple yet effective data augmentation method called guided random mask (GRM) is proposed to discover the lesions with different characteristics. Visual interpretation of the inference result is used as guidance to generate random-sized masks, forcing the DNNs to learn both the prominent and subtle lesions. One notable difference between GRM and conventional data augmentation methods is the association with the training phase of DNNs. The parameters in vanilla augmentation methods are independent of the training phase, which may limit their effectiveness when the scale and appearance of region-of-interests vary. Nevertheless, the effectiveness of the proposed GRM method evolves with the training of DNNs, adaptively regularizing the DNNs to alleviate the over-fitting problem. Moreover, the GRM is a parameter-free augmentation method that can be incorporated into DNNs without modifying the architecture. The GRM is empirically verified on multiple datasets with different modalities, including optical coherence tomography, X-ray, and color fundus images. Quantitative experimental results show that the proposed GRM method achieves higher classification accuracy than the commonly used augmentation methods in multiple networks. Visualization analysis also demonstrates that the GRM can better localize lesions than the vanilla network.

Keywords: deep neural networks; data augmentation; regularization; medical image analysis

Citation: Yu, X.; Wang, S.; Hu, J. Guided Random Mask: Adaptively Regularizing Deep Neural Networks for Medical Image Analysis by Potential Lesions. *Appl. Sci.* **2022**, *12*, 9099. https://doi.org/10.3390/app12189099

Academic Editor: Cosimo Nardi

Received: 31 July 2022
Accepted: 7 September 2022
Published: 9 September 2022

1. Introduction

Deep neural networks (DNNs) have revolutionized the field of medical image analysis by learning to extract high-level abstract features in a data-driven manner, rather than the conventional hand-crafted ones with limited representative ability. Both the convolutional-based [1] and Transformer-based [2] networks have achieved enormous breakthroughs in multiple medical image analysis tasks, such as disease classification, target volume segmentation, lesion detection, and image reconstruction. Based on DNNs, impressive results on numerous diseases have been reported, including the retinopathy of prematurity (ROP) [3], retinal diseases [4,5], breast cancer [6], lung diseases [7], and stomach diseases [8]. These encouraging results demonstrate that DNNs are promising methods to help design computer-aided diagnosis (CAD) systems.

Despite the progress achieved by leveraging the DNN model, the interpretability of the outputted result is a critical aspect that clinicians interested in. A CAD system that utilizes DNNs may output the classification result (e.g., benign or malignant) based on the patient's medical imaging. However, which part of the input image associates with the output helps explain the classification result. The interpretability may also contribute to reducing the false positive or false negative samples, preventing the DNNs from failing silently when the inputs belong to the type of out-of-distribution samples [9,10]. It has been shown that the DNNs behave as object detectors, even without the supervision of the location of the object [11]. There exist several well-known visual interpretation methods that attempt to bridge the object within the input image and the output of DNNs, e.g., guided backpropagation [12], class activation mapping (CAM) [13], and gradient-weighted class activation mapping (Grad-CAM) [14]. Based on these interpretation methods, recent works show that the DNNs indeed localize the potential lesions in recognition of multiple diseases [3,15].

Figure 1 shows three OCT [4] samples that are diagnosed with CNV, accompanied by the visualization result of CAM. By observing the example in the first row, it can be found most of the lesions are located in the center of the image, and the result of CAM shows that the network accurately locates those abnormalities. Given the masked CAM shown on the right side of the first row, it is hard to determine the diagnosis since most lesions are masked. For the second and third examples, the lesions are scattered in the image, much more complicated than those in the first sample. Moreover, the corresponding CAM results indicate that the model only identified part of the lesions on the image's right side. In the third column, green squares are used to point out the lesions ignored by the model, and the CAM-localized regions are masked. These visualization results reveal that the DNNs may bias toward the most distinguishing features in the input, ignoring other sub-clinical lesions that contain valuable information. We suspect that the above limitation may constrain the DNNs' robustness to the variations of the lesions, causing false-negative predictions when the lesions in the image are not prominent. Ideally, it is preferred for the DNNs to recognize both the principal and subtle lesions in the input as clinicians do.

Faced with the aforementioned limitation, the core idea of the proposed method is to leverage the information contained in the CAM as a guide to discover the potential ignored lesions in the input. The visual interpretation result reveals the areas the DNNs focused on. Therefore, the rest may contains ignored sub-clinical lesions that we are interested in. To discover those lesions, a vanilla approach is to fully mask the areas indicated by the visual interpretation, enforcing the DNNs to give the prediction by utilizing the rest of the regions. This approach sounds reasonable for the second and third rows shown in Figure 1, where the DNNs are possible to predict correctly based on the lesions marked by the green squares. However, it is hard to predict the first sample based on the masked input in Figure 1, since critical information in the input is not given. A desirable approach is to moderately mask the potential lesions with the help of visual interpretations without the complete loss of valuable information. Thus, the DNNs can adapt to the inputs that contain either simplified or complicated lesions. Based on this consideration, this paper proposes a simple yet effective data augmentation method called guided random mask (GRM), which randomly masks the areas indicated by the visual interpretation during the training phase. The term "guided" in GRM refers to the information provided by the visual interpretation, and "random mask" implies the stochasticity that produces the effect of regularization.

The proposed GRM is a data augmentation method that can prevent the DNNs from focusing only on prominent input lesions and can better utilize spatial contextual information. Notably, the GRM is a parameter-free method that can accommodate lesions with different scales and complexity. It is known that data augmentation plays a vitally important role in the training of DNNs to mitigate the problem of over-fitting. One inspiration of the proposed GRM is the cutout [16], which randomly masks the input with a fixed size area and helps the DNNs achieve state-of-the-art performance on CIFAR [17] and SVHN [18] datasets. However, compared with the CIFAR and SVHN with relatively small image sizes (32 × 32), it is much harder to apply the cutout to high-resolution medical images because of the hyper-parameter tuning. The hyper-parameter in the cutout is the size of the mask, which can be regarded as the strength of regularization in training DNNs. Its optimal value is task-dependent and requires grid search to achieve the best performance, which may limit its effectiveness in practice. On the contrary, the proposed GRM eliminates the difficulty in hyper-parameter tuning by using the guidance provided by the visual interpretation. Unlike the fixed mask size in cutout, the one in the proposed GRM is adaptively adjusted along with the training of DNNs, making it applicable to recognition tasks with varied scales and complexity of lesions.

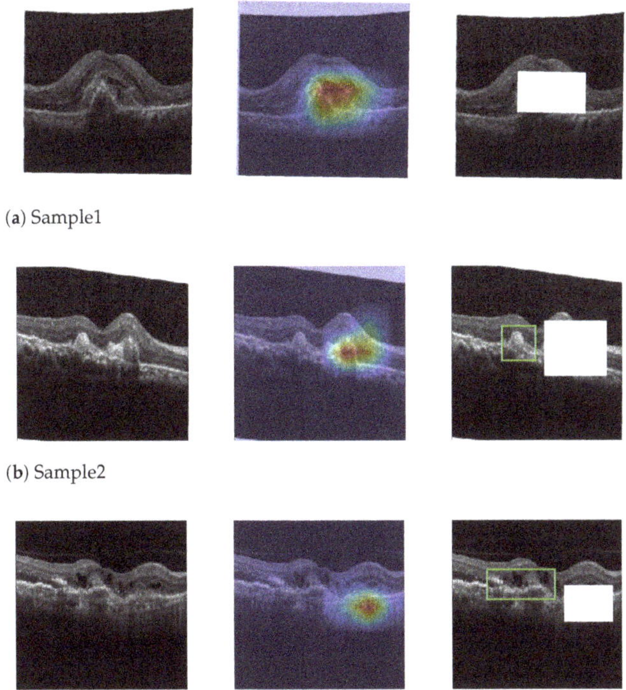

(**a**) Sample1

(**b**) Sample2

(c) Sample3

Figure 1. Three choroidal neovascularization (CNV) samples containing a neovascular membrane. Each row represents an optical coherence tomography (OCT) sample. The three columns denote the original OCT image, the visualization result of CAM, and the image with CAM masked, respectively. For the right-most image in all three rows, the white masks represent the result area of CAM, and the green squares denote the lesions ignored by DNNs. Detailed illustration of dataset and computation principle of CAM can be found in Section 3.

The contributions of the paper can be summarized as follows:

(i)   We found that the DNNs may bias toward the most prominent features and ignore the sub-clinical ones when the input image contains complicated lesions.
(ii)   A parameter-free data augmentation method called GRM is proposed, which utilizes visual interpretation of the prediction result to regularize the training of DNNs adaptively.
(iii)   Visual interpretation demonstrates that DNNs coupled with GRM can more effectively utilize the contextual information than the vanilla models.

(iv)   Ablation studies on multiple datasets, including OCT, X-ray, and ultrasound images, empirically show that the GRM substantially surpasses the benchmark method on various tasks.

The rest of the paper is organized as follows. Section 2 summarizes the related works about the applications in medical imaging and common augmentation methods used for the training of DNNs. Section 3 first illustrates the three types of medical imaging datasets used, followed by a detailed explanation of the proposed method. Section 4 shows the results of extensive experiments, including the comparison of the baseline model and other well-known related augmentation methods. Visualization analysis is also used to verify the effectiveness of the GRM. Finally, Section 5 summarizes and concludes the GRM. The source code is available at https://github.com/hujunjiescu/GRM, accessed on 1 January 2022.

2. Related Works

2.1. DNNs in Medical Image Analysis

DNNs have become ubiquitous methods in the field of medical image analysis, where Deep Convolutional Neural Networks (DCNNs) [1,19,20] and the recently emerged Vision Transformer (ViT) [2] are the two most prevalent paradigms. The following two paragraphs briefly demonstrate their applications in medical image analysis tasks.

For the DCNNs, starting from 2012 when AlexNet [19] won the ILSVRC-2012 competition [21], many breakthroughs in vision-related tasks have been achieved using DCNNs. Several key factors contribute to the success of DCNNs, including massive annotated high-quality datasets, powerful computation capability by utilizing graphics processing units (GPUs), and novel architectures. Multiple architectures of DCNNs proposed in the natural image field have also been successfully applied in medical images. For example, Inception-V3 [22] has been used to identify the retinal diseases in OCT images [4]. Experimental results demonstrate that DCNNs outperform some human experts and can aid in expediting the diagnosis in clinical practice. A three-stage DCNNs-based architecture is proposed in [3] to recognize the existence of ROP based on the fundus images, where multiple popular architectures including VGG [23], GoogLeNet [20], and ResNet [1] delivered promising performance. A novel network architecture called U-Net is proposed in [24] to accomplish biomedical segmentation tasks in an end-to-end manner, surpassing the compared methods by a large margin. This tremendous success makes the U-Net a benchmark in biomedical segmentation tasks. Lots of U-Net's variations have lately been proposed by incorporating attention mechanism [15,25,26], residual convolution blocks [27], etc. Besides the applications in disease diagnoses, DCNNs have also achieved remarkable progress in image reconstruction [28], denoising [29,30], etc.

Transformer [31] is an attention-based model that was initially proposed to solve machine translation tasks. It achieves better performance than the conventional recurrent models, raising expectations that it may also be applicable to the image field. Many researchers attempt to bridge the gap between natural language processing (NLP) and vision, and ViT [2] is one of the well-known Transformer-based models that achieves promising results on the natural image classification task. In addition to the natural image-related tasks, Transformer has also been gaining attention in medical image analysis. Hatamizadeh et al. [32] proposes a Transformer-based segmentation architecture called UNETR that combines the U-Net [24] with Transformer to accomplish the volumetric segmentation task. It achieves the state-of-the-art performance on the dataset of Multi-Atlas Labeling Beyond The Cranial Vault [33] and Medical Segmentation Decathlon (MSD) [34]. A relation Transformer network (RTNet) is proposed in [35] that leverage the Transformer to exploit and interact with the relationships between the lesions and vessels. TransMed is proposed in [36] to incorporate the advantages of DCNNs and Transformer to perform the classification task of multi-model medical image. By combining the feature extraction ability of DCNNs and the spatial relationship modeling capacity of Transformer, TransMed achieves better accuracy than conventional DCNNs-based models.

Even the Transformer delivered competitive performance compared with DCNNs: [37] recently showed that a pure DCNN can surpass the state-of-the-art Transformer by deliberately designing the DCNN's component and architecture. It is hard to say which one is more overwhelming than another since both the DCNNs and Transformers have unique advantages in object recognition.

2.2. Augmentation Methods for Training DCNNs

A massive annotated dataset is an indispensable factor for the success of DNNs since both the DCNNs and Transformers typically have millions of parameters, implying the potential over-fitting problem when the amount of training dataset is limited. During the training phase, it is common to utilize regularization methods to alleviate the over-fitting risk, thus improving the generalization ability. Data augmentation, which aims to increase the diversity of training data, is a frequently used regularization method that contributes to many state-of-the-art results on both natural and medical image analysis tasks.

Current data augmentation methods mainly focus on the domains of spatial and intensity. In the spatial domain, the random crop and flip for the CIFAR [17] dataset have become the standard operations during the training phase [1,16]. U-Net [24] shows that excessive data augmentation by applying elastic deformation to the training dataset is critical for biomedical segmentation, particularly when the number of training samples is limited. For the intensity domain, common augmentation methods include brightness enhancement [38], color transformation [39], noise injection [40], blurring [41] etc. In addition to the domains of spatial and intensity, another type of effective augmentation method is mixup [42], which generates new training samples through the convex combination of random paired examples and their labels. Mixup and its variant [43] have also been applied to medical image segmentation tasks.

Perhaps one of the closest works to ours is the cutout approach [16], which augments the spatial domain by randomly masking squared regions in the input image. Cutout can be regarded as a variant of dropout [44] that randomly drops neurons during the training phase to reduce co-adaptations. Instead of dropping neurons in the dropout, cutout randomly drops squared pixels in the input image. Despite the progress brought by the cutout, one of its limitations lies in the difficulty of determining the optimal masked size (suppose is r) in the input image. The target size varies from the task, indicating that the optimal value of r is task-dependent and can only be determined by trial-and-error. The main reason behind this issue is the separation between the data augmentation and the training phase of DNNs. The proposed GRM method eliminates the problem by leveraging the guidance from the visual interpretation to determine the augmentation parameters, thus bridging the gap between the data augmentation and the training phase. The proposed GRM method has two significant advantages over the cutout. First, the GRM can adaptively adjust the size of the mask with the guidance of visual interpretation without specifying the hyper-parameter r. Second, the GRM can efficiently utilize the contextual information and discover the potential sub-clinical lesions by masking the target region. Essentially, the GRM can be regarded as a regularizer that alleviates the over-fitting problem.

3. Data and Methodology

3.1. Data

Three types of medical imaging datasets are used in the experiments, including OCTs of retinal diseases, X-rays of pneumonia, and color fundus images of glaucoma. Table 1 summarizes the subset of these three datasets used in the experiments. The retinal diseases dataset comprises four classes, i.e., normal, choroidal neovascularization (CNV), diabetic macular edema (DME), and drusen. The pneumonia dataset contains two classes: normal and pneumonia. Similar to the pneumonia dataset, the glaucoma dataset includes normal and glaucoma as the two classes. All three datasets are open source. The retinal diseases and pneumonia dataset can be downloaded at https://data.mendeley.com/datasets/rscbjbr9sj/3, accessed on 1 January 2022, and the glaucoma dataset can be downloaded at https://doi.org/10.5281/zenodo.5793241, accessed on 1 January 2022. Illustrations of these three datasets can be found in Figure 2. Each dataset is manually split into three parts, including training, validation, and test datasets. The experimental results are reported on the test dataset by using the model that achieves the best metric on the validation part.

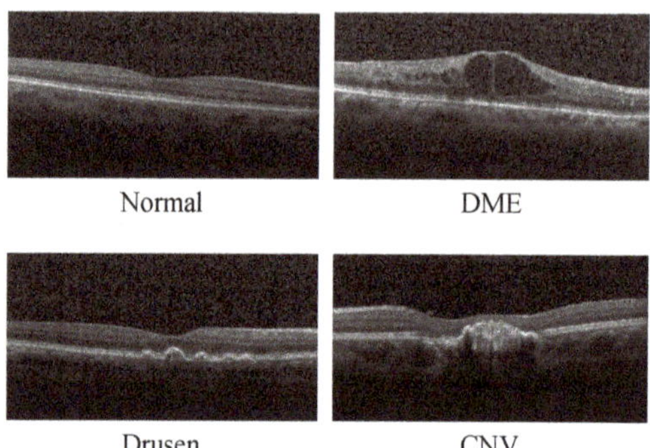

(**a**) Retinal diseases in the modality of OCT.

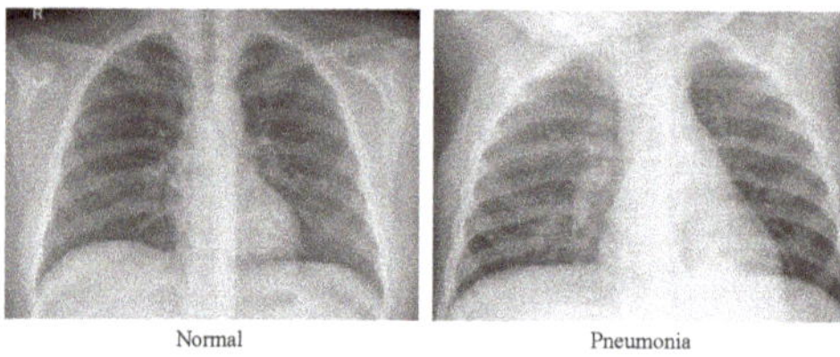

(**b**) Pituitarium in the modality of X-ray.

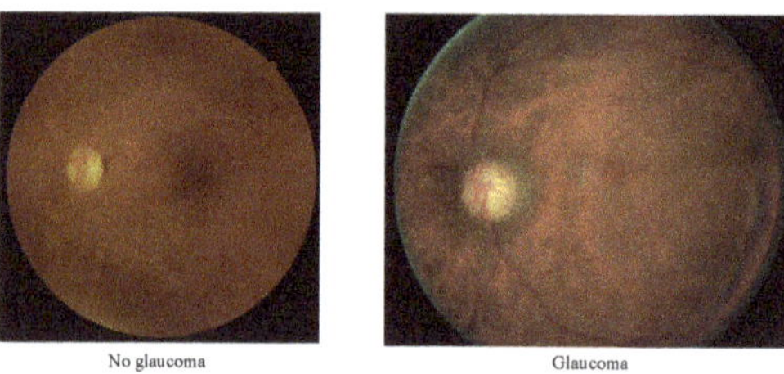

(**c**) Glaucoma in the modality of color image.

Figure 2. Characteristics of different classes in the three datasets.

Table 1. Statistics of the used three medical imaging datasets.

	Part	Modality	Task	Classes	Training Samples	Validation Samples	Test Samples
Retinal diseases	Eyes	OCT	Classification	4	4000	1000	1000
Pneumonia	Chest	X-ray	Classification	2	4632	600	624
Glaucoma	Eyes	Color fundus image	Classification	2	5232	744	744

3.2. Methodology

The proposed GRM is a data augmentation method that bridges the augmentation characteristics with the training phase of the model to identify the ignored sub-clinical lesions adaptively. Two problems need to be solved to achieve the adaptive regularization effect, that is (1) how to discover the region of interest (ROI) that indicates the location of the potential lesions and (2) how to utilize the information contained in the ROI. The corresponding solutions to the two problems are demonstrated in the following subsections.

3.2.1. Localizing Potential Lesions

Consider the C-classes classification task based on DNNs, including the DCNNs and Transformers. Given the input image x, the DNNs denoted as $F(x; W)$ would output the prediction result a^L after the layer-by-layer forward computation, where L denotes the number of layers in the DNNs. a^L is a vector in the length of C, whose elements indicate the probability of each class. Generally, the largest component in a^L, suppose a_c^L, would be the category assigned to the input x. What we are interested in is which part in the x contributes to the class c.

There are multiple ways to solve the above problem. Here, the CAM [13] is utilized for its computational efficiency and simplicity. An essential component in the CAM is the global average pooling (GAP), which is first proposed in the NIN [45] architecture to reduce the use of fully connected (FC) layers. The GAP average the feature maps along the dimension of the channel to reduce the features from a three-dimensional tensor to a one-dimensional vector. In the modern architecture of DNNs, it is common to use the GAP in the penultimate layer to get the global representation of the input, followed by an FC layer whose dimension is C. The core idea of CAM can be regarded as the reverse computation of the above steps, where the learnable weight in the last FC layer (which can be considered as the importance of feature per channel) is used to weight the extracted features to indicate which part in the input is associated with the prediction.

Formally, suppose the features fed into the GAP are denoted as a^{L-1} in the shape of $[K, W, H]$ that indicate the number of channels, width, and height, respectively. W^{L-1} represents the learnable weight within the last FC layer in the shape of $[C, K]$. The probability of class c is then given by the softmax equation $a_c^L = \frac{exp(z_c^L)}{\sum_{i=1}^{C} exp(z_i^L)}$. The scalar variable z_c^L is computed as:

$$z_c^L = \sum_{k=1}^{K} W_{c,k}^{L-1} \cdot \sum_{w=1}^{W} \sum_{h=1}^{H} a_{k,w,h}^{L-1}. \tag{1}$$

The summation on the right side of the above equation represents the GAP, which can be regarded as the feature's context representation along the dimension of the channel. The parameter W^{L-1} thus indirectly represents the contribution of each channel in the a^{L-1} to the predicted score. By leveraging the information contained in W^{L-1} to integrate a^{L-1} in a channel-wise way, it is then possible to highlight the probable areas corresponding to the predicted class. This computation process can be formulated as:

$$M_c = \sum_{k=1}^{K} W_{c,k}^{L-1} \cdot a_k^{L-1}, \tag{2}$$

where M_c denotes the CAM for class c. Each channel in a^{L-1} represents a visual pattern discovered by DNNs. Therefore the CAM can be considered as a weighted summation of the presence of each visual pattern at different spatial locations. Note that the CAM represents the visual pattern in the feature-level's spatial resolution, which is much smaller than the input image. The CAM is required to be upsampled to the resolution of the input image in order to identify the image regions corresponding to the predicted category.

3.2.2. Guided Random Mask

Having identified the possible lesions indicated by the CAM, the next problem to be tackled is how to utilize it to regularize the training of DNNs. We aim to realize moderate regularization effectiveness, that is, to avoid entirely masking the lesions that may cause strong regularization or mask regions with a fixed size that introduce an extra hyper-parameter. Based on this consideration, we propose mask regions with a random size guided by CAM. Figure 3 illustrates the overall computation steps of the proposed GRM method. First, the inference of the input image is required in order to identify its category and locate the potential lesions, which is indicated by the procedures of 1, 2, 3. The next step is generating the bounding box of the CAM, which embodies the majority of regions with high values in CAM. The bounding box of CAM is computed from the binarized CAM, which is accomplished by using the 90th percentile of the original CAM (i.e., a pixel larger than the 90th percentile is 1, otherwise it 0). Then we randomly choose the central point in the bounding box and allocate the width and height with their maximum value the same as that in the bounding box. The bounding box of CAM and randomly generated mask are shown as the white and red boxes in Figure 3, respectively. Finally, the random mask is applied to the raw input by setting the area in the input to 0, later used to train the DNNs. Note that the GRM method is only used in the training phase, not including the test phase.

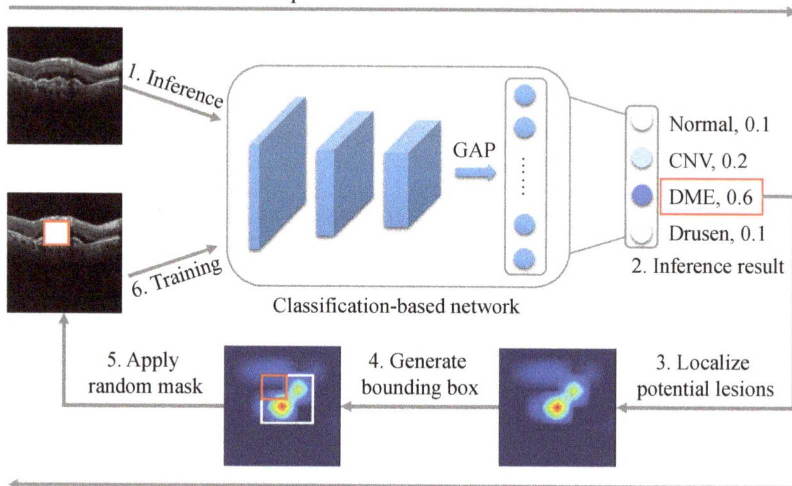

Figure 3. Computation procedures of the proposed GRM method for the classification task.

The above computation processes are designed for the classification task, where its application to segmentation tasks is straightforward. The inference computation to compute CAM can be omitted for the segmentation task since the target is precisely described in the label. It only requires finding the bounding box of the target and generating a corresponding random mask within it. The introduced computational cost is negligible and can be implemented as the preprocessing procedure.

The computational process of the proposed GRM is summarized in Algorithm 1.

Algorithm 1: Algorithm of the proposed GRM method for the classification task.

Input : raw input image a^0
Output: random masked image
for *layer l from 1 to L − 1* **do**
   | // forward computation
   | $a^l = F(a^{l-1}; W^{l-1})$;
end
$z^L = \sum_{k=1}^{K} W_k^{L-1} \cdot \sum_{w=1}^{W} \sum_{h=1}^{H} a_{k,w,h}^{L-1}$;
// find the category
$c = argmax(z^L)$;
// compute and upsample CAM
$\tilde{M}_c = upsample(\sum_{k=1}^{K} W_{c,k}^{L-1} \cdot a_k^{L-1})$;
// find bounding box
$(X, Y, W, H) = BBox(\tilde{M}_c)$;
// uniform sampling center point
$x = uniform(X, X + W)$;
$y = uniform(Y, Y + H)$;
// uniform sampling width and height
$w = uniform(1, W)$;
$h = uniform(1, H)$;
// mask input a^0
$a^0[x - w : x + w, y - h : y + h] = 0$

---

4. Experimental Setup and Results

4.1. Experimental Setup

Multiple modern network architectures including Inception-V3 [22], ResNet-50 [1], DenseNet-121 [46], and ViT [2] are used to verify the generalization of the GRM method. The cross-entropy is used as the cost function for the classification tasks. For the CNNs (ResNet-50, Inception-V3, and DenseNet-121), Adadelta [47] is used as the optimizer to minimize the cost function, where the learning rate is set to 1.0. For the ViT, AdamW [48] is used as the optimizer, coupled with a cosine decay learning rate scheduler and 20 epochs of linear warm-up. The learning rate is set to 0.0001. The size of the image is fixed as 224 for all the experiments.

The number of training epochs is set to 300, which is long enough for the convergence of training. All networks are implemented by using PyTorch [49]. The experiments are carried out on a server with Linux OS and CPU Intel Xeon E5-2620 @2.4GHz, four NVIDIA TITAN RTX GPUs, and 64 GB of RAM.

4.2. Ablation Studies of GRM

To verify the effectiveness of the proposed GRM method, we first quantitatively compare the network with and without the GRM. Table 2 shows the accuracy of multiple networks on the three tasks. For the vanilla network, it can be found that the Inception-V3 achieves the highest accuracy among all the tasks. For example, the accuracy of Inception-V3 on retinal diseases is 94.9, far beyond the 89.9 of the ViT. The inferiority of ViT can be attributed to the difficulty in hyper-parameter tuning and the limited size of the medical image dataset, which significantly increase the risk of over-fitting. By comparing the vanilla network with the one with GRM applied, it can be observed that the accuracy of GRM is unanimously improved among all the tasks, regardless of the network architecture. The highest improvement is from 90.0 to 94.2 in the ResNet-50 on the pneumonia task. A varying degree of improvement is also obtained for the ViT in the three datasets. These encouraging results illustrate that the proposed GRM is broadly applicable to datasets composed of varied modalities and class numbers.

Table 2. Comparison of accuracy (%) between the vanilla networks and the one applied with GRM on the three medical image analysis tasks.

Task	Network	Vanilla	GRM
Retinal diseases	Inception-V3	94.9	**96.7**
	ResNet-50	93.7	**96.3**
	DenseNet-121	93.6	**96.0**
	ViT	89.9	**92.6**
Pneumonia	Inception-V3	90.4	**92.8**
	ResNet-50	90.0	**94.2**
	DenseNet-121	88.8	**92.1**
	ViT	90.2	**91.8**
Glaucoma	Inception-V3	89.2	**91.6**
	ResNet-50	87.5	**90.6**
	DenseNet-121	88.9	**90.0**
	ViT	87.5	**88.5**

One of the reasons for the effectiveness of GRM is the adaptive regularization, which helps the network better extract the context information and alleviate the over-fitting issue. To delve into the training procedure, Figure 4 summarizes the training and validation loss of ResNet-50 in the three tasks. It can be observed the regularization effectiveness brought by GRM in the task of retinal diseases in Figure 4a, where the training loss of GRM (red dotted line) decreases slower than the one in the vanilla network (red solid line), implying the GRM helps to mitigate over-fitting to the training dataset. On the contrary, the validation loss of GRM (green dotted line) is distinctly lower than the one of baseline (green solid line), demonstrating that the GRM increases the network's generalization capacity. Similar convergence results can be found in the pneumonia task. The effectiveness of GRM can also be notably found in the glaucoma task, where the validation loss of the vanilla ResNet-50 increases rapidly from the 50th epoch, and its ascending speed goes faster along the training epochs. This convergence behavior can be commonly observed in the training of networks. By adding GRM to the network, the stability of validation loss is significantly improved, as shown in the green dotted line in Figure 4c. These convergence results confirmed the regularization impact brought by the GRM, which helps combat the over-fitting issue on the training dataset and boosts the generalization ability on the validation dataset.

4.3. Comparison between GRM with Other Augmentation Methods

To further validate the effectiveness of the GRM, we also compare it with cutout [16] and mixup [42]. As shown in Table 3, the GRM is superior to the cutout in improving the diagnosis accuracy. The most significant improvement happens in the Inception network for retinal diseases, which raises the accuracy from 94.3 to 96.7. For pneumonia and glaucoma diseases, different degrees of improvement can also be found in various networks. Experimental results in Table 3 also demonstrate the advantage of GRM over mixup. It can be observed that the GRM outperforms mixup in most tasks except the ViT of retinal diseases, where the accuracy of mixup is 92.8, marginally higher than the 92.6 of GRM. One significant advantage of the GRM and cutout lies in the adaptivity to the inputs. The mask size in the cutout is fixed, whereas the GRM can adaptively adjust the size and location of the mask according to the input. For the Mixup, it combines paired inputs and labels convexly to alleviate the overfitting problem, which can be used together with GRM to increase the capacity of the networks.

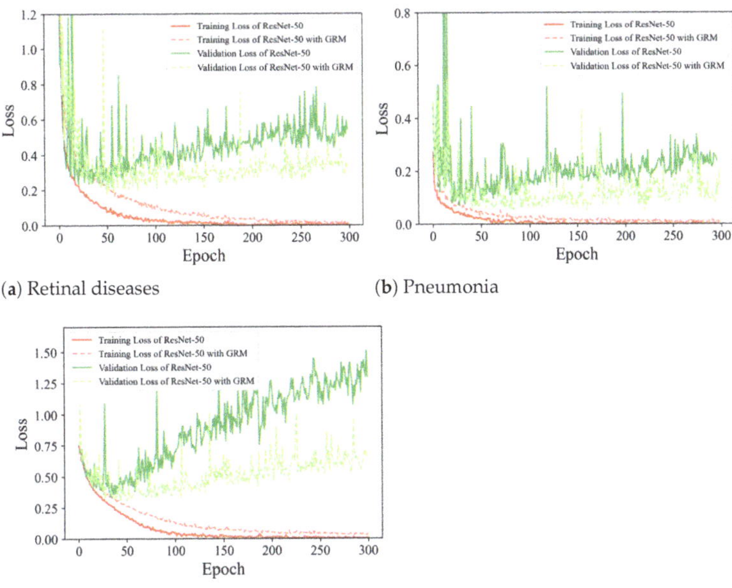

(**a**) Retinal diseases   (**b**) Pneumonia

(**c**) Glaucoma

Figure 4. Convergence comparison between the vanilla ResNet-50 and ResNet-50 applied with the proposed GRM on the retinal disease, pneumonia, and glaucoma tasks, respectively.

Table 3. Comparison of accuracy (%) between the GRM and related methods on the three medical image analysis tasks.

Task	Network	GRM	Cutout	Mixup
Retinal diseases	Inception-V3	**96.7**	94.3	95.8
	ResNet-50	**96.3**	94.1	92.8
	DenseNet-121	**96.0**	95.1	95.6
	ViT	92.6	92.0	**92.8**
Pneumonia	Inception-V3	**92.8**	89.2	91.2
	ResNet-50	**94.2**	89.7	91.5
	DenseNet-121	**92.1**	92.0	91.0
	ViT	**91.8**	91.1	89.1
Glaucoma	Inception-V3	**91.6**	89.0	91.4
	ResNet-50	**90.6**	87.3	89.4
	DenseNet-121	**90.0**	89.6	88.8
	ViT	**88.5**	86.2	86.1

4.4. Visualization Analysis

The motivation of GRM roots in the potential bias of the vanilla network, which attempts to capture the most prominent characteristics of lesions and may ignore the sub-clinical ones. To demonstrate whether the GRM can remit the issue or not, Figure 5 compares the visualization results between the vanilla network and the one applied with GRM on five retinal diseases cases. The first row represents a relatively simple sample that contains abnormals in the center of the image. It can be found that both of the two networks have precisely identified the lesions. For the sample shown in the second row, it can be seen that the vanilla network biases to the right-most lesions and neglects the abnormals located in the center. On the contrary, the GRM has accurately discovered most of the lesions. Similar results can be observed in the third sample. For the fourth sample containing complicated lesions, the vanilla network biases the right-bottom areas, while the GRM has precisely identified intricate lesions. In the fifth sample, both networks found the most distinguished lesion on the right side, whereas only the GRM has identified the nearby subtle lesions.

These visualization results demonstrate two points. First, the DNNs are object detectors that attempt to discover the abnormalities in the input image. It performs well in those images that contain prominent characteristics, such as the sample shown in the first row in Figure 5. Second, the vanilla DNNs may fail to capture the prominent and subtle lesions simultaneously for the image comprised of complicated features. With the help of GRM, the DNNs can efficiently utilize the context information and show much better performance than the vanilla network.

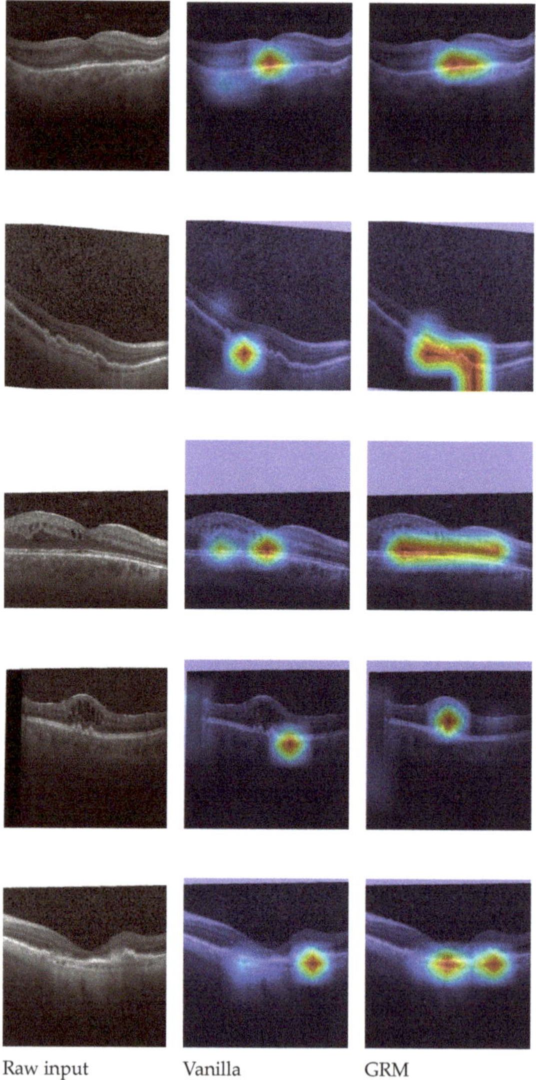

Figure 5. Qualitative comparison between the vanilla ResNet-50 and the ResNet-50 applied with the proposed GRM on five retinal disease cases. Each row denotes a case.

5. Conclusions

This paper proposes a simple yet effective data augmentation method named GRM that aims to discover the potential sub-clinical lesions ignored by the DNNs. The visual interpretation results are used as guidance to help locate the ROIs. Random masking of those ROIs enforces the DNNs to better utilize the context information. Moreover, it also increases the DNNs' robustness to the input since the model is required to predict the category from the incomplete input. Conventional data augmentation method (e.g., cutout) requires to specify the size of the mask, which increases the difficulty during practice when the size of the target varies. On the contrary, the proposed GRM adaptively changes the size and location of the mask according to the characteristics of the target.

Ablation experiments on multiple network architectures are carried out to validate the effectiveness of GRM. The GRM can substantially increase the networks' recognition accuracy on different tasks compared to the vanilla network. The network applied with GRM exhibits evident lower loss on the validation dataset, implying that the GRM helps to increase the networks' generalization capacity. Visualization experiments further demonstrate that the GRM contributes to exploit the sub-clinical lesions and helps reduce the false predictions during practice. In the training phase, the GRM leverages the CAM of the inference result as guidance to randomly mask the input, which is later used to train the network. From a more general point of view, the GRM can be applied iteratively, i.e., the inference and training of the sample can be repeated multiple times till the stability of CAM is achieved. The iterative method may contribute to the learning process of the network because of the enhanced regularization effectiveness. The exploration of the iterative version of GRM is left as a future work.

Author Contributions: Methodology, J.H.; writing—original draft, X.Y.; writing—review and editing, S.W. and J.H. All authors have read and agreed to the published version of the manuscript.

Funding: This work was supported by the National Natural Science Foundation of China under Grant 62106162, China Postdoctoral Science Foundation under Grant 2021M692269, and Sichuan University Postdoctoral Science Foundation under Grant 2022SCU12080.

Conflicts of Interest: The authors declare no conflict of interest.

References

1. He, K.; Zhang, X.; Ren, S.; Sun, J. Deep residual learning for image recognition. In Proceedings of the IEEE Conference on Computer Vision and Pattern Recognition, Las Vegas, NV, USA, 27–30 June 2016; pp. 770–778.
2. Dosovitskiy, A.; Beyer, L.; Kolesnikov, A.; Weissenborn, D.; Zhai, X.; Unterthiner, T.; Dehghani, M.; Minderer, M.; Heigold, G.; Gelly, S.; et al. An Image is Worth 16 × 16 Words: Transformers for Image Recognition at Scale. In Proceedings of the International Conference on Learning Representations, Virtual, 26 April–1 May 2020.
3. Hu, J.; Chen, Y.; Zhong, J.; Ju, R.; Yi, Z. Automated analysis for retinopathy of prematurity by deep neural networks. *IEEE Trans. Med. Imaging* **2018**, *38*, 269–279. [CrossRef] [PubMed]
4. Kermany, D.S.; Goldbaum, M.; Cai, W.; Valentim, C.C.; Liang, H.; Baxter, S.L.; McKeown, A.; Yang, G.; Wu, X.; Yan, F.; et al. Identifying medical diagnoses and treatable diseases by image-based deep learning. *Cell* **2018**, *172*, 1122–1131. [CrossRef] [PubMed]
5. Hu, J.; Chen, Y.; Yi, Z. Automated segmentation of macular edema in OCT using deep neural networks. *Med. Image Anal.* **2019**, *55*, 216–227. [CrossRef] [PubMed]
6. Wang, Z.; Zhang, L.; Shu, X.; Lv, Q.; Yi, Z. An end-to-end mammogram diagnosis: A new multi-instance and multiscale method based on single-image feature. *IEEE Trans. Cogn. Dev. Syst.* **2020**, *13*, 535–545. [CrossRef]
7. Anthimopoulos, M.; Christodoulidis, S.; Ebner, L.; Christe, A.; Mougiakakou, S. Lung pattern classification for interstitial lung diseases using a deep convolutional neural network. *IEEE Trans. Med. Imaging* **2016**, *35*, 1207–1216. [CrossRef]
8. Lu, Y.; Chen, Y.; Zhao, D.; Liu, B.; Lai, Z.; Chen, J. CNN-G: Convolutional neural network combined with graph for image segmentation with theoretical analysis. *IEEE Trans. Cogn. Dev. Syst.* **2020**, *13*, 631–644. [CrossRef]
9. DeVries, T.; Taylor, G.W. Learning confidence for out-of-distribution detection in neural networks. *arXiv* **2018**, arXiv:1802.04865.
10. Jiang, H.; Kim, B.; Guan, M.; Gupta, M. To trust or not to trust a classifier. In *Advances in Neural Information Processing Systems*; The MIT Press: Cambridge, MA, USA, 2018; pp. 5541–5552.
11. Zhou, B.; Khosla, A.; Lapedriza, A.; Oliva, A.; Torralba, A. Object detectors emerge in deep scene cnns. *arXiv* **2014**, arXiv:1412.6856.
12. Springenberg, J.T.; Dosovitskiy, A.; Brox, T.; Riedmiller, M. Striving for simplicity: The all convolutional net. *arXiv* **2014**, arXiv:1412.6806.

13. Zhou, B.; Khosla, A.; Lapedriza, A.; Oliva, A.; Torralba, A. Learning deep features for discriminative localization. In Proceedings of the IEEE Conference on Computer Vision and Pattern Recognition, Las Vegas, NV, USA, 27–30 June 2016; pp. 2921–2929.

14. Selvaraju, R.R.; Cogswell, M.; Das, A.; Vedantam, R.; Parikh, D.; Batra, D. Grad-cam: Visual explanations from deep networks via gradient-based localization. In Proceedings of the IEEE International Conference on Computer Vision, Venice, Italy, 27–29 October 2017; pp. 618–626.

15. Yang, H.; Kim, J.Y.; Kim, H.; Adhikari, S.P. Guided soft attention network for classification of breast cancer histopathology images. *IEEE Trans. Med. Imaging* **2019**, *39*, 1306–1315. [CrossRef]

16. DeVries, T.; Taylor, G.W. Improved regularization of convolutional neural networks with cutout. *arXiv* **2017**, arXiv:1708.04552.

17. Krizhevsky, A.; Hinton, G. *Learning Multiple Layers of Features from Tiny Images*; Citeseer: State College, PA, USA, 2009.

18. Netzer, Y.; Wang, T.; Coates, A.; Bissacco, A.; Wu, B.; Ng, A.Y. Reading digits in natural images with unsupervised feature learning. In Proceedings of the NIPS Workshop on Deep Learning and Unsupervised Feature Learning, Granada, Spain, 16 December 2011.

19. Krizhevsky, A.; Sutskever, I.; Hinton, G.E. Imagenet classification with deep convolutional neural networks. In *Advances in Neural Information Processing Systems*; The MIT Press: Cambridge, MA, USA, 2012; pp. 1097–1105.

20. Szegedy, C.; Liu, W.; Jia, Y.; Sermanet, P.; Reed, S.; Anguelov, D.; Erhan, D.; Vanhoucke, V.; Rabinovich, A. Going deeper with convolutions. In Proceedings of the IEEE Conference on Computer Vision and Pattern Recognition, Boston, MA, USA, 7–12 June 2015; pp. 1–9.

21. Deng, J.; Dong, W.; Socher, R.; Li, L.J.; Li, K.; Li, F.F. ImageNet: A large-scale hierarchical image database. In Proceedings of the IEEE Conference on Computer Vision and Pattern Recognition, Miami, FL, USA, 20–25 June 2009; pp. 248–255.

22. Szegedy, C.; Vanhoucke, V.; Ioffe, S.; Shlens, J.; Wojna, Z. Rethinking the inception architecture for computer vision. In Proceedings of the IEEE Conference on Computer Vision and Pattern Recognition, Las Vegas, NV, USA, 27–30 June 2016; pp. 2818–2826.

23. Simonyan, K.; Zisserman, A. Very deep convolutional networks for large-scale image recognition. *arXiv* **2014**, arXiv:1409.1556.

24. Ronneberger, O.; Fischer, P.; Brox, T. U-net: Convolutional networks for biomedical image segmentation. In Proceedings of the International Conference on Medical Image Computing and Computer-Assisted Intervention, Munich, Germany, 5–9 October 2015; Springer: Berlin/Heidelberg, Germany, 2015; pp. 234–241.

25. Oktay, O.; Schlemper, J.; Folgoc, L.L.; Lee, M.; Heinrich, M.; Misawa, K.; Mori, K.; McDonagh, S.; Hammerla, N.Y.; Kainz, B.; et al. Attention u-net: Learning where to look for the pancreas. *arXiv* **2018**, arXiv:1804.03999.

26. Roy, A.G.; Navab, N.; Wachinger, C. Concurrent spatial and channel 'squeeze & excitation' in fully convolutional networks. In Proceedings of the International Conference on Medical Image Computing and Computer-Assisted Intervention, Granada, Spain, 16–20 September 2018; Springer: Berlin/Heidelberg, Germany, 2018; pp. 421–429.

27. Alom, M.Z.; Hasan, M.; Yakopcic, C.; Taha, T.M.; Asari, V.K. Recurrent residual convolutional neural network based on u-net (r2u-net) for medical image segmentation. *arXiv* **2018**, arXiv:1802.06955.

28. Shan, H.; Padole, A.; Homayounieh, F.; Kruger, U.; Khera, R.D.; Nitiwarangkul, C.; Kalra, M.K.; Wang, G. Competitive performance of a modularized deep neural network compared to commercial algorithms for low-dose CT image reconstruction. *Nat. Mach. Intell.* **2019**, *1*, 269–276. [CrossRef]

29. Yang, Q.; Yan, P.; Zhang, Y.; Yu, H.; Shi, Y.; Mou, X.; Kalra, M.K.; Zhang, Y.; Sun, L.; Wang, G. Low-dose CT image denoising using a generative adversarial network with Wasserstein distance and perceptual loss. *IEEE Trans. Med. Imaging* **2018**, *37*, 1348–1357. [CrossRef]

30. Shan, H.; Zhang, Y.; Yang, Q.; Kruger, U.; Kalra, M.K.; Sun, L.; Cong, W.; Wang, G. 3-D convolutional encoder-decoder network for low-dose CT via transfer learning from a 2-D trained network. *IEEE Trans. Med. Imaging* **2018**, *37*, 1522–1534. [CrossRef]

31. Vaswani, A.; Shazeer, N.; Parmar, N.; Uszkoreit, J.; Jones, L.; Gomez, A.N.; Kaiser, Ł.; Polosukhin, I. Attention is all you need. In *Advances in Neural Information Processing Systems*; The MIT Press: Cambridge, MA, USA, 2017; Volume 30.

32. Hatamizadeh, A.; Tang, Y.; Nath, V.; Yang, D.; Myronenko, A.; Landman, B.; Roth, H.R.; Xu, D. UNETR: Transformers for 3D Medical Image Segmentation. In Proceedings of the IEEE/CVF Winter Conference on Applications of Computer Vision, Waikoloa, HI, USA, 4–8 January 2022; pp. 574–584.

33. Landman, B.; Xu, Z.; Igelsias, J.; Styner, M.; Langerak, T.; Klein, A. MICCAI multi-atlas labeling beyond the cranial vault–workshop and challenge. In Proceedings of the MICCAI Multi-Atlas Labeling Beyond Cranial Vaul—Workshop Challenge, Munich, Germany, 5–9 October 2015; Volume 5, p. 12.

34. Simpson, A.L.; Antonelli, M.; Bakas, S.; Bilello, M.; Farahani, K.; Van Ginneken, B.; Kopp-Schneider, A.; Landman, B.A.; Litjens, G.; Menze, B.; et al. A large annotated medical image dataset for the development and evaluation of segmentation algorithms. *arXiv* **2019**, arXiv:1902.09063.

35. Huang, S.; Li, J.; Xiao, Y.; Shen, N.; Xu, T. RTNet: Relation Transformer Network for Diabetic Retinopathy Multi-lesion Segmentation. *IEEE Trans. Med. Imaging* **2022**, *41*, 1596–1607. [CrossRef]

36. Dai, Y.; Gao, Y.; Liu, F. Transmed: Transformers advance multi-modal medical image classification. *Diagnostics* **2021**, *11*, 1384. [CrossRef]

37. Liu, Z.; Mao, H.; Wu, C.Y.; Feichtenhofer, C.; Darrell, T.; Xie, S. A ConvNet for the 2020s. *arXiv* **2022**, arXiv:2201.03545.

38. Dong, H.; Yang, G.; Liu, F.; Mo, Y.; Guo, Y. Automatic brain tumor detection and segmentation using u-net based fully convolutional networks. In Proceedings of the Annual Conference on Medical Image Understanding and Analysis, Edinburgh, UK, 11–13 July 2017; Springer: Berlin/Heidelberg, Germany, 2017; pp. 506–517.

39. Liskowski, P.; Krawiec, K. Segmenting retinal blood vessels with deep neural networks. *IEEE Trans. Med. Imaging* **2016**, *35*, 2369–2380. [CrossRef] [PubMed]

40. Christ, P.F.; Elshaer, M.E.A.; Ettlinger, F.; Tatavarty, S.; Bickel, M.; Bilic, P.; Rempfler, M.; Armbruster, M.; Hofmann, F.; D'Anastasi, M.; et al. Automatic liver and lesion segmentation in CT using cascaded fully convolutional neural networks and 3D conditional random fields. In Proceedings of the International Conference on Medical Image Computing and Computer-Assisted Intervention, Athens, Greece, 17–21 October 2016, Springer: Berlin/Heidelberg, Germany, 2016; pp. 415–423.

41. Sirinukunwattana, K.; Pluim, J.P.; Chen, H.; Qi, X.; Heng, P.A.; Guo, Y.B.; Wang, L.Y.; Matuszewski, B.J.; Bruni, E.; Sanchez, U.; et al. Gland segmentation in colon histology images: The glas challenge contest. *Med. Image Anal.* **2017**, *35*, 489–502. [CrossRef]

42. Zhang, H.; Cisse, M.; Dauphin, Y.N.; Lopez-Paz, D. mixup: Beyond empirical risk minimization. *arXiv* **2017**, arXiv:1710.09412.

43. Bdair, T.; Navab, N.; Albarqouni, S. ROAM: Random Layer Mixup for Semi-Supervised Learning in Medical Imaging. *arXiv* **2020**, arXiv:2003.09439.

44. Srivastava, N.; Hinton, G.; Krizhevsky, A.; Sutskever, I.; Salakhutdinov, R. Dropout: A simple way to prevent neural networks from overfitting. *J. Mach. Learn. Res.* **2014**, *15*, 1929–1958.

45. Lin, M.; Chen, Q.; Yan, S. Network in network. In Proceedings of the International Conference on Learning Representations, Scottsdale, AZ, USA, 2–4 May 2013.

46. Huang, G.; Liu, Z.; Van Der Maaten, L.; Weinberger, K.Q. Densely connected convolutional networks. In Proceedings of the IEEE Conference on Computer Vision and Pattern Recognition, Honolulu, HI, USA, 21–26 July 2017; pp. 4700–4708.

47. Kingma, D.; Ba, J. ADADELTA: An Adaptive Learning Rate Method. *arXiv* **2012**, arXiv:1212.5701.

48. Loshchilov, I.; Hutter, F. Decoupled Weight Decay Regularization. In Proceedings of the International Conference on Learning Representations, Vancouver, BC, Canada, 30 April–3 May 2018.

49. Paszke, A.; Gross, S.; Chintala, S.; Chanan, G.; Yang, E.; DeVito, Z.; Lin, Z.; Desmaison, A.; Antiga, L.; Lerer, A. Automatic differentiation in PyTorch. In Proceedings of the 31st Conference of Advances in Neural Information Processing Systems, Long Beach, CA, USA, 4–9 December 2017. Avialble online: ttps://openreview.net/forum?id=BJJsrmfCZ (accessed on 30 July 2022).

applied
sciences

MDPI

Article

Towards Explainable Deep Neural Networks for the Automatic Detection of Diabetic Retinopathy

Hanan Saleh Alghamdi

Information Systems Department, Faculty of Computing and Information Technology, King Abdulaziz University, P.O. Box 80200, Jeddah 21589, Saudi Arabia; hsaalghamdi@kau.edu.sa

Featured Application: The proposed approach can be applied to any of the Convolutional Neural Networks-based architecture to explain, evaluate and validate the model's decisions.

Abstract: Diabetic Retinopathy (DR) is a common complication associated with diabetes, causing irreversible vision loss. Early detection of DR can be very helpful for clinical treatment. Ophthalmologists' manual approach to DR diagnoses is expensive and time-consuming; thus, automatic detection of DR is becoming vital, especially with the increasing number of diabetes patients worldwide. Deep learning methods for analyzing medical images have recently become prevalent, achieving state-of-the-art results. Consequently, the need for interpretable deep learning has increased. Although it was demonstrated that the representation depth is beneficial for classification accuracy for DR diagnoses, model explainability is rarely analyzed. In this paper, we evaluated three state-of-the-art deep learning models to accelerate DR detection using the fundus images dataset. We have also proposed a novel explainability metric to leverage domain-based knowledge and validate the reasoning of a deep learning model's decisions. We conducted two experiments to classify fundus images into normal and abnormal cases and to categorize the images according to the DR severity. The results show the superiority of the VGG-16 model in terms of accuracy, precision, and recall for both binary and DR five-stage classification. Although the achieved accuracy of all evaluated models demonstrates their capability to capture some lesion patterns in the relevant DR cases, the evaluation of the models in terms of their explainability using the Grad-CAM-based color visualization approach shows that the models are not necessarily able to detect DR related lesions to make the classification decision. Thus, more investigations are needed to improve the deep learning model's explainability for medical diagnosis.

Keywords: explainable deep networks; diabetic retinopathy; deep learning; Grad-CAM; convolutional neural networks; ResNet; DenseNet

Citation: Alghamdi, H.S. Towards Explainable Deep Neural Networks for the Automatic Detection of Diabetic Retinopathy. *Appl. Sci.* **2022**, *12*, 9435. https://doi.org/10.3390/app12199435

Academic Editor: Dimitris Mourtzis

Received: 16 August 2022
Accepted: 9 September 2022
Published: 21 September 2022

Publisher's Note: MDPI stays neutral with regard to jurisdictional claims in published maps and institutional affiliations.

1. Introduction

Diabetes is a major cause of life-threatening systemic vascular complications, including stroke, heart attacks, kidney failure, and blindness. According to the International Diabetes Federation [1], around 463 million people had diabetes in 2019. The number of people with diabetes had increased to 422 million in 2014 [2] and is estimated to rise to 700 million by 2045. Diabetic Retinopathy (DR) is a common complication of diabetes, found in a third of diabetes patients, and remains the primary cause of avoidable vision loss in working-aged people [3]. DR is caused by damage to the retinal blood vessels; however, it might not have symptoms until it advances to the vision-threatening stage. Early detection of DR is essential to reduce the avoidable vision loss threat of DR. DR screening is performed through the examinations of fundus photographs by a trained clinician to determine DR presence and severity. The severity of DR is determined by the presence of DR lesions, including microaneurysms, hemorrhages, cotton wool spots, and exudates, as demonstrated in Figure 1. Given the limited number of retina specialists and the increased number of diabetes patients worldwide, in-person assessments are impractical and unsustainable. These examinations could result in too-late detection of DR when the treatment is not as effective as in the early stages of the disease. Thus, the necessity of an automated DR screening approach has long been recognized. Significant progress has been made in computer vision, pattern recognition, and machine learning. The automatic detection of DR began to appear in 2010, and since then, analyzing fundus images for DR detection has been performed using numerous approaches. These methods have been applied at different levels of analysis, ranging from general image classification, lesion detection, anatomical structure segmentation, and DR severity determination.

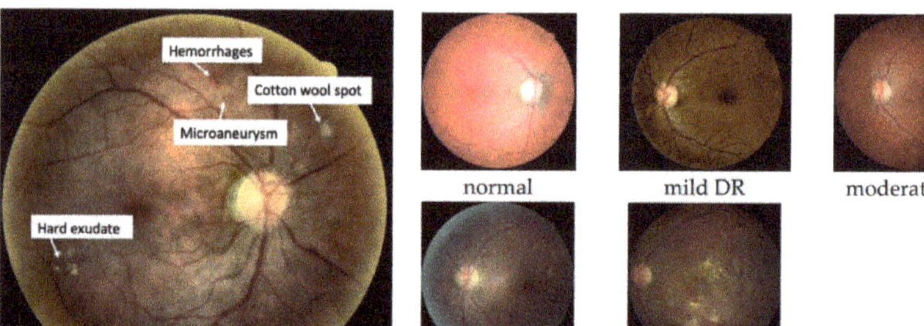

Figure 1. Different DR lesions (**left**) and DR stages (**right**).

Early methods were based on classical computer vision techniques and thresholds [4]. Later, traditional machine learning algorithms had also been applied to DR detection. For example, in [5], Chowdhury et al. trained a random forest classifier on the DIARETDB1 dataset to detect abnormalities in fundus images. The experiment showed that the random forest achieved a better classification accuracy of 93.58% than the Naïve Bayes classifier, which reached 83.63%. Bourouis et al. in [6] developed a hybrid model for DR classification using three kernels for an SVM-based classifier, including the Fisher, Kullback–Leibler, and Bhattacharyya kernels. The experiments were conducted on multiple public DR datasets and achieved 91.33% accuracy on the DRIVE dataset with the Bhattacharyya kernel. Emon et al. in [7] evaluated eight different machine learning models on a dataset consisting of 1151 instances and contained features extracted from the Messidor image set. According to the study, the logistic regression algorithm resulted in the best performance of 75% accuracy.

Artificial Intelligence (AI) algorithms, particularly deep learning (DL), have shown great potential in almost all domains. In the medical imaging field, DL has demonstrated effectiveness for various tasks such as pathologies detection, diagnosis, and prognosis of diseases, for example, brain tumors, lung infections, and retinal disorders. DL is a subcategory of machine learning consisting of a hierarchical, multilayer neural network model for automatic feature extraction. CNNs are the most common DL approach for image classification. CNNs are well-known DL architecture in which neurons are organized in two-dimensional planes to extract basic features from overlapping regions at the lower layers. Then, at the higher layers, these features are combined to form more complex and comprehensive features. However, despite the wide application of DL in automatic diagnosis systems, most DL algorithms remain as black boxes to medical experts.

A fully automated method with a lack of human verification would be unconscionable and potentially dangerous in a clinical setting. The lack of transparency in such systems and the inability to explain the rationale behind the DL models' decisions could prevent the clinical acceptance of integrating such components into the healthcare systems. Domain experts, especially in the medical area, often require insights into the DL model's decision-making process to ensure the reasonableness of the predictions. The increasing demand for explainability by both the end-users and the researchers has led to some noteworthy innovations in the last years. Thus, explainable AI (XAI) has experienced a surge in medical imaging literature. However, how these explanation methods can be used to evaluate and compare DL architectures is still not well explored [8].

The automatic detection of DR began to appear in 2010, and the early methods were based on classical computer vision techniques and thresholds [4]. Later, traditional machine learning algorithms were also applied to DR detection. For example, in [5], Chowdhury et al. trained a random forest classifier on the DIARETDB1 dataset to detect abnormalities in fundus images. The experiment showed that the random forest achieved a better classification accuracy of 93.58% than the Naïve Bayes classifier, which reached 83.63%. Bourouis et al. [6] developed a hybrid model for DR classification using three kernels for an SVM-based classifier, including the Fisher, Kullback–Leibler, and Bhattacharyya kernels. The experiments were conducted on multiple public DR datasets and achieved 91.33% accuracy on the DRIVE dataset with the Bhattacharyya kernel. Emon et al. [7] evaluated eight different machine learning models on a dataset consisting of 1151 instances and contained features extracted from the Messidor image set. According to the study, the logistic regression algorithm resulted in the best performance of 75% accuracy.

However, deep learning and CNNs have proved their superiority over other traditional machine learning algorithms for object detection and image classification tasks. Thus, deep learning and CNNs have been applied and evaluated for the diagnosis of DR [9]. Authors in [10] used the Kaggle DR dataset [11] to train a CNN model to classify referable and nonreferable DR images. They achieved 98.2% accuracy in the Messidor-2 dataset [12]. Transfer learning, which has demonstrated promising results in medical image diagnosis, uses state-of-the-art CNN models pretrained on a large general image dataset. The knowledge learned on a primary task is utilized and transferred to a secondary task. Transfer learning mitigates the need for a vast amount of data and substantial computational resources.

Thus, many recent studies also utilized transfer learning with CNN architectures. The authors in [13] trained AlexNet, VggNet, GoogleNet, and ResNet on the publicly available Kaggle platform and achieved 95.68% accuracy for the best model. The researchers in [14,15] used a dataset provided by APTOS and Kaggle. In [14], the researchers trained ResNet50, Xception Nets, DenseNets, and VGG, all pretrained on ImageNet, and the best model achieved an accuracy of 81.3%, while in [15], the authors tried fine-tuning a pretrained Inception-V3 model for five-class classification. They subsampled a smaller version of the Kaggle DR classification challenge dataset for model training and achieved an accuracy of 90.9%. Table 1 summarizes the related approaches for the DR automatic detection task.

Table 1. Summary of some DR automatic detection approaches applied by other related works.

Reference	Dataset	Approach	Accuracy
[5]	DIARETDB1	Random Forest	93.58%
[6]	Public DR datasets	SVM with Bhattacharyya kernel	91.33%
[7]	Messidor	Logistic Regression	75.00%
[10]	Kaggle DR dataset, Messidor-2	CNN	98.20%
[13]	Kaggle DR dataset	AlexNet, VggNet, GoogleNet, and ResNet	95.68%
[14]	Kaggle DR dataset	ResNet50, Xception Nets, DenseNets, and VGG	81.30%
[15]	Kaggle DR dataset	Inception-V3	90.90%

In this paper, we pursue to evaluate some state-of-the-art DL models for the task of DR detection fundus photographs in terms of their accuracy, sensitivity, and specificity. Additionally, we aim to compare these algorithms based on their explainability. This would increase the expert insights and help decide the most reasonable and trustworthy models for DR detection from retinal photographs.

The key contributions of this paper are three-fold:

1. Evaluate three state-of-the-art deep transfer learning algorithm models using color fundus images for automatic DR detection;
2. Optimize the proposed transfer learning deep learning architectures through early stopping and dropout techniques to control the models' overfitting tendency.
3. Perform Grad-CAM analysis to provide human-interpretable explanations of the deep architectures' predictions of DR.

2. Materials and Methods

This section discusses our approach in detail, covering the dataset and the evaluated deep learning models, followed by the prediction explainability techniques, performance evaluation metrics, and proposed explainability measure.

2.1. Dataset

In this work, we used a publicly available dataset at Kaggle [16]. This would allow further investigation and benchmarking comparison. This dataset consists of a wide variety of retinal photographs as it was collected from multiple clinics using different cameras and over an extended period of time. The images were rated by a clinician for the severity of DR on a scale of 0 to 4: 0 for normal, 1 for mild, 2 for moderate, 3 for severe, and 4 for proliferative DR [16]. Images in this dataset may contain artifacts or are out of focus. The level of variation in this dataset introduced complexity and difficulty for any classifier model, and thus, it is very important to validate the models' decisions. However, the dataset was originally imbalanced, and most samples belong to the normal healthy retina class. In addition, there were no samples dedicated to a validation set. Figure 2 shows the original dataset distribution consisting only of training and test sets of 28,103 and 7022 samples. Indeed, imbalanced training samples would generally lead to a naïve behavior classifier, which tends to classify the samples according to the majority class to minimize the cost function over all training samples.

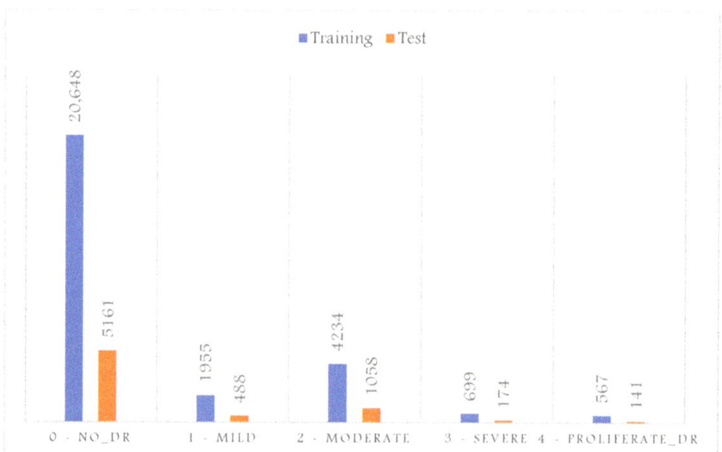

Figure 2. Original DR dataset distribution.

Moreover, the evaluation of imbalanced test samples is biased and misleading. Thus, to overcome this challenge, we sample three sets of training, validation, and testing to contain the same number of samples per class shown in Figure 3. The proliferative DR class contains the least number of samples; thus, the sampling for the training, validation, and test sets was based on the number of samples available for this class. This results in a total number of images in the training set of 2200, 600 for the validation, and 700 for the test set. We also converted the task into a binary classification to detect all abnormal cases in one category; thus, as shown in Figure 4, all abnormal categories were grouped.

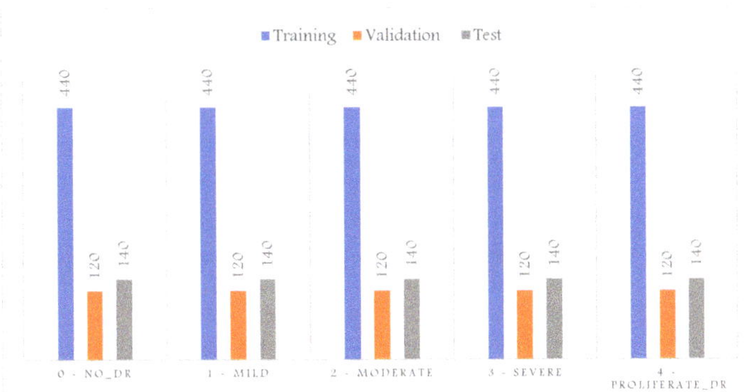

Figure 3. DR dataset distribution after balancing the five categories.

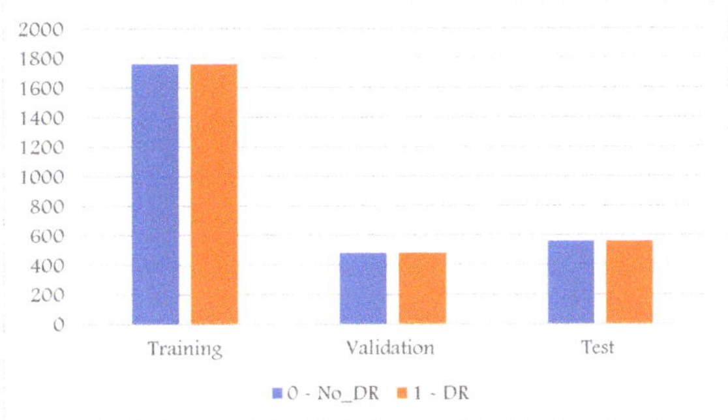

Figure 4. DR dataset distribution after converting it into binary categories (normal, abnormal).

2.2. CNN Models

The CNN models employed in this study were pre-trained on the large-scale ImageNet dataset [17], which includes 1000 categories of different objects. These models normally perform highly on general classification tasks, especially for the objects presented in the training dataset. However, their performance can be lowered when applied to specific domains, such as DR detection. In the following subsections, we start by describing the basic architecture of the CNN model. Then, we briefly describe the three pre-trained models used in this work and highlight their main characteristics.

2.2.1. Convolutional Neural Networks

CNNs are the most common artificial neural networks used for performing computing vision tasks such as image classification, object detection, and segmentation. The advantage of CNNs over other machine learning algorithms such as Support Vector Machine, K-Nearest Neighbors, Random Forest, among others, is that the CNNs can automatically learn representative features from the images and has a higher generalization capacity [18]. A CNN is typically divided into three main components: the convolutional, pooling, and dense layers. A convolutional layer learns the features and passes the features to a pooling layer to perform downsampling. A dense layer learns how to classify the extracted features into different categories. The output layer usually uses the softmax activation function to generate the probability distribution of each category in the problem domain.

2.2.2. Visual Geometry Group

The authors in [18] proposed the architecture of the Visual Geometry Group (VGG) network in 2013 and submitted their model for the 2014 ImageNet Challenge. VGG model uses a small receptive field of size 3×3 throughout the entire network with a 1-pixel stride. It is worth noting that the two consecutive 3×3 convolutional filter layers, without spatial pooling in between, provide a receptive field of size 5×5, and the three 3×3 convolutional layers filters result in a receptive field of 7×7. This unique characteristic allows the network to converge faster, makes the decision functions more discriminative, and reduces the number of weight parameters.

2.2.3. The Residual Network

ResNet architecture is one of the most popular and successful deep learning models for computer vision tasks. The residual network has multiple variants, including ResNet-16, ResNet-18, ResNet-34, ResNet-50, ResNet-101, ResNet-110, ResNet-152, ResNe-t164, ResNet-1202, and so forth. The residual unit is the main building block of the ResNet. The intuition behind the residual unit is to ease the costly training of the very deep networks by using a direct connection that skips some layers in between [19]. This connection is called a 'skip connection' or 'shortcut connection' and is the core of residual blocks. With the introduction of skip connection, the output is changed to F(x) + x instead of F(x) in the other layers. The skip connections in ResNet solve the vanishing gradient in deep neural networks by allowing the gradient to flow through this alternate shortcut pathway [19]. The deep ResNet is a stack of residual units seen as small neural networks with a skip connection. ResNet18 is a 72-layer architecture with 18 deep layers. The input size to the network is $224 \times 224 \times 3$, which is predefined.

2.2.4. DenseNet-121

DenseNet is another type of CNN that uses dense connections between layers through the Dense Blocks [20]. Dense Blocks connect all layers directly with each other. However, each layer obtains additional inputs from all previous layers and passes its feature maps to all subsequent layers in a feed-forward process. DenseNets alleviate the vanishing gradient problem, encourage feature reuse, and reduce the number of learnable parameters [20]. DenseNet-121 is the simple DenseNet network designed for the ImageNet dataset. It consists of multiple dense and transition blocks. Transition Block performs as a 1×1 convolution with 128 filters, followed by a 2×2 pooling with a stride of 2, resulting in dividing the size of the volume by dividing volume size and the number of feature maps in half.

2.3. Models' Explainability Using Grad-CAM

Deep architectures take in more than a million parameters of complex, convoluted operations. Thus, the interpretability of such algorithms is challenging. Class Activation Mapping (CAM) is one technique proposed to enhance the explainability of deep learning models. The basic idea behind CAM is to localize the deep discriminative features and visualize the object parts detected by the CNN [21].

The study in [22] inspired this idea and demonstrates that convolutional layers of CNNs behave as object detectors despite no supervision of the object's location. To generate the CAMs, the predicted class weights are projected back to the activation maps of the previous convolutional layer to highlight class-specific discriminative regions. This approach provides visual explanations as each activation map contains different spatial information about the input, and when the convolutional layer is close to the classification layer, its activations are sufficiently high-level to provide a visual localization to explain the final decision. Let f be a CNN-based classification model and c a target category. Given an input image x and a convolutional layer of f, the CAM with respect to c can be defined as a linear combination of the activation maps of the convolutional layer, as follows [23]:

$$\mathrm{CAM}_c(x) = \sum_{k=1}^{N_f} w_k^c A_k \tag{1}$$

where N_f denotes the number of filters of the convolutional layer, A_k is the kth filter of the activation, and w_k^c are weight coefficients indicating the importance of the activation maps with respect to the target class c. However, CAM is restricted to having a global average pooling (GAP) layer after the final convolutional layer and then a dense linear layer. The GAP computes the average of each feature map for each corresponding class, and the resulting vector is fed into the softmax activation layer, which outputs the class probabilities. If the CNN-based model does not have a GAP in the final layer, CAM requires removing the fully connected layer before the final output and replacing it with the GAP [21]. Gradient-weighted Class Activation Mapping Grad-CAM has been suggested as a generalization version of CAM, as it can be applied to any CNN-based models without modifying their architectures [23]. Similar to the CAM, Grad-CAM employs the spatial information preserved through convolutional layers to highlight the parts of an input image that are important for the classifier decisions. However, Grad-CAM uses class-specific gradient information produced by the feature maps of the last convolutional layer to generate a class-discriminative localization heatmap corresponding to a particular class [23]. The importance of feature map k for the target class c is computed using the gradient of the logits of class c with respect to the activation maps of the final convolutional layer, and the gradients are averaged across each feature map, a ReLU nonlinearity is applied to only consider the pixels that have a positive influence on the score of the target class [23]:

$$L_{Grad-CAM}^c = ReLU\left(\sum_{k=1}^{N_f} w_k^c A_k\right) \tag{2}$$

2.4. Performance Evaluation Metrics

In this work, five evaluation metrics were employed to provide complete coverage and unbiased analysis of the results. This includes the following:

- Accuracy: calculated as the percentage of the correctly classified images by:

$$Accuracy = \frac{TP + TN}{N} \tag{3}$$

where N is the total number of images in the evaluated set, TP is the true positive, i.e., detected abnormal cases, and TN is the true negative, i.e., normal cases not detected as abnormal.

- Precision: calculated as the number of TP divided by the sum of TP and false positives, normal cases detected as abnormal.

$$Precision = \frac{TP}{TP + FP} \tag{4}$$

- Sensitivity/Recall: calculated as the number of, divided by the sum of TP and false negatives FN, abnormal cases detected as normal:

$$Recall = \frac{TP}{TP + FN} \tag{5}$$

- F1-Score: defined as the harmonic mean of precision and recall:

$$F1 - Score = 2*\frac{Precision * Recall}{Precision + Recall} \tag{6}$$

- Confusion matrix: A confusion matrix is a table used for summarizing a classifier's performance. The number of correctly and incorrectly classified samples are summarized with count values and broken down by each category.

2.5. Proposed Explainability Evaluation Metric

In this paper, we propose a novel explanatory metric to validate a deep learning model's decision, a model's conformity that measures the proportion of model attention to DR-related lesions. To calculate a model's conformity over the whole test set, we average the conformity of each instance in that set. We visualize the evaluated deep learning models' Grad-CAM and evaluate the results using the conformity measure as follows:

$$\text{conformity} = \begin{cases} \frac{1}{N} \sum_{i=1}^{N} 1 - \frac{FDL_i}{1+FDL_i} & \text{when } N_l = 0 \\ \frac{1}{N} \sum_{i=1}^{N} \frac{TDL_i}{TDL_i+FUL_i+FDL_i} & \text{when } N_l > 0 \end{cases} \tag{7}$$

where N is the total number of images in the evaluated set, N_l. is the number of DR lesion regions present in image i, TDL_i is the number of correctly detected lesions, FUL_i. is the number of undetected lesions, and FDL_i is the number of incorrectly detected lesions. When $N_l = 0$, i.e., in the case of normal images, we assume that the whole image should contribute to the classifier prediction. Thus, no specific region should be highly activated and highlighted using Grad-CAM. Therefore, the conformity of a model, when tested on image i, equals one in this case. In contrast, if the model highlights many irrelevant regions, the conformity approaches zero. When $N_l > 0$, i.e., in case of abnormal images, all lesions' regions should be highlighted using Grad-CAM. The conformity would equal one if all lesions' regions were highlighted and approach zero if the classifier either detects false regions or misses relevant DR signs regions.

3. Results

In this study, we first compared the performance of the three models on the test set for both five classes and binary classification to see how well each model differentiates abnormal and abnormal fundus photos in these two tasks. Then, we evaluated each model's explainability as measured by our proposed conformity metric to validate the models' performance. We visualized the Grad-CAM outputs to compute the conformity of normal and abnormal retinal photographs. Finally, we discussed the correlation between explainability and the models' performance.

3.1. Model Performance on the Test Set

3.1.1. Binary Classification

Table 2 presents the three models' performance evaluation on the test and train sets for the binary classification of retinal images, i.e., whether normal or contains DR-related signs. As shown, VGG-16 resulted in the highest accuracy on the test set with the least variance between train and test set accuracies. On the other hand, Dense-Net121 clearly overfits, resulting in much lower test accuracy than training accuracy.

Figure 5 shows the three models' confusion matrices on the test and train sets also for the binary classification of retinal images. As can be seen, VGG-16 resulted in the lowest number of false positives and the highest number of true positives, while ResNet-18 has the highest number of false positives and the lowest number of true positives.

Table 2. Performance evaluation on the test and train sets for DR detection.

Model	Precision [1]	Recall [1]	F1-Score [1]	Train Accuracy	Test Accuracy
VGG16	**0.87**	0.52	0.65	78.07%	**73.04%**
ResNet-18	0.67	0.68	0.67	78.44%	67.14%
DenseNet-121	0.74	**0.71**	**0.73**	**91.11%**	72.95%

[1] Calculated for the test set.

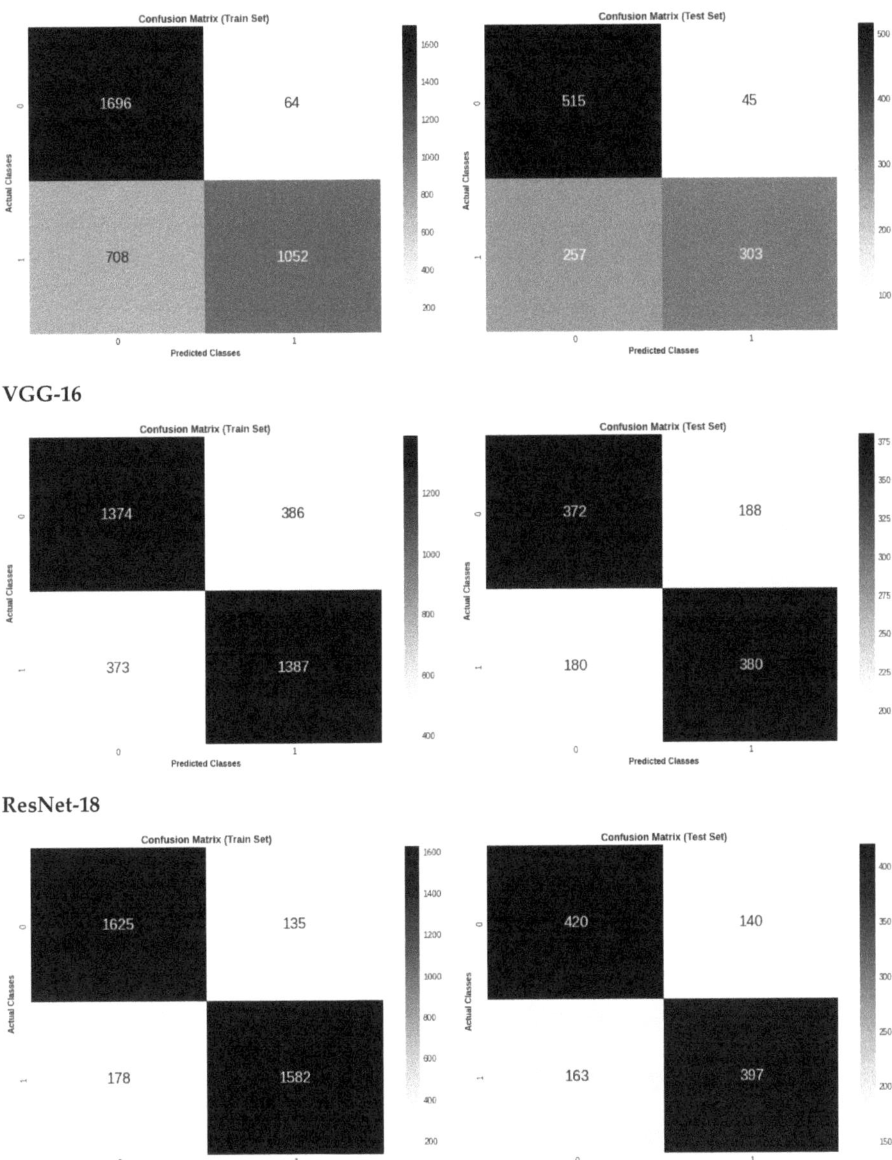

VGG-16

ResNet-18

DenseNet-121

Figure 5. Confusion matrix evaluation of the three models on the test and train sets for binary classes classification.

3.1.2. Multiple Classification

Table 3 presents the three models' performance evaluation on the test and train sets for the five DR stages classification of retinal images. Again, as shown, VGG-16 resulted in the highest accuracy on the test set with the least variance between train and test set accuracies, and Dense-Net121 overfitted the train set, resulting in much lower test accuracy than training accuracy.

Table 3. Performance evaluation on the test and train sets for DR stages classification.

Model	Precision [1]	Recall [1]	F1-Score [1]	Train Accuracy	Test Accuracy
VGG16	**0.45**	**0.48**	**0.47**	64.27%	**48.43%**
ResNet-18	0.44	**0.48**	0.46	76.18%	47.86%
DenseNet-121	0.42	0.46	0.44	**83.05%**	45.57%

[1] Calculated for the test set.

Figure 6 shows the three models' confusion matrices on the test and train sets for the five DR stages classification of retinal images. It is worth noting that VGG-16 demonstrated the highest classification accuracy between the two early stages of DR, which might mean the ability to capture some DR lesions not seen by other models.

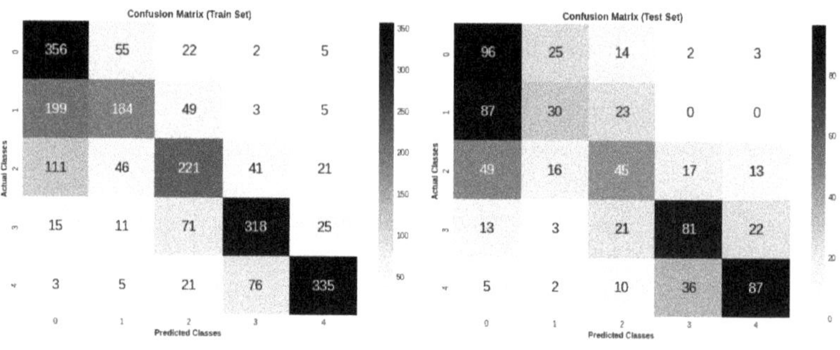

VGG-16

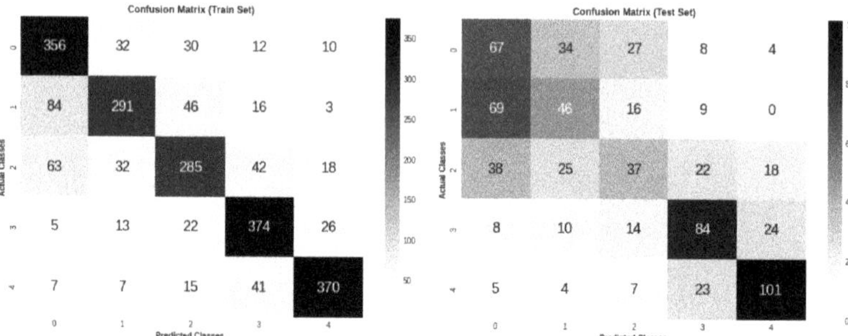

ResNet-18

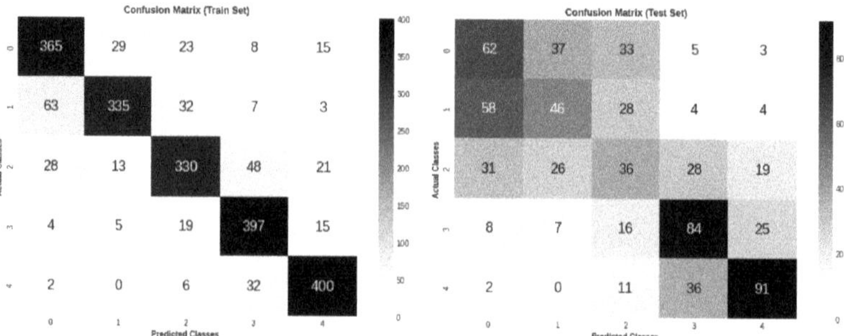

DenseNet-121

Figure 6. Confusion matrix evaluation of the three models on the test and train sets for five classes classification.

3.2. Models Explainability on the Test Set

Table 4 demonstrates some examples of Grad-CAM outputs along with the original fundus photos. Table 5 presents the results of the proposed explainability metric, conformity. As can be seen from Table 4, classifiers might activate irrelevant background regions or normal retinal structures such as the optic disc or the macula. Additionally, the deep learning classifiers do not capture some clear lesion sign regions. For example, as shown in Table 4, image b, ResNet-18 model decisions were based on the background regions. Furthermore, in image c, ResNet-18 and DenseNet-121 models emphasized some irrelevant regions and failed to find DR-related signs even though the DR lesions are distinctive. Lastly, ResNet-18 and DenseNet-121 models are confused by the normal retina structure, which caused false classification by these models.

Table 4. Examples of Grad-CAM output of the evaluated deep learning models.

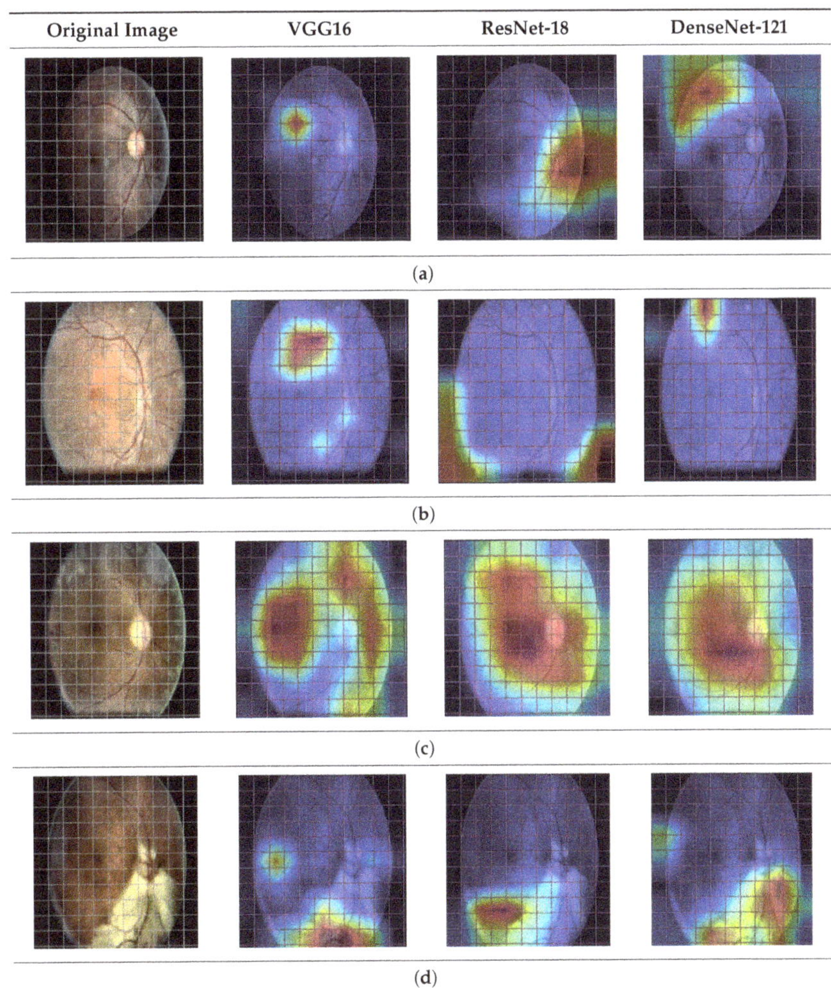

Original Image	VGG16	ResNet-18	DenseNet-121

(a)

(b)

(c)

(d)

Table 5. Model conformity measures.

Model	Conformity with Normal Retinal Photos	Conformity with Abnormal Retinal Photos	Average Conformity
VGG16	**0.2000**	**0.2414**	**0.2207**
ResNet-18	0.0294	0.0645	0.0469
DenseNet-121	0.0385	0.0286	0.0336

4. Discussion

In recent technological advancements, the diffusion of deep learning architectures allows for more promising results corresponding to various applications, including medical imaging and DR diagnosis. Despite achieving remarkable results in terms of model accuracies, deep learning-based methods have not achieved a significant deployment in clinical settings. One major reason is the lack of tools to inspect the decisions of deep learning models, as these models might make the right decision due to wrong reasoning. This is a serious issue, which makes it essential to give more attention to analyzing the black box nature of deep learning models. Another issue related to the performance evaluation of deep learning models in the medical field is the skewness of the data used for training and testing. This is usually due to the domination of normal over abnormal cases. Highly skewed data means the data are not evenly distributed. Machine learning models are designed to improve accuracy by reducing error and tend to produce biased and inaccurate results when faced with imbalanced datasets. Evaluation of an imbalanced dataset using accuracy metric, for example, can also be misleading as the minority class is normally the class of interest, i.e., the disease cases.

In this work, we started by creating a balanced DR dataset by obtaining the same number of instances for all classes. The main objective of dataset balancing is to train unbiased models and to have an accurate and valid evaluation. In this work, balancing data experiments reveal that the deep learning models tend to overfit the training set and do not necessarily perform well on unseen fundus photographs. This highlights the importance of giving more attention to this difficulty before feeding the algorithms with skewed data and validating the experimental results.

To overcome the challenge of unexplained predictions, we proposed a new metric that measures the models' attention to the DR symptoms. We conducted two experiments to classify the fundus images into two and five classes. We fine-tuned three state-of-the-art deep learning architectures in both cases and visualized their decisions using Grad-CAM techniques. Our conformity metric is designed to demonstrate the models' capability to generate a valid rationale for the classification decision. The conformity values range between 1 if all DR signs regions are highlighted by the attention techniques and approach zero if the classifier either detects false regions or misses relevant DR relevant regions. Analyzing the three fine-tuned models results in their conformity and discloses some interesting characteristics of these models and the attention methods. First, Grad-CAM, as a class-discriminative localization technique, can generate visual explanations for all three CNN-based models without requiring architectural changes or re-training. However, visualizations lend insight into the failures of these models to capture the region of interest related to the DR diagnosis task. Second, as shown in Tables 2, 3 and 5, the VGG-16 model manifests the lowest generalization error and the highest conformity and explainability capabilities. This could be due to the small receptive field size used throughout the entire network and the lack of skip connections.

Third, as seen in Tables 2, 3 and 5, DenseNet-121 led to the highest generalization error and overfitting of the training. Interestingly, the conformity metric of this model is the lowest compared to the other models. This emphasizes the necessity for both the data balancing step and the regularization of these models. Additionally, it highlights the correlation between the models' performance and our proposed explainability metric.

5. Conclusions

In this paper, we evaluated three state-of-the-art models for DR binary and five-stage classification using a fundus images dataset. First, we created balanced training, validation, and test sets to ensure the validity of the evaluation results. Evaluating imbalanced sets can be misleading, especially with a skewed dataset and the domination of one class over another. Second, we optimized and fine-tuned the three models and evaluated their performance. The results show that the complexity and depth of these models make them prone to overfitting. Thus, their performance on the test degrades significantly. However, VGG-16 resulted in the least gap between training and test set accuracies and achieved the best generalization among the other models. Third, we proposed a new metric to compare the classification performance of the three models from the explainability aspect, conformity. The proposed metric utilizes the Grad-CAM technique to measure the proportion of model attention to DR-related signs. The superiority of VGG-16 was further demonstrated when evaluating the models using conformity metrics. VGG-16 achieved significantly higher conformity and showed much more justified decisions than the other models. In the future, we aim to evaluate other deep learning models' explainability using our proposed metric and to incorporate lesion detectors with a general classifier to achieve more interpretable classification decisions.

Funding: This project was funded by the Deanship of Scientific Research (DSR) at King Abdulaziz University, Jeddah, Saudi Arabia, under Grant No. J:4-612-1441.

Institutional Review Board Statement: Not applicable.

Informed Consent Statement: Not applicable.

Data Availability Statement: Publicly available datasets were used in this study. These data can be found at https://www.kaggle.com/ratthachat/aptos-eye-preprocessing-in-diabetic-retinopathy (accessed on 4 July 2022).

Acknowledgments: The author acknowledges with thanks the DSR for technical and financial support. The author also thanks Kaggle for making the DR dataset used in this work available.

Conflicts of Interest: The author declares no conflict of interest.

References

1. Arcadu, F.; Benmansour, F.; Maunz, A.; Willis, J.; Haskova, Z.; Prunotto, M. Deep learning algorithm predicts diabetic retinopathy progression in individual patients. *NPJ Digit. Med.* **2019**, *2*, 92. [CrossRef]
2. WHO. Diabetes. 2021. Available online: https://www.who.int/news-room/fact-sheets/detail/diabetes (accessed on 16 July 2021).
3. Retinopathy, D.; Understanding, D. Diabetic Retinopathy—Epidemiology Forecast to 2029. 2021, pp. 1–5. Available online: https://www.reportlinker.com/p05961707/Diabetic-Retinopathy-Epidemiology-Forecast-to.html?utm_source=GNW (accessed on 26 August 2022).
4. Abràmoff, M.D.; Reinhardt, J.M.; Russell, S.R.; Folk, J.C.; Mahajan, V.B.; Niemeijer, M.; Quellec, G. Automated Early Detection of Diabetic Retinopathy. *Ophthalmology* **2010**, *117*, 1147–1154. [CrossRef]
5. Chowdhury, A.R.; Chatterjee, T.; Banerjee, S. A Random Forest Classifier-Based Approach in the Detection of Abnormalities in the Retina. *Med. Biol. Eng. Comput.* **2019**, *57*, 193–203. [CrossRef] [PubMed]
6. Bourouis, S.; Zaguia, A.; Bouguila, N.; Alroobaea, R. Deriving Probabilistic SVM Kernels from Flexible Statistical Mixture Models and its Application to Retinal Images Classification. *IEEE Access* **2019**, *7*, 1107–1117. [CrossRef]
7. Emon, M.U.; Zannat, R.; Khatun, T.; Rahman, M.; Keya, M.S. Performance Analysis of Diabetic Retinopathy Prediction using Machine Learning Models. In Proceedings of the 2021 6th International Conference on Inventive Computation Technologies (ICICT), Coimbatore, India, 20–22 January 2021; pp. 1048–1052. [CrossRef]
8. Zhou, J.; Gandomi, A.H.; Chen, F.; Holzinger, A. Evaluating the quality of machine learning explanations: A survey on methods and metrics. *Electronics* **2021**, *10*, 593. [CrossRef]
9. Anoop, B.K. Binary Classification of DR-Diabetic Retinopathy using CNN with Fundus Colour Images. *Mater. Today Proc.* **2022**, *58*, 212–216. [CrossRef]
10. Pires, R.; Avila, S.; Wainer, J.; Valle, E.; Abramoff, M.D.; Rocha, A. A data-driven approach to referable diabetic retinopathy detection. *Artif. Intell. Med.* **2019**, *96*, 93–106. [CrossRef] [PubMed]
11. Dataset, K. Diabetic Retinopathy Detection. 2015. Available online: https://www.kaggle.com/c/diabetic-retinopathy-detection (accessed on 30 May 2022).

12. Decencière, E.; Zhang, X.; Cazuguel, G.; Lay, B.; Cochener, B.; Trone, C.; Gain, P.; Ordonez, R.; Massin, P.; Erginay, A.; et al. Feedback on a publicly distributed database: The Messidor database. *Image Anal. Stereol.* **2014**, *33*, 231–234. [CrossRef]
13. Wan, S.; Liang, Y.; Zhang, Y. Deep convolutional neural networks for diabetic retinopathy detection by image classification. *Comput. Electr. Eng.* **2018**, *72*, 274–282. [CrossRef]
14. Sarki, R.; Michalska, S.; Ahmed, K.; Wang, H.; Zhang, Y. Convolutional neural networks for mild diabetic retinopathy detection: An experimental study. *bioRxiv* **2019**, 763136. [CrossRef]
15. Hagos, M.T.; Kant, S. Transfer Learning based Detection of Diabetic Retinopathy from Small Dataset. 2019. Available online: http://arxiv.org/abs/1905.07203 (accessed on 26 August 2022).
16. Chatpatanasiri, R. APTOS: Eye Preprocessing in Diabetic Retinopathy. 2019. Available online: https://www.kaggle.com/ratthachat/aptos-eye-preprocessing-in-diabetic-retinopathy (accessed on 26 August 2022).
17. Russakovsky, O.; Deng, J.; Su, H.; Krause, J.; Satheesh, S.; Ma, S.; Huang, Z.; Karpathy, A.; Khosla, A.; Bernstein, M.; et al. "ImageNet Large Scale Visual Recognition Challenge. *Int. J. Comput. Vis.* **2015**, *115*, 211–252. [CrossRef]
18. Simonyan, K.; Zisserman, A. Very deep convolutional networks for large-scale image recognition. In Proceedings of the 3rd International Conference on Learning Representations, Conference Track Proceedings, San Diego, CA, USA, 7–9 May 2015; pp. 1–14.
19. He, K.; Zhang, X.; Ren, S.; Sun, J. Deep residual learning for image recognition. In Proceedings of the IEEE Conference on Computer Vision and Pattern Recognition, Las Vegas, NV, USA, 27–30 June 2016; pp. 770–778. [CrossRef]
20. Huang, G.; Liu, Z.; van der Maaten, L.; Weinberger, K.Q. Densely connected convolutional networks. In Proceedings of the 30th IEEE Conference on Computer Vision and Pattern Recognition, Honolulu, HI, USA, 21–26 July 2017; pp. 2261–2269. [CrossRef]
21. Zhou, B.; Khosla, A.; Lapedriza, A.; Oliva, A.; Torralba, A. Learning Deep Features for Discriminative Localization. In Proceedings of the IEEE Conference on Computer Vision and Pattern Recognition, Las Vegas, NV, USA, 27–30 June 2016; pp. 2921–2929. [CrossRef]
22. Zhou, B.; Khosla, A.; Lapedriza, A.; Oliva, A.; Torralba, A. Object detectors emerge in deep scene CNNs. In Proceedings of the 3rd International Conference on Learning Representations, Conference Track Proceedings, San Diego, CA, USA, 7–9 May 2015.
23. Selvaraju, R.R.; Cogswell, M.; Das, A.; Vedantam, R.; Parikh, D.; Batra, D. Grad-CAM: Visual Explanations from Deep Networks via Gradient-Based Localization. *Int. J. Comput. Vis.* **2020**, *128*, 336–359. [CrossRef]

*applied
sciences*

MDPI

Article

Deep 3D Convolutional Neural Network for Facial Micro-Expression Analysis from Video Images

Kranthi Kumar Talluri [†], Marc-André Fiedler [†] and Ayoub Al-Hamadi *

Neuro-Information Technology Group, Otto von Guericke University Magdeburg, 39106 Magdeburg, Germany
* Correspondence: ayoub.al-hamadi@ovgu.de
† These authors contributed equally to this work.

Abstract: Micro-expression is the involuntary emotion of the human that reflects the genuine feelings that cannot be hidden. Micro-expression is exhibited by facial expressions that last for a short duration and have very low intensity. Because of these reasons, micro-expression recognition is a challenging task. Recent research on the application of 3D convolutional neural networks (CNNs) has gained much popularity for video-based micro-expression analysis. For this purpose, both spatial as well as temporal features are of great importance to achieve high accuracies. The real possibly suppressed emotions of a person are valuable information for a variety of applications, such as in security, psychology, neuroscience, medicine and many other disciplines. This paper proposes a 3D CNN model architecture which is able to extract spatial and temporal features simultaneously. Thereby, the selection of the frame sequence plays a crucial role, since the emotions are only distinctive in a subset of the frames. Thus, we employ a novel pre-processing technique to select the Apex frame sequence from the entire video, where the timestamp of the most pronounced emotion is centered within this sequence. After an extensive evaluation including many experiments, the results show that the train–test split evaluation is biased toward a particular split and cannot be recommended in case of small and imbalanced datasets. Instead, a stratified K-fold evaluation technique is utilized to evaluate the model, which proves to be much more appropriate when using the three benchmark datasets CASME II, SMIC, and SAMM. Moreover, intra-dataset as well as cross-dataset evaluations were conducted in a total of eight different scenarios. For comparison purposes, two networks from the state of the art were reimplemented and compared with the presented architecture. In stratified K-fold evaluation, our proposed model outperforms both reimplemented state-of-the-art methods in seven out of eight evaluation scenarios.

Keywords: micro-expression analysis; 3D CNN; Apex frame sequence; stratified K-fold; intra-dataset and cross-dataset evaluation

Citation: Talluri K.K.; Fiedler, M.-A.; Al-Hamadi, A. Deep 3D Convolutional Neural Network for Facial Micro-Expression Analysis from Video Images. *Appl. Sci.* **2022**, *12*, 11078. https://doi.org/10.3390/app122111078

Academic Editor: Cosimo Nardi

Received: 19 August 2022
Accepted: 27 October 2022
Published: 1 November 2022

1. Introduction

Micro-expressions are the type of expression that occurs when a person tries to suppress their true feelings. Micro-expressions are spontaneous and usually last for less than 0.5 s [1], resulting in tiny facial muscles' movements. As per [2], micro-expressions are generated when someone attempts to hide their true intentions, as these expressions can neither be mimicked nor concealed. In 1966, micro-expressions were first discovered by Haggard and Isaacs [3], assuming that they are associated with defense mechanisms and conveyed feelings.

Later, in 1969, when Ekman and Friesen [4] conducted some experiments, they inadvertently encountered micro-expression. In the experiment, they examined the video of a person with depression who tried to lie about their suicide plan. During ordinary watching of the patient's video, they did not observe anything suspicious, but when they watched the same clip at reduced video speed, they noticed that after the doctor asked the patient about his future, there was an expression of pain on the patient's face. These expressions lasted for a short time and were only present in two frames of the video, so they named them micro-expressions [2].

Micro-expression recognition is applicable in many domains, e.g., doctors can recognize if a patient is suffering from pain [5], or it can be used with criminals during interrogations or at court to find out if they are lying. In addition, the real possibly suppressed emotions of a person are valuable information for many more applications in the fields of security, psychology, neuroscience, and medicine. Traditionally, micro-expression recognition was performed using a handcrafted descriptor that was manually adjusted to extract features from video clips or images. Local Binary Patterns (LBP) [6], optical flow features [7], and Local Binary Pattern histograms from Three Orthogonal Planes (LBP-TOP) [8] were some of these famous manual techniques used for feature extraction. However, the main drawback of the handcrafted methods was that they extract the manually created features and do not provide a generic data representation. Deep learning has recently gained much popularity, and convolutional neural networks (CNNs) are extensively used in solving computer vision problems. By using CNNs [9–11], the results in the field of micro-expression analysis have also been outperformed compared to traditional approaches, which is why this field continues to be of great interest for the research community.

Most of the recently proposed architectures for micro-expression recognition are based on CNN, long short-term memory (LSTM), or a combination of both network types. The disadvantage of standard CNNs is that they are only able to gather spatial features and are unable to capture motion over several frames. Some work [12] attempted to capture temporal information with the optical flow, but this often involves detecting unwanted background motion. Usually, the spatial information was extracted with CNNs, and subsequently, the features obtained were given into an LSTM to analyze the temporal information [13,14]. However, simultaneous extraction of the spatio-temporal features is not possible with this workflow. To overcome all the limitations of existing methods, a 3D CNN model is proposed in this work. It is a new custom architecture for directly recognizing the persons' micro-expressions from the video sequences. In addition, we employ a novel pre-processing technique by selecting the Apex frame sequence from the entire video. The advantage of this approach is that the Apex event with the highest emotion shown by the subject is located in the middle of this sequence. This yields better results in classification compared to using the initial sequence. Moreover, extensive experimental evaluation is performed with intra-dataset as well as cross-dataset experiments on the three benchmark datasets CASME II, SMIC, and SAMM. This comparison of cross-dataset results is unprecedented in the state of the art for micro-expression recognition, and we are the first to perform this kind of evaluation. It clearly shows that our new 3D CNN architecture outperforms other state-of-the-art models in terms of recognition performance.

The main contributions of this paper can be summarized as follows:

- We have developed a 3D CNN model architecture for micro-expression recognition which is able to extract spatial and temporal features simultaneously;
- A novel pre-processing technique is employed by selecting the Apex frame sequence from the entire video, where the timestamp of the most pronounced emotion is centered within this sequence;
- Stratified K-fold was applied for model evaluation because it is suitable for small datasets with imbalanced class distribution as in our case;
- Comprehensive experimental validation was performed by comparing the proposed model with two reimplemented state-of-the-art methods in intra-dataset as well as cross-dataset evaluations in a total of eight different scenarios. To the best of our knowledge, such an extensive evaluation in this or comparable manner has not been conducted for micro-expression recognition so far.

The paper is organized in the following way: The related works proposed so far are presented in Section 2. Section 3 is about the spontaneous micro-expression datasets. In Section 4, the pre-processing steps are examined, and in Section 5, our network architecture is outlined in detail. Section 8 reports the results and discussions. Finally, in Section 10, a conclusion is drawn and future work is outlined.

2. Related Works

In micro-expression recognition, feature extraction is the most crucial task [15]. The classification accuracy of micro-expressions is directly proportional to the feature extraction efficiency. In recent years, many methods have been proposed to effectively extract features from facial images [15]. These approaches are mainly classified into two types: those based on handcrafted methods and those based on deep learning.

2.1. Handcrafted Methods

In the past decade, micro-expression recognition has been based entirely on handcrafted approaches. In [16], a feature extraction based on 3D histogram-oriented gradients (3DHOG) was proposed to detect motion in the smaller regions of the face. Polikovsky et al. [17] presented a 3DHOG descriptor capable of capturing temporal characteristics and sudden changes on the facial surface. A classifier such as k-means and the voting method were used to classify the micro-expressions. Pfister et al. [18] utilized the temporal interpolation method to normalize the length of the video sequence to deal with the problem of short video samples. In addition, temporal and spatial features were extracted using the LBP-TOP descriptor and used for the classification with support vector machine (SVM) or random forest. In [19], new feature extraction techniques were proposed using a fusion approach. This is based on a histogram of the motion boundaries, where the vertical and horizontal components of the differential optical flow are fused. The extracted features are classified by SVM. Shreve et al. [20] proposed a technique to spot the temporal information from long videos. During expressions in the facial regions, strains were produced. The optical flow uses these strains in the mouth, chin, forehead, and cheek areas of the face to identify temporal features. Moreover, the extended video sequence consists of both the micro- and macro-expression. In [21], a new approach was presented that automatically recognizes facial expressions. Instead of an expression, this approach concentrated more on clues related to identity. Hence, video frames were used to generate registration points or landmarks on the face; then, the local regions and LBP were extracted for feature representation, and finally, the expressions were classified by SVM.

2.2. Deep Learning-Based Methods

Recent developments in GPUs extended the potential of deep learning models in various domains. In the last few years, deep learning has become popular in solving computer vision problems, leading to more advanced algorithms for micro-expression recognition. Liong et al. [22] designed a network that uses the on-set and Apex frame of each sample to compute optical flow features. Three such optical flow information, namely the vertical, horizontal and optical strain, were used, which is why this network is called shallow triple-stream 3D CNN. Thus, a compact and discriminative feature representation is learned according to the authors. In [23], a 3D convolutional neural network capable of extracting spatial and temporal features was proposed, taking advantage of three-dimensional kernels. The authors used a frame sequence to classify the micro-expression, but in the entire paper, the authors have not discussed the use of Apex frame or the selection of an Apex frame sequence. In this paper, the author considered only initial starting frames of the samples. Furthermore, two different networks were designed, one using the whole image as input and the other obtaining only cropped parts of the face. For this purpose, experiments were performed on the intermediate and late fusion of eyes and mouth regions based on 3D CNNs.

The 3D CNN-based network by Zhang et al. [24] proposed a multi-scale fusion network that fuses local regions (such as eyes or mouth) along with the global region (full face) in order to select the appropriate region of interest to focus on for micro-expression recognition. These features were later passed to an LSTM to better process the temporal features. Thus, they presented a multi-channel fusion model. In [25], the VGG network was used for micro-expression detection. Since VGG is a deep network model, batch normalization and dropout techniques were applied to avoid the problem of overfitting. Furthermore, the insufficient dataset problem was tackled by choosing a random starting frame, fixing the sequence length, and repeating the process for entire samples to amplify the data. Xu et al. [26] proposed a method using optical flow to extract motion information between only two frames (on-set and Apex) in the entire sequence of samples. These features were inputted to a pre-trained MobileNetV2 and the micro-expressions classified by SVM. They conclude that micro-expressions involve only local areas of the face, and there are some irrelevant muscle movements. In [27], the temporal sample deformation method was introduced to preserve the temporal information, since normalizing the length of the video sequence is an essential aspect in the case of micro-expressions. The new sequence is randomly sampled by a normal distribution. The authors developed a three-branched architecture combining 2D and 3D CNNs. They were able to show the advantages of single 3D kernel sizes and multiple 3D kernel combination. The Eulerian video magnification (EVM) technique adopted by Wang et al. [28] for magnifying the motion in micro-expressions was performed to extract the spatio-temporal features for the CNNs. Therefore, Eulerian motion feature maps were extracted by employing a spatial scale temporal filtering approach. Next, these feature maps are fed into a 3D CNN for final recognition. In contrast to the previous approaches, where the optical flow information was obtained from on-set and Apex frames, Chen et al. [29] utilized a novel method by dividing the optical flow image into small blocks and then processed them by CNN. A weighted loss function of implicit semantic data augmentation is applied for augmentation of the training data in the deep features space. In [30], the authors used only two frames (Apex frame and on-set frame) from the entire frame sequence of micro-expression samples. Later, feature extraction was performed using Bi-Weighted Oriented Optical Flow (Bi-WOOF) based on only those two frames instead of a complete sequence. Thereby, discriminately weighted motion features are intended to be captured and are weighted by their own magnitudes. The facial regions are weighted by the magnitude of optical strain. Subsequently, SVM was used to classify the micro-expressions. Li et al. [31] presented a method that consists of two sub-networks. First, a hierarchical convolutional recurrent neural network is utilized for the extraction of spatio-temporal features. Then, a principal-component-analysis-based recurrent neural network is applied, and the features are merged through the fusion of sub-networks.

3. Datasets

There are two types of micro-expressions posed and spontaneous ones. In earlier studies, participants were often asked to pose their facial expressions, so the datasets were generated on this basis. However, the disadvantage of collecting datasets in this way was that the expressions elicited by the participants were not genuine, which lead them to differ from natural micro-expressions. In this paper, three spontaneous datasets are used: Chinese Academy of Sciences Micro-Expression II (CASME II), Spontaneous Actions and Micro-Movements (SAMM), and Spontaneous Micro-Expression Corpus (SMIC).

3.1. CASME II

The dataset [32] consists of 255 samples collected from 26 subjects with the same ethnicity. Action Units (AUs) on the basis of the Facial Action Coding System (FACS) system were employed for labeling the samples. The videos are recorded at a resolution of 640×480 pixels. The samples in the CASME II dataset are distributed among seven classes namely happiness, disgust, fear, sadness, repression, surprise, and others. However, only the three classes happiness, disgust, and surprise are taken for the experiment (see Table 1), as the remaining classes only contained very few samples and the class others was not annotated consistently across the three datasets. The rest of the samples was therefore discarded.

Table 1. Experimental datasets with their class distribution.

Emotions	CASME II	SMIC	SAMM
Happy	32	51	26
Disgust	63	70	15
Surprise	28	43	9
Total	123	164	50

3.2. SMIC

In the SMIC [33] dataset, 16 participants were recorded. Participants were chosen from three different ethnicities to make this dataset more diversified. This dataset consists of 164 samples, which were further classified into three classes, namely the positive class that contains happy emotions, combining the three sad, disgust, and fear emotions to form the negative class and the surprise class.

3.3. SAMM

The SAMM [34] dataset consists of micro-expression samples obtained from 29 participants. This was the first spontaneous high-resolution dataset in which participants belonged to different geographical regions from a total of 13 ethnic groups. AUs were used to label these samples based on FACS. SAMM contains seven emotions that were captured at 200 FPS with a high resolution of 2040×1088. Again, only three classes are further processed because of the small amount of samples in the remaining classes and to ensure a fair comparison between the databases.

4. Pre-Processing

In micro-expression recognition, pre-processing is one of the most critical stages, consisting of necessary steps before extracting the essential and useful features. Face detection and face alignment are used to bring all the frames to a common reference. Next, the landmarks are identified to discard unwanted background information or noise that negatively affects the model's accuracy. The final step is to capture the Apex frame sequence where the emotion is high. Most of the pre-processing is performed by using algorithms from the Dlib library.

4.1. Face Detection and Alignment

Detecting the face and aligning the faces to a common reference were major steps which ensure that the extracted features belong to same location corresponding to each face. Both these steps were performed by calling a function in Dlib named `dlib.get_frontal_face_detector()`. This is a face detector that receives as input the image and the up-scaling factor for the two arguments. By increasing the up-scaling value, even smaller faces can be detected in the image, but this also increases the computation time [35]. The output of the face detection function returns x, y, w, and h values, which are the coordinates of the diagonal corners forming a bounding box around the detected face.

4.2. Facial Landmark Detection

After detecting the face, the next task was to obtain the facial landmarks. Based on these landmarks, the face can be cropped properly ignoring the background as shown in Figure 1. In Dlib, there is a pre-trained model called `shape_predictor_68_face_landmarks.dat` that is used to generate the landmarks on the detected face. With the help of this pre-trained model, a function called `dlib.shape_predictor()`. acts as a landmark detector. The output of this model returns 68 points on the face [35]. After determining the landmarks, all micro-expression frames were cropped according to the requirements of each model.

Figure 1. The workflow of pre-processing steps.

4.3. Apex Frame Spotting

In our work, Apex frame spotting is not applied, since the information is already provided in the datasets. The purpose of this section is to give an overview of existing approaches for this task, which in particular has high relevance for real-world applications. In real applications, the recorded data do not contain information about the Apex, which indicates the point of time in the sequence with the highest micro-expression present. However, this information can help to effectively reveal the genuine emotions behind a particular video sample [36]. Few works have been proposed to detect the Apex frame from a video sequence. In [37], the authors used an LBP descriptor to extract the features from each frame and computed the difference by subtracting the features of the on-set frame with the remaining frames. Then, they divided the sequence into two halves and calculated the sum of each subsequence. The subsequence with the largest difference is selected. This binary search process continues as described until an Apex frame is determined from the entire sequence. In [38], optical flow and the LBP algorithm were used to determine the Apex frame. Based on the distance of facial movements and their directions, the optical flow was employed to examine the facial changes for identifying the Apex frame. In [39], the authors used Region HOOF to spot the Apex frame with the help of five ROIs.

4.4. Selection of Apex Frame Sequence

In pre-processing, the last task was to choose the best sequence of frames with high emotion or where the facial muscle movement was at its peak. The sequence with high emotion needs to be considered for micro-expression recognition. Details about the frames with high emotional content are obtained from the data annotations by analyzing their AU levels. The CASME II and SMIC datasets also provide information about the on-set frame (where micro-expression starts), the Apex frame (where micro-expression peaks), and the off-set frame (where micro-expression ends). By taking advantage of these details, the selection of frame sequence is performed. The Apex frame sequence consists only of frames surrounding the Apex event with the highest micro-expression intensity. For instance, a fixed frame length of 36 frames is chosen for a video sample from CASME II. Then, 18 frames before and 18 frames after the Apex frame (including it) are considered in the Apex frame sequence. For SMIC, the sequence length is 26 frames.

5. Network Architectures

The application of 3D CNN exploits the spatio-temporal correlation. Our proposed model is named Model-A. The main aim of this work is to improve the recognition performance of micro-expressions by using a 3D CNN. Two state-of-the-art models are reimplemented to compare the performance with our model. These two models are named Model-B [23] and Model-C [40], respectively. Both also employ a 3D CNN in their model architecture. In this section, the networks of Model-A, Model-B, and Model-C are presented.

5.1. Model-A (Proposed 3D CNN Model)

The goal of the proposed 3D CNN model architecture named Model-A is to recognize micro-expressions in video clips. The architecture is demonstrated in Figure 2. The dimensions height, width, and depth were passed as input into the model. Height and width are kept constant with a size of 128×128 pixels, and the depth dimension varies based on the frame sequence length of each dataset (36 frames for CASME II, 26 frames for SMIC, and 30 frames for the Combined dataset). The entire model was created using the Keras framework. Initially, a sequential model was constructed to which the layers were added.

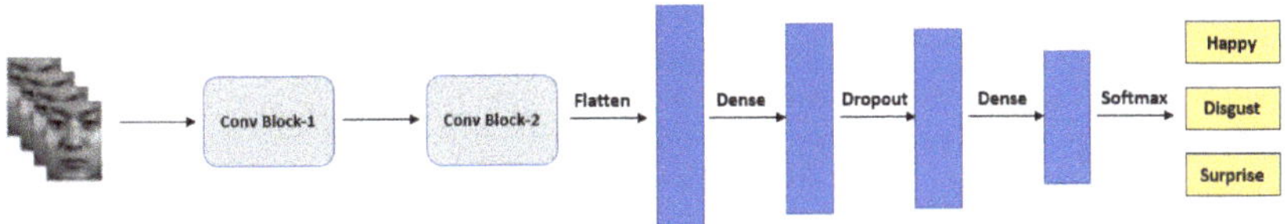

Figure 2. Architecture of our proposed 3D CNN Model-A.

The main block of Model-A consists of two Conv blocks followed by flatten and dense layers. Each Conv block contains a convolution, batch normalization, max pooling, and dropout layer, as shown in Figure 3. For the convolutional layer, a 3D CNN was chosen instead of a 2D CNN because 3D CNNs are capable of capturing temporal information along with spatial information. Each convolutional layer operates with 16 filters and a kernel size of $3 \times 3 \times 3$. The model uses a HeNormal initializer for better initial weights. The padding type was kept the same so that edge information was retained. The output from each convolution was passed to the batch normalization in order to converge the model faster and reduce the problem of overfitting. Then, 3D max-pooling with a $3 \times 3 \times 3$ kernel was applied to decrease the spatio-temporal size and preserve critical features. Furthermore, to avoid overfitting, dropout layers were added to the model with a dropout rate of 0.4.

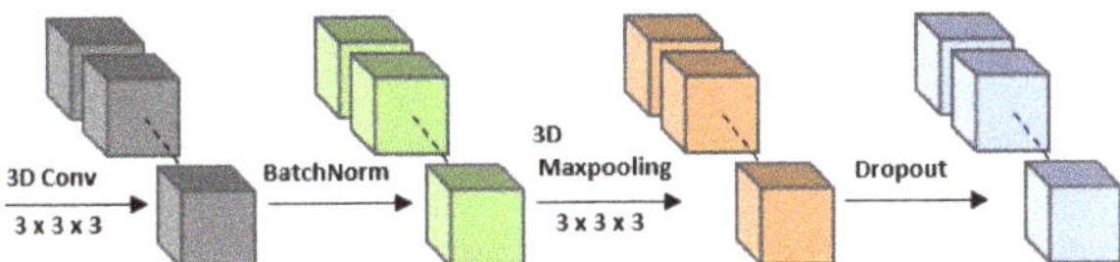

Figure 3. Layers of each Conv block.

As the dimensions for the dense layer need to be one-dimensional, a flatten layer was applied prior to dense. The first dense layer consists of 128 neurons, which is subsequently connected to another dropout layer. Then, a second dense layer with three neurons performs the classification into the three micro-expression classes. Finally, softmax is applied to the output. Rectified linear unit (ReLU) activation functions were used across all feasible layers. A summary of the sequential model is shown in Table 2, indicating the output feature map size of each layer.

Table 2. Summary of our Model-A with CASME II dataset samples as input.

Layer Type	Filter Size	Output Shape
Conv3D-1	$3 \times 3 \times 3$	$36 \times 128 \times 128 \times 16$
BatchNorm-1	-	$36 \times 128 \times 128 \times 16$
3D-MaxPooling-1	$3 \times 3 \times 3$	$12 \times 42 \times 42 \times 16$
Dropout-1	-	$12 \times 42 \times 42 \times 16$
Conv3D-2	$3 \times 3 \times 3$	$12 \times 42 \times 42 \times 16$
BatchNorm-2	-	$12 \times 42 \times 42 \times 16$
3D-MaxPooling-2	$3 \times 3 \times 3$	$4 \times 14 \times 14 \times 16$
Dropout-2	-	$4 \times 14 \times 14 \times 16$
Flatten	-	12,544
Dense-1	-	128
Dropout-3	-	128
Dense-2	-	3

5.2. Split-Model

The architecture of Model-A employed for the train–test split evaluation differs slightly from the proposed one in Section 5.1. The only difference is that the architecture consists of three Conv blocks instead of two for Model-A. The rest of the layers and model parameters remain unchanged. For better distinction, this variant for the intra-dataset train–test split is named Split Model.

5.3. Model-B

This reimplemented model was originally called MicroExpSTCNN [23] in the corresponding paper. To avoid confusion, our reimplementation is called Model-B. It is a 3D CNN with input dimensions for width and height of 64×64, whereby the depth dimension changes depending on the database selection. The model starts with a convolutional layer with 32 filters and a kernel size of $3 \times 3 \times 15$. Then, max-pooling is applied with a size of $3 \times 3 \times 3$ to retain important features. To prevent model overfitting, a dropout layer with 0.5 dropout rate is employed afterwards. Then, the obtained output was flattened and inputted to a dense layer with 128 neurons, which was followed by another dropout layer. Finally, a dense layer with three neurons and softmax activation are used to classify the samples into the classes anger, disgust, or happy. More details can be found in the original paper [23].

5.4. Model-C

In the original work, the architecture was referred to as the 3D CNN network model [40]. This is the second reimplementation, which will be called Model-C in the rest of this paper. This model is again a 3D CNN, but some model parameters were not listed by the authors, e.g., the input image shape, the number of filters and the kernel size for the convolution, the max-pooling size, the number of neurons in the dense layer, and the dropout rate. To the best of our knowledge, missing values were assumed and empirically selected. An input image size of 128×128 was chosen, and the depth depends as with Model-B on the database. A convolution with 32 filters and a $3 \times 3 \times 3$ kernel was picked; additionally, batch normalization was added for better convergence. For the two max-pooling blocks, a pooling size of $3 \times 3 \times 3$ was taken. Moreover, a dropout rate of 0.5 was implemented, and for the dense layer, 128 neurons were chosen. See [40] for more information.

6. Model Training Parameters

In this section, the parameter specifications for each model are specified.

6.1. Model-A

Our proposed model was implemented with the Keras framework and TensorFlow backend. The system specification involved in all experimental setups includes a Intel® Core™ i7-4770S CPU @ 3.10GHz, with 16 GB RAM and a NVIDIA GeForce RTX 2080 Ti GPU with 12 GB. As hyperparameters during model training, a learning rate of 0.001 was chosen with a batch size of 4. Categorical cross-entropy was used as the loss function, and stochastic gradient descent (SGD) was used as the optimizer with a momentum of 0.9. Accuracy was the metric utilized for measuring the performance of the model. In total, the model was trained for 100 epochs.

6.2. Model-B

In the paper [23], the model was implemented through the Keras framework via TensorFlow. The model training was performed along with an NVIDIA Tesla K80 GPU server with 24 GB dedicated GDDR5 graphics processor. For the experiment, SGD was the optimizer used along with categorical cross-entropy as the loss function. The model was trained with a batch size of eight for 100 epochs.

6.3. Model-C

The experiment conducted in the paper [40] also employed the Keras framework with TensorFlow backend. The experimental setup was equipped with an Intel® Core™ i7-9700 CPU 3.70 GHz with 32 GB RAM and a NVIDIA GeForce GTX 3090 GPU. The model parameters included a learning rate of 0.01, an SGD optimizer with a momentum 0.9, and categorical cross-entropy as the loss function. The Nesterov accelerated gradient is employed for SGD, and the learning rate decay is 1×10^{-6}. The model was trained with a batch size of four for 800 epochs.

Note that the learning rate of 0.01 was changed to 0.001 and the number of training epochs was changed from 800 to 100 for experimental purposes. Many model parameters were not mentioned in the paper and therefore had to be assumed. As a result, some given hyperparameters had to be adjusted, as they were no longer suitable for model training due to the assumptions made. The initial assumptions on the parameters were made based on the proposed Model-A, since Model-C has a similar architecture. Subsequently, these model parameters were fine-tuned and empirically optimized to improve the results of Model-C.

7. Experimental Analysis

In this section, an experiment is conducted to observe the difference in the performance of the proposed model by selecting the initial frames or the frames around the Apex frame. For experimental purposes, 36 frames were used in CASME II and 26 frames in SMIC. The number of frames for SMIC is less because the average video length is shorter at 34 frames compared to 68 frames in CASME II. From the results comparison in Table 3, it is clearly visible that the performance of the proposed model achieves 9.6% higher accuracy when using the Apex frame sequence instead of the original frame sequence for the CASME II dataset. A similar observation is made for the SMIC dataset, where Model-A performs 9.4% better in accuracy using the Apex frame sequence compared to the initial frame sequence. More details about the process of selecting the Apex frame sequence can be found in Section 4.4.

Table 3. Performance of Model-A using initial frames and the Apex frame sequence.

	Type of Frame Sequence	
Dataset	**Initial Frame Sequence**	**Apex Frame Sequence**
CASME II	46.9%	56.5%
SMIC	34.3%	43.7%

From the results of this experiment, the overall superiority of our choice of the Apex frame sequence is clearly demonstrated. Therefore, in all experiments for the eight scenarios in the next section, the Apex frame sequence is used instead of the initial frames.

8. Results And Discussions

The results for the two evaluation techniques and eight scenarios are given and discussed in this section. All the evaluation results are stated in the terms of accuracy. The accuracy metric is used for model evaluation, as it is applied in two state-of-the art models [23,40] as well. For better comparison, this paper applies the accuracy as the main evaluation measure for all three models in the eight scenarios.

8.1. Train–Test Split

In the experiments, the Split Model was used for train–test split evaluation. It was trained on the CASME II dataset and achieved an accuracy of 85.2% compared to 80.3% by Model-C.

The results obtained seemed good at first sight, but a major problem was encountered, as they are not reproducible for randomly shuffled splits and differ drastically in their performance. In an experiment of five random splits, Run-1 achieved the highest accuracy of 85.2%, whereas Run-3 achieved only 45%. The results of the other three splits are in between them. The validation curves for all runs are shown in Figure 4.

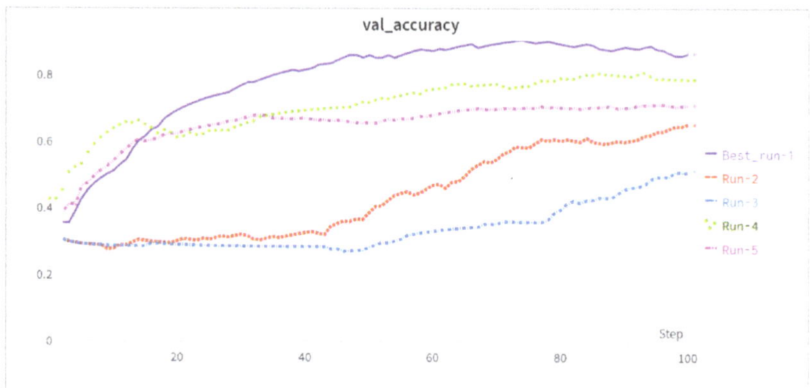

Figure 4. Evaluation accuracies for our Split Model with train–test split for multiple runs.

This clearly illustrates that a train–test split is not suitable for the small and imbalanced datasets in micro-expression recognition. In other papers in the state of the art, a split was then accordingly selected allowing the generation of the best possible results. However, from our point of view, this does not correspond to a fair comparison and is not a good scientific practice. To address this common practice in the field of micro-expression recognition and to provide a fair comparison, we compared our model with two state-of-the-art models in eight evaluation scenarios for both intra-dataset and cross-dataset. The splits were generated using stratified K-fold validation. More details can be found in the next subsection. To the best of our knowledge, we are the first to perform such a comparison in this manner in micro-expression recognition.

8.2. Stratified K-Fold

The approach of stratified K-fold validation ensures that in every split, there will be always the same ratio of class samples. Therefore, this approach was deployed for all experiments. Overall, stratified K-fold validation was employed for eight different evaluation scenarios. The number of K was always chosen to be five. Three of these scenarios are intra-dataset experiments and five are cross-dataset experiments. For this purpose, varying combinations of the three datasets CASME II, SMIC, and SAMM (see Section 3) were applied. The combined dataset is a merging of CASME II and SMIC. As the number of samples in SAMM with 50 was insufficient for training, it is only used for testing purposes in the cross-dataset comparisons.

The results of the three methods Model-A, Model-B, and Model-C across all eight scenarios are summarized in Table 4. It can be observed that our proposed Model-A outperforms the state-of-the-art architectures Model-B and Model-C in seven out of eight scenarios, with the exception of Scenario-8, where Model-C performs best. This clearly shows the superiority of our architecture compared to the methods presented so far. Detailed results for each scenario with further explanations and discussions can be found in the following subsections.

8.2.1. Scenario-1

Scenario-1 is an intra-dataset experiment where both training and validation were performed on the SMIC dataset. Our Model-A reached an accuracy of 43.7% compared to 33.5% and 37.3% of Model-B and Model-C, respectively. Thus, the performance difference is 10.2% and 6.4%. The advantage of Model-A is that it is deeper and consists of more convolutional layers. Therefore, it is able to better identify spatio-temporal features related to micro-expressions. Another benefit is provided by our pre-processing technique, which involves selecting an Apex sequence and interpolating the first and last frames of each video, resulting in fewer samples being discarded because of too short length.

A overall low accuracy can be noticed for all three models. The main reason for this is that the videos were recorded at 100 FPS, which is low compared to the other benchmark datasets. Thus, there is a lag in the models' ability to capture the temporal information, which is an essential factor in micro-expression recognition. Consequently, poor temporal feature quality leads to low accuracies. Additionally, the average sample length in SMIC is only 34 frames, which is shorter than for the other databases. It is likely that this only allows the extraction of features with lower quality from the samples.

Table 4. Accuracy comparison of the three models across all eight evaluation scenarios.

Scenario	Train	Test	Model-A	Model-B	Model-C
01	SMIC	SMIC	**43.7%**	33.5%	37.3%
02	CASME II	CASME II	**56.5%**	45.4%	48.1%
03	Combined	Combined	**88.2%**	85.4%	80.4%
04	SMIC	SAMM	**44.3%**	31.1%	42.0%
05	CASME II	SAMM	**24.8%**	24.3%	23.1%
06	SMIC	CASME II	**44.7%**	43.7%	39.1%
07	CASME II	SMIC	**37.7%**	35.4%	36.5%
08	Combined	SAMM	27.1%	23.1%	**36.9%**

8.2.2. Scenario-2

In Scenario-2, CASME II was applied for training and testing. Again, Model-A performs best with a margin of 8.4% to the second best Model-C. An accuracy of 56.5% was obtained for Model-A, which is 12.8% higher than for the SMIC dataset in Scenario-1. Model-B and Model-C were also able to increase by 11.9% and 10.8%, respectively. This can be mainly attributed to the higher frame rate in the video recordings, which is with 200 FPS twice as high as with SMIC. Thus, the networks are capable of detecting even more minor motion information. As mentioned above, a higher frame rate can help produce higher accuracies in micro-expression recognition. Another advantage of the CASME II dataset is that the on-set frame, Apex frame, and off-set frame information is provided by the authors. As the Apex frame was known in the entire video, the pre-processing was executed easily in a correct manner.

Even though the accuracy achieved with CASME II is higher than with SMIC, it is still not very high for micro-expression recognition. There are some reasons for this: The major problem in CASME II was the data imbalance. So, for the experiments, only three classes are appropriate: namely, happiness, disgust, and surprise. These emotions were mainly considered in order to be consistent with SMIC and SAMM. Even though the dataset has 255 samples, only 123 samples were utilized in the experiments. Thus, the low overall performance in CASME II can be attributed to insufficient and imbalanced samples.

8.2.3. Scenario-3

In Scenario-3, SMIC and CASME II are combined by joining their samples. Discussing the performance of the individual models, all three models achieved more than 80% accuracy. The performance of Model-A with an accuracy of 88.2% is the best over all eight scenarios. Model-B and Model-C rank behind with 85.4% and 80.4%, respectively.

The higher performance for this scenario can be explained by two reasons: First, the combined dataset contains an amount of 230 more samples for training. Second, the model benefits from the mixture of higher and lower frame rates, making it more robust.

8.2.4. Scenario-4

In this cross-dataset scenario, the SMIC dataset was used for model training, whereas SAMM was used for evaluation. Model-A achieved an accuracy of 44.3%. This can be regarded as relatively high when it is compared to the intra-dataset evaluation results of SMIC in Scenario-1, where the performance of Model-A is lowered by 0.6%. In cross-dataset experiments, the ethnicities of the subjects play a crucial role, as they will lead to better generalization of the model. It is all the more astonishing that the model trained on SMIC, which includes only three ethnicities (Africa, Asia, and Caucasian) in the training samples, performs that well on SAMM.

Compared to Model-C, Model-A performs better by 2.3%. A relatively large drop can be seen for Model-B with a difference of over 13% to Model-A.

8.2.5. Scenario-5

In Scenario-5, the models were trained on CASME II and evaluated on SAMM. Compared to Scenario-4, where SAMM was also employed for evaluation, there is a significant decrease in the accuracies of between 7% and 20% for the three models. Reasons for this may be that CASME II contains only Asian subjects and therefore generalizes less accurately for other people. In addition, CASME II has a high data imbalance.

All three models perform almost identically in Scenario-5 with a slight edge: Model-A is 0.5% higher than Model-B and 1.7% higher than Model-C.

8.2.6. Scenario-6

The SMIC dataset was used for training in Scenario-6 and CASME II was used for evaluating the model. Model-A again outperforms the two others with 44.7% accuracy, which is 1.0% and 5.6% higher in comparison with Model-B and Model-C, respectively.

In contrast to Scenario-2, where intra-dataset evaluation was performed for CASME II, the model trained on SMIC has to cope with a drop of 11.8%.

It can be observed that all models trained on SMIC perform similarly well on the different test sets. The accuracy of Model-A on the intra-dataset in Scenario-1 is 43.7%, while in Scenario-4 on SAMM, it is 44.3%. Such balanced results are found only with the model trained on SMIC. Due to the comparatively high number of samples of 164, the different ethnicities of the subjects, and a relatively balanced class distribution, the networks trained on SMIC show the best generalization and do not vary strongly across the datasets, which would be an indicator for a bias in the training data.

8.2.7. Scenario-7

In this scenario, the training was performed on CASME II and evaluation was performed on SMIC. The performances of all three models are comparable with only a difference of 2.3% between the worst and the best. Model-A ranks again first with an accuracy of 37.7%.

The drop of Model-A in Scenario-7 is kept within limits at 6% compared to the model trained on SMIC (Scenario-1). Nevertheless, if we examine all three models trained on CASME II, it is noticeable that the results differ heavily depending on the test set. Here, again, the lack of diversity and the class imbalance problem of CASME II become obvious. The intra-dataset experiment (Scenario-2) performs about 20 % better than testing on SMIC (Scenario-7), which is in turn around 13 % better than on SAMM (Scenario-5). Overall, it appears that CASME II contains high levels of bias, and therefore, the models trained on it have limited ability to cope with other data.

8.2.8. Scenario-8

Scenario-8 along with Scenario-5 contain the worst results across all eight experiments. Here, again, testing is completed on SAMM, but here, the training is carried out on the combined dataset of SMIC and CASME II. This scenario forms an exception, because it is the only scenario not dominated by Model-A. Model-C clearly performs better here with 36.9% compared with Model-A's 27.1%.

Comparing the results of Model-A, CASME II seems to have a greater impact on training the model. Through the combination, the accuracy compared to training solely on CASME II can increase from 24.8% to 27.1%. However, this still leaves a large gap with the SMIC trained model in Scenario-4, where the accuracy is 17.2% higher. This confirms the conclusions drawn in Scenario-7 that CASME II is only limitedly qualified for the implementation of models with a high degree regarding generalization ability.

9. Applications and Use Cases

Micro-expressions play an important role in applications of lie detection, person authentication, and many more. For instance, in an interrogation, facial micro-expressions of the criminal can assist the police in convicting him. Another use case is that border guards can use them to identify unusual behavior of persons during border control [41]. Moreover, knowing a patient's true feelings is considered to be very helpful for psychotherapists while treating their patients. Furthermore, micro-expression can also be used in the field of marketing to understand people's reactions and comments to the company's advertisements, goods and services [41].

10. Conclusions

In this work, we proposed a novel 3D CNN architecture capable of extracting features from both spatial as well as temporal dimensions simultaneously. Especially in micro-expression recognition, these combined spatio-temporal features are crucial for the subsequent performance of the network. In order to improve the input sequence selection, a pre-processing technique was introduced to select the Apex frame sequence from the video, which is the part with the most visible emotions. This Apex frame sequence shows better results than picking the first or a random starting frame from the video. Another benefit of our pre-processing is that for samples with short videos, the first and last frame will be interpolated to meet the fixed-lenght input criteria so that they do not need to be discarded.

In addition, an extensive experimental evaluation was performed in our work, which is unprecedented in the state of the art for micro-expression recognition. To conduct this comparison, two state-of-the-art models were reimplemented, and intra-dataset as well as cross-dataset experiments were executed on the three datasets: CASME II, SMIC, and SAMM across a total of eight different evaluation scenarios. In addition, a stratified K-fold evaluation was proposed and adopted, since in a classical train–test split comparison, the performance varies strongly depending on the split chosen and the samples included for training and testing, making the results not meaningful. Our proposed Model-A was able to outperform the other two state-of-the-art architectures in seven out of eight evaluation scenarios, clearly demonstrating the superiority of our network. The highest accuracy of 88.2% is shown in the intra-dataset validation in Scenario-3, where the CASME II and SMIC datasets were merged.

Overall, in particular, the cross-dataset results of consistently below 50% accuracy exhibit great potential for improvement. Specifically, the CASME II dataset reveals that it is not suitable for achieving a network with a high degree of generalization ability. The lack of diversity in the subjects' origin countries and the class imbalance between the samples lead to a severe performance drop on other datasets. In comparison, SMIC is well suited for achieving consistent results on a variety of data, but results also need to be further improved. Especially, the different ethnicities (Africa, Asia, and Caucasian) of the subjects and the class distribution are good, but the frame rate of 100 FPS and the total number of 164 samples are not high enough. Even though the results seem comparatively poor and irrelevant at first glance, they show an unvarnished, fair and reproducible comparison which allows existing challenges and future work to be identified for the field of micro-expression recognition. In this regard, the main task for the future is the acquisition of new large databases with a high level of annotation quality. Thereby, SMIC can provide good preliminary knowledge on how to set up and accomplish this dataset recording. In addition, generative adversarial networks might be able to contribute to the collection of these additionally required amounts of data. In recent years, these could be successfully exploited in many areas of deep learning to enhance the available data.

Author Contributions: Conceptualization, K.K.T. and M.-A.F.; methodology, K.K.T. and M.-A.F.; software, K.K.T. and M.-A.F.; validation, K.K.T. and M.-A.F.; formal analysis, K.K.T. and M.-A.F.; investigation, K.K.T. and M.-A.F.; resources, K.K.T. and M.-A.F.; data curation, K.K.T. and M.-A.F.; writing—original draft preparation, K.K.T. and M.-A.F.; writing—review and editing, K.K.T., M.-A.F. and A.A.-H.; visualization, K.K.T. and M.-A.F.; supervision, A.A.-H.; project administration, A.A.-H.; funding acquisition, A.A.-H. All authors have read and agreed to the published version of the manuscript.

Funding: This work was supported by the German Research Foundation (DFG) under grants AL 638/13-1 and AL 638/14-1.

Institutional Review Board Statement: Ethical review and approval were waived for this study due to the use of public databases, which were conducted according to the guidelines of the Declaration of Helsinki and approved by the relevant review boards. We complied with the terms of use of the databases regarding the publication of data.

Informed Consent Statement: According to the documentation of the used public databases, informed consent was obtained from all subjects involved.

Data Availability Statement: The SAMM, SMIC, and CASME II datasets are publicly not available. To obtain access to these datasets, please write an email to the corresponding authors, and the access procedures to these datasets are mentioned in the respective links: the SAMM dataset can be obtained at http://www2.docm.mmu.ac.uk/STAFF/m.yap/dataset.php (accessed on 23 December 2021); the SMIC datasets can be obtained at https://www.oulu.fi/en/university/faculties-and-units/faculty-information-technology-and-electrical-engineering/center-machine-vision-and-signal-analysis (accessed on 22 December 2021); the CASME II dataset can be obtained at http://fu.psych.ac.cn/CASME/casme2-en.php (accessed on 23 February 2022).

Conflicts of Interest: The authors declare no conflict of interest.

Abbreviations

The following abbreviations are used in this manuscript:

3D CNN	3D Convolutional Neural Networks
3DHOG	3D Histogram-Oriented Gradients
ANN	Artificial Neural Networks
AU	Action Unit
CASME	Chinese Academy of Sciences Micro-Expression
CASME II	Chinese Academy of Sciences Micro-Expression II
CNN	Convolutional Neural Networks
EVM	Eulerian Video Magnification
FACS	Facial Action Coding System
FPS	Frames per second
HOG	Histogram of Gradients
LBP	Local Binary Pattern
LBP-TOP	Local Binary Pattern histograms from Three Orthogonal Planes
LSTM	Long Short-Term Memory
MMOD	max-margin object-detection algorithm
ReLU	Rectified Linear Unit
ROI	Region of Interest
SAM	Self-Assessment Manikins
SAMM	Spontaneous Actions and Micro-Movements
SGD	Stochastic Gradient Descent
SMIC	Spontaneous Micro-Expression Corpus
SVM	Support Vector Machine
TIM	Temporal interpolation model

References

1. Yan, W.J.; Wu, Q.; Liang, J.; Chen, Y.H.; Fu, X. How fast are the leaked facial expressions: The duration of micro-expressions. *J. Nonverbal Behav.* **2013**, *37*, 217–230. [CrossRef]
2. Ekman, P. Darwin, deception, and facial expression. *Ann. N. Y. Acad. Sci.* **2003**, *1000*, 205–221. [CrossRef]
3. Haggard, E.A.; Isaacs, K.S. Micromomentary facial expressions as indicators of ego mechanisms in psychotherapy. In *Methods of Research in Psychotherapy*; The Century Psychology Series; Springer: Boston, MA, USA, 1966; pp. 154–165. [CrossRef]
4. Ekman, P.; Friesen, W.V. Nonverbal leakage and clues to deception. *Psychiatry* **1969**, *32*, 88–106. [CrossRef] [PubMed]
5. Endres, J.; Laidlaw, A. Micro-expression recognition training in medical students: A pilot study. *BMC Med. Educ.* **2009**, *9*, 1–6. [CrossRef] [PubMed]
6. Huang, X.; Wang, S.J.; Zhao, G.; Piteikainen, M. Facial micro-expression recognition using spatiotemporal local binary pattern with integral projection. In Proceedings of the 2015 IEEE International Conference on Computer Vision Workshop (ICCVW), Santiago, Chile, 7–13 December 2015; pp. 1–9.
7. Liu, Y.J.; Zhang, J.K.; Yan, W.J.; Wang, S.J.; Zhao, G.; Fu, X. A main directional mean optical flow feature for spontaneous micro-expression recognition. *IEEE Trans. Affect. Comput.* **2015**, *7*, 299–310. [CrossRef]
8. Wang, Y.; See, J.; Phan, R.C.W.; Oh, Y.H. Lbp with six intersection points: Reducing redundant information in lbp-top for micro-expression recognition. In *ACCV 2014: Computer Vision—ACCV 2014*; Lecture Notes in Computer Science Book Series; Springer: Cham, Switzerland, 2015; pp. 525–537.
9. Guo, J.; Zhou, S.; Wu, J.; Wan, J.; Zhu, X.; Lei, Z.; Li, S.Z. Multi-modality network with visual and geometrical information for micro emotion recognition. In Proceedings of the 2017 12th IEEE International Conference on Automatic Face & Gesture Recognition (FG 2017), Washington, DC, USA, 30 May–3 June 2017; pp. 814–819.
10. Liong, S.T.; Gan, Y.S.; Yau, W.C.; Huang, Y.C.; Ken, T.L. Off-apexnet on micro-expression recognition system. *arXiv* **2018**, arXiv:1805.08699.
11. Patel, D.; Hong, X.; Zhao, G. Selective deep features for micro-expression recognition. In Proceedings of the 2016 23rd International Conference on Pattern Recognition (ICPR), Cancun, Mexico, 4–8 December 2016; pp. 2258–2263.
12. Li, J.; Wang, Y.; See, J.; Liu, W. Micro-expression recognition based on 3D flow convolutional neural network. *Pattern Anal. Appl.* **2019**, *22*, 1331–1339. [CrossRef]
13. Khor, H.Q.; See, J.; Phan, R.C.W.; Lin, W. Enriched long-term recurrent convolutional network for facial micro-expression recognition. In Proceedings of the 2018 13th IEEE International Conference on Automatic Face & Gesture Recognition (FG 2018), Xi'an, China, 15–19 May 2018; pp. 667–674.
14. Kim, D.H.; Baddar, W.J.; Jang, J.; Ro, Y.M. Multi-objective based spatio-temporal feature representation learning robust to expression intensity variations for facial expression recognition. *IEEE Trans. Affect. Comput.* **2017**, *10*, 223–236. [CrossRef]
15. Wang, L.; Jia, J.; Mao, N. Micro-Expression Recognition Based on 2D-3D CNN. In Proceedings of the 2020 39th Chinese Control Conference (CCC), Shenyang, China, 27–29 July 2020; pp. 3152–3157.
16. Polikovsky, S.; Kameda, Y.; Ohta, Y. Facial micro-expressions recognition using high speed camera and 3D-gradient descriptor. In Proceedings of the 3rd International Conference on Imaging for Crime Detection and Prevention (ICDP 2009), London, UK, 3 December 2009.
17. Polikovsky, S.; Kameda, Y.; Ohta, Y. Facial micro-expression detection in hi-speed video based on facial action coding system (FACS). *IEICE Trans. Inf. Syst.* **2013**, *96*, 81–92. [CrossRef]
18. Pfister, T.; Li, X.; Zhao, G.; Pietikäinen, M. Recognising spontaneous facial micro-expressions. In Proceedings of the 2011 International Conference on Computer Vision, Barcelona, Spain, 6–13 November 2011; pp. 1449–1456.
19. Lu, H.; Kpalma, K.; Ronsin, J. Motion descriptors for micro-expression recognition. *Signal Process. Image Commun.* **2018**, *67*, 108–117. [CrossRef]
20. Shreve, M.; Godavarthy, S.; Goldgof, D.; Sarkar, S. Macro-and micro-expression spotting in long videos using spatio-temporal strain. In Proceedings of the 2011 IEEE International Conference on Automatic Face & Gesture Recognition (FG), Santa Barbara, CA, USA, 21–25 March 2011; pp. 51–56.
21. Wang, Y.; Yu, H.; Stevens, B.; Liu, H. Dynamic facial expression recognition using local patch and LBP-TOP. In Proceedings of the 2015 8th International Conference on Human System Interaction (HSI), Warsaw, Poland, 25–27 June 2015; pp. 362–367.
22. Liong, S.T.; Gan, Y.S.; See, J.; Khor, H.Q.; Huang, Y.C. Shallow Triple Stream Three-dimensional CNN (STSTNet) for Micro-expression Recognition. In Proceedings of the 2019 14th IEEE International Conference on Automatic Face & Gesture Recognition (FG 2019), Lille, France, 14–18 May 2019; pp. 1–5.
23. Reddy, S.P.T.; Karri, S.T.; Dubey, S.R.; Mukherjee, S. Spontaneous facial micro-expression recognition using 3D spatiotemporal convolutional neural networks. In Proceedings of the 2019 International Joint Conference on Neural Networks (IJCNN), Budapest, Hungary, 14–19 July 2019; pp. 1–8.
24. Zhang, H.; Liu, B.; Tao, J.; Lv, Z. Facial Micro-Expression Recognition Based on Multi-Scale Temporal and Spatial Features. In Proceedings of the ICMI '21: International Conference on Multimodal Interaction, Montreal, QC, Canada, 18–22 October 2021; pp. 80–84.
25. Jin, W.; Meng, X.; Wei, D.; Lei, W.; Xinran, W. Micro-expression recognition algorithm based on the combination of spatial and temporal domains. *High Technol. Lett.* **2021**, *27*, 303–309.

26. Xu, W.; Zheng, H.; Yang, Z.; Yang, Y. Micro-Expression Recognition Base on Optical Flow Features and Improved MobileNetV2. *KSII Trans. Internet Inf. Syst. (TIIS)* **2021**, *15*, 1981–1995.
27. Wu, C.; Guo, F. TSNN: Three-Stream Combining 2D and 3D Convolutional Neural Network for Micro-Expression Recognition. *IEEJ Trans. Electr. Electron. Eng.* **2021**, *16*, 98–107. [CrossRef]
28. Wang, Y.; Ma, H.; Xing, X.; Pan, Z. Eulerian Motion Based 3DCNN Architecture for Facial Micro-Expression Recognition. In *International Conference on Multimedia Modeling*; Lecture Notes in Computer Science; Springer: Cham, Switzerland, 2020; pp. 266–277.
29. Chen, B.; Liu, K.H.; Xu, Y.; Wu, Q.Q.; Yao, J.F. Block Division Convolutional Network with Implicit Deep Features Augmentation for Micro-Expression Recognition. *IEEE Trans. Multimed.* 2022, *early access*. [CrossRef]
30. Liong, S.T.; See, J.; Wong, K.; Phan, R.C.W. Less is more: Micro-expression recognition from video using apex frame. *Signal Process. Image Commun.* **2018**, *62*, 82–92. [CrossRef]
31. Li, J.; Wang, T.; Wang, S.J. Facial micro-expression recognition based on deep local-holistic network. *Appl. Sci.* **2022**, *12*, 4643. [CrossRef]
32. Yan, W.J.; Li, X.; Wang, S.J.; Zhao, G.; Liu, Y.J.; Chen, Y.H.; Fu, X. CASME II: An improved spontaneous micro-expression database and the baseline evaluation. *PLoS ONE* **2014**, *9*, e86041. [CrossRef]
33. Li, X.; Pfister, T.; Huang, X.; Zhao, G.; Pietikäinen, M. A spontaneous micro-expression database: Inducement, collection and baseline. In Proceedings of the 2013 10th IEEE International Conference and Workshops on Automatic Face and Gesture Recognition (FG), Shanghai, China, 22–26 April 2013; pp. 1–6.
34. Davison, A.K.; Lansley, C.; Costen, N.; Tan, K.; Yap, M.H. Samm: A spontaneous micro-facial movement dataset. *IEEE Trans. Affect. Comput.* **2016**, *9*, 116–129. [CrossRef]
35. King, D.E. Dlib-ml: A Machine Learning Toolkit. *J. Mach. Learn. Res.* **2009**, *10*, 1755–1758.
36. Oh, Y.H.; See, J.; Le Ngo, A.C.; Phan, R.C.W.; Baskaran, V.M. A survey of automatic facial micro-expression analysis: Databases, methods, and challenges. *Front. Psychol.* **2018**, *9*, 1128. [CrossRef]
37. Liong, S.T.; See, J.; Wong, K.; Le Ngo, A.C.; Oh, Y.H.; Phan, R. Automatic apex frame spotting in micro-expression database. In Proceedings of the 2015 3rd IAPR Asian Conference on Pattern Recognition (ACPR), Kuala Lumpur, Malaysia, 3–6 November 2015; pp. 665–669.
38. Yan, W.J.; Chen, Y.H. Measuring dynamic micro-expressions via feature extraction methods. *J. Comput. Sci.* **2018**, *25*, 318–326. [CrossRef]
39. Ma, H.; An, G.; Wu, S.; Yang, F. A region histogram of oriented optical flow (RHOOF) feature for apex frame spotting in micro-expression. In Proceedings of the 2017 International Symposium on Intelligent Signal Processing and Communication Systems (ISPACS), Xiamen, China, 6–9 November 2017; pp. 281–286.
40. Jiao, Y.; Jing, M.; Hu, Y.; Sun, K. Research on a Micro-Expression Recognition Algorithm based on 3D-CNN. In Proceedings of the 2021 3rd International Conference on Intelligent Control, Measurement and Signal Processing and Intelligent Oil Field (ICMSP), Xi'an, China, 23–25 July 2021; pp. 221–225.
41. Takalkar, M.; Xu, M.; Wu, Q.; Chaczko, Z. A survey: Facial micro-expression recognition. *Multimed. Tools Appl.* **2018**, *77*, 19301–19325. [CrossRef]

Article

Lightweight Dual Mutual-Feedback Network for Artificial Intelligence in Medical Image Super-Resolution

Beibei Wang [1], Binyu Yan [1,*], Gwanggil Jeon [2,*], Xiaomin Yang [1], Changjun Liu [1] and Zhuoyue Zhang [1]

[1] College of Electronics and Information Engineering, Sichuan University, Chengdu 610065, China
[2] Department of Embedded Systems Engineering, Incheon National University, Incheon 22012, Republic of Korea
* Correspondence: yanby@scu.edu.cn (B.Y.); gjeon@inu.ac.kr (G.J.)

Abstract: As a result of hardware resource constraints, it is difficult to obtain medical images with a sufficient resolution to diagnose small lesions. Recently, super-resolution (SR) was introduced into the field of medicine to enhance and restore medical image details so as to help doctors make more accurate diagnoses of lesions. High-frequency information enhances the accuracy of the image reconstruction, which is demonstrated by deep SR networks. However, deep networks are not applicable to resource-constrained medical devices because they have too many parameters, which requires a lot of memory and higher processor computing power. For this reason, a lightweight SR network that demonstrates good performance is needed to improve the resolution of medical images. A feedback mechanism enables the previous layers to perceive high-frequency information of the latter layers, but no new parameters are introduced, which is rarely used in lightweight networks. Therefore, in this work, a lightweight dual mutual-feedback network (DMFN) is proposed for medical image super-resolution, which contains two back-projection units that operate in a dual mutual-feedback manner. The features generated by the up-projection unit are fed back into the down-projection unit and, simultaneously, the features generated by the down-projection unit are fed back into the up-projection unit. Moreover, a contrast-enhanced residual block (CRB) is proposed as each cell block used in projection units, which enhances the pixel contrast in the channel and spatial dimensions. Finally, we designed a unity feedback to down-sample the SR result as the inverse process of SR. Furthermore, we compared it with the input LR to narrow the solution space of the SR function. The final ablation studies and comparison results show that our DMFN performs well without utilizing a large amount of computing resources. Thus, it can be used in resource-constrained medical devices to obtain medical images with better resolutions.

Keywords: attention module; dual mutual feedback; lightweight; medical image super-resolution; unity feedback

Citation: Wang, B.; Yan, B.; Jeon, G.; Yang, X.; Liu, C.; Zhang, Z. Lightweight Dual Mutual-Feedback Network for Artificial Intelligence in Medical Image Super-Resolution. *Appl. Sci.* **2022**, *12*, 12794. https://doi.org/10.3390/app122412794

Academic Editor: Cosimo Nardi

Received: 28 October 2022
Accepted: 9 December 2022
Published: 13 December 2022

Publisher's Note: MDPI stays neutral with regard to jurisdictional claims in published maps and institutional affiliations.

1. Introduction

The aim of SR is to learn a mapping function from input low-resolution (LR) images to output high-resolution (HR) images. High-resolution medical images are very important for doctors in terms of making accurate diagnoses of lesions; thus, SR for medical images has recently received a great deal of attention. However, image super-resolution remains a challenge, as LR images lose a certain amount of information as compared to HR images [1]. Many researchers have tried to find a solution to this critical issue [2–5].

On the basis of deep learning, Dong et al. proposed the SR convolutional neural network (SRCNN) [2], which utilizes the convolutional neural network (CNN) architecture and is vastly superior to other traditional methods. Thereafter, Dong et al. proposed fast SR convolutional neural networks (FSRCNNs) [5], which up-sample feature maps using deconvolution in the last layer of the network and provide more accurate estimates with less computation. The deconvolutional layer generates HR features by enlarging feature maps. Then, the subpixel convolutional layer was proposed by Shi et al. [6], which expands the number of feature channels in order to store more pixels and rearrange them to generate HR features. The Laplacian pyramid super-resolution network (LapSRN) [7] up-samples LR feature maps progressively, which enables it to reconstruct multi-scale SR images in one training session.

To further improve SR performance, deep networks were introduced into SR. The very deep SR convolutional neural network (VDSR) [8] proposed by Kim et al. is the first deep multiple-scale model. It bypasses interpolated LR images to the end by residual learning. Then, on the basis of VDSR [8], the authors proposed a deeply recursive convolutional network (DRCN) [9], which trains the network using a recursive-supervised strategy and achieves a similar performance to VDSR [8] with fewer parameters. Deep dense SR (DDSR) [10] was proposed for the SR of medical images, which uses densely connected hidden layers to obtain informative high-level features.

However, it remains a challenge for deep neural networks to go deeper because of the various difficulties associated with training, such as gradient vanishing/exploding problems. Rresidual learning was proposed to solve these problems. The deep residual network (ResNet) [11] is a representative model, which achieves a remarkable performance based on residual learning. Tai et al. [12] used residual learning and recursive learning to realize a very deep network without an enormous amount of parameters. The SR network using dense skip connections (SRDenseNet) [13] is another representative model based on residual learning. It bypasses all previous features to latter layers in blocks and densely concatenates all blocks. The enhanced deep SR network (EDSR) [14] proposed by Lim et al. removes the use of batch normalization (BN), which is harmful to the final performance in SR tasks. EDSR also employs a pretraining strategy and residual scaling techniques to improve the final performance. On the basis of residual learning, for the SR of three-dimensional (3D) brain MRI images, Pham et al. [15] proposed a deep 3D CNN.

Above classical SR methods are all feedforward SR methods; low-frequency information is directly passed to the following layer or bypassed to the latter layers through skip connections. The feedback mechanism enables the previous layers to perceive high-frequency information from the latter layers, but no new parameters are introduced. It is widely used in the domain of computer vision [16–19]. Recently, Haris et al. [20] proposed error feedback for image SR, which was used in two back-projection units. Thereafter, the SR feedback network (SRFBN) [21] was proposed, which contains a feedback block that functions in a self-feedback manner. For the SR of medical images, the feedback adaptive weighted dense network (FAWDN) [22] was proposed based on an adaptive weighted dense block and feedback connection.

Although the feedback mechanism is used in some SR methods, it is rarely used in lightweight SR methods. The feedback mechanism enables the previous layers to perceive high-frequency information from latter layers, but no new parameters are introduced. Therefore, it is very applicable for lightweight networks. Moreover, most medical devices are resource-constrained, so lightweight feedback SR networks with good performance are desired. In order to meet the demand, a lightweight dual mutual-feedback network (DMFN) is proposed for artificial intelligence in medical image super-resolution. The DMFN feeds the HR features generated by the up-projection unit back into the down-projection unit, and feeds LR features generated by the down-projection unit back into the up-projection unit, which forms a dual mutual-feedback architecture, as shown in Figure 1. Our method that was trained using natural images is named DMFN, and our method that was trained using medical images is named DMFN+. They were tested on MRI13 from [22] and compared with other state-of-the-art SR methods, as shown in Figure 2. Our method performs very well with little computational cost.

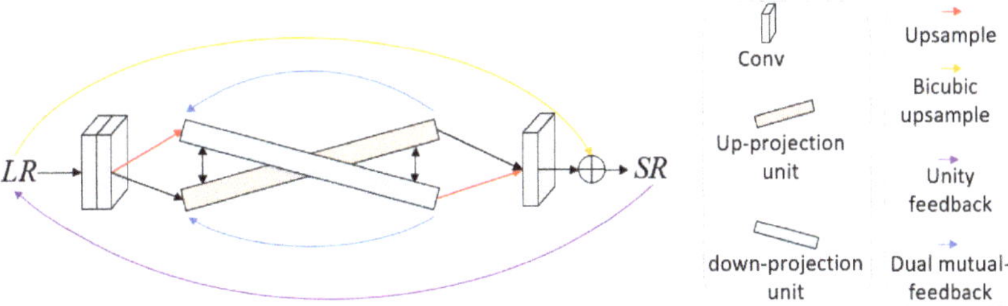

Figure 1. The structure of DMFN.

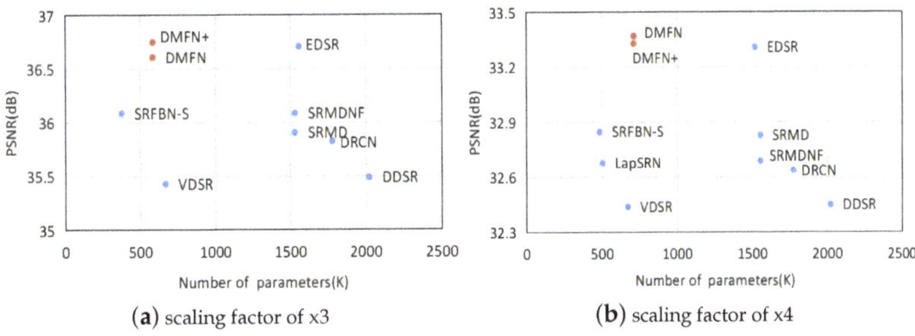

Figure 2. PSNR vs. parameters on MRI13 in [22].

Our contributions are summarized as follows:

- To better perceive the high-level information from each other, we designed a dual mutual-feedback structure. The HR features generated by the up-projection unit are fed back into the down-projection unit, and the LR features generated by the down-projection unit are fed back into the up-projection unit.
- To boost the expressive ability of the network, we propose a contrast-enhanced residual block (CRB) for use as each cell block in the projection units. CRB uses the contrast-enhanced channel and spatial attention within residual learning. The contrast-enhanced channel attention module learns the pixel contrast of each feature map to restore the textures, structures, and edges of images. The contrast-enhanced spatial attention module learns the pixel contrast in the same spatial location along the channel dimension to infer finer spatial-wise information.
- To narrow the search domain of the SR function, we designed a unity feedback. We down-sampled the SR result to LR image as the inverse process of SR. We then compared it with the input LR to calculate the unity feedback loss. The proposed unity feedback is helpful in terms of learning a better SR function with very few introduced parameters, which can be applied as a module to other SR networks.

2. Related Work

In this study, we designed a feedback network, which is inspired by SRFBN [21]. Moreover, inspired by [20], we used two back-projection units working in a dual mutual-feedback manner. Furthermore, we propose an attention-based module CRB for use as each cell block in the two back-projection units.

2.1. Attention Mechanism

The attention mechanism helps the networks perceive more informative features. Previously, the attention mechanism was used for image classification tasks [23] in RNN. Recently, inspired by the non-local means method, [24] learned the relationship between pixels with weighted sum t using long-range dependencies acquisition. Then, Hu et al. [25] learned the dependencies between channels with very little computational cost. The residual channel attention block (RCAB) proposed in [26] first used channel attention within the residual block. The convolutional block attention module (CBAM) [27] enhanced the discriminate learning ability of the network with the help of channel and spatial attention. Hui et al. [28] proposed the contrast-enhanced channel attention (CCA) and argued that channel attention with standard deviation can better learn the interdependencies between feature channels.

Inspired by [28], we designed contrast-enhanced spatial attention, which learns the contrast of pixels in the spatial dimension to infer finer spatial-wise information in feature maps. Then, we used contrast-enhanced channel attention and spatial attention successively within the residual block, which is named the contrast-enhanced residual block (CRB).

2.2. Back-Projection

Irani et al. [29] used back-projection for image enhancements, which confirmed that iterative updates and down-sampling can minimize reconstruction error. Dai et al. [30] proposed bilateral back-projection for SR networks with a single LR input. Then Dong et al. [31] used iterative back-projection and incorporated non-local information to improve reconstruction performance. Timofte et al. [32] enhanced the reconstruction capabilities of learning-based SISR with the refinement of back-projection. Hairs et al. [20] learned the errors after up- and down-sampling to refine the intermediate features, which was used to realize up-projection and down-projection. The up- and down-projection units were then learned iteratively to further improve reconstruction performance.

Inspired by [20], we argue that mutual learning between two back-projection units will improve their performance, as it enhances the information exchange between the two. Further experimental results indicate that mutual learning between two back-projection units performs better than the existing independent learning methods.

2.3. Feedback Mechanism

In feedforward SR methods, the low-frequency information is directly passed to the following layer or is bypassed to the latter layers through skip connections. The feedback mechanism enables the previous layers to perceive the high-level information of latter layers, which is widely used in the domain of computer vision [16–19]. Recently, Hairs et al. [20] used error feedback in back-projection units to correct intermediate features. Then, Han et al. [33] designed a dual-state structure with delayed feedback to exchange signals between states. SRFBN [21] is a feedback network with a feedback block, which iteratively feeds the output features back to itself as the input.

Inspired by the above feedback methods, we used dual mutual feedback on two back-projection units, which feeds the HR features generated by the up-projection unit back into the down-projection unit, and feeds the LR features generated by the down-projection unit back into the up-projection unit. Our dual mutual feedback performs better than dual self-feedback and single feedback manners.

3. Method or Methodology

In this section, we present the overall architecture of DMFN, including the dual mutual-feedback component, the contrast-enhanced residual block (CRB) that is used as each cell block in the dual mutual-feedback component, and the loss function.

3.1. Architecture of DMFN

Similar to SRFBN [21], our DMFN can be unfolded into several iterations because of the feedback manner, and the iteration t is set from 1 to T. The back-projection units feed back their output results to each other iteratively in a dual mutual-feedback manner. As shown in Figure 3, two convolutional layers are firstly used to obtain shallow features, which are then up-sampled. Then, the shallow features and the up-sampled shallow features are learned by the dual mutual-feedback component. In the dual mutual-feedback component, the HR features generated by the up-projection unit are fed back into the down-projection unit, and the LR features generated by the down-projection unit are fed back into the up-projection unit in the next iteration, which forms a dual mutual-feedback structure. Then, the outputs of dual mutual feedback from all iterations are concatenated for image reconstruction by fusing them with the bicubic interpolated results. Finally, we down-sampled the SR results to LR images as the inverse process of SR in the unity feedback component. Then, we compared it with the input LR to calculate the unity feedback loss.

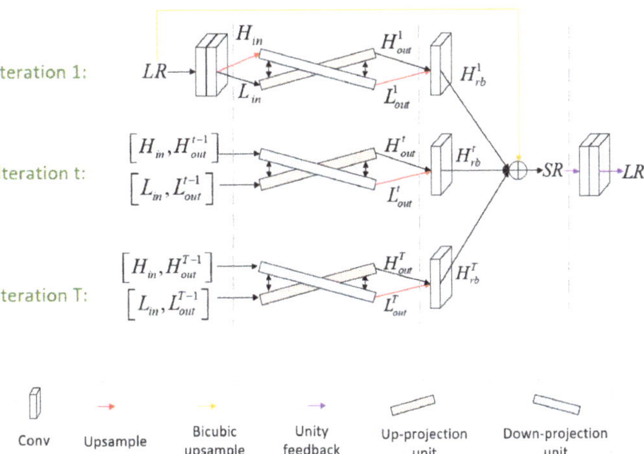

Figure 3. The unfolded DMFN.

We define L_{in} and H_{in} as the shallow features learned by the first component, which can be obtained by

$$\left\{ \begin{array}{l} L_{in} = f_c(LR) \\ H_{in} = f_{up}(L_{in}) \end{array} \right\}, \tag{1}$$

where f_c contains two convolutional layers to obtain shallow LR features. f_{up} is a deconvolutional upsampling operation.

In the dual mutual-feedback component of the t-th iteration, we use L^t_{out} to represent the LR features generated by the down-projection unit, and H^t_{out} to represent the HR features generated by the up-projection unit. The functions are as follows:

$$L^t_{out} = \left\{ \begin{array}{ll} f_{d-p}(H_{in}) & t = 1 \\ f_{d-p}([H_{in}, H^{t-1}_{out}]) & t \geq 2 \end{array} \right\}, \tag{2}$$

$$H^t_{out} = \left\{ \begin{array}{ll} f_{u-p}(L_{in}) & t = 1 \\ f_{u-p}([L_{in}, L^{t-1}_{out}]) & t \geq 2 \end{array} \right\}, \tag{3}$$

where f_{d-p} are the operations of the down-projection unit, which contains some features from the up-projection unit because of themutual learning between the two back-projection units. f_{u-p} is the operations of the up-projection unit, which also contains some features from the down-projection unit. [] is the concat function.

For reconstruction, we up-scale the LR features generated by the down-projection unit, which are then fused with the HR features generated by the up-projection unit. We define the final HR feature results of the t-th iteration as follows:

$$H^t_{rb} = H^t_{out} + f_{up}(L^t_{out}). \tag{4}$$

Since the final HR features of all iterations are fused and then added to the bicubic interpolated result of LR input, the final SR result is as follows:

$$SR = f_{cm}([H_{rb}^1, H_{rb}^2, ..., H_{rb}^T]) + f_{BC}(LR),\tag{5}$$

where f_{cm} is a conv-3 compression layer, and f_{BC} represents the bicubic up-sample function.

Finally, we down sample the SR result to the LR image named LR' by the down-sampling function f_{down}, which contains two convolutional layers for down sampling and channel transformation. The unity feedback loss calculated by LR' and LR is used to narrow the search domain of the SR function.

$$LR' = f_{down}(SR),\tag{6}$$

3.2. Dual Mutual-Feedback Component

The dual mutual-feedback component of the t-th iteration is shown in Figure 4. Pink represents the up-projection unit, and blue represents the down-projection unit. Then, we unfold the two back-projection units. The upward arrows represent the up-sampling operation, and the downward arrows represent the down-sampling operation. The pink arrows connect to an up-projection unit, and the blue arrows connect to a down-projection unit. Then, we use mutual learning (black arrows) between the two back-projection units to exchange information. Finally, the outputs of the two units are fed back into each other in the next iteration to realize dual mutual feedback.

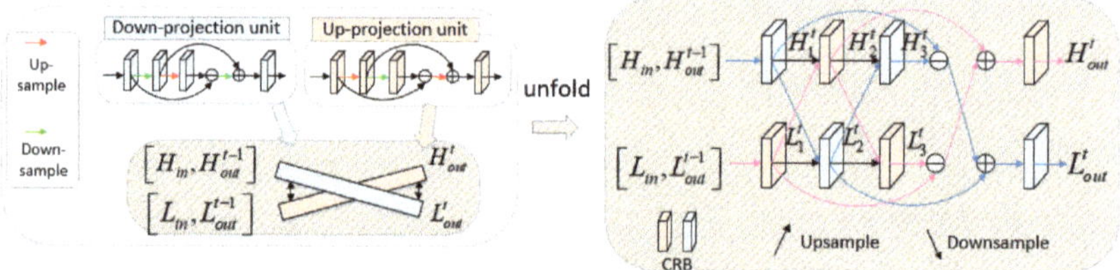

Figure 4. Dual mutual-feedback component of the t-th iteration in DMFN.

In the dual mutual-feedback component of the t-th iteration, we define the HR features as H_1^t, H_2^t and H_3^t, and the LR features as L_1^t, L_2^t and L_3^t. We use f_{CRB} to represent the operations of CRB. The dual mutual-feedback procedure is as follows:

$$L_1^t = \begin{cases} f_{CRB}(L_{in}) & t = 1 \\ f_{CRB}([L_{in}, L_{out}^{t-1}]) & t \geq 2 \end{cases},\tag{7}$$

$$H_1^t = \begin{cases} f_{CRB}(H_{in}) & t = 1 \\ f_{CRB}([H_{in}, H_{out}^{t-1}]) & t \geq 2 \end{cases},\tag{8}$$

$$\begin{cases} L_2^t = f_{CRB}([L_1^t, f_{down}(H_1^t)]) \\ H_2^t = f_{CRB}([H_1^t, f_{up}(L_1^t)]) \end{cases},\tag{9}$$

$$\begin{cases} L_3^t = f_{CRB}([L_2^t, f_{down}(H_2^t)]) \\ H_3^t = f_{CRB}([H_2^t, f_{up}(L_2^t)]) \end{cases},\tag{10}$$

$$\begin{cases} L_{out}^t = f_{CRB}(L_2^t + f_{down}(H_3^t - H_1^t)) \\ H_{out}^t = f_{CRB}(H_2^t + f_{up}(L_3^t - L_1^t)) \end{cases}.\tag{11}$$

3.3. Contrast-Enhanced Residual Block (CRB)

To further boost the expressive ability of our network, we propose a contrast-enhanced residual block (CRB), which is used as each cell block of the dual mutual-feedback component, as shown in Figure 5. CRB uses contrast-enhanced channel attention and spatial attention within the residual block. Contrast-enhanced channel attention assigns different weights to channels, and contrast-enhanced spatial attention assigns different weights to spatial locations. Therefore, the feature learning ability of residual blocks is enhanced. As shown in Figure 5, the input features F_{in} are learned by a multi-layer perceptron f_{mlp}

(Conv-ReLU-Conv) and then are learned by contrast-enhanced channel attention. The input of contrast-enhanced channel attention X is shown below:

$$X = f_{mlp}(F_{in}).\tag{12}$$

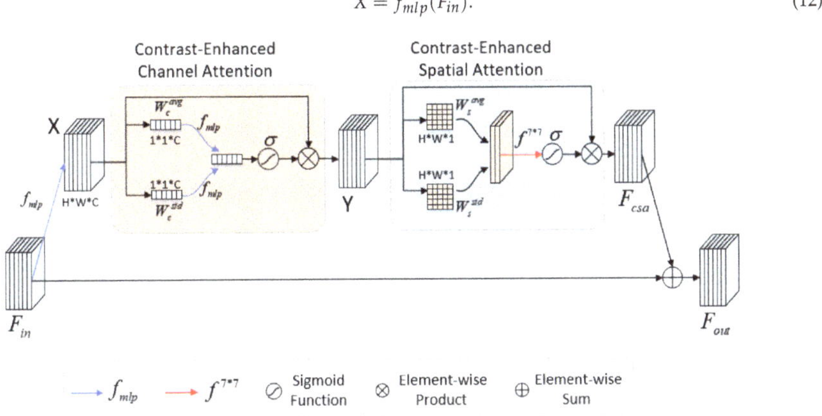

Figure 5. Contrast-enhanced residual block (CRB).

3.3.1. Contrast-Enhanced Channel Attention

As is the case for CCA [28], both standard deviation and average pooling are used to describe the context of each channel. Standard deviation enables the network to perceive more channels with a greater pixel contrast, as it represents image details related to structures, textures, and edges. Average pooling enables the network to perceive more informative channels. The size of feature maps X is $H \times W \times C$ and we use $c \in (1, \ldots\ldots C)$ to represent the channel number. We use $i \in (1, \ldots\ldots H)$ and $j \in (1, \ldots\ldots W)$ to represent the pixel location in each feature map. The weights of each channel calculated by average pooling and standard deviation are shown below:

$$w_c^{avg} = \frac{1}{HW} \sum_{(i,j) \in x_c} x_c^{i,j},\tag{13}$$

$$w_c^{std} = \sqrt{\frac{1}{HW} \sum_{(i,j) \in x_c} (x_c^{i,j} - \frac{1}{HW} \sum_{(i,j) \in x_c} x_c^{i,j})^2},\tag{14}$$

Then, we use $W_c^{avg} \in R^{1 \times 1 \times C}$ and $W_c^{std} \in R^{1 \times 1 \times C}$ to represent the average-pooled and standard deviation results of X on all channels. They are learned by f_{mlp}, and normalized with the application of the sigmoid function. Finally, the input feature maps X are rescaled by the element-wise product. The features learned by contrast-enhanced channel attention are shown below:

$$Y = X * \sigma(f_{mlp}(W_c^{avg}) + f_{mlp}(W_c^{std})).\tag{15}$$

3.3.2. Contrast-Enhanced Spatial Attention

We argue that the standard deviation value in the spatial dimension indicates the pixel contrast in the same spatial location along the channel dimension. The pixels with a higher standard deviation value must have a higher information value in some channels, which should be given more attention. Average pooling enables the network to perceive more informative spatial locations along the channel dimension. Therefore, both standard deviation and average pooling are used to describe the pixel weights in the spatial dimension, which enhances the image details. The size of feature maps Y is $H \times W \times C$ and we use $c \in (1, \ldots\ldots C)$ to represent the channel number. We use $i \in (1, \ldots\ldots H)$ and

$j \in (1, \ldots\ldots W)$ to represent the pixel location in each feature map. The weights of each spatial location calculated by average pooling and standard deviation are shown below:

$$w_{i,j}^{avg} = \frac{1}{C} \sum_{c=1}^{C} y_c^{i,j}. \tag{16}$$

$$w_{i,j}^{std} = \sqrt{\frac{1}{C} \sum_{c=1}^{C} (y_c^{i,j} - \frac{1}{C} \sum_{c=1}^{C} y_c^{i,j})^2}. \tag{17}$$

Then, we use $W_s^{avg} \in R^{H \times W \times 1}$ and $W_s^{std} \in R^{H \times W \times 1}$ to represent the average-pooled and standard deviation results of Y across the channel. They are learned using $f^{7 \times 7}$ (conv-7 layer), which is helpful to identify important spatial locations, and normalized with the application of the sigmoid function. Finally, it rescales the input Y using the element-wise product. The features learned by contrast-enhanced spatial attention are shown below:

$$F_{csa} = Y * \sigma(f^{7 \times 7}[W_s^{avg}, W_s^{std}]) \tag{18}$$

Finally, because CRB is a residual block, F_{out}, as the output of CRB, can be obtained by

$$F_{out} = F_{csa} + F_{in}. \tag{19}$$

3.4. Loss Function

We designed a unity feedback method that down-samples the SR result SR to LR images LR', as the inverse process of SR. Then, we compared it with the input LR to obtain the unity feedback loss, which can be used to narrow the search domain of the SR function. We chose the L1 loss function and used w to represent the weight of the unity feedback loss. Accordingly, our loss function is as follows:

$$Loss = \|SR, HR\|_1 + w\|LR', LR\|_1, \tag{20}$$

4. Experimental Results

In this section, we first introduce the setting of our experiments. Then, we present the experiments and analyze the results to prove the effectiveness of our methods, which include unity feedback, dual mutual-feedback feedback, mutual learning, the concat function for SR reconstruction, and CRB.

4.1. Setting

4.1.1. Datasets

First, we trained our DMFN with the DIV2k [34] dataset and validated it with Set5 [35]. This was used to compare it with other SR methods trained using natural images. Moreover, the corresponding ablation models were trained with the DIV2k [34] dataset and validated with Set5 [35]. Then, as is the case with FAWDN [22], we trained our network with the medical image dataset MRIMP and validated it with MRI13, which was used in [22]. This was named DMFN+. Finally, all the comparison results were tested with three medical image datasets: the MRI13 dataset in [22], ADNI100 [36] dataset and OASIS100 [37] dataset.

The DIV2k [34] dataset contains 800 training images, which have a resolution of 2K. We increased the number of images 10-fold through rotation and cropping. The medical image dataset MRIMP in FAWDN [22] contains 1444 training images. These were obtained using GoogleMR by crawling the keywords IXI [38], ADNI [36], KneeMR [39], and LSMRI [40]. LR images come from the bicubic down-sampling of HR images. All experimental results were obtained on a GPU under the PyTorch framework.

4.1.2. Implementation

The Adam optimizer was employed to train our network. We set the initial learning rate (lr) to 0.0005, the epochs to 1000 (halved every 200 epochs), the batch size to 16, and the base filter number to 32. As is the case for SRFBN-S [21], we performed the dual mutual-feedback procedure four times.

4.2. Effectiveness of Unity Feedback

In our study, we designed a unity feedback to narrow the search domain of the SR function, so our loss function contains two parts, as shown in Equation (20). We used the SR loss and unity feedback loss to train our DMFN, and we used w to represent the weight assigned to the unity feedback loss. In this experiment, we increase the weight of unity feedbak loss w from 0 to 1 to obtain the best trade-off. Then, we compare the SR results of DMFN with different weights in Table 1. The unity feedback improved the performance of our DMFN, and performed best when $w = 0.1$. Therefore, we set the weight of unity feedback loss to 0.1 to supervise the training of our methods.

Table 1. Comparison of different weights assigned to unity feedback loss on DMFN.

w	Scale	Params	MRI13	ADNI100	OASIS100
			PSNR/SSIM	PSNR/SSIM	PSNR/SSIM
0		583 k	36.58/0.9480	30.13/0.8716	34.22/0.9475
0.01		583 k	36.61/0.9482	30.15/0.8718	34.36/0.9479
0.1	×3	583 k	36.72/0.9485	30.17/0.8719	34.35/0.9478
1		583 k	36.54/0.9480	30.14/0.8715	34.33/0.9477

4.3. Effectiveness of Dual Mutual Feedback

For evaluating the effectiveness of dual mutual feedback, we performed ablation studies for dual feedback, mutual feedback, and feedback. The ablation architectures are shown in Figure 6. First, to realize the ablation of dual feedback, we used feedback with down sampling, as shown in Figure 6b. Therefore, it perfomed two more up-sampling steps and one more down-sampling step than the proposed DMFN in every feedback session. Second, to realize the ablation of mutual feedback, the architecture has one more up-sampling and down-sampling step than the proposed DMFN in each feedback, as shown in Figure 6c. Finally, to realize the ablation of feedback, we used a down-sampling operation after the up-projection unit to build the feedforward architecture, and up-sampled all features at the last layer, as shown in Figure 6d. All the ablation architectures have more parameters and a higher computational complexity than the proposed DMFN.

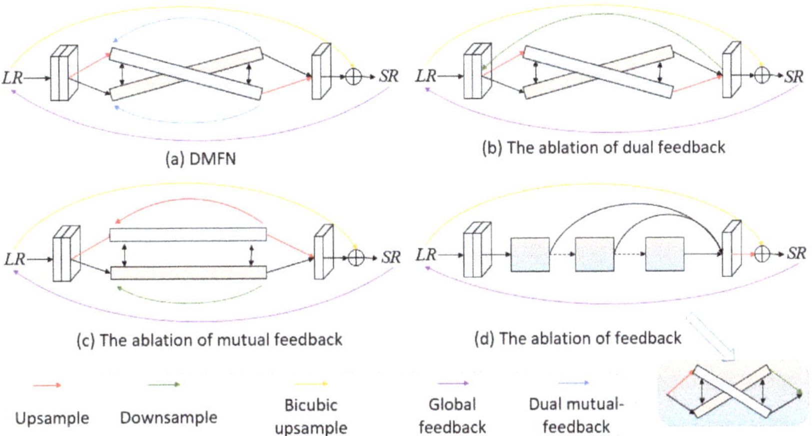

Figure 6. The ablation architectures. (**a**) is the DMFN. (**b**) represents the ablation of dual feedback. (**c**) represents the ablation of mutual feedback. (**d**) represents the ablation of feedback.

The experimental results are shown in Table 2. First, the ablation of dual feedback did not perform well. This is because the input of the two units is so similar that the information exchange between them cannot function adequately. Second, the ablation of mutual feedback also exhibited a poorer performance than DMFN. This is because it has less information exchange for its self-feedback architecture, and the additional up sampling and down sampling are not directly used for SR reconstruction. Finally, DMFN with feedback manner has fewer parameters but performs better than the feedforward manner. This is because the feedback mechanism enables the previous layers to perceive high-level information of latter layers.

Table 2. The ablation studies of dual feedback, mutual feedback, and feedback on DMFN.

Dual	Mutual	Feedback	Scale	Params	MRI13	ADNI100	OASIS100
					PSNR/SSIM	PSNR/SSIM	PSNR/SSIM
✓	✓	✓		583 k	36.72/0.9485	30.17/0.8719	34.35/0.9478
✗	✓	✓	×3	631 k	36.49/0.9479	30.13/0.8716	34.24/0.9475
✓	✗	✓		687 k	36.58/0.9480	30.14/0.8716	34.31/0.9477
✓	✓	✗		633 k	36.53/0.9479	30.14/0.8714	34.35/0.9477

4.4. Ablation Study of Mutual Learning between Two Back-Projection Units

Inspired by [20], we argue that mutual learning between the two back-projection units will improve their performance, as it facilitates information exchange. We performed an ablation study for the mutual-learning method, in which two back-projection units were learned independently, as shown in Figure 7. As illustrated in Table 3, the mutual-learning method performed better than the independent-learning method.

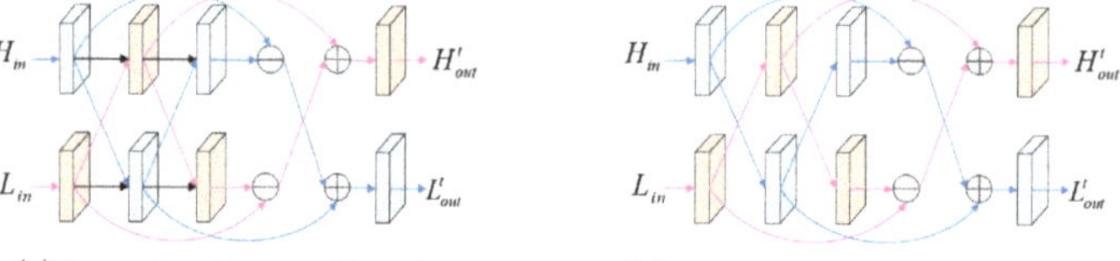

(**a**) Two units with mutual learning   (**b**) Two units with independent learning

Figure 7. The comparison of our mutual-learning method and the independent-learning method between the two back-projection units. The black arrows show the mutual learning between two units, which are not used in independent-learning.

Table 3. Comparisons of mutual learning and independent learning on DMFN.

Architecture	Scale	Params	MRI13	ADNI100	OASIS100
			PSNR/SSIM	PSNR/SSIM	PSNR/SSIM
DMFN with mutual learning	×3	583 k	36.72/0.9485	30.17/0.8719	34.35/0.9478
DMFN with independent-learning		574 k	36.55/0.9479	30.13/0.8713	34.31/0.9476

4.5. Ablation Study of the Concat Function for SR Reconstruction

Certain multi-branch methods [7,9,21] reconstruct the SR image using multi-prediction, such as SRFBN [21], which generates a prediction in each feedback procedure. However, we argue that previous feedback procedures cannot produce a meaningful prediction as a result of their very shallow HR features. Accordingly, we concatenated the HR features of all feedback procedures to obtain the final SR result. We compared our concat function and the multi-prediction method using the DMFN, as shown in Figure 8. As illustrated in Table 4, the concat function performed better than the multi-prediction method on the DMFN. Therefore, the concat function was shown to be effective and applicable to feedback networks.

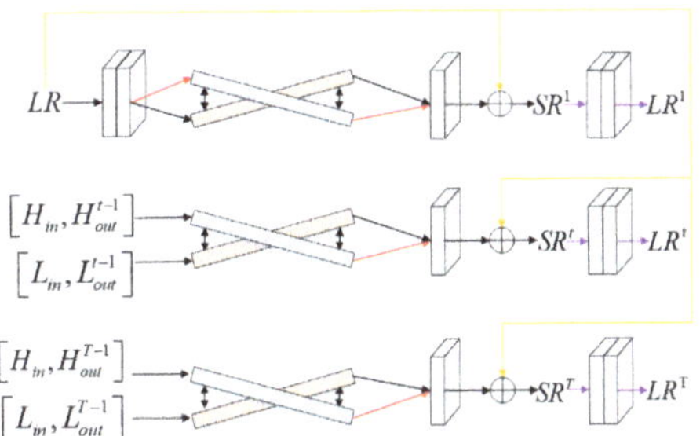

Figure 8. DMFN without the concat function, which degrades into the multi-prediction method.

Table 4. Comparison of the concat function and multi-prediction used on DMFN.

Reconstruction Method	Scale	Params	MRI13	ADNI100	OASIS100
			PSNR/SSIM	PSNR/SSIM	PSNR/SSIM
concat function	×3	583 k	36.72/0.9485	30.17/0.8719	34.35/0.9478
multi-prediction		582 k	36.24/0.9468	30.08/0.8708	34.07/0.9465

4.6. Improvement of CRB

A contrast-enhanced residual block (CRB) was designed by fusing both the contrast-enhanced channel and spatial attention within residual learning, which is used in the dual mutual-feedback component as each cell block. In this experiment, we replaced the CRB with several attention-based modules used in existing methods to evaluate their effectiveness, such as CBAM [27], RCAB [26], and CCA [28]. Figure 9 shows a comparison of these attention models. Our CRB and CBAM [27] contain both channel and spatial attention, while CCA [28] and RCAB [26] are models based on the channel attention. We used the above attention models on our DMFN to compare their performance. They are denoted as DMFN-CBAM, DMFN-RCAB and DMFN-CCA. As illustrated in Table 5, our CRB performed better than the above attention-based models. Therefore, our CRB was shown to be efficient and to improve SR performance.

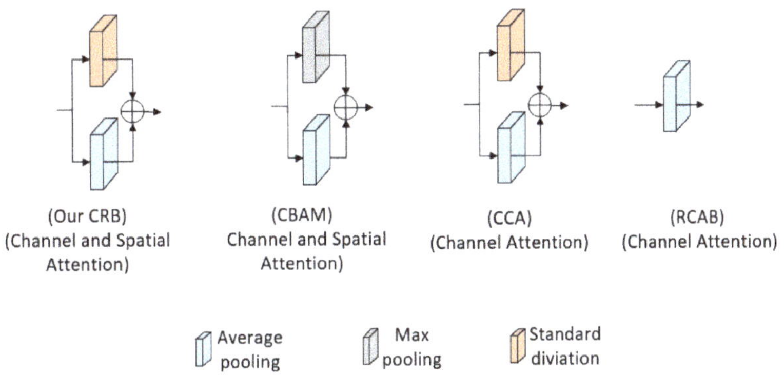

Figure 9. Existing attention models [26–28].

Table 5. Comparison of existing attention models and our CRB on DMFN.

Attention Module	Scale	Params	MRI13	ADNI100	OASIS100
			PSNR/SSIM	PSNR/SSIM	PSNR/SSIM
DMFN		583 k	36.72/0.9485	30.17/0.8719	34.35/0.9478
DMFN-CBAM	×3	583 k	36.40/0.9476	30.13/0.8721	34.21/0.9472
DMFN-RCAB		582 k	36.47/0.9478	30.13/0.8718	34.24/0.9473
DMFN-CCA		582 k	36.56/0.9479	30.14/0.8717	34.29/0.9475

4.7. Comparison with Classical SR Methods

Our DMFN was trained using DIV2K [34].Therefore, we compared it with other classical SR methods trained using natural images, such as SRCNN [2], FSRCNN [5], VDSR [8], DRCN [9], LapSRN [7], SRDenseNet [13], DDSR [10], EDSR [14], SRMD [41], SRMDNF [41], SRFBN-S [21] and FAWDN [22]. Moreover, we compared our DMFN+ with FAWDN+ [22], which was trained using the images of MRIMP and part of DIV2K [34], as the network suffered from overfitting when trained with MRIMP. Our DMFN+ was trained with MRIMP, and there was no overfitting, which demonstrates the stability of our method. As illustrated in Table 6, we compared their PSNR and SSIM values, and our DMFN was shown to perform better than the other natural image SR methods with fewer parameters. Furthermore, our DMFN+ demonstrated a better performance than the medical image SR methods with fewer parameters.

Finally, a visual comparison of SR medical images was performed, as shown in Figure 10. For ADNI100 [36], the performance of DMFN+ was the best, followed by EDSR [14]. For OASIS100 [37], the performance of DMFN+ was the best, followed by DMFN. However, DMFN+ requires fewer than half the parameters of EDSR [14], so our DMFN and DMFN+ provide a better trade-off. In summary, our methods recover image details and textures better than most of the other methods.

Table 6. Comparison of PSNR/SSIM for differentscale factors on the MRI13 [22], ADNI100 [36], and OASIS100 [37] datasets. The red and blue represent the best and second-best results, respectively.

Methods	Scale	Params	MRI13	ADNI100	OASIS100
			PSNR/SSIM	PSNR/SSIM	PSNR/SSIM
Bicubic		-	37.95/0.9677	30.74/0.8943	33.54/0.9585
SRCNN [2]		8 K	40.76/0.9820	32.49/0.9258	35.52/0.9717
FSRCNN [5]		13 K	40.90/0.9809	40.90/0.9809	34.47/0.7778
VDSR [8]		666 K	41.78/0.9835	33.09/0.9328	36.70/0.9759
DRCN [9]		1774 K	42.15/0.9838	33.22/0.9336	37.19/0.9779
LapSRN [7]		251 K	41.99/0.9840	32.96/0.9317	36.98/0.9772
SRDenseNet [13]		7160 K	42.97/0.9846	33.33/0.9348	37.69/0.9788
DDSR [12]	×2	2020 K	41.92/0.9833	33.00/0.9318	36.97/0.9766
EDSR [14]		1370 K	43.61/0.9853	33.50/0.9359	38.18/0.9797
SRMD [41]		1511 K	42.26/0.9841	33.15/0.9335	37.27/0.9780
SRMDNF [41]		1511 K	42.76/0.9848	33.30/0.9348	37.69/0.9794
SRFBN-S [21]		282 K	42.77/0.9843	33.29/0.9342	38.88/0.9806
FAWDN [22]		7170 K	43.35/0.9850	33.41/0.9352	37.91/0.9791
FAWDN+ [22]		7170 K	43.59/0.9851	33.87/0.9400	38.10/0.9798
DMFN(ours)		475 K	43.38/0.9850	33.41/0.9353	39.27/0.9813
DMFN+(ours)		475 K	43.57/0.9851	33.84/0.9397	39.43/0.9818

Table 6. *Cont.*

Methods	Scale	Params	MRI13	ADNI100	OASIS100
			PSNR/SSIM	PSNR/SSIM	PSNR/SSIM
Bicubic	×3	-	32.70/0.9126	27.95/0.8063	28.97/0.8919
SRCNN [2]		8 K	35.03/0.9393	29.29/0.8527	31.27/0.9225
FSRCNN [5]		13 K	35.35/0.9406	29.48/0.8562	30.91/0.7568
VDSR [8]		666 K	35.43/0.9449	29.83/0.8661	31.60/0.9308
DRCN [9]		1774 K	35.83/0.9449	29.92/0.8672	32.18/0.9371
SRDenseNet [13]		7160 K	36.25/0.9469	30.06/0.8704	32.64/0.9477
DDSR [12]		202 0K	35.49/0.9433	29.68/0.8638	31.72/0.9330
EDSR [14]		1555 K	36.71/0.9484	30.27/0.8735	33.26/0.9444
SRMD [41]		1528 K	35.91/0.9456	29.91/0.8676	32.29/0.9385
SRMDNF [41]		1528 K	36.09/0.9465	30.02/0.8698	32.52/0.9405
SRFBN-S [21]		375 K	36.09/0.9459	29.99/0.8691	33.79/0.9453
FAWDN [22]		7170 K	36.60/0.9481	30.16/0.8719	33.00/0.9429
FAWDN+ [22]		7170 K	36.73/0.9479	30.75/0.8839	33.19/0.9450
DMFN(ours)		583 K	36.72/0.9485	30.17/0.8719	34.35/0.9478
DMFN+(ours)		583 K	36.75/0.9482	30.76/0.8843	34.41/0.9493
Bicubic	×4	-	29.90/0.8591	26.37/0.7298	26.37/0.7298
SRCNN [2]		8 K	31.75/0.8914	27.49/0.7838	28.47/0.8621
FSRCNN [5]		13 K	32.17/0.8934	27.66/0.7876	28.49/0.6900
VDSR [8]		666 K	32.44/0.9027	28.00/0.8028	28.96/0.8748
DRCN [9]		1774 K	32.64/0.9034	28.04/0.8029	29.29/0.8830
LapSRN [7]		502 K	32.68/0.9072	27.99/0.8027	29.26/0.8837
SRDenseNet [13]		7160 K	32.97/0.9075	28.18/0.8079	29.65/0.8900
DDSR [12]		2020 K	32.45/0.9029	27.91/0.8011	29.15/0.8798
EDSR [14]		1518 K	33.31/0.9107	28.42/0.8136	30.48/0.9003
SRMD [41]		1552 K	32.83/0.9060	28.05/0.8044	29.66/0.8884
SRMDNF [41]		1552 K	32.69/0.9062	28.13/0.8080	28.13/0.8080
SRFBN-S [21]		483 K	32.85/0.9069	28.20/0.8086	31.01/0.9006
FAWDN [22]		7170 K	33.22/0.9098	28.30/0.8117	30.05/0.8957
FAWDN+ [22]		7170 K	33.21/0.9086	28.81/0.8259	30.38/0.8895
DMFN(ours)		707 K	33.37/0.9104	28.30/0.8117	31.47/0.9051
DMFN+(ours)		707 K	33.33/0.9094	28.94/0.8310	31.54/0.9079

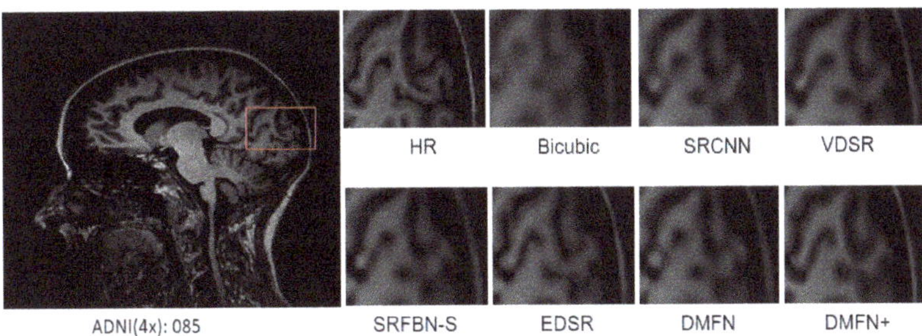

Figure 10. *Cont.*

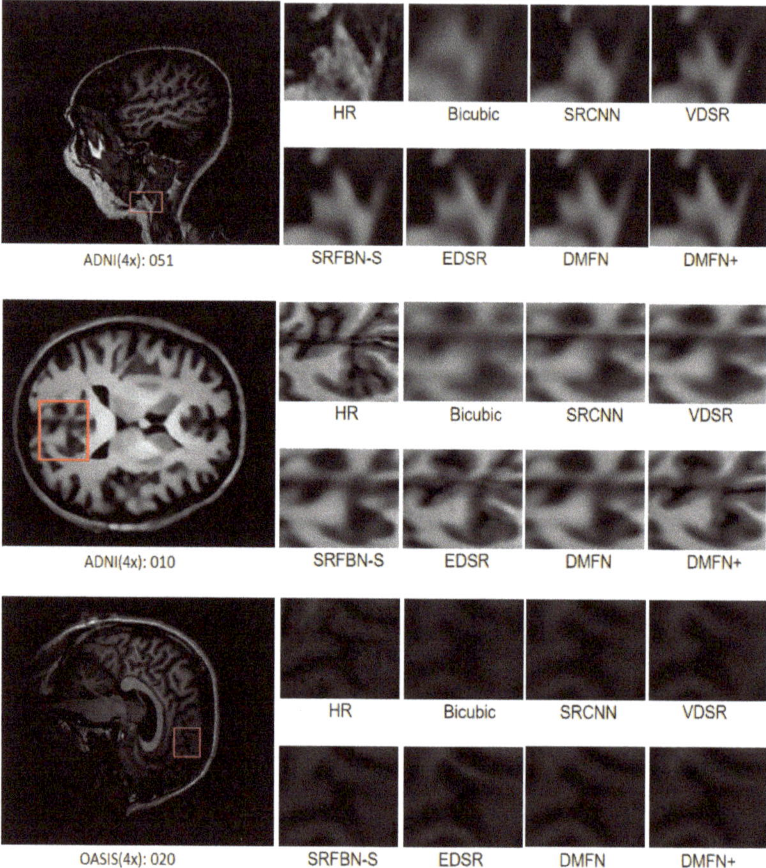

Figure 10. Comparison of visualization results on ADNI100 [36] and OASIS100 [37] datasets. Images on the right are the recovery details of the red box.

5. Conclusions

In this paper, a lightweight dual mutual-feedback network (DMFN) is proposed for use in artificial intelligence in medical image super-resolution. It contains two back-projection units working in a dual mutual-feedback manner. We propose a contrast-enhanced residual block (CRB), which is used in the back-projection units as each cell block. The CRB uses the contrast-enhanced channel and spatial attention within residual learning to enhance its ability to express details. We used the concat function for SR image reconstruction. Finally, a unity feedback method was designed to supervise the process of SR, which down-sampled the SR result to LR images as the inverse process of SR. As illustrated in the experimental results, our DMFN outperformed the other methods with very little computational cost. Accordingly, our method can help doctors to make accurate diagnoses by improving the resolution of medical images. The DMFN introduces a feedback mechanism into medical image SR and was shown to exhibit good performance on synthetic datasets. However, we are not sure whether it will perform well in real-world medical image SR, as the degradation of real-world images is complicated. In the future, we will focus our attention on real-world medical image SR, as it is possible that the feedback mechanism can also be used to improve performance in this scenario.

Author Contributions: B.W.: Conceptualization, Methodology, Software, Writing—Original draft; B.Y.: Supervison, Validation; G.J.: Supervison, Formal analysis; C.L.: Data Curation, Resources; X.Y.: Writing—Review, Editing, Funding acquisition; Z.Z.: Investigation, Visualization. All authors have read and agreed to the published version of the manuscript.

Funding: This research was funded by Sichuan University and Yibin Municipal People's Government University and City strategic cooperation special fund project (Grant No. 2020CDYB-29); Science and Technology plan transfer payment project of Sichuan province (2021ZYSF007); The Key Research and Development Program of Science and Technology Department of Sichuan Province (No. 2020YFS0575, No.2021KJT0012-2021YFS0067); The funding from Science Foundation of Sichuan Science and Technology Department (2021YFH0119); The funding from Sichuan University (Grant No. 2020SCUNG205) and National Natural Science Foundation of China under Grant No. 62201370.

Institutional Review Board Statement: Not applicable.

Informed Consent Statement: Not applicable.

Data Availability Statement: Not applicable.

Conflicts of Interest: The authors declare no conflict of interest.

References

1. Dmitry Ulyanov, A.V.; Lempitsky, V. Deep image prior. In Proceedings of the 2018 IEEE Conference on Computer Vision and Pattern Recognition (CVPR), Salt Lake City, UT, USA, 18–22 June 2018.
2. Dong, C.; Loy, C.C.; He, K.; Tang, X. Learning a deep convolutional network for image super-resolution. In Proceedings of the European Conference on Computer Vision (ECCV), Zurich, Switzerland, 6–12 September 2014.
3. Tai, Y.; Yang, J.; Liu, X.; Xu, C. MemNet: A Persistent Memory Network for Image Restoration. In Proceedings of the 2017 IEEE International Conference on Computer Vision (ICCV), Venice, Italy, 22–29 October 2017.
4. Fu, S.; Lu, L.; Li, H.; Li, Z.; Wu, W.; Paul, A.; Jeon, G.; Yang, X. A Real-Time Super-Resolution Method Based on Convolutional Neural Networks. *Circuits Syst. Signal Process.* **2020**, *39*, 805–817. [CrossRef]
5. Dong, C.; Loy, C.C.; Tang, X. Accelerating the Super-Resolution Convolutional Neural Network. In Proceedings of the European Conference on Computer Vision (ECCV), Amsterdam, The Netherlands, 8–16 October 2016.
6. Shi, W.; Caballero, J.; Huszár, F.; Totz, J.; Aitken, A.P.; Bishop, R.; Rueckert, D.; Wang, Z. Real-Time Single Image and Video Super-Resolution Using an Efficient Sub-Pixel Convolutional Neural Network. In Proceedings of the 2016 IEEE Conference on Computer Vision and Pattern Recognition (CVPR), Las Vegas, NV, USA, 27–30 June 2016.
7. Lai, W.S.; Huang, J.B.; Ahuja, N.; Yang, M.H. Deep Laplacian Pyramid Networks for Fast and Accurate Super-Resolution. In Proceedings of the 2017 IEEE Conference on Computer Vision and Pattern Recognition (CVPR), Honolulu, HI, USA, 21–26 July 2017.
8. Kim, J.; Lee, J.; Lee, K. Accurate Image Super-Resolution Using Very Deep Convolutional Networks. In Proceedings of the 2016 IEEE Conference on Computer Vision and Pattern Recognition (CVPR), Las Vegas, NV, USA, 27–30 June 2016.
9. Kim, J.; Lee, J.K.; Lee, K.M. Deeply-Recursive Convolutional Network for Image Super-Resolution. In Proceedings of the 2016 IEEE Conference on Computer Vision and Pattern Recognition (CVPR), Las Vegas, NV, USA, 27–30 June 2016.
10. Wei, S.; Wu, W.; Jeon, G.; Ahmad, A.; Yang, X. Improving resolution of medical images with deep dense convolutional neural network. *Concurr. Comput. Pract. Exp.* **2018**, *32*, e5084. [CrossRef]
11. He, K.; Zhang, X.; Ren, S.; Sun, J. Deep Residual Learning for Image Recognition. In Proceedings of the 2016 IEEE Conference on Computer Vision and Pattern Recognition (CVPR), Las Vegas, NV, USA, 27–30 June 2016.
12. Tai, Y.; Yang, J.; Liu, X. Image Super-Resolution via Deep Recursive Residual Network. In Proceedings of the 2017 IEEE Conference on Computer Vision and Pattern Recognition (CVPR), Honolulu, HI, USA, 21–26 July 2017.
13. Tong, T.; Li, G.; Liu, X.; Gao, Q. Image Super-Resolution Using Dense Skip Connections. In Proceedings of the 2017 IEEE International Conference on Computer Vision (ICCV), Venice, Italy, 22–29 October 2017.
14. Lim, B.; Son, S.; Kim, H.; Nah, S.; Lee, K.M. Enhanced Deep Residual Networks for Single Image Super-Resolution. In Proceedings of the 2017 IEEE Conference on Computer Vision and Pattern Recognition (CVPR), Honolulu, HI, USA, 21–26 July 2017.
15. Pham, C.H.; Ducournau, A.; Fablet, R.; Rousseau, F. Brain MRI super-resolution using deep 3D convolutional networks. In Proceedings of the 2017 IEEE 14th International Symposium on Biomedical Imaging (ISBI), Melbourne, Australia, 18–21 April 2017.
16. Gilbert, C.D.; Sigman, M. Brain States: Top-Down Influences in Sensory Processing. *Neuron* **2007**, *54*, 677–969. [CrossRef] [PubMed]
17. Stollenga, M.; Masci, J.; Gomez, F.; Schmidhuber, J. Deep Networks with Internal Selective Attention through Feedback Connections. In Proceedings of the Advances in Neural Information Processing Systems (NIPS), Montreal, QC, Canada, 8–13 December 2014.
18. Zamir, A.R.; Wu, T.L.; Sun, L.; Shen, W.; Shi, B.E.; Malik, J.; Savarese, S. Feedback Networks. In Proceedings of the 2016 IEEE Conference on Computer Vision and Pattern Recognition (CVPR), Las Vegas, NV, USA, 27–30 June 2016.

19. Carreira, J.; Agrawal, P.; Fragkiadaki, K.; Malik, J. Human Pose Estimation with Iterative Error Feedback. In Proceedings of the 2016 IEEE Conference on Computer Vision and Pattern Recognition (CVPR), Las Vegas, NV, USA, 27–30 June 2016.
20. Haris, M.; Shakhnarovich, G.; Ukita, N. Deep Back-Projection Networks for Super-Resolution. In Proceedings of the 2018 IEEE Conference on Computer Vision and Pattern Recognition (CVPR), Salt Lake City, NT, USA, 18–22 June 2018.
21. Li, Z.; Yang, J.; Liu, Z.; Yang, X.; Jeon, G.; Wu, W. Feedback Network for Image Super-Resolution. In Proceedings of the 2019 IEEE Conference on Computer Vision and Pattern Recognition (CVPR), Long Beach, CA, USA, 16–20 June 2019.
22. Chen, L.; Yang, X.; Jeon, G.; Anisetti, M.; Liu, K. A Trusted Medical Image Super-Resolution Method based on Feedback Adaptive Weighted Dense Network. *Artif. Intell. Med.* **2020**, *106*, 101857. [CrossRef] [PubMed]
23. Mnih, V.; Heess, N.; Graves, A.; Kavukcuoglu, K. Recurrent models of visual attention. In Proceedings of the Advances in Neural Information Processing Systems (NIPS), Montreal, QC, Canada, 8–13 December 2014.
24. Wang, X.; Girshick, R.; Gupta, A.; He, K. Non-local Neural Networks. In Proceedings of the 2018 IEEE Conference on Computer Vision and Pattern Recognition (CVPR), Salt Lake City, NT, USA, 18–22 June 2018.
25. Hu, J.; Shen, L.; Sun, G. Squeeze-and-Excitation Networks. In Proceedings of the 2018 IEEE Conference on Computer Vision and Pattern Recognition (CVPR), Salt Lake City, NT, USA, 18–22 June 2018.
26. Zhang, Y.; Li, K.; Li, K.; Wang, L.; Zhong, B.; Fu, Y. Image Super-Resolution Using Very Deep Residual Channel Attention Networks. In Proceedings of the European Conference on Computer Vision (ECCV), Munich, Germany, 8–14 September 2018.
27. Woo, S.; Park, J.; Lee, J.Y.; Kweon, I. CBAM: Convolutional Block Attention Module In Proceedings of the European Conference on Computer Vision (ECCV), Munich, Germany, 8–14 September 2018.
28. Hui, Z.; Gao, X.; Yang, Y.; Wang, X. Lightweight image super-resolution with information multi-distillation network. In Proceedings of the 27th ACM International Conference on Multimedia, Nice, France, 21–25 October 2019.
29. Irani, M.; Peleg, S. Motion Analysis for Image Enhancement: Resolution, Occlusion, and Transparency. *J. Vis. Commun. Image Represent.* **2002**, *4*. [CrossRef]
30. Dai, S.; Han, M.; Wu, Y.; Gong, Y. Bilateral Back-Projection for Single Image Super Resolution. In Proceedings of the IEEE International Conference on Multimedia and Expo (ICME), Beijing, China, 2–5 July 2007.
31. Dong, W.; Zhang, L.; Shi, G.; Wu, X. Nonlocal back-projection for adaptive image enlargement. In Proceedings of the 2009 16th IEEE International Conference on Image Processing (ICIP), Cairo, Egypt, 7–10 November 2009.
32. Timofte, R.; Rothe, R.; Van Gool, L. Seven Ways to Improve Example-Based Single Image Super Resolution. In Proceedings of the 2016 IEEE Conference on Computer Vision and Pattern Recognition (CVPR), Las Vegas, NV, USA, 27–30 June 2016.
33. Han, W.; Chang, S.; Liu, D.; Yu, M.; Witbrock, M.; Huang, T.S. Image Super-Resolution via Dual-State Recurrent Networks. In Proceedings of the 2018 IEEE Conference on Computer Vision and Pattern Recognition (CVPR), Salt Lake City, UT, USA, 18–22 June 2018.
34. Agustsson, E.; Timofte, R. NTIRE 2017 Challenge on Single Image Super-Resolution: Dataset and Study. In Proceedings of the 2017 IEEE Conference on Computer Vision and Pattern Recognition (CVPR), Honolulu, HI, USA, 21–26 July 2017.
35. Bevilacqua, M.; Roumy, A.; Guillemot, C.; Alberi-Morel, M.L. Low-Complexity Single Image Super-Resolution Based on Nonnegative Neighbor Embedding. In Proceedings of the BMVC 2012 - Electronic Proceedings of the British Machine Vision Conference, Bristol, UK, 3–7 September 2012.
36. Adni. Available online: http://adni.loni.usc.edu/ (accessed on 22 May 2019).
37. Oasis. Available online: https://www.oasis-brains.org/ (accessed on 15 May 2019).
38. Ixi-Dataset. Available online: http://brain-development.org/ixi-dataset/ (accessed on 16 May 2019).
39. Bien, N.; Rajpurkar, P.; Ball, R.; Irvin, J.; Park, A.; Jones, E.; Bereket, M.; Patel, B.; Yeom, K.; Shpanskaya, K.; et al. Deep-learning-assisted diagnosis for knee magnetic resonance imaging: Development and retrospective validation of MRNet. *PLoS Med.* **2018**, *15*, e1002699. [CrossRef] [PubMed]
40. Sudirman, S; Al Kafri, A; Natalia, F; Meidia, H; Afriliana, N; Al-Rashdan, W; Bashtawi, M; Al-Jumaily, M Lumbar spine mri dataset. *Mendeley Data* **2019**, *2*. [CrossRef]
41. Zhang, K.; Zuo, W.; Zhang, L. Learning a Single Convolutional Super-Resolution Network for Multiple Degradations. In Proceedings of the 2018 IEEE Conference on Computer Vision and Pattern Recognition(CVPR), Salt Lake City, UT, USA, 18–22 June 2018.

 *applied sciences*

Article

Mixed-Sized Biomedical Image Segmentation Based on U-Net Architectures

Priscilla Benedetti [1,2], **Mauro Femminella** [1,3,*] **and Gianluca Reali** [1,3]

1   Department of Engineering, University of Perugia, Via G. Duranti 93, 06125 Perugia, Italy
2   Department of Electronics and Informatics (ETRO), Vrije Universiteit Brussel, Pleinlaan 2, 1050 Brussels, Belgium
3   Consorzio Nazionale Interuniversitario per le Telecomunicazioni (CNIT), 43124 Parma, Italy
*   Correspondence: mauro.femminella@unipg.it

Abstract: Convolutional neural networks (CNNs) are becoming increasingly popular in medical Image Segmentation. Among them, U-Net is a widely used model that can lead to cutting-edge results for 2D biomedical Image Segmentation. However, U-Net performance can be influenced by many factors, such as the size of the training dataset, the performance metrics used, the quality of the images and, in particular, the shape and size of the organ to be segmented. This could entail a loss of robustness of the U-Net-based models. In this paper, the performance of the considered networks is determined by using the publicly available images from the 3D-IRCADb-01 dataset. Different organs with different features are considered. Experimental results show that the U-Net-based segmentation performance decreases when organs with sparse binary masks are considered. The solution proposed in this paper, based on automated zooming of the parts of interest, allows improving the performance of the segmentation model by up to 20% in terms of Dice coefficient metric, when very sparse segmentation images are used, without affecting the cost of the learning process.

Keywords: Image Segmentation; convolutional neural networks; biomedical image analysis

1. Introduction

Technological progress and advanced tools for medical analysis have significantly contributed to reducing waiting times for the diagnosis of various diseases. In particular, in oncology, the increase in the number of screening tests associated with the drastic decrease in diagnosis times has contributed to significantly reducing the mortality rate of diseases. Since the late 1980s, diagnostic imaging has been essential to visualizing organs and tissues in detail in order to detect tumors even in their early stages [1,2]. However, the tomographic images of the human body, obtained by CT-Scan (Computed Tomography), require a specialist to manually identify and segment the area of interest.

In the field of image processing, segmentation is defined as the process of decomposing an image into its constituent regions or into the objects that compose it [3]. Since the manual approach to segmentation, still used by a large part of the medical staff, is time-consuming and tedious, some techniques have been proposed to make it automatic. This can be done on the basis of certain criteria concerning the pixels belonging to a region. This is a complex objective, also because the accuracy of the result depends on the type of information to be extracted from the image.

This paper shows a solution, based on Artificial Intelligence (AI) technologies, to automate, speed up and possibly improve the analysis of the images compared to what a human operator could do. This tool incorporates and enhances some known results based on deep learning. In particular, it is based on the use of a deep convolutional neural network (CNN) that allows one to automatically process and analyze multi-scale digital images, known in the literature as U-Net [4]. Our solution includes image processing techniques that improve visualization in terms of quality, such as increasing contrast and provide a valuable tool

Citation: Benedetti, P.; Femminella, M.; Reali, G. Mixed-Sized Biomedical Image Segmentation Based on U-Net Architectures. *Appl. Sci.* **2023**, *13*, 329. https://doi.org/10.3390/app13010329

Academic Editor: Cosimo Nardi

Received: 21 November 2022
Revised: 14 December 2022
Accepted: 22 December 2022
Published: 27 December 2022

for the visual analysis of specific morphological characteristics of objects present within the image, such as their perimeter and area. In more detail, the main focus of our research is the definition of an appropriate automatic segmentation method based on the current state-of-the-art and the comparison of its performance across multiple types of images at different scales. The developed neural network allows processing and identifying a wide typology of anatomical organs, following an adequate training of the model using the CT-Scans of the patients. This feature determines a very versatile algorithm behavior, extensible to any organ. However, if the loss function used to drive the training process cannot suitably capture the information present in the mask of the organs at different scales, the resulting prediction could be unsatisfactory in operation.

The main contribution of this paper is to show how to improve performance in the presence of target images characterized by very sparse signals without significantly improving the cost of the learning process and without introducing different learning algorithms. Furthermore, by making slight changes to the model parameters, the algorithm could be used not only in the medical field, but also in other fields that require semantic segmentation of images. In fact, the applied techniques have universal value, although in this paper they have been treated in relation to the prefixed purpose. The contribution of this paper is twofold:

- *First*, a performance analysis of the baseline segmentation model on different types of organs is shown. In some case studies, the model provides good results. However, the performance is often unsatisfactory when small organs or restricted regions of interest, which are important in the diagnosis of serious pathologies, are considered [5].
- *Then*, a proposal for improving multi-scale segmentation, based on lightweight image preprocessing is shown. This proposal leads to 20% improvement in the score evaluated by using a metric based on the Dice coefficient [6], also known as F1-score [7]. This method can be generally applied to any target image characterized by a sparse binary mask.

The remainder of this paper is organized as follows: Section 2 presents some insights on the background and challenges motivating this work, along with the related works. Then, Section 3 presents the implemented model and the relevant parameters. Section 4 includes the results obtained by using the proposed methodology. Some final remarks are reported in Section 5.

2. Background and Related Works

2.1. Segmentation of Medical Images

In radiology, a CT-Scan is a diagnostic technique used for reproducing sectional and three-dimensional images of the human body. Images are obtained from the computerized analysis of the information present in the X-ray scans. Since each image is the projection of the object itself from one of multiple angles, it is possible to reconstruct three-dimensional objects.

In general, medical Image Segmentation can be useful for multiple purposes [8], such as:

- To diagnose conditions, including damage and injury to internal organs and bones, stroke, cancer and problems with blood flow;
- To guide tests and treatments. For example, before radiotherapy treatment, a CT scan is performed to determine the location, size and shape of the tumor to be treated;
- To monitor the evolution of patients' conditions, such as the presence and size of the tumor during and after specific treatments.

In order to quantitatively describe radiodensity, the Hounsfield scale, or Hounsfield Unit (HU), is typically regarded as the reference unit of measurement [9]. The HU values give preliminary information on the nature of the observed tissues.

Image Segmentation is a technique of partitioning an image into distinct and meaningful parts, called segments. The purpose of this process is to simplify and change the representation of images, for identifying and extracting some objects of particular interest and making it easier to analyze individually. In fact, it is particularly useful in applications of computer vision, such as object recognition, image compression and analysis of digital image content. In the medical area, segmentation is useful for many purposes, such as to locate and identify tumor cells, measure tissue volumes, perform virtual surgery simulations and intra-surgical navigation and integrate slower and more subjective manual human labor.

It is possible to categorize segmentation techniques into three classes: clustering, edge detection and region extraction [3]. The two macro problems to avoid are sub-segmentation, which means merging semantically different objects in the same area and over-segmentation, or the subdivision of the same object into multiple areas. In order to make algorithms autonomous, they should not assume any prior knowledge of the image to be available; otherwise, it would be difficult to ensure that results are satisfactory for any type of analysis. It is also possible to obtain the segmentation of the image in a number of regions, such that each of them is homogeneous and coherent with respect to a certain criterion and at the same time their union returns the original image. To this end, the following formal definition of the segmentation problem is given [10]: let X be the image domain and let P be a homogeneity predicate, that is a feature extracted from the image and associated with each pixel, defined on a set S of connected pixels of X. The segmentation of X consists of the partition of X into a number of n sub-images or regions S_i, with $i = 1, \cdots, n$, such that:

$$\bigcup_{i=1}^{n} S_i = X \tag{1}$$

$$S_i \cap S_j = \emptyset \qquad \forall (i,j)(i \neq j)$$

$$P(S_i) = true \qquad \forall i = 1, \cdots, n$$

$$P(S_i \cup S_j) = false \qquad \forall (i,j)(i \neq j)$$

However, the partition of an image into homogeneous regions with respect to certain characteristics does not guarantee the correct subdivision into semantic objects, especially in particularly complex images. In this regard, it is not certain that the automatic segmentation of images admits a single solution that is also robust. Nevertheless, it is possible to reach the expected result with a good approximation by implementing a multi-level convolutional model [11].

2.2. Convolutional Neural Networks

CNN is an artificial neural network architecture widely used in deep learning for image processing. It analyzes images through artificial neurons placed in three dimensions, called channels: height, width and depth [12]. It is specialized to detect and classify images and extract their features, such as corners and edges. For example, its ability to recognize objects allows the detection of tumors, in the medical field, and of obstacles, in the field of autonomous car driving. A convolutional neural network has a structure consisting of multiple levels of feature detectors:

- Convolutional layer;
- Non-linear activation functions;
- Pooling;
- Fully connected network (optional);
- Dropout layer (optional).

Our proposal makes use of the results presented in the pioneering paper [4], in which the author proposes a particular CNN, called U-Net, to solve the problem of automatic semantic segmentation of biomedical images. While a CNN is implemented in order to learn the feature map of an image and exploit it to create a more nuanced mapping by converting the image itself into a vector, when segmenting an image it is also necessary to reconstruct the image from this vector obtained. This task is particularly difficult as it is more complex to convert a vector to an image rather than the other way around. The basic idea of a U-Net is to take advantage of the functionality mapping, learned during the conversion of an image into a vector, and use it to reconstruct the output image. In this way, the integrity of the image structure is preserved and distortion is greatly reduced.

The resulting structure includes elements typical of deep learning tools, such as convolutional layers and max pooling. Figure 1 shows the structure of the U-Net, with the parameters used in this work. Details about the configuration that has been actually used in this work are given in Section 3.2.

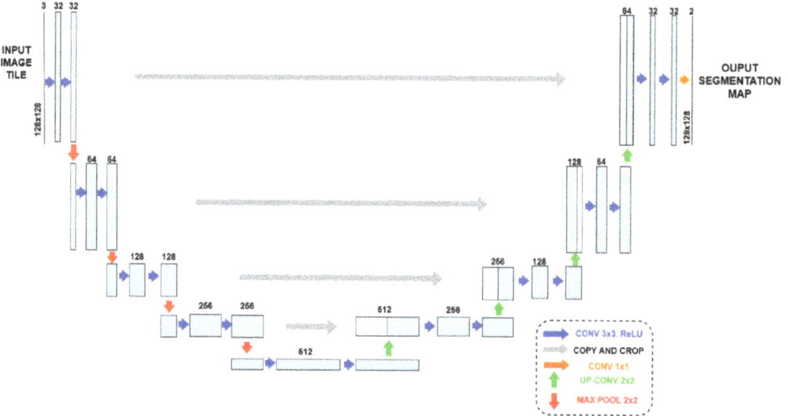

Figure 1. Model of the U-Net architecture used in this work. Each blue rectangle represents a multi-channel feature map. The number of channels is written above each box and represents the value used in the experiments. The white boxes represent the copied feature maps. The arrows denote the type of operation performed.

2.3. Related Works

In [13], the authors present a deep learning based segmentation model that relies on image-specific fine tuning. The presented model performs bounding based binary segmentation with a P-Net [14], a structure adapted from VGG-16 [15], a well known CNN model. The proposed model shows good performance, but it requires an extended number of training iterations and samples to reach good performance. Ref. [16] makes use of a modified U-Net network [4] for Image Segmentation. After an initial phase of image augmentation, the proposed network shows a good performance, but it is focused only on liver images. Other medical images with more sparse binary masks are not considered. An interesting analysis of the impact of the different parts of the U-Net architecture on segmentation accuracy is presented in [17]. In such a paper, the authors propose a reduced version of the U-Net network that sensibly reduces the number of operations required for segmentation. Nevertheless, this model does not always reach the same performance results of the standard U-Net network. Ref. [18] presents an exhaustive survey of the state-of-the-art U-Net-based Image Segmentation, with its numerous application fields. In [18], the need for improvement techniques without relying on extended datasets is mentioned as a research challenge.

For what concerns the segmentation of images with a sparse target signal, ref. [19] adds the attention mechanism ECA-Net (Efficient Channel Attention Neural Networks) into the standard U-Net architecture in order to improve the ability of the model to segment small items in the target images. This approach was applied on insulator string images and was not tested for biomedical Image Segmentation. In [20], the performance of U-Net for COVID-19 lesion segmentation on lung CT-Scans is compared with the achievable performance of another deep learning model, SegNet. Although SegNet produces better lesion detection accuracy, U-Net turns out to be better for multi-class segmentation. A solution for segmentation tasks dealing with organs of highly varying dimensions is proposed in [21]. The authors consider segmentation in head and neck (HaN) CT images, which is characterized by the presence of big and small organs. The proposed solution consists of combining a standard deep learning network for 3D images, the S-net, with a smaller one specialized for smaller segments. The main network provides the secondary one with the probabilistic location of the small organs in the HaN samples. This ensemble, called FocusNet, produced good performance using the publicly available MICCAI 2015 Head and Neck dataset. Nevertheless, it requires the introduction of a new module, which increases the model complexity and the training effort. In this regard, a lightweight solution based on data pre-processing is proposed. Similarly, ref. [22] shows a solution for small organ segmentation, considering whole-body Magnetic Resonance Imaging (MRI) scans. With the implementation of a two-staged fully CNN, a coarse-scale segmentation is first executed and then refined in order to refine the segmentation of the considered organs. Although the proposed method has a 50% gain with respect to the the state-of-the-art for small organ MRI segmentation, the overall performance is still unsatisfactory, with a 0.56 Dice similarity coefficient. Finally, blood vessel segmentation is investigated in [23,24]. In [23], a U-Net is used to perform coronary artery stenosis detection on X-ray coronary angiograms, including a module that leverages the temporal consistency of consecutive frames to limit the number of false positives. Ref. [24] is focused on the segmentation of coronary angiograms. The proposal relies on a nested encoder–decoder architecture named T-Net, which produced an accuracy of 83%. None of these mentioned proposals were used for CT-Scan blood vessel segmentation.

Finally, it is worth mentioning that a significant effort has gone into using annotated 2D organ sections to provide a three-dimensional representation of organs. In this regard, the 3D U-Net [25] represents a pioneering paper, extending the previous achievements of Olaf Ronneberger et al. obtained with the initial U-Net [4], achieving good results. However, this goal is different from that of our research, which investigates automatic segmentation of 2D CT-Scan slices that, in any case, are fundamental for also building the volumetric segmentation, if needed.

3. Dataset, Model and Parameters Configuration

3.1. Dataset

This paper shows the performance achievable by using the U-Net architecture for medical image segmentation during both the training and test phases. For this purpose, a publicly available dataset, the *3Dircadb1* dataset (https://www.ircad.fr/research/datasets/liver-segmentation-3d-ircadb-01/, accessed on 1 November 2022), was used. It includes 3D CT-Scan images of 10 women and 10 men with liver tumors in 75% of cases. The dataset is anonymized. For each patient, it includes masks of different organs. For our analysis the images of three different organs available in the dataset have been selected. They are characterized by very different shape, size, compactness and mask features: liver, bone and portal vein. The difference between these organs and relevant masks allows performing a multi-scale and multi-shape analysis of the achievable U-Net performance. In fact, while liver binary masks show a considerable amount of information (for instance, with reference to the slice in Figure 2, see the white portion in Figure 3a), in case of bone and portal vein, a gradual and significant decrease of information appear in their masks. In such situations, the shortage of the available white pixels, representing the information needed for training the network, limits the effectiveness of the segmentation process. Thus, processing images of these different organs allowed us not only to test the effectiveness of the U-Net, but also to introduce simple yet effective strategies to overcome such limitations.

The selected three organ images are provided for all the patients of the dataset. Each patient's folder includes four subfolders, in which the DICOM-formatted images, the labelled images, the mask images and the surface meshes can be found.

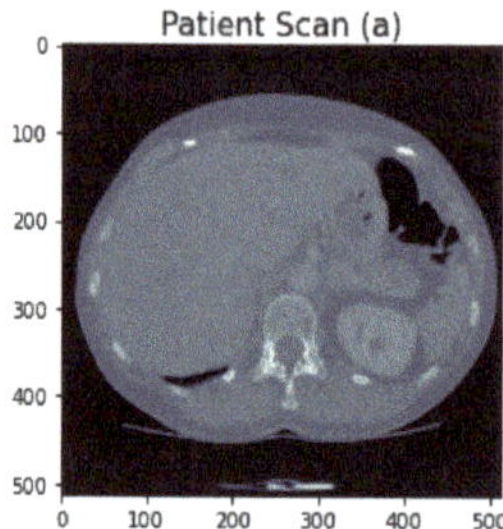
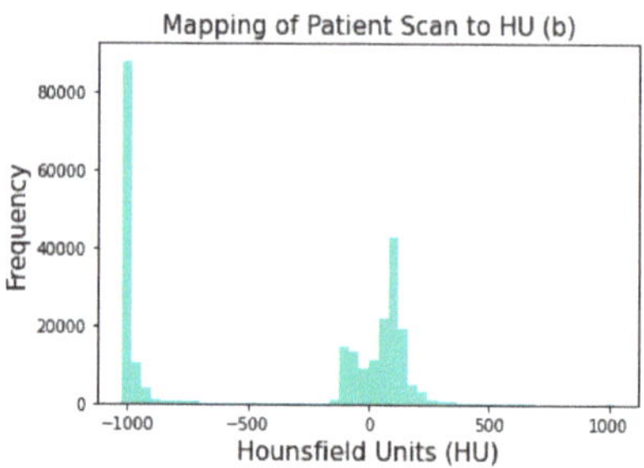

Figure 2. CT-Scan plot from '*PATIENT_DICOM*' (**a**) and corresponding histogram of HU values (**b**).

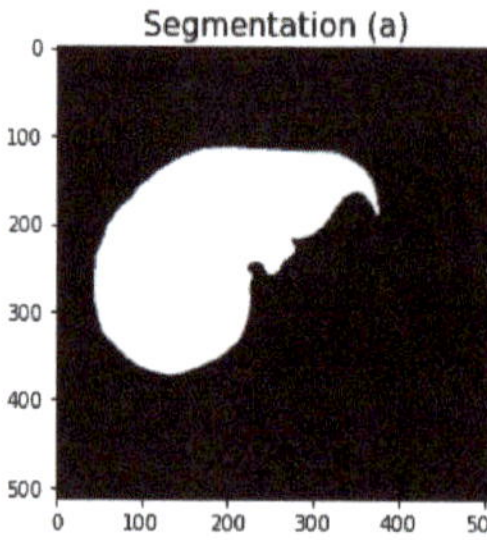
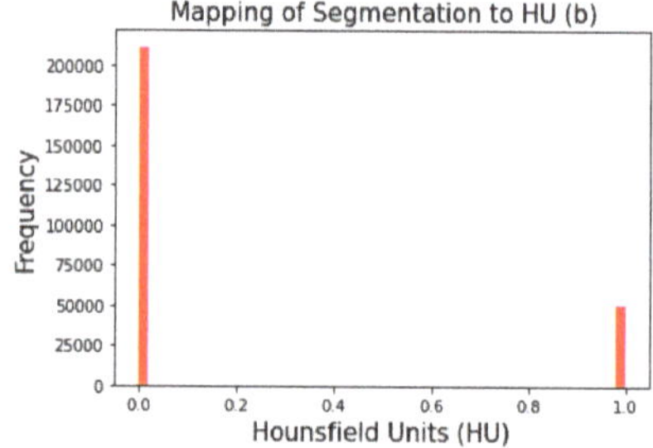

Figure 3. Binary mask plot from '*MASKS_DICOM*' (**a**) and corresponding histogram of HU values (**b**).

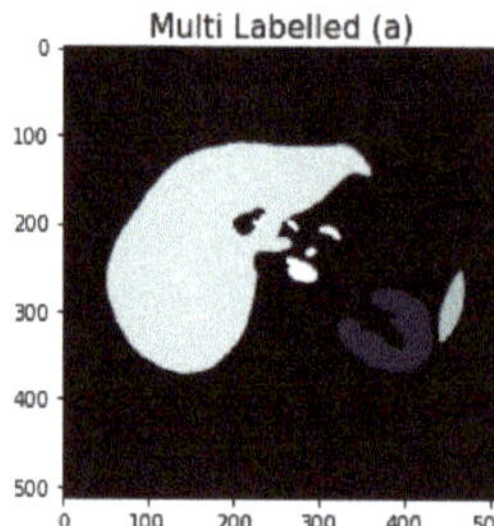
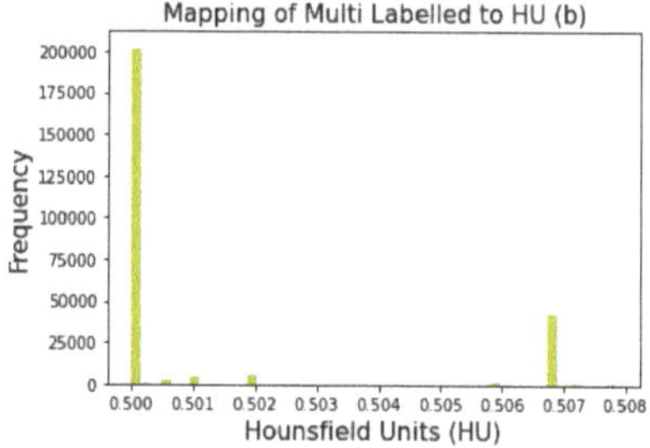

Figure 4. Multi segmented mask plot from '*LABELLED_DICOM*' (**a**) and corresponding histogram of HU values (**b**).

All images are of the same size, equal to 512×512 pixels. To process the dataset, 20 folders have been used, each one corresponding to a different patient. These folders are called '*3Dircadb1.number*', with *number* varying between 01 and 20 and each of them, containing the following four subfolders:

- '*PATIENT_DICOM*': The patient's images in DICOM format. Figure 2a shows an example of a complete CT-Scan slice, with relevant frequency of HU values reported in the companion histogram of Figure 2b;
- '*MASKS_DICOM*': A set of subfolders corresponding to the names of the various segmented organs, containing the binary DICOM image of each original mask. Figure 3a shows an example of a liver mask, with the relevant frequency of HU values (corresponding to black and white in this case) in the histogram of Figure 3b. In this subfolder, the segmentation masks of liver, bone and portal vein used in this work can be found for each patient;
- '*LABELLED_DICOM*': The ensemble of segmented images corresponding to all the patient's analyzed organs, including the ones considered in this work (liver, bone and portal vein), in DICOM format. Figure 4a shows an example of a multi-segmented

mask, with relevant frequency of HU values (one for each mask) shown in a histogram in Figure 4b;

- *'MESHES_VTK'*: All the files corresponding to the surface meshes of the various areas of interest in Visualization Toolkit (VTK) format.

3.2. Model Architecture

Our proposal makes use of the U-Net, a new CNN presented in the pioneering paper [4] to solve the problem of automatic semantic segmentation of biomedical images. U-Net is a CNN whose architecture was adapted in order to make use of a reduced image dataset while continuing to produce fairly precise segmentation. Figure 1 shows the structure of the U-Net used in this work, together with the relevant parameters. The architecture is divided into two *paths*. The path on the left is called *"Contraction Path"* or *"Encoder Path"*, while the one on the right is referred to as *"Expansion Path"*, or *"Decoder Path"*. In the middle data, concatenations are performed, indicated by the grey arrows in Figure 1. They implement acquisition of localized information from the feature maps.

The *Contraction Path* consists of a certain number of contraction blocks followed by a max 2 × 2 pooling. Each contraction block downsamples the input image, received from the previous level, in a feature map, applying two levels of 3 × 3 convolution. The number of kernels, or feature maps, after each block doubles, so that the architecture can effectively learn the complex structure of the input image. In this case, the input image is 128 × 128 pixels and 32 features are used in the first step. Parameters, such as the size of the input image and the number of features, can be changed according to the architecture to implement, as some criteria may not make the network work properly. Moving further and further down, the bottom level averages between the two paths and uses two CNN 3 × 3 levels followed by a level called *"up convolution"* 2 × 2.

Subsequently, the *Expansion Path* section, which also consists of a series of expansion blocks, allows the network to propagate information from the lowest resolution level to higher ones. This way, it amplifies (upsamples) the final image feature map. Each block passes the received input to two 3 × 3 convolutional levels and a 2 × 2 upsampling level. Symmetrically to the left branch, after each block the number of feature maps used by the convolutional layer is halved. However, as shown by the gray arrows, each time the copied feature map of the corresponding contraction level is added to the input. This ensures that the features and information learned while shrinking the image are subsequently used to reconstruct it correctly. Clearly, the initial levels of the encoder contain more information, so they guarantee a significant boost in the up-sampling process, allowing the recovery of details and significantly improving the result. The architecture, being symmetrical, is such that the number of expansion blocks is equal to the number of contraction blocks. Going up to the final level of the expansion path, the resulting map passes through the final level of 3 × 3 convolution, with the number of feature maps equal to the number of desired segments. Hence, the same feature map used for the contraction is then used to expand the vector and obtain the output image, which represents the segmentation of the input image.

3.3. Model Implementation

To train the classification algorithm, implemented with Tensorflow (https://www.tensorflow.org/, accessed on 1 November 2022), the CT-Scan records of 17 patients were used, consisting of 2445 images (and related masks), encapsulated in DICOM files (Digital Imaging and Communications in Medicine, https://www.dicomstandard.org/, accessed on 1 November 2022). The image of a single slice of a CT scan is referred to as a *sample* throughout the paper.

After importing the appropriate packages, the functions for loading and importing DICOM images with related tags are implemented. The function code is shown in Appendix A. By using the `process_path(filename)` function, it is possible to load the DICOM image and the relevant mask for each file path. In the specific example, the one necessary for the *liver* segmentation is shown: It is necessary to convert the sample values to the unit of measurement for CT-Scans, that is the HU, since, by default, the values returned from the first upload (resulting from `decode_dicom_image()`) are not expressed in the HU units. This transformation is linear and, if stored in the DICOM header at the time of image acquisition, the relevant slope and intercept can be recovered by using the tag codes (00281052, 00281053) in a completely automated way. This function is subsequently applied to each TensorFlow dataset containing the list of paths related to images, for each folder, using the `map()` method. The use of tensors is preferred in order to model and process data. In fact, through the representation in computational graphs, they facilitate parallel computing, making full use of the computing resources used (Graphic Processing Units, GPUs). Furthermore, with the use of tensors, the computation of derivatives, fundamental in the learning process of a neural network, is accelerated.

Once all the functions for image processing are defined, these methods are invoked in order to view the data collected so far. For example, Figure 2 shows an example of the images present in the 3Dircadb1.1 [26] folder of patient n.1 (a) and the associated histogram (b). Through the histogram, it is possible to see how many pixels correspond to air and how many of them to tissues. Looking at the histogram, it emerges that a lot of air is present, there is an abundance of soft tissue, mainly muscle, liver and part of the lung, but also fat. Only a small piece of bone is present in the CT scan, which appears as a tiny sliver, difficult to appreciate in the HU histogram (expected values around 400 HU) due to the small number of relevant samples.

Once data are processed, a further operation is required on the arrays of the training and test sets, in order to scale the size of the images from 512×512 pixels to 128×128 pixels and add a dimension to the channel (the gray scale, which is 1), since the built model accepts images with this resolution as input. For what concerns the construction of the actual model, it is made up of 10 convolutional levels that outline the architecture of the U-Net, described in the previous section. In addition, dropout levels are inserted after each pooling level, as the amount of information in the considered dataset is quite limited. In fact, in the absence of dropout, the training of the algorithm would require more effort to converge due to overfitting (overfitting is a phenomenon that happens when a statistical model excessively adapts to its data during the training process, thus losing generality [27]). To limit the risk of overfitting, possible solutions consist of increasing the volume of data, reducing the complexity of the architecture or adding regularization. The latter solution is implemented by adding the dropout after each level of convolution, keeping in mind the loss of information as a consequence: If part of the information is lost in the first level, it is lost for the entire network. Therefore, the final scheme of the neural network that we present starts with a low dropout rate in the first few levels to limit the loss of information, which gradually increases to limit overfitting.

The model structure is implemented by using the Keras library (https://keras.io/, accessed on 1 November 2022). In particular, the `get_model()` function is created, which receives as arguments the optimizer, the loss function, the metric and the learning rate selected. The code (Listing 1) below illustrates the first level of convolution of the Contraction Path:

Listing 1. Keras model creation: Encoder Path's first level.

```
1 conv1 = Conv2D(32, (3, 3), activation='relu', padding='same')
2                  (inputs)
3 conv1 = Conv2D(32, (3, 3), activation='relu', padding='same')
4                  (conv1)
5 pool1 = MaxPooling2D(pool_size=(2, 2))(conv1)
6 drop1 = Dropout(0.5)(pool1)
```

The first two lines of code create a convolution kernel, which processes the received input to produce an output tensor. First, the set of images coming from the outside is used as input, then the output of the first convolution becomes the input of the second. The constructor arguments used are as follows: Number of filters (integer power of 2), from which they learn the convolutional levels, kernel size (an integer or a tuple of two odd integers), which specifies the height and the width of the 2D convolution window, `activation`, which specifies the activation function to apply after performing the convolution and padding value (`padding`), which specifies, based on the set value, whether the size of the input volume was changed. In this work, the ReLU (Rectified Linear Activation Unit) activation function is used, while the padding value is the 'same', which means that the spatial dimensions of the input are maintained, so that the volume of the output has the same size. Therefore, reliance on padding, image shrinkage or information loss is avoided.

ReLU is a simple function with better performance than other activation functions, such as Sigmoid and Tanh. Its equation is as follows [28]:

$$R(x) = max(0, x). \tag{2}$$

It returns 0 if it receives a negative input, while for any other positive input value x, it returns the same value. In fact, the function does not perform any complex calculations and therefore the model takes little time to train and converges very quickly. Another advantage is sparsity, which implies better predictive power and less model overfitting, as it increases the likelihood that neurons are actually processing meaningful data. This occurs since in neural networks, such as the one under consideration, matrices have many zero cells and for this reason they are called 'sparce matrices'. Therefore ReLU, by providing zero output for negative inputs, makes sure that the network is sparse and that neurons are not turned on to process unmeaningful data. However, the phenomenon of 'dying ReLU' could occur. A ReLU neuron is dead if it is stuck on the negative side and always returns 0, i.e., once it goes negative, it is unlikely to recover. This problem can occur when the learning rate is too high and a lower rate could solve this issue. The learning rate used in our experiments is equal to 10^{-3}.

In the first level, the aim is to reduce the spatial dimensions of the output volume through MaxPooling, applying it to the output of convolution. MaxPooling is a pooling operation that selects the maximum element from the area of the feature map covered by the filter, sub-sampling the input along its spatial dimensions. The window size is set with the `pool_size` argument, which is an integer or a tuple of two integers (if only one integer is specified, it is used for both the height and width of the window). Therefore, the output after this level is a feature map containing the most important characteristics of the previous input feature map.

Finally, as already mentioned, a dropout level is added, which receives the result of the MaxPooling as input. Inputs to dropout not set to 0 are scaled by 1 / (1 − rate), so that the sum across all inputs is unchanged [28]. The `rate` value is the fraction of the input units to be released. It is in the range between 0 and 1; in particular, it is set to `rate = 0.5`, since, as mentioned above, it is good to lose only a small amount of information in the first level.

To complete the Contraction Path, the other four levels of convolution are implemented in the same way as the first one, except for the input of the first convolution of each level, with the input, in this case, being the output of the previous level dropout. In level 5 of the Contraction Path, both pooling and dropout are not necessary as they are the final level of the structure (see Figure 1). The Decoder Path is built symmetrically, with the same number of levels as the Encoder Path. The syntax of the code (Listing 2) is as follows:

Listing 2. Keras model creation: Decoder Path's sixth level.

```
1  up6 = concatenate([Conv2DTranspose(256, (2, 2), strides=(2, 2),
2                            padding='same')(conv5), conv4],
3                            axis=3)
4      conv6 = Conv2D(256, (3, 3), activation='relu', padding='same')
5                (up6)
6      conv6 = Conv2D(256, (3, 3), activation='relu', padding='same')
7                (conv6)
```

In level 6, upsampling takes place first. It consists of a concatenation which takes a list of tensors as input, all of the same shape except for the concatenation axis (`axis = 3`). It returns a single tensor, which is the concatenation of all inputs [28]. The first item in the list is the result of the transposed convolution made on the output of the level 5 convolution, while the second is the output of the level 4 convolution. The transposed convolution performs an inverse transformation to the normal convolution made in the descent path, since in this way the output begins to be generated which has the same shape as the original input, while maintaining some connectivity with the shape of the convolution output. The parameters of this operation are the number of filters, the kernel size, padding, just like a normal Conv2D, and also the number of steps (`strides`) of the convolution along the height and width (an integer or a tuple of two integers). In the code `strides = (2, 2)` is set; that, is the filter is moved 2 pixels horizontally for each reading from left to right, then down 2 pixels for the next row, establishing the outputs in the feature map. During this operation, Conv2DTranspose learns during training and attempts to fill in the details as part of the upsampling process to resample the original input. Then two normal Conv2D convolutions are executed: the first takes the result of the upsamplig concatenation as input. Its output becomes the input of the second convolution. At the end of the first level of the Decoder Path, the U-Net structure continues with three other levels that have the same shape as the one just described. Finally, the tenth and final level of the network consists of a single convolution, in which the activation function (Listing 3) is the 'Sigmoid' [29] and no padding is done:

Listing 3. Keras Model creation: level 10.

```
1 conv10 = Conv2D(1, (1, 1), activation='sigmoid')(conv9)
```

The final step to complete the definition of the `get_model` () function is the configuration of the model for training (Listing 4), which is also used for its validation:

Listing 4. Model configuration.

```
1 model.compile(optimizer=optimizer(learning_rate=learning_rate),
2                       loss=loss_metric, metrics=metrics)
```

3.4. Parameters Configuration

This section analyzes the specific arguments that are passed to the `get_model` () function and the results obtained.

As mentioned above, the hyperparameters that allow controlling the model training process are the optimizer, the loss function and the metric. Following an experimental optimization process to identify an optimal configuration from a performance point of view, which minimizes a predefined loss function on test data, the following parameters were selected:

- Optimizer = 'Adam';
- Learning Rate = '1e-3';
- Loss Metric = 'dice_coef_loss';
- Metric = 'dice_coef'.

3.4.1. Adam Optimizer and Learning Rate

The optimizer selected implements the Adam algorithm [30], acronym for Adaptive Moment Estimation, for the optimization of the descent of the stochastic gradient, applying the principles of the RMSProp and AdaGrad optimizers [31].

Gradient descent is a technique used for determining the global maximum and minimum points of a function of several variables. The stochastic approximation is applied when the cost function is too expensive at each iteration and breaks down the addends at each iteration into a sum. In the context of artificial neural networks, the descent of the gradient evaluates the model by using an input corresponding to a known output and corrects each parameter of the model in a proportional quantity (but of opposite sign) with respect to its contribution to the result error.

The Adam algorithm is computationally efficient, requires little memory and is suitable for handling large data volumes or numbers of parameters. The fundamental parameter that Adam receives is the learning rate, that is, a floating point value that indicates the size of the passage at each iteration and that influences the criterion for evaluating whether the newly acquired information is more important than the past information item. Therefore, the learning rate must be neither too high—otherwise, learning will jump above the minima of the loss function; nor too low—otherwise the convergence will happen too slowly, with the possibility of getting stuck in an undesired local minimum. During the training phase, it is advisable to adjust and adapt the learning rate in the right way as it does not change, but remains unchanged throughout the execution of the model. For the proposed model, the value of the learning rate was set at the default value recommended for the Adam algorithm, equal to 10^{-3}. Then, the optimal value was determined experimentally, by varying it in the range $[10^{-4} \div 10^{-2}]$, using a logarithmic spacing for the search process.

From the relevant literature, the good robustness and speed of Adam, compared to other optimizers, emerges [32]. However, performance comparisons of various optimizers are strongly dependent on the specific workload and hyperparameter tuning. Hence, Adam has been selected after a preliminary experimental comparison with alternative optimizers such as SGD and Adagrad. In our analysis, Adam provided the best performance in a limited training time. Moreover, coherently with our results, the Adam optimizer is widely used in neural networks for Image Segmentation [17,33–35], since it provides better results compared with some alternatives even with varying network architectures [33].

3.4.2. Metric and Loss

A metric function based on the Sørensen–Dice coefficient [6] is used in the model. It is often referred to as the Dice coefficient and is equivalent to the F1-score [7]. It is a statistical index that measures the similarity between two sets of data. In particular, in the context of Image Segmentation, it compares the estimated output of the algorithm with the known reference masks, measuring the affinity between two binary images. The `dice_coef()`, used as accuracy metric in List 4 (model configuration), is based on the following equation:

$$DSC = \frac{2 \mid X \cap Y \mid + smooth}{\mid X \mid + \mid Y \mid + smooth}. \tag{3}$$

where $|X|$ and $|Y|$ are the sizes (expressed as number of elements) of the two sets X and Y. In this case, X and Y represent the sets of white pixels in the masks generated by the U-Net and in the reference one, respectively. The DSC is the ratio of the double of the number of elements common to both sets (size of intersection set) and the sum of the size of the two sets. The expression returns a value between 0 and 1. A Dice coefficient equal to 1 denotes a perfect and complete overlap. In our code, these sets are obtained by flattening each image by using the `flatten()` method, which returns a one-dimensional array. At this point, Equation (3) is used to implement the Dice metric. For the evaluation of the coefficient on the expected masks, the numerator is approximated by the sum of the elements of the matrix obtained by the element-wise product between the elements of the forecast and those of the input mask. The advantage of using the Dice coefficient is that it maintains sensitivity in more heterogeneous data sets and is less sensitive to outliers. In the DSC Formula (3), the parameter *smooth* indicates the smoothing coefficient, which is a value between 0 and 1 that prevents the occurrence of a zero denominator [36,37]. In our model, the value *smooth=1* is set.

Finally, the `dice_coef_loss()` loss function is defined as the opposite of the Dice coefficient.

4. Performance Evaluation

The used processing environment is the well known Google Colab (https://colab.research.google.com/, accessed on 1 November 2022). This tool allows easy repeatability of experiments and code sharing. The free version of Google Colab was used. It offers remote execution on virtual machines (VMs) with limited lifetime (12 h) [38]. The available computing resources for each VM are 12 GB of RAM, 78 GB of disk space and GPU-accelerated computing on Nvidia Tesla K80 [39]. The Tesla K80 GPU combines two graphics processors to increase performance. It is characterized by 2496 shading units, 208 texture mapping units and 48 render output units per GPU.

The comparison done in this work focused on three different organs, with very different features: liver, bone and portal vein. Each patient of the considered dataset contains segmentation images for each of these three organs.

For each organ, a target U-Net was trained and tested. From the original set of 20 patients, 3 patients were discarded due to anomalies in the images that could negatively influence the training process. Figure 5 shows two examples of altered CT-Scans from *3Dircadb1*. In the figure, it is possible to see the presence of an additional circular area that encloses the body section. This area alters the sample's HU values, blurring the difference among the body section and the external area. Hence, these images can increase the probability of unreliable segmentation in the model.

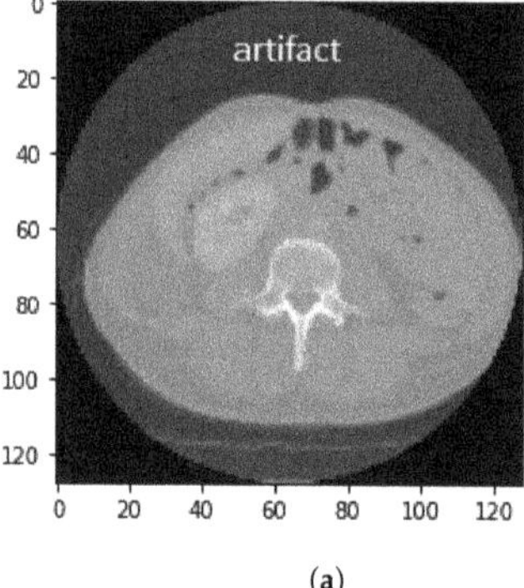

(a)

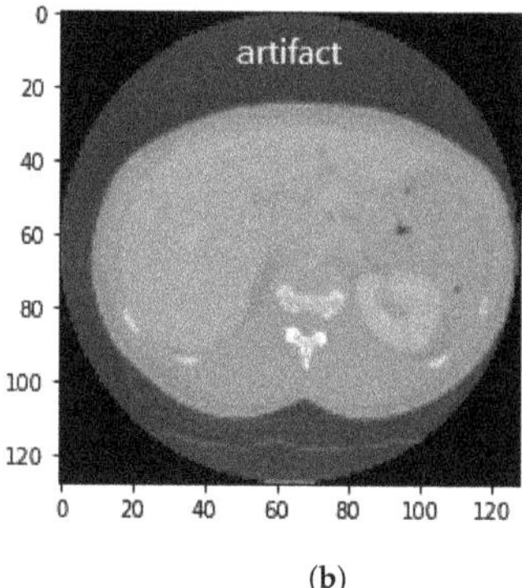

(b)

Figure 5. Two different samples of altered CT-Scans from the *3Dircadb1* dataset: presence of an additional grey circular area that encloses the body section, marked by the word 'artifact' to better highlight it.

During the data processing phase presented in Section 3, the CT-Scan images were resized to 128×128 pixels. This choice was motivated by the availability of limited computing resources in the Colab VM. In fact, the training of the considered U-Net with images of the original sizes (512×512 pixels) or on a 256×256 resized version is not feasible in Google Colab, leading to memory error even when considering strongly limited batches (5 or 10 images for a batch). Hence, our training phase consisted of 90 iterations for each organ on 80% of the dataset, processing 128×128 CT-Scan images in batches of 32 images. The CT-Scan images are sent as input to the network after a shuffling step. This procedure increases the robustness of the network to both image variability and overfitting. Processing of randomly sorted images is common in Image Segmentation tasks [20,24,40].

As mentioned above, the U-Net was implemented by Keras and Tensorflow, with the following configuration, which is referred to as 'base configuration' in this paper:

- Optimizer: Adam [30], learning rate 1×10^{-3}.
- Performance evaluation metric: Dice coefficient [6].

The first comparison was carried out with the same configuration for the three organs. The U-Net with the configuration described in the previous section provides the results reported in Table 1, tested on the remaining 20% of the samples, consisting of 489 images. The results are expressed in terms of DSC and Accuracy [7], defined as in (4). Nevertheless, due to the nature of the problem, the simple Accuracy metric provides unreliable results.

$$Accuracy = \frac{|X \cap Y| + |\overline{X} \cap \overline{Y}|}{|X| + |\overline{X}|}. \tag{4}$$

In fact, while the DSC decreases with the increasing sparsity of the binary masks, as expected, the corresponding increasing imbalance between 0 s and 1 s in the masks leads to a very high Accuracy. Hence, this high Accuracy does not reflect the actual quality of the segmentation. In fact this quality significantly decreases for small image segments, such as the bone and, in particular, the portal vein, as shown in the following analysis. For this reason, DSC will hereafter be considered as the evaluation metric of the segmentation. Samples for organ images, original masks and masks predicted by the U-Net are reported in Figures 6–9, for liver, bones and two vein samples, respectively.

Table 1. Performance comparison results on the test set.

Organ	Samples	DSC	Accuracy
Liver	489	97.85%	88.65%
Bone	489	81.35%	95.93%
Portal Vein	489	58.53%	98.15%

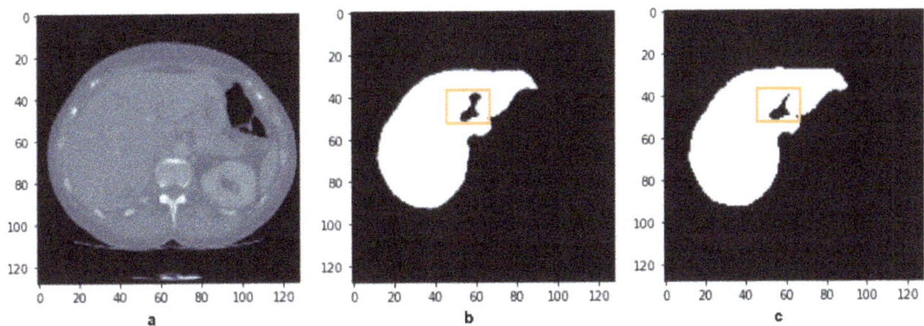

Figure 6. Sample image (**a**), original mask (**b**) and predicted mask (**c**) for the liver.

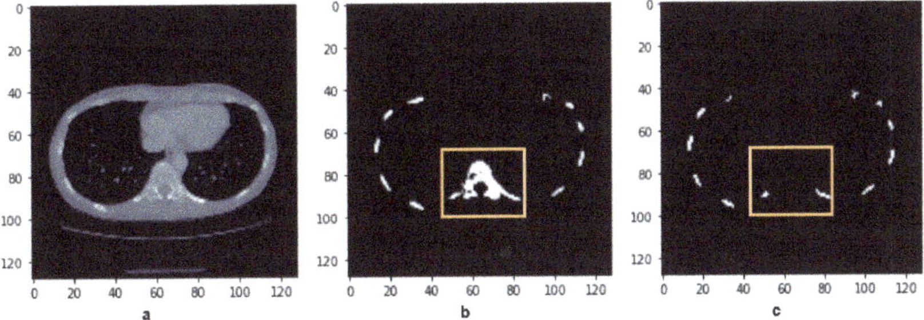

Figure 7. Sample image (**a**), original mask (**b**) and predicted mask (**c**) for the bone.

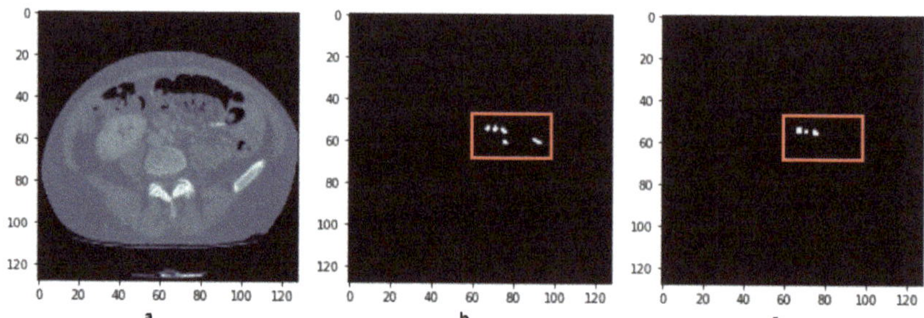

Figure 8. Portal vein sample 1: (**a**) image, (**b**) original mask, and (**c**) predicted mask without zooming in the pre-processing.

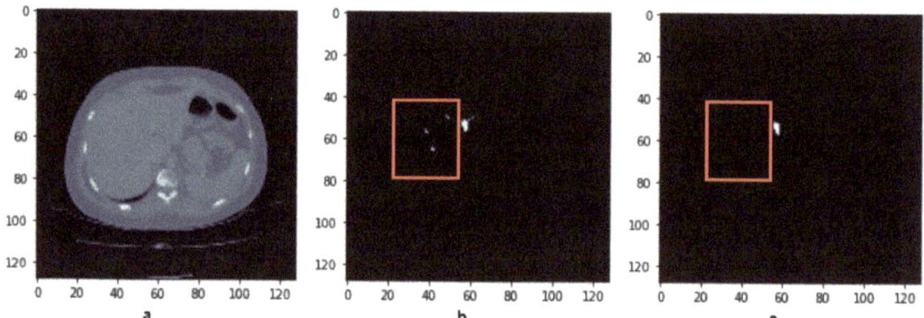

Figure 9. Portal vein sample 2: (**a**) image, (**b**) original mask and (**c**) predicted mask without zooming in the pre-processing.

As shown in Figure 6, the network produces good results (97.85%) for the liver on the test set. The overall parenchyma tissue is well defined in the predicted mask (Figure 6c), with a good precision in the segment border. The internal sections still suffer from some inaccuracy, as shown by the orange boxes in the figure. However, these results are fairly good, considering the size of the training dataset available.

The results for bone segmentation indicate a significant decrease in the DSC value, which is equal to 81.35%. This is still borderline acceptable for not very small organs [13,21,22]. This is confirmed by the analysis of the sample images shown in Figure 7. The predicted mask (Figure 7c) shows how the trained model correctly intercepted the location and perimeter of the bone structure. Nevertheless, it can be seen that in cases of increased sparsity of the target signal, the percentage of error for the same model is increasing. This causes a reduction in the segmentation accuracy of more than the 15% in terms of the DSC metric. In particular, some segments in the predicted mask are predicted with a significantly reduced area with respect to the original mask (Figure 7b), while the spinal section is missing. Even if the signal of the spinal section is not so sparse compared to the other bone sections in the original mask, its rather small dimensions lead to an unreliable segmentation.

This phenomenon becomes more evident when the model is tested by using the portal vein samples. In this case, the prediction performance drops significantly (58.53%) and becomes not acceptable at all. This can be attributed to the increased and significant sparsity of the signal in the portal vein binary masks, due to the specific features of the vasculature of the torso. Both Figures 8 and 9 depict examples of prediction outputs. As can be observed in Figure 8, the trained model cannot detect the vascularization sparse spots in the segmentation mask (Figure 8b) correctly, as it can only reproduce in the output mask the biggest segment detected from the original sample (Figure 8c). This behaviour is further confirmed in Figure 9. In the presence of two different segment regions, one with a bigger area, the other with sparse vascularization spots, the model correctly detects the first one, while it ignores most of the segments in the latter, as shown in the red box (Figure 9c).

A known approach to solve class imbalance in machine learning is the usage of a weighted cross entropy loss. The introduction of weights to penalize the misclassification of minority classes is present in medical image analysis and segmentation as well, in particular for medical Image Segmentation [41]. Hence, the application of a weighted cross entropy loss in our model was evaluated to improve the performance of portal vein images segmentation. The class weights have been configured with the *balanced* option of the Python module `compute_class_weight` (https://scikit-learn.org/stable/modules/generated/sklearn.utils.class_weight.compute_class_weight.html, accessed on 1 November 2022). Nevertheless, considering the same number of training epochs, the introduction of the weighted cross entropy as loss metric significantly degrades the segmentation performance on portal vein images. The evaluation on the test set provides a DSC value equal to 27.8%, as reported in Table 2.

Table 2. DSC metric for different percentages of cropping of the original images (portal vein).

Approach	DSC
Base configuration (no zoom)	58.53%
40% image cropping	62.04%
50% image cropping	65.23%
60% image cropping	76.39%
70% image cropping	81.45%
Weighted cross entropy loss	27.8%

Hence, in order to improve the DSC metric for the portal vein segmentation and, in general, for vascularization samples, the following solution is proposed. A preprocessing step, consisting of magnifying the portion of interest of the binary masks and of the corresponding areas in slice images, is introduced as follows:

- Zooming the image and the corresponding mask in the target area of vascularization, in order to increase the size of the vascularization segments. The zoom was configured to enhance the body section of interest, by cropping with the Tensorflow function `tf.image.central_crop()` up to 70% of the original image. Hence, in this case, the resulting image is reduced to a size of 154×154 pixels from the original size of 512×512 pixels. This resulted in the optimal percentage for the best DSC metric, found experimentally, as shown in Table 2.
- Resizing the samples to the 128×128 format, as in the previous experiments.

The rationale of this strategy is that the higher resolution of original images is leveraged to provide the U-Net with additional information, in order to enlarge the target area. In fact, with the original setting, the amount of the target area in the mask is too small to drive the Dice metric towards a correct recognition of it. This is due to the metric definition itself, which is based on the percentage amount of pixels of the target image (see (3)). This causes, as mentioned in the comments to Figures 8 and 9, the missing recognition of small, sparse spots of the vascularization. Although the resolution in the original setting cannot be handled by the available computing resources, it offers the opportunity to gain additional information via zooming the interested portion of the image before resolution rescaling. The advantages are multiple: not only is the information overlooked in base configuration (i.e., the one without any zoom) leveraged, but the training time and the amount of necessary computing resources are kept unmodified.

After pre-processing, a similar U-Net was trained by using the new images for 90 iterations. The new model leads to an increase in the DSC metric up to 22% for portal veins, corresponding to 81.45% on the test set with the largest and optimal zoom level. Figures 10 and 11 depict two sample outputs obtained by using the new model, specially tuned to segment vascularization in medical CT scans. The positive impact of preprocessing can also be observed on the predicted mask quality. Even in the presence of sparse signals, the model is good at detecting all the present segments, although some minor discrepancies in the vascularization segment shapes (orange box) are present (Figures 10c and 11c). Nevertheless, through the proposed approach, the model is able to fetch significantly small segments that were previously ignored, as shown in another sample in Figure 12 on the green box on the right side of the binary mask (Figure 12c). This is further confirmed by the sample in Figure 13, replicating the same two distinct segments highlighted in the green box (Figure 13c).

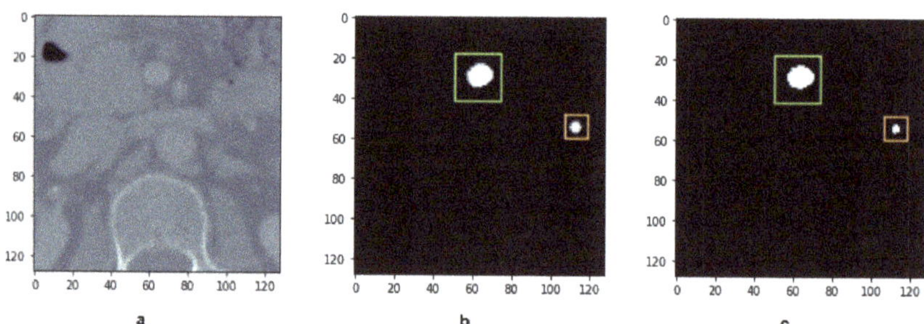

Figure 10. Portal vein sample 3: image (**a**), original mask (**b**) and predicted mask (**c**) after pre-processing with zoom.

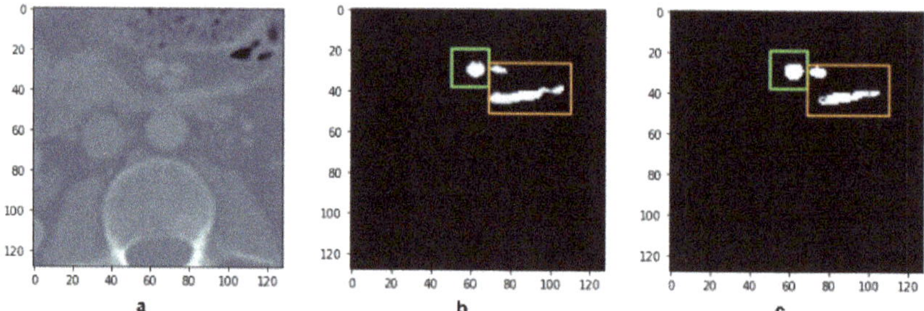

Figure 11. Portal vein sample 4: image (**a**), original mask (**b**) and predicted mask (**c**) after pre-processing with zoom.

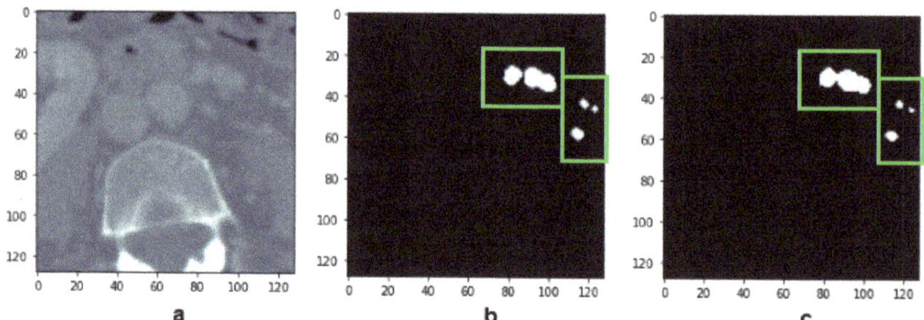

Figure 12. Portal vein sample 5: image (**a**), original mask (**b**) and predicted mask (**c**) after pre-processing with zoom.

Although results are appreciable, some margin for further improvement exists. In particular, some research is still necessary to refine the model in order to identify even the smallest segments, such as the ones missing in the prediction output of the sample shown in Figure 13c. This cannot be done by further zooming the available images, thus different techniques have to be designed in the future, or images with higher resolution are necessary.

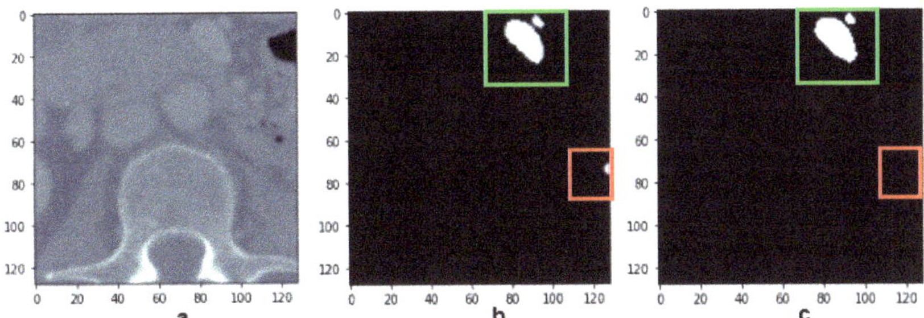

Figure 13. Portal vein sample 6: image (**a**), original mask (**b**) and predicted mask (**c**) after pre-processing with zoom.

The optimal zoom level was found by testing different configurations, as shown in Table 2. Figure 14 shows the trend of the DSC metric as a function of the zoom level. The saturation effect is evident for values higher than a zoom level of 3×. This effect can be explained by the fact that the maximum lossless zoom that can be performed on these images is 4×, since the original resolution of the images is 512×512 and the target one used to feed the U-Net is 128×128. Beyond this zoom value, the cropped image will have a size inferior to 128×128; thus, further improvements for larger magnification levels are not expected. In this case, it is not possible to use a zoom value of 4×, since this would also crop part of the white portion of the mask. In addition, the figure shows that the proposed zoom procedure does not impact the Accuracy metric, which remains stable in the range 98–99%, as expected. In fact, our procedure does not involve any loss of useful data for the mask of the considered organ; thus, the Accuracy is not affected by it.

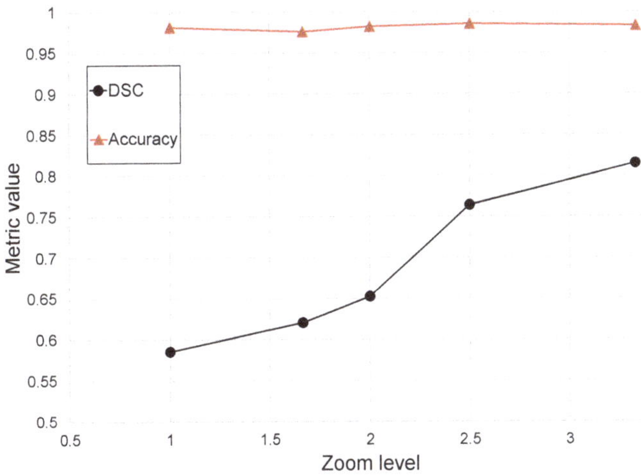

Figure 14. DSC and Accuracy metrics as a function of the zoom level.

Another evolutionary step is in the automatization of the whole process. In the current approach, the area of interest must be first identified by the medical staff. Then, the automatic zoom on the selected area is performed on the CT Scans and the corresponding masks, obtaining images in a 128×128 format. Although most of the operations are already automatic, it would be possible to leverage the first phase in order to identify the area deserving more attention and using the first output to automatically focus the zoom operation on the right portion of the image. This will be an objective of future investigations. In this regard, Figure 15 presents a general overview of the segmentation procedure detailed in this section. After the preliminary image preprocessing, consisting in the rescaling to a resolution of 128 × 128, the U-Net is trained on a portion of the dataset. The performance is then evaluated on the remainder of the samples. In the presence of small organs, as shown in what follows, a significant DSC decrease can be observed. Hence, if this decrease produces DSC values lower than a predefined threshold, corresponding to an 80% decrease in this paper, samples are further processed with an incremental zoom of images. This procedure stops in cases of data loss, i.e., if part of the organ signal is cut away by the zoom, or when the U-Net is able to achieve the desired performance score in terms of DSC.

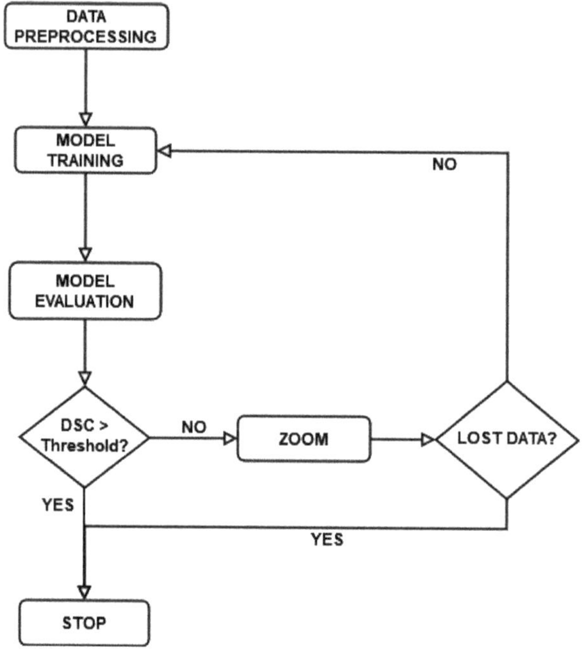

Figure 15. Flow diagram of the training of the mixed size medical segmentation procedure via U-Net.

Small organ segmentation has been previously investigated in other papers, such as [34,35]. FocusNetv2, presented in [34], was developed to segment small organs in the head and neck (HaN) area. This network is composed of various parts. They include a main segmentation network based on Snet, a U-Net variant, a small-organ localization network, to localize the center locations of small organs and a small-organ segmentation network, which refines the Snet's segmentation with the information provided by the small-organ localization network. The proposed ensemble is tested on various organs of different size, providing a maximum DSC of 82.45% on small organs. Hence, the performance obtained with our approach is comparable with the DSC obtained with more complex and computationally expensive models, at a reduced cost. A modular network is used in [35] as well, to segment pancreas images, a small organ, from the NIH dataset. The proposed network operates with two levels, a coarse-segmentation stage and a fine-segmentation stage, with the introduction of a saliency transformation module. This module converts the previous iteration's segmentation probability map into spatial weights for the current iteration. The proposed model produces a DSC of 84.5% on the pancreas NIH dataset. Hence, the performance obtained with the approach presented in this paper is comparable with the state-of-the-art, without requiring additional complexity in the network, increasing neither the computational cost nor the needed volume of training data.

In summary, with the proposed approach, the segmentation of binary masks with sparse signals, such as in the case of vascularization or small organ scans, is improved. In particular, an increased accuracy evaluated through the DSC metric is of great importance in the medical field, especially for vascularization images. In fact, the analysis of these images is fundamental to accelerating the diagnosis of various cancer types, due to the tumor angiogenesis process [5]. It is worth stressing that the achieved performance is reached with the base version of Google Colab, without the need for additional, paid resources. This is an important result, since it demonstrates that testing with U-Net can be carried out easily in a standard developing environment available to everyone, without requiring one to set up a dedicated cluster with multi-core CPUs or GPUs. Thus, the entry level of this approach is represented only by the skill of the programmers, without infrastructure costs. The price to pay is a small variability in execution times, due to the fact that resources in Google Colab are not reserved exclusively for free users and the fact that activity cannot last more than 12 consecutive hours. However, this is definitely acceptable for a developing work. Obviously, it is possible to set up a local cluster offering the Jupyter Notebook service or pay for dedicated cloud resources in order to not experience the limitations of the free account, but this is more suitable in operation, when service discontinuity is not acceptable. In any case, the software to run is exactly the same developed on the free version of Google Colab.

Finally, it is interesting to evaluate the time taken by the overall procedure to complete the test phase, i.e., in operation. The results of the evaluation campaign are reported in Table 3. The evaluation included four cases, each one repeated several times to evaluate both mean time and standard deviation, as follows: (i) single image; (ii) set of images of a single patient, with 129 slices; (iii) set of images of a single patient, with 172 slices; (iv) set of images of a single patient, with 200 slices. The usage of the dataset of different patients, each with a different number of slices, allows evaluating the scalability of the procedure. The used images are already loaded in Google Colab, to avoid including the effects of the upload time in the evaluation, since it depends on external factors, such as speed and reliability of network connection. It results that the segmentation of a single image taken from a batch of the test set by Tensorflow requires 3.286 s on average. In addition, in cases in which the zoom of images is not necessary, the initial pre-processing (rescaling operation) requires a mean time of 21 ms. In cases in which the zooming procedure is completed, it takes a slightly higher mean time, equal to 23 ms. Both of them are negligible with respect to the processing time. When the processing is executed on a larger set of images, processed in batches of 10 images, both processing and pre-processing times scale very well. In fact, they exhibit a sub-linear behaviour, with a processing time for 200 images well below 20 s and pre-processing always below 100 ms, with an impact of zooming operation that is essentially negligible. This phenomenon can be explained by the high parallelism that can be achieved with Tensorflow when batches of data are processed, that allows to fully exploit the capabilities of multi-core CPUs and GPUs. A second comment is that, in general, the variability of the processing time, evaluated by means of the standard deviation, is always quite limited, which is highly desirable. When execution times are very small, such as in the case of pre-processing, results could be slightly different from expected, due to free resource allocation in Google Colab. In short, the time taken to execute the overall procedure is negligible with respect to the time taken by a human operator to complete the same task, which makes the proposed approach definitely affordable.

Table 3. Processing and pre-processing times for different sets of images.

Number of Images	Processing Time (Avg ± Std)	Pre-Processing Time (Avg ± Std)	
		without Zoom	with Zoom
1	3.286 s ± 0.046 s	0.021 s ± 0.001 s	0.023 s ± 0.003 s
129	11.669 s ± 0.854 s	0.028 s ± 0.004 s	0.051 s ± 0.011 s
172	14.918 s ± 1.045 s	0.027 s ± 0.005 s	0.081 s ± 0.073 s
200	17.189 s ± 0.132 s	0.055 s ± 0.041 s	0.095 s ± 0.017 s

5. Conclusions

In this study, the performance of a state-of-the-art segmentation model, based on U-Net [4] is first analyzed by using CT-Scans of different organs, namely liver, bones and portal vein. Our results show that the standard U-Net network can provide very good results for segmenting large organs, approaching 98% Accuracy values in terms of DSC metric, but it exhibits poor performance when organ images characterized by small and sparse segmentation training masks are used. In particular, this happened in the case of vascularization images. In fact, segmentation of small organs is a recurring and challenging issue for automated medical Image Segmentation [22,34,35]. In order to overcome this problem, a novel approach producing a significant improvement of the DSC metric in the most critical cases is proposed. This approach does not require the use of additional data samples, nor a significant additional computational burden. In fact, for all our analyses, the free computing infrastructure made available by Google Colab was enough.

In particular, the suitable working conditions of the baseline U-Net are determined for segmenting sparse and/or small sections through image pre-processing. Since a quite general approach and metric (the DSC) to drive the segmentation is used, the solution can be used in other situations based on similar images and metrics. In the most critical case, relevant to portal vein images, the DSC improvement obtained is 20%. This comes basically at a very small cost, leaving untouched the training time and computing requirements in comparison with the baseline U-Net processing.

Our approach is based on training the U-Net by using differentiated zoom levels in different areas of test images. Thus, it follows that resorting to higher resolution images can bring further significant benefits in terms of segmentation accuracy. Clearly, an increase in resolution would require additional storage space and computing power. However, determination of the achievable performance improvement, if anything, needs further research. In fact, the observed improvement is due to the usage of the Dice metric and the impact of additional resolution, and a higher scaling factor on images cannot be easily determined; a deep investigation is necessary.

Due to its flexibility and small computational cost, the application of differentiated zoom levels, associated with the Dice metric, used to enhance the segmentation quality of small entities, can potentially be applied to other use-cases characterized by sparse segmentation signals. Nevertheless, the application of the proposed approach to other use-cases is beyond the scope of this paper. Finally, the proposed approach can have a significant impact in operation, since a correct segmentation of a vasculature or any other small organs and disease areas is essential for helping medical diagnosis.

Author Contributions: Conceptualization, methodology and validation, G.R.; software, G.R. and P.B.; writing—original draft preparation, P.B. and M.F.; writing—review and editing, M.F. and G.R.; supervision, G.R.; funding acquisition, G.R. All authors have read and agreed to the published version of the manuscript.

Funding: This research was funded by the company Molecular Horizon srl to the Department of Engineering of the University of Perugia.

Institutional Review Board Statement: Not applicable.

Informed Consent Statement: Not applicable.

Data Availability Statement: This reasearch has been conducted on the publicy availabe Liver segmentation 3D-IRCADb-01 from IRCAD institute, available at https://www.ircad.fr/research/data-sets/liver-segmentation-3d-ircadb-01/ (accessed on 1 November 2022).

Conflicts of Interest: The authors declare no conflict of interest.

Abbreviations

The following abbreviations are used in this manuscript:

CNN	Convolutional Neural Network
CT	Computed Tomography
HU	Hounsfield Unit
MRI	Magnetic Resonance Imaging
DICOM	Digital Imaging and Communications in Medicine
VTK	Visualization Toolkit
GPU	Graphics Processing Unit
VM	Virtual Machine

Appendix A

Listing A1. Loading CT-Scan into the related tensors starting from the file paths.

```
1  def process_path(filename):
2      patient_bytes = tf.io.read_file(filename)
3      patient_image = tfio.
4                  image.
5                  decode_dicom_image(patient_bytes,
6                                  color_dim=False,
7                                  on_error='skip',
8                                  scale='preserve',
9                                  dtype=tf.uint16,
10                                 name=None)
11     tf.cast(patient_image, tf.int32, name=None)
12     patient_image = tf.image.resize(patient_image, (D,D))
13     patient_image = tf.squeeze(patient_image, axis=0)
14     mask_path = tf.strings.regex_replace(filename,
15                                 'PATIENT_DICOM',
16                                 'MASKS_DICOM/liver')
17     mask_bytes = tf.io.read_file(mask_path)
18     mask_image = tfio.image.decode_dicom_image(mask_bytes,
19                                 scale='auto',
20                                 on_error='lossy',
21                                 dtype=tf.uint8)
22     mask_image = tf.squeeze(mask_image, axis=0)
23     intercept_tag = tfio.
24                  image.
25                  decode_dicom_data(patient_bytes,
26                             tags=np.uint32(int("00281052",
27                                 16)))
28     intercept = tf.strings.to_number(intercept_tag, tf.float32)
29     slope_tag = tfio.
30                  image.
31                  decode_dicom_data(patient_bytes,
32                             tags=np.uint32(int("00281053",
33                                 16)))
34     slope = tf.strings.to_number(slope_tag, tf.float32)
35     patient_image = tf.
36                  math.
37                  add(tf.math.multiply(patient_image, slope),
38                      intercept) /100.-10.
39     return patient_image, mask_image
```

References

1. Kapoor, L.; Thakur, S. A survey on brain tumor detection using image processing techniques. In Proceedings of the 2017 7th International Conference on Cloud Computing, Data Science & Engineering—Confluence, Noida, India, 12–13 January 2017; pp. 582–585. [CrossRef]
2. Jaume, S.; Ferrant, M.; Macq, B.; Hoyte, L.; Fielding, J.; Schreyer, A.; Kikinis, R.; Warfield, S. Tumor detection in the bladder wall with a measurement of abnormal thickness in CT scans. *IEEE Trans. Biomed. Eng.* **2003**, *50*, 383–390. [CrossRef] [PubMed]
3. Fu, K.; Mui, J. A survey on Image Segmentation. *Pattern Recognit.* **1981**, *13*, 3–16. [CrossRef]
4. Ronneberger, O.; Fischer, P.; Brox, T. U-Net: Convolutional Networks for Biomedical Image Segmentation. *arXiv* **2015**, arXiv:1505.04597[cs].
5. Ehling, J.; Theek, B.; Gremse, F.; Baetke, S.; Möckel, D.; Maynard, J.; Ricketts, S.A.; Grüll, H.; Neeman, M.; Knuechel, R.; et al. Micro-CT imaging of tumor angiogenesis: Quantitative measures describing micromorphology and vascularization. *Am. J. Pathol.* **2014**, *184*, 431–441. [CrossRef]
6. Carass, A.; Roy, S.; Gherman, A.; Reinhold, J.; Jesson, A.; Arbel, T.; Maier, O.; Handels, H.; Ghafoorian, M.; Platel, B.; et al. Evaluating White Matter Lesion Segmentations with Refined Sørensen-Dice Analysis. *Sci. Rep.* **2020**, *10*, 8242. [CrossRef]
7. Powers, D.M.W. Evaluation: From precision, recall and F-measure to ROC, informedness, markedness and correlation. *arXiv* **2020**, arXiv:2010.16061.
8. Wang, R.; Lei, T.; Cui, R.; Zhang, B.; Meng, H.; Nandi, A.K. Medical Image Segmentation using deep learning: A survey. *IET Image Process.* **2022**, *16*, 1243–1267. [CrossRef]
9. International Atomic Energy Agency. *Diagnostic Radiology Physics—A Handbook for Teachers and Students*; Non-Serial Publications, IAEA: Wien, Austria, 2014 .
10. Horowitz, S.L.; Pavlidis, T. Picture Segmentation by a Tree Traversal Algorithm. *J. ACM* **1976**, *23*, 368–388. [CrossRef]
11. Zaitoun, N.M.; Aqel, M.J. Survey on Image Segmentation Techniques. *Procedia Comput. Sci.* **2015**, *65*, 797–806.
12. Albawi, S.; Mohammed, T.A.; Al-Zawi, S. Understanding of a convolutional neural network. In Proceedings of the 2017 International Conference on Engineering and Technology (ICET), Antalya, Turkey, 21–23 August 2017; pp. 1–6. [CrossRef]
13. Wang, G.; Li, W.; Zuluaga, M.; Aughwane, R.; Patel, P.; Aertsen, M.; Doel, T.; David, A.; Deprest, J.; Ourselin, S.; et al. Interactive Medical Image Segmentation Using Deep Learning With Image-Specific Fine Tuning. *IEEE Trans. Med Imaging* **2018**, *37*, 1562–1573. [CrossRef]
14. Wang, G.; Zuluaga, M.A.; Li, W.; Pratt, R.; Patel, P.A.; Aertsen, M.; Doel, T.; David, A.L.; Deprest, J.A.; Ourselin, S.; et al. DeepIGeoS: A Deep Interactive Geodesic Framework for Medical Image Segmentation. *IEEE Trans. Pattern Anal. Mach. Intell.* **2019**, *41*, 1559–1572. [CrossRef] [PubMed]
15. Simonyan, K.; Zisserman, A. Very Deep Convolutional Networks for Large-Scale Image Recognition. In Proceedings of the 3rd International Conference on Learning Representations, ICLR 2015, San Diego, CA, USA, 7–9 May 2015.
16. Li, X.; Qian, W.; Xu, D.; Liu, C. Image Segmentation Based on Improved Unet. *J. Phys. Conf. Ser.* **2021**, *1815*, 012018. [CrossRef]
17. Lu, H.; She, Y.; Tie, J.; Xu, S. Half-UNet: A Simplified U-Net Architecture for Medical Image Segmentation. *Front. Neuroinform.* **2022**, *16*. [CrossRef] [PubMed]
18. Siddique, N.; Paheding, S.; Elkin, C.P.; Devabhaktuni, V. U-Net and Its Variants for Medical Image Segmentation: A Review of Theory and Applications. *IEEE Access* **2021**, *9*, 82031–82057. [CrossRef]
19. Han, G.; Zhang, M.; Wu, W.; He, M.; Liu, K.; Qin, L.; Liu, X. Improved U-Net based insulator Image Segmentation method based on attention mechanism. *Energy Rep.* **2021**, *7*, 210–217.
20. Saood, A.; Hatem, I. COVID-19 lung CT Image Segmentation using deep learning methods: U-Net versus SegNet. *BMC Med. Imaging* **2021**, *21*, 19. [CrossRef]
21. Gao, Y.; Huang, R.; Chen, M.; Wang, Z.; Deng, J.; Chen, Y.; Yang, Y.; Zhang, J.; Tao, C.; Li, H. FocusNet: Imbalanced Large and Small Organ Segmentation with an End-to-End Deep Neural Network for Head and Neck CT Images. In *Proceedings of the Medical Image Computing and Computer Assisted Intervention—MICCAI 2019: 22nd International Conference, Shenzhen, China, 13–17 October 2019, Proceedings, Part III*; Springer: Berlin/Heidelberg, Germany, 2019; pp. 829–838. [CrossRef]
22. Valindria, V.V.; Lavdas, I.; Cerrolaza, J.J.; Aboagye, E.O.; Rockall, A.G.; Rueckert, D.; Glocker, B. Small Organ Segmentation in Whole-body MRI using a Two-stage FCN and Weighting Schemes. In *Proceedings of the Machine Learning in Medical Imaging: 9th International Workshop, MLMI 2018, Held in Conjunction with MICCAI 2018, Granada, Spain, 16 September 2018*; Springer: Berlin/Heidelberg, Germany, 2018.
23. Wu, W.; Zhang, J.; Xie, H.; Zhao, Y.; Zhang, S.; Gu, L. Automatic detection of coronary artery stenosis by convolutional neural network with temporal constraint. *Comput. Biol. Med.* **2020**, *118*, 103657. [CrossRef]
24. Jun, T.J.; Kweon, J.; Kim, Y.H.; Kim, D. T-Net: Nested encoder–decoder architecture for the main vessel segmentation in coronary angiography. *Neural Netw.* **2020**, *128*, 216–233. [CrossRef]
25. Çiçek, O.; Abdulkadir, A.; Lienkamp, S.S.; Brox, T.; Ronneberger, O. 3D U-Net: Learning Dense Volumetric Segmentation from Sparse Annotation. In *International Conference on Medical Image Computing and Computer-Assisted Intervention*; Springer: Cham, Switzerland, 2016. [CrossRef]
26. Soler, L.; Hostettler, A.; Agnus, V.; Charnoz, A.; Fasquel, J.; Moreau, J.; Osswald, A.; Bouhadjar, M.; Marescaux, J. 3D image reconstruction for comparison of algorithm database: A patient specific anatomical and medical image database. *Tech. Rep. IRCAD* **2010**. Available online: http://www-sop.inria.fr/geometrica/events/wam/abstract-ircad.pdf (accessed on 28 September 2022).

27. Bishop, C.M. Pattern Recognition and Machine Learning. In *Information Science and Statistics*; Springer: New York, NY, USA, 2006; Volume 4, p. 738. [CrossRef]
28. Keras API. Available online: https://keras.io/api/ (accessed on 28 September 2022).
29. Han, J.; Moraga, C. The influence of the sigmoid function parameters on the speed of backpropagation learning. In *Proceedings of the From Natural to Artificial Neural Computation*; Mira, J., Sandoval, F., Eds.; Springer: Berlin/Heidelberg, Germany, 1995; pp. 195–201.
30. Kingma, D.P.; Ba, J. Adam: A Method for Stochastic Optimization. In Proceedings of the 3rd International Conference on Learning Representations, ICLR 2015, San Diego, CA, USA, 7–9 May 2015.
31. Ruder, S. An overview of gradient descent optimization algorithms. *arXiv* **2016**, arXiv:1609.04747.
32. Mustapha, A.; Lachgar, M.; Ali, K. Comparative study of optimization techniques in deep learning: Application in the ophthalmology field Comparative study of optimization techniques in deep learning: Application in the ophthalmology field. *J. Phys. Conf. Ser.* **2021**, *1743*, 012002. [CrossRef]
33. Manugunta, R.K.; Maskeliūnas, R.; Damaševičius, R. Deep Learning Based Semantic Image Segmentation Methods for Classification of Web Page Imagery. *Future Int.* **2022**, *14*, 277. [CrossRef]
34. Gao, Y.; Huang, R.; Yang, Y.; Zhang, J.; Shao, K.; Tao, C.; Chen, Y.; Metaxas, D.N.; Li, H.; Chen, M. FocusNetv2: Imbalanced large and small organ segmentation with adversarial shape constraint for head and neck CT images. *Med. Image Anal.* **2021**, *67*, 101831. [CrossRef]
35. Yu, Q.; Xie, L.; Wang, Y.; Zhou, Y.; Fishman, E.K.; Yuille, A.L. Recurrent Saliency Transformation Network: Incorporating Multi-stage Visual Cues for Small Organ Segmentation. In *Proceedings of the 2018 IEEE/CVF Conference on Computer Vision and Pattern Recognition (CVPR)*; IEEE Computer Society: Los Alamitos, CA, USA, 2018; pp. 8280–8289. [CrossRef]
36. Wang, L.; Wang, C.; Sun, Z.; Chen, S. An Improved Dice Loss for Pneumothorax Segmentation by Mining the Information of Negative Areas. *IEEE Access* **2020**, *8*, 167939–167949. [CrossRef]
37. Jadon, S. A survey of loss functions for semantic segmentation. In Proceedings of the 2020 IEEE Conference on Computational Intelligence in Bioinformatics and Computational Biology (CIBCB), Via del Mar, Chile, 27–29 October 2020; pp. 1–7. [CrossRef]
38. Google Colab FAQs. Available online: https://research.google.com/colaboratory/faq.html (accessed on 28 September 2022).
39. Tesla K80 | NVIDIA. Available online: https://www.nvidia.com/en-gb/data-center/tesla-k80/ (accessed on 28 September 2022).
40. Weng, Y.; Zhou, T.; Li, Y.; Qiu, X. NAS-Unet: Neural Architecture Search for Medical Image Segmentation. *IEEE Access* **2019**, *7*, 44247–44257. [CrossRef]
41. Ben naceur, M.; Akil, M.; Saouli, R.; Kachouri, R. Fully automatic brain tumor segmentation with deep learning-based selective attention using overlapping patches and multi-class weighted cross-entropy. *Med. Image Anal.* **2020**, *63*, 101692. [CrossRef]

applied
sciences

Article

Binary Starling Murmuration Optimizer Algorithm to Select Effective Features from Medical Data

Mohammad H. Nadimi-Shahraki [1,2,3,*], Zahra Asghari Varzaneh [4], Hoda Zamani [1,2] and Seyedali Mirjalili [3,5,*]

1 Faculty of Computer Engineering, Najafabad Branch, Islamic Azad University, Najafabad 8514143131, Iran
2 Big Data Research Center, Najafabad Branch, Islamic Azad University, Najafabad 8514143131, Iran
3 Centre for Artificial Intelligence Research and Optimisation, Torrens University Australia, Brisbane, QLD 4006, Australia
4 Department of Computer Science, Faculty of Mathematics and Computer, Shahid Bahonar University of Kerman, Kerman 7616914111, Iran
5 Yonsei Frontier Lab, Yonsei University, Seoul 03722, Republic of Korea
* Correspondence: nadimi@iaun.ac.ir (M.H.N.-S.); ali.mirjalili@torrens.edu.au (S.M.)

Abstract: Feature selection is an NP-hard problem to remove irrelevant and redundant features with no predictive information to increase the performance of machine learning algorithms. Many wrapper-based methods using metaheuristic algorithms have been proposed to select effective features. However, they achieve differently on medical data, and most of them cannot find those effective features that may fulfill the required accuracy in diagnosing important diseases such as Diabetes, Heart problems, Hepatitis, and Coronavirus, which are targeted datasets in this study. To tackle this drawback, an algorithm is needed that can strike a balance between local and global search strategies in selecting effective features from medical datasets. In this paper, a new binary optimizer algorithm named BSMO is proposed. It is based on the newly proposed starling murmuration optimizer (SMO) that has a high ability to solve different complex and engineering problems, and it is expected that BSMO can also effectively find an optimal subset of features. Two distinct approaches are utilized by the BSMO algorithm when searching medical datasets to find effective features. Each dimension in a continuous solution generated by SMO is simply mapped to 0 or 1 using a variable threshold in the second approach, whereas in the first, binary versions of BSMO are developed using several S-shaped and V-shaped transfer functions. The performance of the proposed BSMO was evaluated using four targeted medical datasets, and results were compared with well-known binary metaheuristic algorithms in terms of different metrics, including fitness, accuracy, sensitivity, specificity, precision, and error. Finally, the superiority of the proposed BSMO algorithm was statistically analyzed using Friedman non-parametric test. The statistical and experimental tests proved that the proposed BSMO attains better performance in comparison to the competitive algorithms such as ACO, BBA, bGWO, and BWOA for selecting effective features from the medical datasets targeted in this study.

Keywords: disease diagnosis; medical data; feature selection; binary metaheuristic algorithms; starling murmuration optimizer (SMO); transfer function

Citation: Nadimi-Shahraki, M.H.; Asghari Varzaneh, Z.; Zamani, H.; Mirjalili, S. Binary Starling Murmuration Optimizer Algorithm to Select Effective Features from Medical Data. *Appl. Sci.* **2023**, *13*, 564. https://doi.org/10.3390/app13010564

Academic Editor: Cosimo Nardi

Received: 20 November 2022
Revised: 20 December 2022
Accepted: 29 December 2022
Published: 31 December 2022

1. Introduction

With recent advancements in medical information technology, a huge volume of raw medical data is rapidly generated from different medical resources such as medical examinations, radiology, laboratory tests, mobile health applications, and wearable health-care technologies [1–3]. Extracting informative knowledge from these medical data using artificial intelligence and machine learning algorithms can help in faster treatment and significantly reduce patient mortality rates [4,5]. Application of these algorithms in some diseases such as Diabetes, Heart problems, Hepatitis, and Coronavirus is more common than others due to their high epidemic and mortality rates, expensive tests, and the requirement of special experience [6–8]. One of the main challenges in such disease datasets is the existence of redundant and irrelevant features [9], which can decrease the effectiveness of disease diagnosis systems. In medical data mining and machine learning [10,11], one of the most crucial preprocessing steps is feature selection, which eliminates redundant and irrelevant features to uncover effective ones. Since there are 2^N distinct feature subsets in a dataset with N features, the feature selection problem is NP-hard [12,13]. Therefore, evaluating all feature subsets to find effective features is very costly, and if each feature is added to the dataset, then the complexity will be doubled [13,14].

Filter-based, wrapper-based, and embedded methods are the three main categories of feature selection techniques [15,16]. The classification algorithm is not involved in filter-based methods, which typically operate based on feature ranking. Wrapper-based methods use a classifier algorithm to evaluate individual candidate subsets of features as opposed to filter-based methods [17,18]. Embedded methods combine the qualities of filter and wrapper methods, and the feature selection algorithm is integrated as part of the learning algorithm [16]. Many wrapper feature selection methods based on metaheuristic algorithms have been proposed [15,16] that can effectively solve feature selection problems as an NP-hard problem in a reasonable response time [19,20]. The main goal of using metaheuristic algorithms is to search the feature space and find near-optimal solutions effectively. Metaheuristic algorithms are recognized as robust problem solvers to solve a variety of problems with different types, such as continuous [21], discrete [22–24], and constraint [25,26]. Particle swarm optimization (PSO) [27], ant colony optimization (ACO) [28], differential evolution (DE) [29], cuckoo optimization algorithm (COA) [30], krill herd (KH) [31], social spider algorithm (SSA) [32], crow search algorithm (CSA) [33], grasshopper optimization algorithm (GOA) [34], quantum-based avian navigation optimizer algorithm (QANA) [35] and African vultures optimization algorithm (AVOA) [36] are some of the successful metaheuristic algorithms that are promisingly developed to solve feature selection problems.

Many metaheuristic-based methods have been proposed to select features from medical data [37–39]. However, a few of them can select effective features that may provide acceptable accuracy in diagnosing all the targeted diseases in this study, including Diabetes, Heart problems, Hepatitis, and Coronavirus [40]. The main reason for this drawback is generating and storing many irrelevant and redundant features in the medical processes, which reduces the efficiency of classification algorithms used in disease diagnosis systems. Therefore, a metaheuristic algorithm is needed to select useful and effective features from medical datasets by striking a proper balance between local and global search strategies. Responding to this need, particularly for the datasets targeted in the scope of this study, is our motivation to introduce binary versions of the newly proposed starling murmuration optimizer (SMO) algorithm [41], which can balance between its search strategies efficiently. The SMO algorithm uses a dynamic multi-flock construction and three search strategies: separating, diving, and whirling. Starlings in large flocks turn, dive, and whirl across the sky in SMO. The separating search strategy enriches population diversity by employing the quantum harmonic oscillator. With the help of a quantum random dive operator, the diving search strategy enhances the exploration. In contrast, the whirling search strategy significantly uses cohesion force in the vicinity of promising regions. The SMO algorithm has shown a high ability to solve different complex and engineering problems, but it was not yet developed for solving feature selection problems. The binary version of SMO or BSMO is expected to effectively solve the feature selection problem.

The BSMO algorithm generates candidate subsets of features using two different approaches. The first approach develops binary versions of BSMO using several S-shaped and V-shaped transfer functions. In contrast, in the second approach, BSMO maps each dimension in a continuous solution generated by SMO to 0 or 1 using a variable threshold method. The scope of this study is limited to selecting effective features from four targeted datasets consisting of Diabetes, Heart, Hepatitis, and Coronavirus. The performance of the

BSMO's variants is assessed on targeted datasets in terms of fitness, accuracy, sensitivity, specificity, precision, and error. The results are contrasted with competing binary algorithms like the ant colony optimization (ACO) [28], binary bat algorithm (BBA) [42], binary grey wolf optimization (bGWO) [43], and binary whale optimization algorithm (BWOA) [39]. The main contributions of this study can be summarized as follows.

- Developing the BSMO algorithm as a binary version of the SMO algorithm.
- Transferring the continuous solutions to binary ones effectively using two different approaches, including S-shaped and V-shaped transfer functions and value threshold method.
- Evaluating BSMO on medical datasets targeted in this study and comparing its performance with other popular feature selection algorithms.
- Finding satisfactory results in selecting effective features from the targeted medical datasets.

The rest of this paper is organized as follows. The related works are reviewed in Section 2. A description of the standard SMO algorithm is presented in Section 3. The details of the proposed BSMO algorithm are presented in Section 4. Section 5 includes the experimental evaluation and the comparison between the proposed BSMO and contender algorithms. Section 6 concludes this study and its finding, and suggests some future works.

2. Related Works

Real-world optimization problems have different properties and involve various intricacies, creating critical challenges for optimization algorithms in solving them. Generally, optimization problems in mechanical and engineering applications are mostly faced with multiple properties, such as linear and non-linear constraints in decision variables, non-differentiable objectives, and constraint functions. Therefore, many constraint-handling methods, such as penalty functions, static, dynamic, annealing, adaptive, co-evolutionary, and the death penalty, are developed to cope with such challenges [44]. The other optimization problems, especially in feature selection applications, mostly involve different intricacies such as discrete search spaces, existing irrelevant and redundant features, and high dimensionality feature space. Feature selection is a common way in preprocessing phase to cope with such intricacies by selecting only a small subset of relevant features from the original dataset [45,46]. Feature selection reduces the feature space's dimensionality, speeds up the learning process, simplifies the learned model, and boosts classifier performance by eliminating redundant and irrelevant features [47–49].

The topic of feature selection is presented as a binary optimization problem with the conflicting objectives of reducing the number of features and enhancing classification accuracy. Each solution is presented by a D-dimensional binary vector that only has the two values 0 and 1, where 0 signifies that the corresponding feature is not selected, and 1 indicates that it is selected. The number of dimensions in this binary vector corresponds to the number of features in the initial feature dataset. In many machine learning and data mining tasks, including intrusion detection [50–53], spam detection [54,55], financial problem prediction [56], and classification [57–59]. Particularly, finding an optimal subset of features from medical datasets is a challenging problem that many researchers have recently considered. Metaheuristic algorithms are recognized as prominent problem-solver to solve optimization problems especially feature selection. Based on the source of their inspiration, metaheuristic algorithms may be divided into eight groups: physical-based, biology-based, swarm-based, social-based, mu-sic-based, sport-based, chemistry-based, and math-based [60–62]. Since most metaheuristic algorithms are proposed for continuous problems, many binarization methods such as logical operators, variable threshold methods and transfer functions, are developed to map the continuous feature space to the binary one. In the literature, the most famous transfer functions are S-shaped [63], V-shaped [64–66], U-shaped [67,68], X-shaped transfer function [69], and Z-shaped [70]. This section presents an overview of the most recent related works on metaheuristics for the wrapper feature selection problem in medical data classification.

Nadimi-Shahraki et al. [40] proposed an improved whale optimization algorithm called BE-WOA. In BE-WOA, a pooling mechanism and three effective search strategies, migration, preferential selection, and surrounded prey, are used to improve the WOA to select effective features from medical datasets. BE-WOA also applied to predict Coronavirus 2019 disease or COVID-19. The obtained results prove the efficiency of the BE-WOA algorithm. The gene selection technique is used for high-dimensional datasets where the number of samples is small, and the number of features is large. Finding the best feature subset in a dataset is the process of gene selection [71]. For gene selection, Alirezanejad et al. [72] developed two Xvariance heuristics against mutual congestion. This approach involves ranking the features first. Then, using Monte's cross-validation, ten subsets of features are chosen based on forward feature selection (FFS). To enhance the results, majority voting is applied to the features selected in the prior stage to calculate accuracy, sensitivity, specificity, and matthews correlation coefficient.

Asghari Varzaneh et al. [73] proposed a new COVID-19 intubation prediction strategy using the binary version of the horse herd optimization algorithm to select the effective features. The results of the tests showed that the proposed feature selection method is better than other methods. Pashaei et al. [74] introduced two binary variations of the chimp optimization algorithm using S-shaped and V-shaped transfer functions for biomedical data classification. In a recent study, Nadimi-Shahraki et al. [75] proposed the binary version of the quantum-based avian navigation optimizer algorithm (BQANA) to select the optimal feature subset from high-dimensional medical datasets. The reported results show that the BQANA using a threshold method can dominate all contender algorithms. Alweshah et al. [76] proposed the greedy crossover (GC) operator strategy to boost the exploration capability of the coronavirus herd immunity optimizer (CHIO). Then, some medical datasets were used to evaluate the performance of the proposed algorithm in addressing the feature selection problem in the field of medical diagnosis. The results indicated that the GC operator strikes a balance between the search strategies of the CHIO algorithm.

For challenges involving medical feature selection, Anter et al. [77] proposed a hybrid crow search optimization algorithm combined with chaos theory and a fuzzy c-means algorithm (CFCSA). The suggested algorithm avoids local optima and improves the CSA's convergence using chaos theory and the global optimization method. The test results show the efficiency and stability of CFCSA for solving medical data and real problems. Singh et al. [78] proposed a hybrid ensemble-filter wrapper feature selection algorithm to improve the performance of classifiers in medical data applications. In this algorithm, first, the filter-based method is used based on the weight points to produce the ranking of the features. Then, the sequential forward selection algorithm is used as a wrapper-based feature selection to generate an optimal feature subset. To propose the binary version of the atom search optimization algorithm (ASO), Too et al. [79] applied four S-shaped and four V-shaped transfer functions to solve the feature selection problem. Among the eight presented binary versions, BASO based on the S1–shaped transfer function has the highest performance. Moreover, Mirjalili et al. [67] proposed a new binary version of the PSO algorithm using a U-shaped transfer function to transform continuous velocity values into binary values. The results show that U-shaped transfer functions significantly increase the performance of BPSO.

Elgamal et al. [80] enhanced the reptile search optimization algorithm (RSA) by employing the chaotic map and simulated annealing algorithm to tackle feature selection issues for high-dimensional medical datasets. Applying chaos theory to RSA improves its exploration ability, and hybridizing RSA with the simulated annealing algorithm can avoid local optima trapping. Many metaheuristic algorithms have been proposed to solve feature selection problems, such as binary ant lion optimizer (BALO) [81], return-cost-based binary firefly algorithm (Rc-BBFA) [82], chaotic dragonfly algorithm (CDA) [83], binary chimp optimization algorithm (BChOA) [84], altruistic whale optimization algorithm (AltWOA) [85], binary African vulture optimization algorithm (BAVOA) [86], and binary dwarf mongoose optimization algorithm (BDMSAO) [87].

Studying related works shows that various metaheuristic algorithms have been used to select effective features from medical data. However, most of them cannot find effective features for providing an acceptable diagnosis of important diseases such as Diabetes, Heart, Hepatitis, and Coronavirus. To respond to this weakness, the BSMO algorithm is introduced to develop a new wrapper feature selection method for these diseases in this study.

3. Starling Murmuration Optimizer (SMO)

SMO is a population-based metaheuristic algorithm recently developed by Zamani et al. [41]. The SMO algorithm is modeled the starlings' behavior during their stunning murmuration using three new search strategies, separating, diving, and whirling. The starling's population is denoted by $S = \{s_1, s_2, \ldots, s_N\}$ where N is the population size. The position of each starling s_i at iteration t is denoted using a vector $X_i(t) = (x_{i,1}, x_{i,2}, \ldots, x_{i,D})$ and its fitness value is expressed by $F_i(t)$. In first iteration, each $X_i(t)$ is initiated by a uniform random distribution in a D-dimensional search space using Equation (1), where X^L and X^U are lower and upper bounds of the search space, respectively and $rand\,(0,1)$ is a random value between 0 and 1.

$$X_i(t) = X^L + rand(0,1) \times (X^U - X^L), \quad i = 1, 2, \ldots, N \tag{1}$$

For the rest of the iterations, the population of starlings is moved using the separating, diving, and whirling search strategies. The details of these search strategies are discussed in the following sections.

3.1. Separating Search Strategy

The separation search strategy is promoted diversity throughout the population. In this strategy, first, a portion of starlings with size P_{sep} are randomly selected to separate from population S using Equation (2). Then, some dimensions of the selected starlings are updated using Equation (3), where $X_G(t)$ is the global best position, and $X_r(t)$ is randomly selected from a population S. In each iteration, the best position obtained so far is stored, then these positions are joined with the separated positions with size P_{sep}, ultimately $X_{r'}(t)$ is randomly selected from these sets. $Q_1(y)$ is a separation operator which is calculated using Equation (4), where α is the quantum harmonic oscillator, parameters m and k are the particle's mass and strength, respectively and the parameter h is Planck's constant. Moreover, the function H_n is the Hermite polynomial with integer index n, and y is a random number.

$$P_{sep} = \frac{\log(t + D)}{\log(MaxIt) \times 2} \tag{2}$$

$$X_i(t+1) = X_G(t) + Q_1(y) \times (X_{r'}(t) - X_r(t)) \tag{3}$$

$$Q_1(y) = \left(\frac{\alpha}{2^n \times n! \times \pi^{\frac{1}{2}}}\right)^{\frac{1}{2}} H_n(\alpha \times y) \times e^{-0.5 \times \alpha^2 \times y^2}, \ \alpha = \left(\frac{m \times k}{\hbar}\right)^{\frac{1}{4}} \tag{4}$$

The rest of the starlings with a size of $\acute{N}$ $(N - P_{sep})$ is flocked using dynamic multi-flock construction to search the problem space using either diving or whirling search strategies. Each iteration creates a dynamic multi-flock using k non-empty flocks $f_1 \ldots f_k$. First, k best starlings are separated from the population $\acute{N}$ and stored in matrix R, then the rest of the population $(\acute{N}\text{-}R)$ is divided among the k flocks. Finally, each position of R assigns to each flock such that $f_{1\leftarrow} \{R_1 \cup f_1\}, \ldots, f_{k\leftarrow} \{R_k \cup f_k\}$.

As shown in Equation (6), the diving and whirling search strategies are assigned to the flocks based on the quality of each flock. The quality of each flock ($Q_q\,(t)$) is evaluated using Equation (5), where k is the number of flocks, $sf_{ij}\,(t)$ is the fitness value of the starling

s_i in the flock f_j, and n is the number of starlings in each flock. The parameter $\mu_Q(t)$ in Equation (6) denotes the average of all flock's quality.

$$Q_q(t) = \frac{\sum_{i=1}^{k} \frac{1}{n} \sum_{j=1}^{n} sf_{ij}(t)}{\frac{1}{n} \sum_{i=1}^{n} sf_{qi}(t)} \tag{5}$$

$$X_i(t+1) = \begin{cases} \text{Diving search strategy} & Q_q(t) \leq \mu_Q(t) \\ \text{Whirling search strategy} & Q_q(t) > \mu_Q(t) \end{cases} \tag{6}$$

3.2. Diving Search Strategy

The diving search strategy is encouraged the selected flocks ($Q_q(t) \leq \mu_Q(t)$) to explore the search space effectively. The starlings are moved using upward and downward quantum random dives (QRD). The starlings of a flock switch among these quantum dives using two quantum probabilities shown in Equation (7), where $\left|\psi^{Up}(X_i)\right|$ and $\left|\psi^{Down}(X_i)\right|$ are the upward and downward probabilities that are computed using Equations (8) and (9). Parameters φ and θ are set by the user, and $|\psi(\delta_2)\rangle$ is an inverse-Gaussian distribution that is computed using Equation (10), where the values of λ and μ are set by the user, and y is a random number.

$$QRD = \begin{cases} \text{Upward quantum dive} & \left|\psi^{Up}(X_i)\right| > \left|\psi^{Down}(X_i)\right| \\ \text{Downward quantum dive} & \left|\psi^{Up}(X_i)\right| \leq \left|\psi^{Down}(X_i)\right| \end{cases} \tag{7}$$

$$\left|\psi^{Up}(X_i)\right\rangle = e^{i\varphi}\cos\theta \times |\psi(\delta_2)\rangle - e^{-i\varphi}\sin\theta \times |\psi(\delta_2)\rangle \tag{8}$$

$$\left|\psi^{Down}(X_i)\right\rangle = e^{i\varphi}\sin\theta \times |\psi(\delta_2)\rangle + e^{-i\varphi}\cos\theta \times |\psi(\delta_2)\rangle \tag{9}$$

$$|\psi(\delta_2)\rangle = \sqrt{\frac{\lambda}{2 \times \pi \times y^3}} \times e\left[-\frac{\lambda(y-\mu)^2}{2 \times \mu^2 \times y}\right] \tag{10}$$

The downward and upward quantum dives are computed using Equations (11) and (12), respectively, where $|\psi(R_D)\rangle$ is selected from set R, $|\psi(X_i)\rangle$ is the position of starling s_i in the current iteration, the position of $|\psi(X_r)\rangle$ is randomly selected among flocks assigned for diving strategy, $|\psi(X_j)\rangle$ is randomly selected from the population S and the best starlings set. $|\psi(\delta_1)\rangle$ is a random position selected from the best starlings set obtained from the first iteration so far and the starling population S.

$$|\psi(t+1,\, X_i)\rangle = |\psi(R_D)\rangle - |\psi^{Down}(X_i)\rangle \times (|\psi(X_i)\rangle - |\psi(X_r)\rangle) \tag{11}$$

$$|\psi(t+1,\, X_i)\rangle = |\psi(R_D)\rangle + |\psi^{Up}(X_i)\rangle \times (|\psi(X_i)\rangle - |\psi(X_j)\rangle + |\psi(\delta_1)\rangle) \tag{12}$$

3.3. Whirling Search Strategy

Starlings of a flock exploit the search problem using the whirling search strategy when the quality of the flock is more than the average quality of all flocks ($Q_q(t) > \mu_Q(t)$). The whirling search strategy is denoted in Equation (13), where $X_i(t+1)$ is the next position of starling s_i at iteration t, a position $X_{RW}(t)$ is randomly selected from set R of flocks that are considered for the whirling search strategy, $X_N(t)$ randomly selected from all flocks that want to use the whirling search strategy. $C_i(t)$ is the cohesion operator which is calculated using Equation (14), where $\xi(t)$ is a random number between intervals 0 and 1.

$$X_i(t+1) = X_i(t) + C_i(t) \times (X_{RW}(t) - X_N(t)) \tag{13}$$

$$C_i(t) = cos(\xi(t)) \tag{14}$$

The pseudocode of the SMO algorithm is shown in Algorithm 1.

Algorithm 1: Starling Murmuration Optimizer (SMO)

Input: N (Population size), k (Flocks size), and *MaxIt* (Maximum iterations).
Output: Global best solution.

1:   **Begin**
2:   Randomly distributed N starlings in the search space.
3:   **Set** $t = 1$.
4:   **While** $t \leq MaxIt$
5:      Separating a portion of starlings with size P_{sep} from the population using Equation (2).
6:      The rest of the population is flocked into k flocks using the dynamic multi-flock construction.
7:      Computing the quality of each flock (fq) using Equation (5).
8:      **For** $q = 1: k$
9:        **If** $Q_q \, (t) \leq \mu_Q \, (t)$
10:         Moving starlings of the flock fq using the diving strategy.
11:        **Else**
12:         Moving starlings of the flock fq using the whirling strategy.
13:        **End if**
14:      **End for**
15:      Update the position of starlings and global best solution.
16:      $t = t + 1$.
17:   **End while**
18:   Return position of best starling as a global best solution.
19:   **End**

---

4. Binary Starling Murmuration Optimizer (BSMO)

SMO is a new metaheuristic algorithm that effectively solves various engineering and complex problems. However, the ability of the SMO algorithm to solve feature selection problems has not been studied yet, which is the motivation of this study. In this study, a binary starling murmuration optimizer (BSMO) is proposed to select effective features from the datasets of four important targeted diseases consisting Diabetes, Heart problems, Hepatitis, and Coronavirus. The proposed BSMO is developed using two different approaches. The first approach uses S-shaped and V-shaped transfer functions, whereas the second approach maps the continuous search space to 0 or 1 using a threshold value.

Suppose matrix X is to represent the population of starlings in the BSMO, then Figure 1 shows the representation scheme of the proposed BSMO algorithm in solving the feature selection problem. Figure 1a–c show starling S_i, binary vector B_i, and the selected feature set SF_i. Each starling S_i is transformed using different transform functions to the binary vector B_i in which the value of 1 for each element means the corresponding feature should be selected to form the selected feature set SF_i. Accordingly, the BSMO algorithm uses the fitness function defined in Equation (15) [83,88].

$$Fit_i = \alpha E + \beta \frac{|SF_i|}{D} \tag{15}$$

where E determines the error rate of the classification algorithm, $|SF_i|$ and D are the number of the selected feature in a subset of SF_i, and the total features in the dataset, respectively. α and $\beta = 1 - \alpha$ are two constant values to control the significance of the classification accuracy and feature subset reduction, respectively. Since the accuracy is more important of the number of features, usually β is very smaller than α, in this study, $\alpha = 0.99$ and $\beta = 0.01$, according to [89].

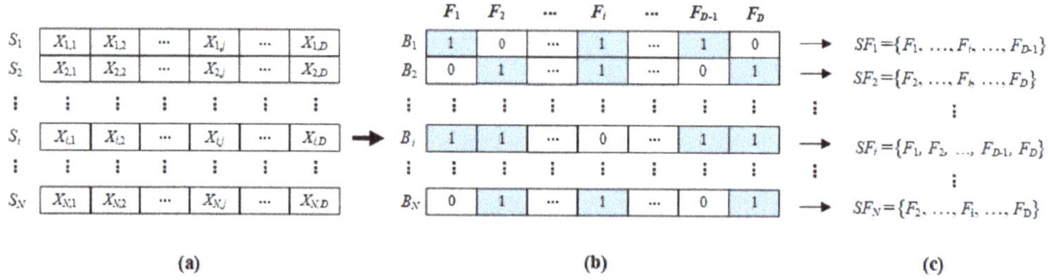

Figure 1. The representation scheme used by BSMO, (**a**) Starling population (Matrix X), (**b**) Binary population (Matrix B), and (**c**) Selected features (SF).

4.1. BSMO Using S-Shaped Transfer Function (S-BSMO)

This method uses the sigmoid transfer function (S-shape) to map the continuous to the binary version of the SMO algorithm. Therefore, updating the position of the starlings by the transfer functions S will cause them to be in a binary search space, and their position vector will only take the values of "0" or "1". The sigmoid function $S2$ formulated in Equation (16) first used in BPSO to develop a binary PSO [89,90].

$$S\left(x_i^d(t+1)\right) = \frac{1}{1+e^{-x_i^d(t)}} \tag{16}$$

where $x_i^d(t)$ and $S\left(x_i^d(t+1)\right)$ show the position and probability of changing the binary position value of the search agent i^{th} in dimension d in the tth iteration, respectively. Since the calculated value of S is still in continuous mode, it must be compared with a threshold value to create binary mode. Therefore, the new position of the search agent is updated using Equation (17), where $b_i^d(t+1)$ is a binary position of ith search agent in dimension d, and r is a random value between 0 and 1.

$$b_i^d(t+1) = \begin{cases} 0 & if & r < S\left(x_i^d(t+1)\right) \\ 1 & if & r \geq S\left(x_i^d(t+1)\right) \end{cases}, \tag{17}$$

In addition to the transfer function $S2$ introduced in Equation (16), three other types of S-shaped transfer functions, including $S1$, $S3$, and $S4$ have been used. All four transfer functions are formulated in Table 1. Moreover, all these transfer functions are shown visually in Figure 2. According to the figure, as the slope of the transfer function S increases, the probability of changing the position value increases. Therefore, $S1$ obtains the highest probability, and $S4$ obtains the lowest probability, effectively updating agents' position and finding the optimal solution.

Table 1. The formulation of S-shaped and V-shaped transfer functions.

Name	S-Shaped Transfer Functions	Name	V-Shaped Transfer Functions
S1-shaped	$T(x) = \frac{1}{1+e^{-2x}}$	V1-shaped	$T(x) = \left\lvert \mathrm{erf}\left(\frac{\sqrt{\pi}}{2}x\right) \right\rvert$
S2-shaped	$T(x) = \frac{1}{1+e^{-x}}$	V2-shaped	$T(x) = \lvert \tanh(x) \rvert$
S3-shaped	$T(x) = \frac{1}{1+e^{\frac{-x}{2}}}$	V3-shaped	$T(x) = \left\lvert \frac{x}{\sqrt{1+x^2}} \right\rvert$
S4-shaped	$T(x) = \frac{1}{1+e^{\frac{-x}{3}}}$	V4-shaped	$T(x) = \left\lvert \frac{2}{\pi}\arctan\left(\frac{\pi}{2}x\right) \right\rvert$

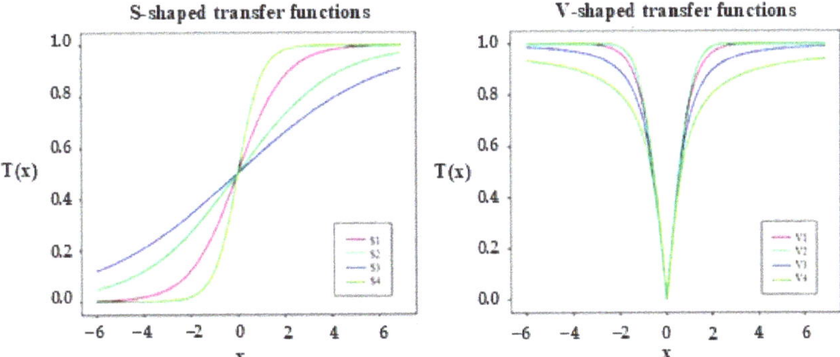

Figure 2. The S-shaped and V-shaped transfer functions [89].

4.2. BSMO Using V-Shaped Transfer Function (V-BSMO)

In this approach, the V-shaped transfer function is used to calculate the probability of changing the position of the agents in the SMO algorithm. Probability values are calculated using the V-shaped (hyperbolic) transfer function by Equation (18) [64], where $x_i^d(t)$ indicates the position value of the ith search agent in dimension d at iteration t.

$$V\left(x_i^d(t+1)\right) = \left|tanh\left(x_i^d(t)\right)\right| \tag{18}$$

Considering that the V-shaped transfer function is different from the S-shaped transfer function, after calculating the probability values, the Equation (19) [64] is used to update the position of each search agent.

$$b_i^d(t+1) = \begin{cases} x_i^d(t)^{-1} & if \quad r < V\left(x_i^d(t+1)\right) \\ \\ x_i^d(t) & if \quad r \geq V\left(x_i^d(t+1)\right) \end{cases} \tag{19}$$

where, $b_i^d(t+1)$ indicates the binary position of the ith search agent at iteration $t + 1$ in dimension d. Moreover, $x_i^d(t)^{-1}$ indicates the complement of $x_i^d(t)$. In addition, r is a random number in [0,1]. Unlike the S-shaped transfer function, the V-shaped transfer function does not force the search agents into 0 or 1. According to Equation (19), if the value of V is small and less than the value of r, the binary position of the search agents in dimension d will not change. On the other hand, if the calculated value of the transfer function is greater than or equal to the value r, the position of the search agents is changed to the complement of the current binary position. Table 1 formulates the mathematical equations of transfer functions $V1$, $V2$, $V3$, and $V4$, and Figure 2 represents transfer functions visually. According to Figure 2, $V1$ has the highest probability, and $V2$, $V3$, and $V4$ have lower probability values for moving the positions of search agents, respectively [89].

4.3. BSMO Using Variable Threshold Method (Threshold-BSMO)

In this section, the SMO transforms the continuous solutions into the binary form using the variable threshold method defined in Equation (20), where $b_i^d(t+1)$ is a new binary position of the ith search agent, and a variable threshold θ is 0.5 that is set by the user.

$$b_i^d(t+1) = \begin{cases} 1 & if \quad x_i^d(t+1) > \theta \\ \\ 0 & if \quad x_i^d(t+1) \leq \theta \end{cases} \tag{20}$$

Figure 3 represents the flowchart of the proposed BSMO algorithm, which is a binary version of the SMO algorithm to solve the feature selection problem. As shown in this figure, the optimization process is started by initializing the input variables, including a maximum number of iterations (*MaxIt*), population size (N), problem size (D), and flocks size (k). First, N starlings are randomly distributed in a D-dimensional search space. Then, a portion of starlings (P_{sep}) using Equation (2) are randomly selected to separate from the population and explore the search space using the separating strategy defined in Equation (3). The rest of the starlings are partitioned between different flocks to exploit the search space using the whirling strategy defined in Equation (13) or explore using the diving strategy defined in Equation (7). The obtained solutions from such search strategies are mapped to binary using two binarization approaches demonstrated in Table 1 and Equation (20). The obtained solutions are restricted to binary values 0 or 1 using Equations (17), (19), and (20). Finally, the solutions are evaluated using Equation (15). The optimization process is repeated until the termination condition, or MaxIt, is satisfied, and the global best solution is reported as the output variable.

4.4. The Computational Complexity of the BSMO Algorithm

Since BSMO has six distinct phases: initialization, separating search strategy, multi-flock construction, diving or whirling search strategy, mapping, and fitness evaluation, its computational complexity can be computed as follows. The initialization phase's computational complexity is O (ND), considering N starlings are randomly allocated in a D-dimensional search space using Equation (1). Then, a portion of the starlings is randomly selected using Equation (2) to explore the search space with computational complexity O (ND). The cost of the multi-flock construction phase to build k flocks by partitioning N starlings is O ($NlogN + k$). In the next phase, the cost of each flock containing n subpopulation for determining its quality utilizing Equation (5) is O (nD), and for moving by either diving or whirling search strategy is also O (nD). Thus, the overall complexity of this phase is O (knD) or O (ND) in the worst case. In the mapping phase, the continuous solutions are transformed into binary ones based on Table 1 and Equation (20) with computational complexity O (ND). Finally, in the fitness evaluation phase, the quality of binary solutions is assessed using Equation (15), consisting of a K-fold cross-validation method, k-NN classifier, and updating. The computational complexity of a K-fold cross-validation method with M samples is O (KM). Since K is a constant value, complexity equals O (M). The k-NN classifier with M samples and D features for training the classifier is O (MD), and the complexity of updating is O (ND). Since these phases are repeated T times, therefor the summation of the computational complexity of BSMO is O ($ND + T$ ($ND + (NlogN+k) + ND + ND + M + MD + ND$)), which is equal to O (TD ($N+M$)).

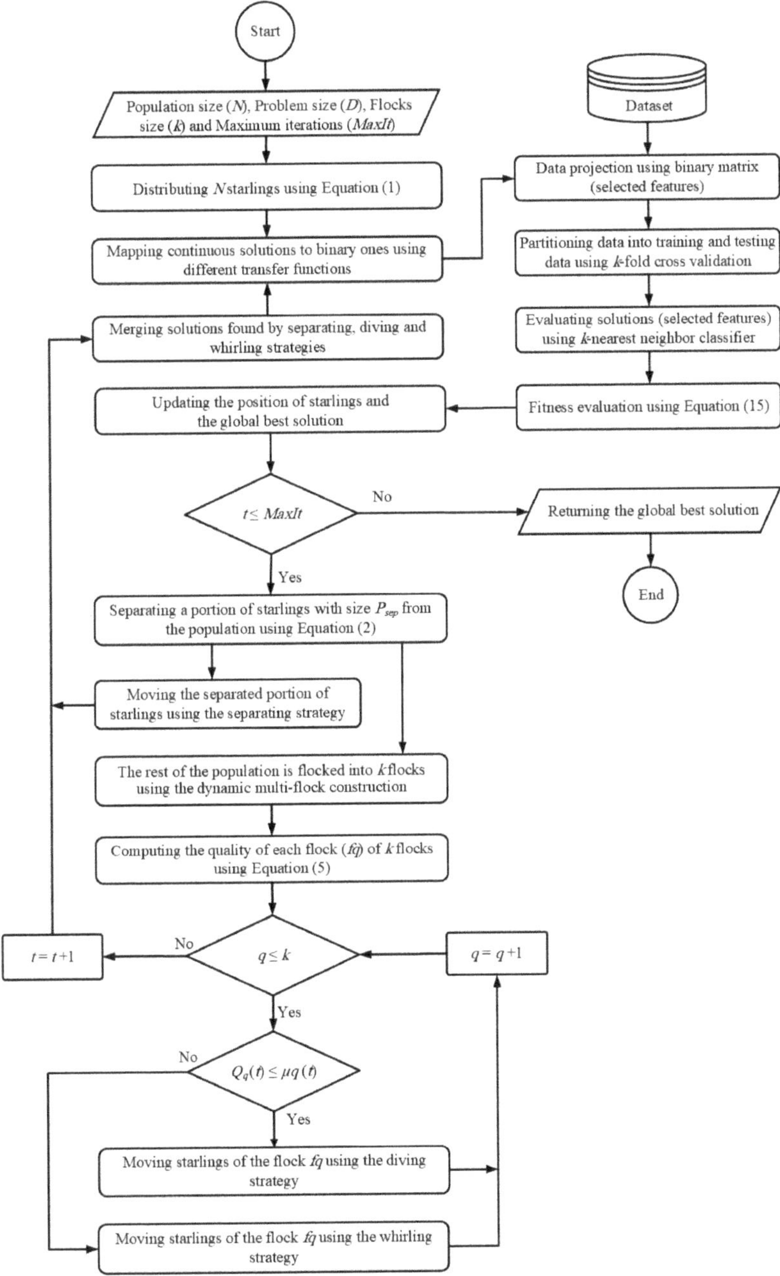

Figure 3. Flowchart of the proposed BSMO algorithm.

5. Experimental Evaluation

The performance of the proposed BSMO algorithm is assessed in finding the optimal feature subset from targeted datasets, Diabetes, Heart, Hepatitis, and Coronavirus diseases 2019, downloaded from [91,92]. Then, the nine BSMO variants' outcomes are then compared with those of competitive algorithms, ACO [28], BBA [42], bGWO [43], and BWOA [39].

All experiments are run under the same experimental conditions. MATLAB R2019b programming language is considered for implementing the BSMO and running all comparative algorithms. All experiments are run using an Intel (R) Core (TM) i5-3770 CPU, 3.4 GHz, 8 GB RAM, and Windows 10 with the 64-bit operating system.

5.1. Parameter Settings of Algorithms and k-NN Classifier

In this study, the k-nearest neighbor (*k-NN*) classifier with $k = 5$ is used to classify the feature subsets in all algorithms [93]. To learn the *k-NN* classifier, each dataset is randomly partitioned using a K-fold cross-validation method into training and testing sets, where K is a constant value equal to 10. One fold is used for the testing set, and the $K-1$ folds are applied for the training set [94,95].

For a fair comparison, all results were obtained under the same experimental conditions. The common parameters in BSMO and comparative algorithms, such as termination criterion and population size (N), are the same. In most optimization algorithms, the termination criterion is defined using the maximum number of iterations (*MaxIt*) or maximum function evaluations (*MaxFEs*), where *MaxIt* = *MaxFEs*/N and it is set to 300 and N is 30. Due to the stochastic nature of the algorithms, all simulations and obtained results are conducted with 15 independent runs. All results are reported using the standard statistical metrics maximum (*Max*), average (*Avg*), and minimum (*Min*) values. In each table, the best result is highlighted in boldface.

Table 2 shows the values of parameters used for BSMO and other comparative algorithms. The parameter values of all contender algorithms were set as same as their original papers. Moreover, a sensitivity analysis on key parameters of the BSMO algorithm, such as flock size (k), and population size (N), is performed to tune the values of these parameters using the offline parameter tuning method. The tuning results were reported in Tables A1–A6 of Appendix A in terms of fitness, error, accuracy, sensitivity, specificity, and precision metrics.

Table 2. Parameters setting.

Algorithms	Parameters
ACO	$\tau = 1, \eta = 1, \rho = 0.2, \alpha = 1$, and $\beta = 0.1$
BBA	$Q_{min} = 0$ and $Q_{max} = 2$
bGWO	a linearly decreases from 2 to 0, C_1, C_2, and C_3 are a random numbers
BWOA	a linearly decreases from 2 to 0, $b = 1$, r_1 and $r_2 \in$ rand (0, 1)
BSMO	$k = 5, \lambda = 20, \mu = 0.5, \theta$ and $\phi \in (0, 1.8)$

5.2. Evaluation Criteria

The performance of proposed BSMO and contender algorithms are assessed using evaluation criteria such as fitness, accuracy, sensitivity, specificity, precision, and error. The fitness evaluation metric is computed using Equation (15). The accuracy, sensitivity, specificity, precision, and error are calculated using Equations (21)–(25) [96,97]. In these equations, parameters *TP* and *TN* specify the number of positive and negative samples that are correctly classified by the classifier, respectively. *FN* is the number of positive samples incorrectly predicted as negative, and FP is the number of negative samples incorrectly predicted as positive using a classifier [98].

$$\text{Accuracy} = \frac{TP + TN}{TP + TN + FP + FN} \tag{21}$$

$$\text{Sensitivity} = \frac{TP}{TP + FN} \tag{22}$$

$$\text{Specificity} = \frac{TN}{TP + FN} \tag{23}$$

$$\text{Precision} = \frac{TP}{TP + FP} \tag{24}$$

The error metric is computed using the mean square error (MSE) denoted in Equation (25), where N is the number of samples, y_i is the observed values and $\hat{y}_i$ is the predicted value. Moreover, evaluating the proposed algorithm does not use any constraint handling methods since no constraints are considered in the feature selection problem.

$$\text{Error} = \frac{1}{N} \sum_{i=1}^{N} (y_i - \hat{y}_i)^2 \tag{25}$$

5.3. Numerical Results and Discussion

In this section, the simulation results of the proposed BSMO algorithm are presented on targeted medical datasets.

5.3.1. Comparison of Algorithms to Detect Diabetes Disease

The Pima Indian Diabetes dataset [91] consists of eight features, 268 samples with diabetes-positive labeling and 500 samples with diabetes-negative. The objective of this dataset is to detect whether or not a patient has diabetes. Table 3 shows that the proposed Threshold-BSMO can achieve the best performance compared to all comparative algorithms.

Table 3. Diabetes disease detection.

Algorithms	Fitness		Accuracy		Sensitivity		Precision		Specificity		Error	
	Avg	Min	Avg	Max	Avg	Max	Avg	Max	Avg	Max	Avg	Min
ACO	0.2384	0.2318	76.5109	77.0865	85.2345	86.6173	60.1832	64.0414	79.9663	82.1451	0.2351	0.2291
BBA	0.2331	0.2281	76.9974	77.4675	86.4089	88.748	79.8734	83.4279	59.365	63.0096	0.23	0.2253
bGWO	0.2295	0.2253	77.3573	77.8725	86.2124	89.3135	80.0664	83.5267	59.8209	65.9114	0.2264	0.2213
BWOA	0.2386	0.2344	76.4744	76.825	85.8432	87.8664	79.8754	82.3961	59.5944	64.961	0.2353	0.2317
S1-BSMO	0.2342	0.2266	76.9719	77.7409	88.5142	89.8454	83.2288	84.242	65.9602	68.2916	0.2504	0.2382
S2-BSMO	0.2352	0.2267	76.8537	77.7341	88.2422	89.2631	83.1426	84.2726	65.7932	68.023	0.2516	0.2369
S3-BSMO	0.2373	0.2291	76.6101	77.4897	88.1787	90.1796	82.925	84.4974	65.5806	67.8662	0.2508	0.2397
S4-BSMO	0.2368	0.2291	76.6654	77.4863	88.3085	89.7088	82.764	83.8476	65.0295	66.8104	0.2533	0.2384
V1-BSMO	0.2344	0.2294	76.889	77.3411	88.2848	89.7132	83.1764	86.1848	65.7345	70.5787	0.2552	0.2422
V2-BSMO	0.2343	0.2266	76.8872	77.6128	88.6261	90.0085	82.9072	83.761	65.7846	67.6503	0.2548	0.2345
V3-BSMO	0.2353	0.2306	76.7716	77.2163	88.245	89.6911	83.1812	84.5091	66.0204	69.1626	0.2547	0.2383
V4-BSMO	0.2335	0.2292	76.9639	77.471	88.1009	89.484	83.2658	84.4214	66.2564	69.1896	0.2534	0.2383
Threshold-BSMO	0.2306	0.2229	77.3077	77.9904	89	89.9871	83.5823	84.7376	66.6321	69.2028	0.253	0.2408

5.3.2. Comparison of Algorithms to Detect Heart Disease

The Statlog (Heart) dataset [91] consists of 13 features and 270 samples without no missing values to detect the absence or presence of heart disease. In this dataset 120 of the samples are labeled with the presence of heart disease and 150 samples are labeled with the absence of this disease. The performance of the proposed BSMO with nine variants is assessed and compared with well-known optimizers to diagnose heart disease. The results in Table 4 show that the proposed Threshold-BSMO can obtain a minimum fitness value of 0.1322 and a maximum accuracy of 87.037 than other algorithms.

Table 4. Heart disease detection.

Algorithms	Fitness		Accuracy		Sensitivity		Precision		Specificity		Error	
	Avg	Min	Avg	Max	Avg	Max	Avg	Max	Avg	Max	Avg	Min
ACO	0.147	0.1387	85.4815	86.2963	88.8186	94.1537	86.8452	89.665	82.7764	86.6325	0.1452	0.137
BBA	0.1414	0.1380	86.0123	86.2963	94.0096	95.4345	89.7266	91.4855	88.5579	91.1526	0.1959	0.1519
bGWO	0.1383	0.1358	86.4198	86.6667	87.4898	93.0586	85.4259	90.2422	80.7175	87.4738	0.1578	0.1444
BWOA	0.1409	0.1387	86.1728	86.2963	89.4656	91.3609	86.9606	90.1189	82.8787	88.087	0.1383	0.137
S1-BSMO	0.151	0.1432	85.1852	85.9259	89.2216	95.26	83.6512	89.9588	78.67	87.0474	0.1481	0.1407
S2-BSMO	0.146	0.1411	85.8148	86.2963	93.6608	95.0876	89.2841	91.4817	86.0433	88.7512	0.1964	0.1593
S3-BSMO	0.1481	0.1424	85.5185	85.9259	93.3517	95.3351	89.403	91.5718	86.2794	88.2128	0.2015	0.1556
S4-BSMO	0.1495	0.1432	85.3333	85.9259	93.1475	94.4033	89.7136	91.6581	87.0123	89.3531	0.1930	0.1556
V1-BSMO	0.1492	0.1403	85.3704	86.2963	93.2132	95.0297	89.4763	91.4379	86.3764	89.284	0.1907	0.1481
V2-BSMO	0.1423	0.1387	85.9383	86.2963	93.8571	96.2621	89.3417	91.9558	89.0497	91.2747	0.1884	0.1593
V3-BSMO	0.1417	0.1380	86.037	86.2963	94.4918	96.2525	89.3503	91.9198	88.4579	91.1828	0.1911	0.1481
V4-BSMO	0.1411	0.1351	86.0741	86.6667	94.1042	95.6443	89.6908	91.8579	88.5817	90.503	0.1956	0.1667
Threshold-BSMO	0.1371	0.1322	86.5432	87.037	89.8998	93.4192	86.7337	90.5212	82.2366	87.3123	0.1346	0.1296

5.3.3. Comparison of Algorithms to Detect Hepatitis Disease

The Hepatitis disease dataset [91] is complex with many missing values that contain occurrences of hepatitis in people. This dataset consists of 19 features with 155 samples, of which 123 samples are categorized in the live class, and 32 are categorized in the die class. The optimization algorithms try to find the best feature set which can detect Hepatitis disease with high accuracy. In this evaluation, the performance of the proposed algorithm is assessed and reported in Table 5. The results show that the BSMO using the variable threshold can obtain the optimum feature set with a minimum fitness value. Additionally, the Threshold-BSMO achieves the highest classification accuracy compared to the contender algorithm.

Table 5. Hepatitis disease detection.

Algorithms	Fitness		Accuracy		Sensitivity		Precision		Specificity		Error	
	Avg	Min	Avg	Max	Avg	Max	Avg	Max	Avg	Max	Avg	Min
ACO	0.1215	0.1074	88.0639	89.625	64.5377	76.411	94.4719	97.8957	75.7176	89.8369	0.1194	0.1037
BBA	0.1116	0.0977	89.1083	90.5	64.4286	80.5122	78.7006	90.214	95.0395	97.9604	0.109	0.095
bGWO	0.1067	0.0932	89.5417	90.9583	63.8564	82.9117	79.1231	85.5983	95.3145	97.5229	0.1046	0.0904
BWOA	0.1209	0.1135	88.1806	88.9583	60.8305	74.3306	78.0184	93.4557	95.2697	98.8117	0.1182	0.1104
S1-BSMO	0.1265	0.1147	87.8319	89	70.6404	80.2298	81.3914	95.0256	99.422	100	0.1659	0.1292
S2-BSMO	0.1218	0.1118	88.1708	89.1667	70.8924	84.532	78.9332	91.5289	99.4674	100	0.1598	0.1171
S3-BSMO	0.1213	0.1051	88.2153	89.9167	71.8705	85.8738	81.3385	96.9048	99.377	100	0.1599	0.1237
S4-BSMO	0.1209	0.1070	88.2306	89.6667	72.851	82.1369	81.9163	93.1111	99.3568	100	0.1603	0.1296
V1-BSMO	0.1109	0.0977	89.1542	90.5	78.8832	85.8624	83.8414	95.5556	99.471	100	0.1587	0.1292
V2-BSMO	0.1106	0.0998	89.2069	90.375	79.3521	87.3972	84.3151	96.3492	99.2964	99.9187	0.1589	0.1342
V3-BSMO	0.1107	0.0994	89.1986	90.375	78.5909	86.1964	85.7139	97.5	99.4433	100	0.1617	0.1412
V4-BSMO	0.1096	0.0990	89.3278	90.375	79.7051	88.4275	84.2503	98.75	99.4127	100	0.1617	0.1425
Threshold-BSMO	0.1081	0.0924	89.5194	91.0417	80.2438	91.3715	85.1981	95.7778	99.4531	100	0.1623	0.1342

5.3.4. Comparison of Algorithms to Detect Coronavirus Disease 2019 (COVID-19)

The COVID-19 pandemic is an infectious disease of severe acute respiratory syndrome Coronavirus 2019 [99] which was initiated in Wuhan, China, in December 2019 and profoundly affected human life [100]. Early detection of Coronavirus disease can reduce the transmission rate and slow the epidemic outbreak. Many optimization algorithms have been developed to alleviate this global crisis [101]. In this section, the performance of the proposed algorithm is evaluated in the Coronavirus disease 2019 (COVID-19) dataset [92]. This dataset consists of two classes, death or recovery, and 13 features, including location, country, gender, age, whether the patients visited Wuhan, whether the patients from Wuhan had fever, cough, cold, fatigue, body pain, malaise, and day's difference between the symptoms being noticed and admission to the hospital. The results reported in Table 6 indicate the proposed Threshold-BSMO outperforms all contender algorithms and BSMO variants to detect COVID-19.

Table 6. Coronavirus disease 2019 (COVID-19) detection.

Algorithms	Fitness		Accuracy		Sensitivity		Precision		Specificity		Error	
	Avg	Min	Avg	Max	Avg	Max	Avg	Max	Avg	Max	Avg	Min
ACO	0.0521	0.0493	95.2805	95.4825	98.3325	99.0601	96.3844	97.4589	74.0994	78.5774	0.0477	0.0452
BBA	0.0508	0.0494	95.3575	95.4838	98.411	98.9281	96.5039	97.1542	74.7778	79.4731	0.0464	0.0452
bGWO	0.0482	0.0455	95.4915	95.7137	98.6061	99.3678	96.1273	97.5426	73.3757	80.9149	0.0451	0.0429
BWOA	0.0518	0.0493	95.2667	95.7164	98.3045	99.0229	96.4153	97.1998	74.626	82.0579	0.0479	0.0428
S1-BSMO	0.0515	0.0493	95.417	95.5988	99.2906	99.7496	97.7266	98.2616	83.0173	87.3208	0.0511	0.0452
S2-BSMO	0.0516	0.049	95.3861	95.5961	99.3947	100	97.5923	97.9672	82.1743	85.5629	0.052	0.0498
S3-BSMO	0.0517	0.0497	95.3308	95.6001	99.3703	100	97.5576	98.2389	81.7173	87.298	0.0521	0.0487
S4-BSMO	0.0516	0.049	95.3347	95.5948	99.4093	100	97.5954	98.126	82.1407	86.2074	0.0532	0.0498
V1-BSMO	0.051	0.0497	95.2469	95.5974	99.8598	100	97.3384	97.924	80.376	84.761	0.0537	0.0476
V2-BSMO	0.0509	0.0489	95.263	95.4812	99.8182	100	97.364	97.8237	80.6954	84.0749	0.053	0.0474
V3-BSMO	0.051	0.0486	95.2695	95.4838	99.7693	100	97.3319	97.9259	80.477	83.9283	0.053	0.0475
V4-BSMO	0.0506	0.0478	95.2692	95.4892	99.7845	100	97.4058	97.956	80.7991	84.4487	0.0532	0.0452
Threshold-BSMO	0.0488	0.0451	95.537	95.8353	99.3774	100	97.7178	98.0502	83.1011	87.2075	0.0518	0.0487

5.4. Convergence Comparison

In addition, to compare the efficiency of BSMO with other comparative algorithms, convergence curves were drawn for each dataset used in the evolution. Figure 4 shows the convergence curves of all algorithms based on the fitness value. According to the figure, Threshold-BSMO has the highest efficiency in diagnosing Diabetes, Hepatitis, Heart, and Coronavirus 2019 diseases with the lowest fitness value compared to competitive algorithms.

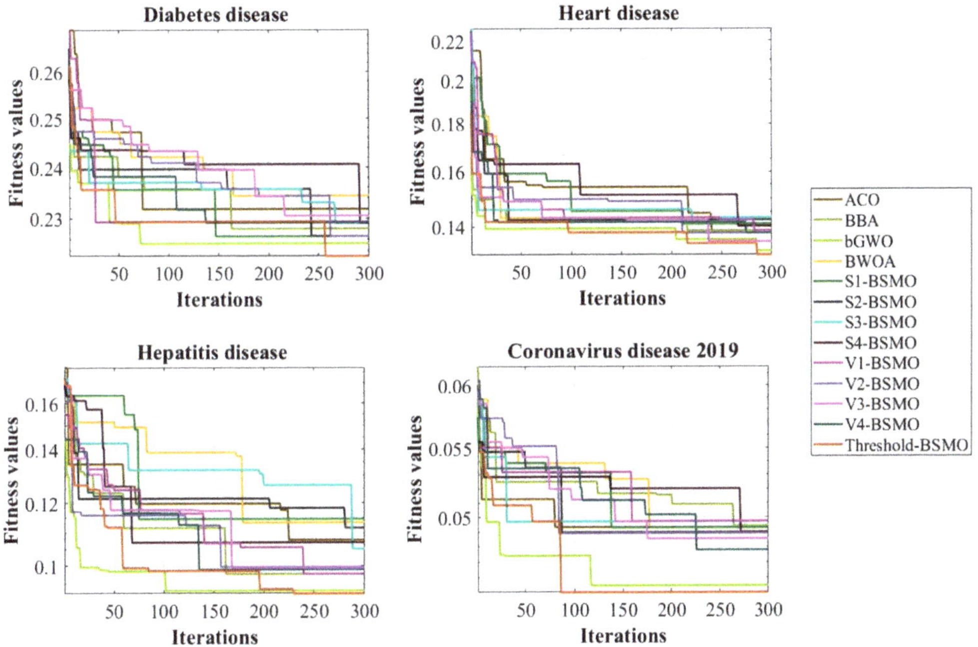

Figure 4. Convergence comparison of the BSMO and comparative algorithms.

5.5. Statistical Analysis

To compare the algorithms fairly and to choose the best transfer function for mapping the continuous solutions to binary ones, Friedman's statistical test was used to rank the algorithms. Table 7 shows the results of Friedman's test according to the fitness values of the algorithms in which the Threshold-BSMO is a great variant to select the effect features from Diabetes, Heart, Hepatic, and Coronavirus diseases.

Table 7. Friedman test.

Algorithms	Medical Problems			
	Diabetes (Rank)	Heart (Rank)	Hepatics (Rank)	COVID-19 (Rank)
ACO	10.37(11)	8.67 (8)	9.23 (8)	9.70 (11)
BBA	10.37 (11)	8.67 (8)	9.23 (8)	9.70 (11)
bGWO	2.80 (2)	2.17 (2)	3.07 (2)	2.27 (2)
BWOA	10.40 (12)	11.23 (12)	9.23 (8)	8.70 (9)
S1-BSMO	5.53 (4)	8.57 (7)	11.87 (11)	7.80 (7)
S2-BSMO	7.43 (8)	9.30 (9)	8.87 (7)	8.97 (10)
S3-BSMO	9.47 (10)	10.73 (11)	9.27 (9)	10.07 (12)
S4-BSMO	9.27 (9)	10.37 (10)	9.43 (10)	8.40 (8)
V1-BSMO	6.13 (5)	5.67 (6)	4.87 (6)	5.53 (3)
V2-BSMO	6.27 (6)	5.13 (5)	4.40 (4)	6.73 (6)
V3-BSMO	6.70 (7)	3.90 (3)	4.60 (5)	6.20 (5)
V4-BSMO	4.67 (3)	4.80 (4)	4.20 (3)	5.60 (4)
Threshold-BSMO	1.60 (1)	1.80 (1)	2.73 (1)	1.33 (1)

6. Conclusions

Many metaheuristic algorithms have been applied in the wrapper-based methods to select effective features from medical data; however, most cannot find those features that can fulfill an acceptable accurate diagnosis of diseases. To deal with this weakness, a new binary metaheuristic algorithm named binary starling murmuration optimization (BSMO) is proposed to select the effective features from different important diseases such as Diabetes, Heart, Hepatitis, and Coronavirus. The proposed BSMO used two different approaches: S-shaped and V-shaped transfer functions and a variable threshold method to convert the continuous solutions to binary ones. Moreover, metrics such as fitness, accuracy, sensitivity, specificity, precision, and error were used to assess the proposed BSMO's performance compared to competing algorithms. Finally, the Friedman non-parametric test was also used to show the proposed algorithm's superiority statistically. The statistical and experimental tests proved that the proposed BSMO algorithm is very competitive in selecting effective features from targeted medical datasets. The proposed Threshold-BSMO can effectively find the optimal feature subset for Diabetes, Heart, Hepatitis, and Coronavirus diseases. Overall, considering the fitness criterion as the main criterion for identifying the most effective binary algorithm in selecting the effective features from the medical datasets targeted in this study, Threshold-BSMO was a superior variant to the contender algorithms.

Although the proposed algorithm can select effective features compared to other comparative algorithms, it was limited to four disease datasets targeted in this study. Therefore, the proposed BSMO algorithm can be applied and improved for other real-world applications. Moreover, a self-adapting parameter tuning method can be applied instead of the try-and-test method used for tuning some parameters of BSMO. The BSMO can be armed by other binarization techniques and transfer functions for selecting effective features in other applications. In addition, the SMO's search strategies can be hybridized with other metaheuristic algorithms to generate better candidate continues solutions.

Author Contributions: Conceptualization, M.H.N.-S., Z.A.V. and H.Z.; methodology, M.H.N.-S., Z.A.V. and H.Z.; software, M.H.N.-S., Z.A.V. and H.Z.; validation, M.H.N.-S., Z.A.V., H.Z. and S.M.; formal analysis, M.H.N.-S., Z.A.V., H.Z. and S.M.; investigation, M.H.N.-S., Z.A.V. and H.Z.; resources, M.H.N.-S. and H.Z.; data curation, M.H.N.-S., Z.A.V., H.Z. and S.M.; writing, original draft preparation, M.H.N.-S., Z.A.V. and H.Z.; writing—review and editing, M.H.N.-S., Z.A.V., H.Z. and S.M.; visualization, M.H.N.-S., Z.A.V. and H.Z.; supervision, M.H.N.-S. and S.M.; project administration, M.H.N.-S. and S.M. All authors have read and agreed to the published version of the manuscript.

Funding: This research received no external funding.

Institutional Review Board Statement: Not applicable.

Informed Consent Statement: Not applicable.

Data Availability Statement: The data and code used in the research may be obtained from the corresponding author upon request.

Conflicts of Interest: The authors declare no conflict of interest.

Appendix A

The metaheuristic optimization algorithms' performance is strongly dependent on selecting the proper values for their parameters. Therefore, in this section, the sensitivity on different values for key parameters of the BSMO algorithm, such as flock size (k) and population size (N), are analyzed and tuned using the offline parameter tuning method. The detailed results of pretests and experiments for tuning the BMSO's parameter values to find its best robustness in solving feature selection problems on targeted medical datasets were reported in Tables A1–A6 in terms of fitness, error, accuracy, sensitivity, specificity, and precision. The Friedman rank in Tables A1 and A2 specifies the highest performance of BSMO when k and N are equal to 5 and 30, respectively.

Table A1. Parameters setting of BSMO algorithm in terms of fitness values.

Algorithms	Metrics	$k = 3, N = 30$				$k = 5, N = 20$			
		Diabetes	Heart	Hepatics	COVID-19	Diabetes	Heart	Hepatics	COVID-19
S1-BSMO	Avg	0.2349	0.1463	0.1235	0.0519	0.2354	0.1482	0.1238	0.0523
	Min	0.2295	0.1382	0.1129	0.0493	0.2305	0.1418	0.1014	0.0490
S2-BSMO	Avg	0.2358	0.1486	0.1222	0.0517	0.2368	0.1492	0.1230	0.0522
	Min	0.2317	0.1411	0.1067	0.0497	0.2292	0.1403	0.1080	0.0505
S3-BSMO	Avg	0.2370	0.1487	0.1196	0.0516	0.2367	0.1506	0.1242	0.0519
	Min	0.2331	0.1403	0.1069	0.0493	0.2241	0.1403	0.1138	0.0485
S4-BSMO	Avg	0.2360	0.1497	0.1225	0.0519	0.2369	0.1517	0.1234	0.0521
	Min	0.2305	0.1432	0.1118	0.0505	0.2319	0.1403	0.1050	0.0505
V1-BSMO	Avg	0.2338	0.1419	0.1103	0.0513	0.2361	0.1418	0.1109	0.0519
	Min	0.2305	0.1358	0.0990	0.0479	0.2319	0.1380	0.0994	0.0510
V2-BSMO	Avg	0.2335	0.1413	0.1096	0.0515	0.2365	0.1428	0.1108	0.0510
	Min	0.2319	0.1380	0.0995	0.0493	0.2345	0.1387	0.1059	0.0497
V3-BSMO	Avg	0.2341	0.1410	0.1091	0.0507	0.2347	0.1423	0.1103	0.0508
	Min	0.2319	0.1351	0.0981	0.0497	0.2305	0.1395	0.1003	0.0475
V4-BSMO	Avg	0.2330	0.1410	0.1092	0.0505	0.2344	0.1422	0.1101	0.0514
	Min	0.2240	0.1380	0.0990	0.0482	0.2319	0.1380	0.0999	0.0486
Threshold-BSMO	Avg	0.2314	0.1375	0.1044	0.0487	0.2324	0.1395	0.1144	0.0497
	Min	0.2268	0.1308	0.0884	0.0463	0.2254	0.1337	0.0978	0.0482
Friedman rank		2				4			

Algorithms	Metrics	$k = 5, N = 30$				$k = 7, N = 30$			
		Diabetes	Heart	Hepatics	COVID-19	Diabetes	Heart	Hepatics	COVID-19
S1-BSMO	Avg	0.2342	0.1460	0.1265	0.0515	0.2352	0.1463	0.1230	0.0517
	Min	0.2266	0.1411	0.1147	0.0493	0.2330	0.1374	0.1146	0.0497
S2-BSMO	Avg	0.2352	0.1481	0.1218	0.0516	0.2344	0.1462	0.1208	0.0518
	Min	0.2267	0.1424	0.1118	0.0490	0.2293	0.1403	0.1088	0.0509
S3-BSMO	Avg	0.2373	0.1495	0.1213	0.0517	0.2360	0.1484	0.1211	0.0517
	Min	0.2291	0.1432	0.1051	0.0497	0.2331	0.1432	0.1128	0.0505
S4-BSMO	Avg	0.2368	0.1492	0.1209	0.0516	0.2367	0.1497	0.1238	0.0518
	Min	0.2291	0.1403	0.1070	0.0490	0.2331	0.1440	0.1120	0.0501
V1-BSMO	Avg	0.2344	0.1423	0.1109	0.0510	0.2343	0.1427	0.1096	0.0509
	Min	0.2294	0.1387	0.0977	0.0497	0.2293	0.1411	0.0990	0.0489
V2-BSMO	Avg	0.2343	0.1417	0.1106	0.0509	0.2339	0.1410	0.1098	0.0508
	Min	0.2266	0.1380	0.0998	0.0489	0.2294	0.1387	0.1046	0.0497
V3-BSMO	Avg	0.2353	0.1411	0.1107	0.0510	0.2354	0.1413	0.1125	0.0515
	Min	0.2306	0.1351	0.0994	0.0486	0.2320	0.1380	0.1073	0.0496
V4-BSMO	Avg	0.2335	0.1414	0.1096	0.0506	0.2330	0.1425	0.1100	0.0507
	Min	0.2292	0.1380	0.0990	0.0478	0.2293	0.1403	0.1049	0.0490
Threshold-BSMO	Avg	0.2306	0.1378	0.1081	0.0488	0.2302	0.1370	0.0920	0.0491
	Min	0.2229	0.1308	0.0924	0.0451	0.2266	0.1308	0.0920	0.0478
Friedman rank		1				3			

Table A2. Parameters setting of BSMO algorithm in terms of error values.

Algorithms	Metrics	$k = 3, N = 30$				$k = 5, N = 20$			
		Diabetes	Heart	Hepatics	COVID-19	Diabetes	Heart	Hepatics	COVID-19
S1-BSMO	Avg	0.2505	0.1954	0.1594	0.0523	0.2541	0.2081	0.1692	0.0538
	Min	0.2422	0.1556	0.1212	0.0498	0.2383	0.1593	0.1412	0.0498
S2-BSMO	Avg	0.2492	0.1949	0.1628	0.0518	0.2556	0.2096	0.1693	0.0531
	Min	0.2370	0.1593	0.1358	0.0475	0.2422	0.1704	0.1342	0.0487
S3-BSMO	Avg	0.2517	0.1956	0.1624	0.0523	0.2541	0.2057	0.1672	0.0535
	Min	0.2408	0.1519	0.1429	0.0476	0.2369	0.1519	0.1421	0.0498
S4-BSMO	Avg	0.2498	0.2016	0.1573	0.0529	0.2546	0.2037	0.1664	0.0536
	Min	0.2371	0.1593	0.1225	0.0487	0.2383	0.1667	0.1358	0.0498
V1-BSMO	Avg	0.2551	0.1930	0.1549	0.0530	0.2589	0.2004	0.1623	0.0527
	Min	0.2461	0.1519	0.1162	0.0486	0.2488	0.1519	0.1346	0.0510
V2-BSMO	Avg	0.2578	0.1946	0.1532	0.0528	0.2560	0.2044	0.1611	0.0542
	Min	0.2461	0.1630	0.1096	0.0464	0.2396	0.1482	0.1300	0.0510
V3-BSMO	Avg	0.2504	0.1906	0.1588	0.0536	0.2590	0.2047	0.1669	0.0546
	Min	0.2357	0.1519	0.1171	0.0486	0.2474	0.1704	0.1479	0.0509
V4-BSMO	Avg	0.2557	0.1878	0.1545	0.0528	0.2554	0.2049	0.1635	0.0538
	Min	0.2474	0.1630	0.1279	0.0487	0.2422	0.1556	0.1417	0.0509
Threshold-BSMO	Avg	0.2514	0.1925	0.1640	0.0525	0.2563	0.2163	0.1629	0.0534
	Min	0.2370	0.1481	0.1233	0.0487	0.2448	0.1593	0.1150	0.0497
Friedman rank		2				4			

Algorithms	Metrics	$k = 5, N = 30$				$k = 7, N = 30$			
		Diabetes	Heart	Hepatics	COVID-19	Diabetes	Heart	Hepatics	COVID-19
S1-BSMO	Avg	0.2504	0.1964	0.1659	0.0511	0.2506	0.2020	0.1612	0.0521
	Min	0.2382	0.1593	0.1292	0.0452	0.2409	0.1630	0.1363	0.0463
S2-BSMO	Avg	0.2516	0.2015	0.1598	0.0520	0.2534	0.1983	0.1702	0.0520
	Min	0.2369	0.1556	0.1171	0.0498	0.2423	0.1593	0.1500	0.0475
S3-BSMO	Avg	0.2508	0.1930	0.1599	0.0521	0.2532	0.1847	0.1598	0.0521
	Min	0.2397	0.1556	0.1237	0.0487	0.2421	0.1630	0.1288	0.0464
S4-BSMO	Avg	0.2533	0.1907	0.1603	0.0532	0.2568	0.1901	0.1650	0.0532
	Min	0.2384	0.1481	0.1296	0.0498	0.2487	0.1519	0.1429	0.0487
V1-BSMO	Avg	0.2552	0.1884	0.1587	0.0537	0.2537	0.1915	0.1572	0.0533
	Min	0.2422	0.1593	0.1292	0.0476	0.2384	0.1593	0.1346	0.0498
V2-BSMO	Avg	0.2548	0.1911	0.1589	0.0530	0.2539	0.1896	0.1637	0.0527
	Min	0.2345	0.1481	0.1342	0.0474	0.2447	0.1630	0.1483	0.0521
V3-BSMO	Avg	0.2547	0.1956	0.1617	0.0530	0.2558	0.2096	0.1531	0.0528
	Min	0.2383	0.1667	0.1412	0.0475	0.2475	0.1889	0.1225	0.0464
V4-BSMO	Avg	0.2534	0.1959	0.1617	0.0532	0.2565	0.1959	0.1472	0.0538
	Min	0.2383	0.1519	0.1425	0.0452	0.2396	0.1556	0.1163	0.0521
Threshold-BSMO	Avg	0.2530	0.1952	0.1623	0.0518	0.2498	0.1986	0.1558	0.0528
	Min	0.2408	0.1519	0.1342	0.0487	0.2395	0.1593	0.1558	0.0487
Friedman rank		1				3			

Table A3. Parameters setting of BSMO algorithm in terms of accuracy values.

Algorithms	Metrics	$k = 3, N = 30$				$k = 5, N = 20$			
		Diabetes	Heart	Hepatics	COVID-19	Diabetes	Heart	Hepatics	COVID-19
S1-BSMO	Avg	76.9228	85.7901	88.1236	95.4467	76.8885	85.5802	88.1083	95.401
	Max	77.3257	86.6667	89.125	95.5961	77.4761	86.2963	90.2917	95.707
S2-BSMO	Avg	76.8251	85.4568	88.1625	95.364	76.717	85.4444	88.0736	95.3321
	Max	77.3582	86.2963	89.75	95.5921	77.4761	86.2963	89.625	95.4852
S3-BSMO	Avg	76.6528	85.4198	88.3667	95.3389	76.677	85.2222	87.9167	95.2899
	Max	77.0899	86.2963	89.6667	95.4852	77.9973	86.2963	89.0417	95.4878
S4-BSMO	Avg	76.7293	85.3086	88.0444	95.3477	76.6684	85.0988	87.9708	95.3
	Max	77.2198	85.9259	89.125	95.4878	77.0933	86.2963	89.875	95.4838
V1-BSMO	Avg	76.9143	86	89.2708	95.2115	76.707	86	89.158	95.197
	Max	77.218	86.6667	90.375	95.4758	77.088	86.296	90.333	95.362
V2-BSMO	Avg	76.9203	86.0617	89.2958	95.2158	76.644	85.926	89.229	95.238
	Max	77.0779	86.2963	90.375	95.3635	76.822	86.296	89.833	95.364
V3-BSMO	Avg	76.9173	86.0741	89.3458	95.2694	76.834	85.951	89.211	95.268
	Max	77.2095	86.6667	90.4583	95.3729	77.227	86.296	90.292	95.595
V4-BSMO	Avg	76.9872	86.0864	89.3708	95.2895	76.876	85.951	89.289	95.204
	Max	78.0041	86.2963	90.375	95.83	77.081	86.296	90.333	95.481
Threshold-BSMO	Avg	77.3771	86.5309	89.8972	95.5124	77.136	86.37	88.919	95.467
	Max	78.1203	87.4074	91.5	95.715	77.862	87.037	90.542	95.599

Algorithms	Metrics	$k = 5, N = 30$				$k = 7, N = 30$			
		Diabetes	Heart	Hepatics	COVID-19	Diabetes	Heart	Hepatics	COVID-19
S1-BSMO	Avg	76.9719	85.8148	87.8319	95.417	76.895	85.778	88.189	95.427
	Max	77.7409	86.2963	89	95.5988	77.216	86.667	88.958	95.599
S2-BSMO	Avg	76.8537	85.5185	88.1708	95.3861	76.927	85.704	88.3	95.389
	Max	77.7341	85.9259	89.1667	95.5961	77.344	86.296	89.542	95.482
S3-BSMO	Avg	76.6101	85.3333	88.2153	95.3308	76.763	85.457	88.192	95.336
	Max	77.4897	85.9259	89.9167	95.6001	76.965	85.926	89.083	95.482
S4-BSMO	Avg	76.6654	85.3704	88.2306	95.3347	76.643	85.333	87.922	95.341
	Max	77.4863	86.2963	89.6667	95.5948	76.96	85.926	89.167	95.484
V1-BSMO	Avg	76.889	85.9383	89.1542	95.2469	76.864	85.963	89.313	95.24
	Max	77.3411	86.2963	90.5	95.5974	77.348	86.296	90.375	95.376
V2-BSMO	Avg	76.8872	86.037	89.2069	95.263	76.918	86.148	89.267	95.227
	Max	77.6128	86.2963	90.375	95.4812	77.353	86.296	89.792	95.365
V3-BSMO	Avg	76.7716	86.0741	89.1986	95.2695	76.774	86.037	89.013	95.23
	Max	77.2163	86.6667	90.375	95.4838	77.075	86.296	89.583	95.607
V4-BSMO	Avg	76.9639	86.0123	89.3278	95.2692	76.992	85.926	89.292	95.293
	Max	77.471	86.2963	90.375	95.4892	77.346	86.296	89.833	95.365
Threshold-BSMO	Avg	77.3077	86.5309	89.5194	95.537	77.328	86.6173	91.0833	95.4834
	Max	77.9904	87.4074	91.0417	95.8353	77.6179	87.4074	91.0833	95.7124

Table A4. Parameters setting of BSMO algorithm in terms of sensitivity values.

Algorithms	Metrics	$k = 3, N = 30$				$k = 5, N = 20$			
		Diabetes	Heart	Hepatics	COVID-19	Diabetes	Heart	Hepatics	COVID-19
S1-BSMO	Avg	88.7609	93.8971	68.6377	99.2675	88.365	93.4924	68.8162	99.2336
	Max	90.2278	96.2467	85.3876	99.7684	89.7842	95.292	78.4811	100
S2-BSMO	Avg	88.3668	93.6035	71.8856	99.3568	88.22	92.9212	69.6659	99.3498
	Max	89.1665	95.3642	84.1441	99.8295	89.7151	95.2571	79.7631	100
S3-BSMO	Avg	88.1191	93.0296	72.7952	99.3912	88.0791	93.0515	69.5805	99.383
	Max	89.0592	95.2235	85.8833	100	89.7059	95.8368	79.1405	100
S4-BSMO	Avg	88.191	93.1775	72.6372	99.4499	88.0248	93.0834	71.6037	99.2982
	Max	89.0733	95.3049	81.5983	100	89.4038	96.6535	88.5142	99.7212
V1-BSMO	Avg	88.3628	94.2365	78.835	99.9088	88.327	94.012	76.679	99.504
	Max	89.4449	95.5444	88.3507	100	89.715	94.952	85.144	100
V2-BSMO	Avg	87.6423	94.536	79.637	99.9252	87.515	93.419	80.879	99.696
	Max	88.1193	96.3431	90.6895	100	88.523	94.47	88.078	100
V3-BSMO	Avg	87.8565	94.0807	79.3312	99.8619	88.05	94.206	75.106	99.811
	Max	88.9304	96.2699	86.7033	100	89.359	95.947	83.572	100
V4-BSMO	Avg	88.4581	94.1059	80.3357	99.8371	88.104	93.948	78.811	99.775
	Max	89.115	95.4339	89.9166	100	89.716	94.769	91.569	100
Threshold-BSMO	Avg	89.0468	94.655	80.116	99.3936	88.503	94.575	75.676	99.348
	Max	91.0607	96.4165	88.838	100	89.44	96.612	84.286	99.548

Algorithms	Metrics	$k = 5, N = 30$				$k = 7, N = 30$			
		Diabetes	Heart	Hepatics	COVID-19	Diabetes	Heart	Hepatics	COVID-19
S1-BSMO	Avg	88.5142	93.6608	70.6404	99.2906	88.514	93.831	71.703	99.26
	Max	89.8454	95.0876	80.2298	99.7496	89.311	95.569	88.021	99.609
S2-BSMO	Avg	88.2422	93.3517	70.8924	99.3947	88.387	93.475	67.965	99.381
	Max	89.2631	95.3351	84.532	100	89.719	94.748	75.613	99.822
S3-BSMO	Avg	88.1787	93.1475	71.8705	99.3703	88.325	92.863	69.775	99.512
	Max	90.1796	94.4033	85.8738	100	90.142	95.51	75.39	100
S4-BSMO	Avg	88.3085	93.2132	72.851	99.4093	88.234	93.467	74.89	99.486
	Max	89.7088	95.0297	82.1369	100	88.447	96.508	90.417	100
V1-BSMO	Avg	88.2848	93.8571	78.8832	99.8598	88.244	94.334	78.981	99.861
	Max	89.7132	96.2621	85.8624	100	88.978	96.082	87.181	100
V2-BSMO	Avg	88.6261	94.4918	79.3521	99.8182	88.217	94.237	76.884	99.87
	Max	90.0085	96.2525	87.3972	100	89.536	96.225	85.649	100
V3-BSMO	Avg	88.245	94.1042	78.5909	99.7693	88.328	94.287	78.435	99.707
	Max	89.6911	95.6443	86.1964	100	89.598	96.606	85.285	100
V4-BSMO	Avg	88.1009	94.0096	79.7051	99.7845	88.377	94.327	77.526	99.802
	Max	89.484	95.4345	88.4275	100	89.576	95.545	84.209	100
Threshold-BSMO	Avg	89	94.6804	80.2438	99.3774	89.1512	94.8397	81.5161	99.4531
	Max	89.9871	96.6626	91.3715	100	90.8328	96.8864	81.5161	100

Table A5. Parameters setting of BSMO algorithm in terms of precision values.

Algorithms	Metrics	$k = 3, N = 30$				$k = 5, N = 20$			
		Diabetes	Heart	Hepatics	COVID-19	Diabetes	Heart	Hepatics	COVID-19
S1-BSMO	Avg	83.3024	89.7759	81.8452	97.6496	82.9358	88.9528	77.9683	97.6013
	Max	84.5112	91.1126	92.5556	98.0808	84.0629	91.3534	92.5714	97.9844
S2-BSMO	Avg	83.1656	89.6198	83.3028	97.5525	82.995	88.9851	77.8085	97.5018
	Max	83.9979	91.6603	96.8889	97.9414	84.226	92.1922	96.6667	98.2263
S3-BSMO	Avg	82.8792	89.3916	80.6788	97.6031	82.7702	89.4225	78.8535	97.4893
	Max	84.0439	92.1316	94	97.9846	84.4498	92.9411	96.6667	98.1287
S4-BSMO	Avg	82.9226	90.0263	82.091	97.5329	82.8662	89.0141	78.5044	97.5208
	Max	83.5297	93.4569	98.3333	98.0334	84.1876	91.7804	92.4643	98.099
V1-BSMO	Avg	83.3034	89.275	82.3873	97.3381	82.916	89.014	82.316	97.303
	Max	84.5946	91.9784	89.9596	97.6875	83.93	90.58	88.611	97.645
V2-BSMO	Avg	82.8698	89.5967	85.3515	97.3886	82.777	90.147	78.368	97.291
	Max	83.2788	91.9132	100	97.8924	83.495	92.77	92.051	97.604
V3-BSMO	Avg	83.0581	89.6372	84.2314	97.4291	82.752	89.063	80.449	97.369
	Max	84.4839	92.3453	94.6429	98.2173	84.24	92.867	96.349	97.735
V4-BSMO	Avg	83.0346	90.0785	84.4764	97.4154	83.013	89.097	81.099	97.38
	Max	83.6545	92.453	100	98.0199	83.788	91.567	93.099	97.932
Threshold-BSMO	Avg	83.7292	91.6759	85.2687	97.7006	83.564	91.301	81.078	97.67
	Max	85.4714	93.7512	97.5	98.3656	84.936	92.869	95.325	98.032

Algorithms	Metrics	$k = 5, N = 30$				$k = 7, N = 30$			
		Diabetes	Heart	Hepatics	COVID-19	Diabetes	Heart	Hepatics	COVID-19
S1-BSMO	Avg	83.2288	89.2841	81.3914	97.7266	83.487	89.175	84.47	97.634
	Max	84.242	91.4817	95.0256	98.2616	84.571	89.756	96.5	98.064
S2-BSMO	Avg	83.1426	89.403	78.9332	97.5923	83.224	89.761	81.574	97.574
	Max	84.2726	91.5718	91.5289	97.9672	84.225	92.71	98	97.908
S3-BSMO	Avg	82.925	89.7136	81.3385	97.5576	82.855	89.229	78.523	97.55
	Max	84.4974	91.6581	96.9048	98.2389	84.549	91.133	86.998	97.873
S4-BSMO	Avg	82.764	89.4763	81.9163	97.5954	83.236	89.538	82.034	97.459
	Max	83.8476	91.4379	93.1111	98.126	84.582	90.938	98.75	97.872
V1-BSMO	Avg	83.1764	89.3417	83.8414	97.3384	82.958	89.327	83.84	97.356
	Max	86.1848	91.9558	95.5556	97.924	84.323	90.731	91	97.672
V2-BSMO	Avg	82.9072	89.3503	84.3151	97.364	82.987	89.725	82.101	97.369
	Max	83.761	91.9198	96.3492	97.8237	84.159	92.147	92.372	97.734
V3-BSMO	Avg	83.1812	89.6908	85.7139	97.3319	83.21	89.99	81.337	97.334
	Max	84.5091	91.8579	97.5	97.9259	84.092	91.817	86.738	97.62
V4-BSMO	Avg	83.2658	89.7266	84.2503	97.4058	83.289	89.8	83.036	97.41
	Max	84.4214	91.4855	98.75	97.956	84.474	92.318	95.238	97.716
Threshold-BSMO	Avg	83.5823	91.4408	85.1981	97.7178	83.5777	91.6578	82.5912	97.6967
	Max	84.7376	93.8631	95.7778	98.0502	84.5587	94.5662	82.5912	97.9926

Table A6. Parameters setting of BSMO algorithm in terms of specificity values.

Algorithms	Metrics	$k = 3, N = 30$				$k = 5, N = 20$			
		Diabetes	Heart	Hepatics	COVID-19	Diabetes	Heart	Hepatics	COVID-19
S1-BSMO	Avg	66.2342	86.542	99.4587	82.4862	65.1298	85.5329	99.2915	82.346
	Max	68.2502	87.9644	100	85.9366	66.4093	88.7883	99.8374	84.4318
S2-BSMO	Avg	65.4575	86.9073	99.5053	81.7373	65.6179	85.9113	99.3399	81.8105
	Max	66.6288	89.656	100	85.0243	67.707	91.2161	99.9187	85.9674
S3-BSMO	Avg	64.8413	86.4002	99.4766	82.3377	65.1382	86.3003	99.2782	81.4987
	Max	66.1969	89.2235	100	84.8969	67.4399	91.9747	100	86.2247
S4-BSMO	Avg	65.7836	87.4361	99.4948	81.9079	65.0415	86.3582	99.4617	81.5022
	Max	66.5935	91.8474	99.9187	84.9453	67.8061	90.3087	100	86.1819
V1-BSMO	Avg	65.75	88.5427	99.368	80.2567	65.413	88.039	99.063	80.501
	Max	67.4069	90.7376	100	83.9446	66.648	89.375	99.399	82.194
V2-BSMO	Avg	66.6065	89.1717	99.4428	80.7396	64.975	89.529	99.286	79.295
	Max	68.4934	91.1563	100	84.5923	66.435	92.307	100	82.191
V3-BSMO	Avg	66.1459	88.7473	99.3145	81.0216	64.927	88.066	99.258	80.392
	Max	67.5985	91.9959	99.8286	85.1101	66.815	90.58	99.829	82.735
V4-BSMO	Avg	65.8974	88.7773	99.4461	81.007	65.451	88.153	99.243	80.932
	Max	67.0376	90.7347	100	85.0112	66.87	91.094	99.837	83.784
Threshold-BSMO	Avg	66.9048	89.1478	99.3309	82.7429	66.572	88.889	99.058	81.727
	Max	69.2527	92.084	100	87.2612	70.241	90.63	100	84.678
Algorithms	Metrics	$k = 5, N = 30$				$k = 7, N = 30$			
		Diabetes	Heart	Hepatics	COVID-19	Diabetes	Heart	Hepatics	COVID-19
S1-BSMO	Avg	65.9602	86.0433	99.422	83.0173	66.509	85.883	99.513	82.554
	Max	68.2916	88.7512	100	87.3208	69.423	86.761	100	84.83
S2-BSMO	Avg	65.7932	86.2794	99.4674	82.1743	66.123	87.242	99.451	82.462
	Max	68.023	88.2128	100	85.5629	68.293	90.414	100	84.963
S3-BSMO	Avg	65.5806	87.0123	99.377	81.7173	66.103	86.476	99.369	81.929
	Max	67.8662	89.3531	100	87.298	69.399	88.543	99.837	84.695
S4-BSMO	Avg	65.0295	86.3764	99.3568	82.1407	65.939	86.436	99.48	81.524
	Max	66.8104	89.284	100	86.2074	66.92	88.336	100	85.479
V1-BSMO	Avg	65.7345	89.0497	99.471	80.376	65.889	88.631	99.324	80.379
	Max	70.5787	91.2747	100	84.761	68.174	89.595	99.919	82.705
V2-BSMO	Avg	65.7846	88.4579	99.2964	80.6954	65.45	89.022	99.279	80.59
	Max	67.6503	91.1828	99.9187	84.0749	66.884	90.656	99.919	82.872
V3-BSMO	Avg	66.0204	88.5817	99.4433	80.477	65.038	88.873	99.174	80.363
	Max	69.1626	90.503	100	83.9283	67.188	90.793	99.473	83.175
V4-BSMO	Avg	66.2564	88.5579	99.4127	80.7991	66.366	88.717	99.438	81.74
	Max	69.1896	91.1526	100	84.4487	67.647	91.444	99.666	83.986
Threshold-BSMO	Avg	66.6321	88.8911	99.4531	83.1011	66.562	89.1274	99.6652	82.6134
	Max	69.2028	92.1136	100	87.2075	68.5097	92.9485	99.6652	85.0483

References

1. Nilashi, M.; bin Ibrahim, O.; Ahmadi, H.; Shahmoradi, L. An analytical method for diseases prediction using machine learning techniques. *Comput. Chem. Eng.* **2017**, *106*, 212–223. [CrossRef]
2. Abdulkhaleq, M.T.; Rashid, T.A.; Alsadoon, A.; Hassan, B.A.; Mohammadi, M.; Abdullah, J.M.; Chhabra, A.; Ali, S.L.; Othman, R.N.; Hasan, H.A.; et al. Harmony search: Current studies and uses on healthcare systems. *Artif. Intell. Med.* **2022**, *131*, 102348. [CrossRef]
3. Qader, S.M.; Hassan, B.A.; Rashid, T.A. An improved deep convolutional neural network by using hybrid optimization algorithms to detect and classify brain tumor using augmented MRI images. *Multimedia Tools Appl.* **2022**, *81*, 44059–44086. [CrossRef]

4. Golestan Hashemi, F.S.; Razi Ismail, M.; Rafii Yusop, M.; Golestan Hashemi, M.S.; Nadimi Shahraki, M.H.; Rastegari, H.; Miah, G.; Aslani, F. Intelligent mining of large-scale bio-data: Bioinformatics applications. *Biotechnol. Biotec. Eq.* **2018**, *32*, 10–29. [CrossRef]
5. Van Woensel, W.; Elnenaei, M.; Abidi, S.S.R.; Clarke, D.B.; Imran, S.A. Staged reflexive artificial intelligence driven testing algorithms for early diagnosis of pituitary disorders. *Clin. Biochem.* **2021**, *97*, 48–53. [CrossRef] [PubMed]
6. Shah, D.; Patel, S.; Bharti, S.K. Heart Disease Prediction using Machine Learning Techniques. *SN Comput. Sci.* **2020**, *1*, 1–6. [CrossRef]
7. Sharma, P.; Choudhary, K.; Gupta, K.; Chawla, R.; Gupta, D.; Sharma, A. Artificial plant optimization algorithm to detect heart rate & presence of heart disease using machine learning. *Artif. Intell. Med.* **2019**, *102*, 101752. [CrossRef]
8. Devaraj, J.; Elavarasan, R.M.; Pugazhendhi, R.; Shafiullah, G.; Ganesan, S.; Jeysree, A.K.; Khan, I.A.; Hossain, E. Forecasting of COVID-19 cases using deep learning models: Is it reliable and practically significant? *Results Phys.* **2021**, *21*, 103817. [CrossRef]
9. Remeseiro, B.; Bolon-Canedo, V. A review of feature selection methods in medical applications. *Comput. Biol. Med.* **2019**, *112*, 103375. [CrossRef]
10. Gokulnath, C.B.; Shantharajah, S.P. An optimized feature selection based on genetic approach and support vector machine for heart disease. *Clust. Comput.* **2018**, *22*, 14777–14787. [CrossRef]
11. Huda, S.; Yearwood, J.; Jelinek, H.F.; Hassan, M.M.; Fortino, G.; Buckland, M. A Hybrid Feature Selection With Ensemble Classification for Imbalanced Healthcare Data: A Case Study for Brain Tumor Diagnosis. *IEEE Access* **2016**, *4*, 9145–9154. [CrossRef]
12. Xue, B.; Zhang, M.; Browne, W.N.; Yao, X. A survey on evolutionary computation approaches to feature selection. *IEEE Trans. Evol. Comput.* **2015**, *20*, 606–626. [CrossRef]
13. García-Torres, M.; Gómez-Vela, F.; Melián-Batista, B.; Moreno-Vega, J.M. High-dimensional feature selection via feature grouping: A Variable Neighborhood Search approach. *Inf. Sci.* **2016**, *326*, 102–118. [CrossRef]
14. Yu, L.; Liu, H. Feature selection for high-dimensional data: A fast correlation-based filter solution. In Proceedings of the 20th International Conference on Machine Learning (ICML-03), Washington, DC, USA, 21–24 August 2003; pp. 856–863.
15. Guyon, I.; Elisseeff, A. An introduction to variable and feature selection. *J. Mach. Learn. Res.* **2003**, *3*, 1157–1182.
16. Gnana, D.A.A.; Balamurugan, S.A.A.; Leavline, E.J. Literature review on feature selection methods for high-dimensional data. *Int. J. Comput. Appl.* **2016**, *136*, 9–17.
17. Kohavi, R.; John, G.H. Wrappers for feature subset selection. *Artif. Intell.* **1997**, *97*, 273–324. [CrossRef]
18. Karegowda, A.G.; Manjunath, A.S.; Jayaram, M.A. Feature Subset Selection Problem using Wrapper Approach in Supervised Learning. *Int. J. Comput. Appl.* **2010**, *1*, 13–17. [CrossRef]
19. Kabir, M.; Shahjahan; Murase, K. A new local search based hybrid genetic algorithm for feature selection. *Neurocomputing* **2011**, *74*, 2914–2928. [CrossRef]
20. Tran, B.; Xue, B.; Zhang, M. Adaptive multi-subswarm optimisation for feature selection on high-dimensional classification. In Proceedings of the Genetic and Evolutionary Computation Conference, Boston, MA, USA, 13–17 July 2019; pp. 481–489. [CrossRef]
21. Zamani, H.; Nadimi-Shahraki, M.H.; Gandomi, A.H. CCSA: Conscious Neighborhood-based Crow Search Algorithm for Solving Global Optimization Problems. *Appl. Soft Comput.* **2019**, *85*, 105583. [CrossRef]
22. Benyamin, A.; Farhad, S.G.; Saeid, B. Discrete farmland fertility optimization algorithm with metropolis acceptance criterion for traveling salesman problems. *Int. J. Intell. Syst.* **2020**, *36*, 1270–1303. [CrossRef]
23. Fard, E.S.; Monfaredi, K.; Nadimi, M.H. An Area-Optimized Chip of Ant Colony Algorithm Design in Hardware Platform Using the Address-Based Method. *Int. J. Electr. Comput. Eng. (IJECE)* **2014**, *4*, 989–998. [CrossRef]
24. Sayadi, M.K.; Hafezalkotob, A.; Naini, S.G.J. Firefly-inspired algorithm for discrete optimization problems: An application to manufacturing cell formation. *J. Manuf. Syst.* **2013**, *32*, 78–84. [CrossRef]
25. Gharehchopogh, F.S.; Nadimi-Shahraki, M.H.; Barshandeh, S.; Abdollahzadeh, B.; Zamani, H. CQFFA: A Chaotic Quasi-oppositional Farmland Fertility Algorithm for Solving Engineering Optimization Problems. *J. Bionic Eng.* **2022**, 1–26. [CrossRef]
26. Nadimi-Shahraki, M.H.; Fatahi, A.; Zamani, H.; Mirjalili, S.; Oliva, D. Hybridizing of Whale and Moth-Flame Optimization Algorithms to Solve Diverse Scales of Optimal Power Flow Problem. *Electronics* **2022**, *11*, 831. [CrossRef]
27. Eberhart, R.; Kennedy, J. A new optimizer using particle swarm theory. In Proceedings of the MHS'95 Proceedings of the sixth International Symposium on Micro Machine and Human Science, Nagoya, Japan, 4–6 October 1995; pp. 39–43.
28. Dorigo, M.; Di Caro, G. Ant colony optimization: A new meta-heuristic. In Proceedings of the 1999 Congress on Evolutionary Computation-CEC99 (Cat No 99TH8406), Washington, DC, USA, 6–9 July 1999; pp. 1470–1477.
29. Storn, R.; Price, K. Differential evolution–a simple and efficient heuristic for global optimization over continuous spaces. *J. Global. Optim.* **1997**, *11*, 341–359. [CrossRef]
30. Rajabioun, R. Cuckoo optimization algorithm. *Appl. Soft. Comput.* **2011**, *11*, 5508–5518. [CrossRef]
31. Gandomi, A.H.; Alavi, A.H. Krill herd: A new bio-inspired optimization algorithm. *Commun. Nonlinear Sci. Numer. Simul.* **2012**, *17*, 4831–4845. [CrossRef]
32. James, J.; Li, V.O. A social spider algorithm for global optimization. *Appl. Soft. Comput.* **2015**, *30*, 614–627. [CrossRef]
33. Askarzadeh, A. A novel metaheuristic method for solving constrained engineering optimization problems: Crow search algorithm. *Comput. Struct.* **2016**, *169*, 1–12. [CrossRef]

34. Saremi, S.; Mirjalili, S.; Lewis, A. Grasshopper optimisation algorithm: Theory and application. *Adv. Eng. Softw.* **2017**, *105*, 30–47. [CrossRef]
35. Zamani, H.; Nadimi-Shahraki, M.H.; Gandomi, A.H. QANA: Quantum-based avian navigation optimizer algorithm. *Eng. Appl. Artif. Intell.* **2021**, *104*, 104314. [CrossRef]
36. Abdollahzadeh, B.; Gharehchopogh, F.S.; Mirjalili, S. African vultures optimization algorithm: A new nature-inspired metaheuristic algorithm for global optimization problems. *Comput. Ind. Eng.* **2021**, *158*, 107408. [CrossRef]
37. Abu Khurmaa, R.; Aljarah, I.; Sharieh, A. An intelligent feature selection approach based on moth flame optimization for medical diagnosis. *Neural Comput. Appl.* **2020**, *33*, 7165–7204. [CrossRef]
38. Moorthy, U.; Gandhi, U.D. A novel optimal feature selection technique for medical data classification using ANOVA based whale optimization. *J. Ambient. Intell. Humaniz. Comput.* **2020**, *12*, 3527–3538. [CrossRef]
39. Zamani, H.; Nadimi-Shahraki, M.H. Feature selection based on whale optimization algorithm for diseases diagnosis. *Int. J. Comput. Sci. Inf. Secur.* **2016**, *14*, 1243.
40. Nadimi-Shahraki, M.H.; Zamani, H.; Mirjalili, S. Enhanced whale optimization algorithm for medical feature selection: A COVID-19 case study. *Comput. Biol. Med.* **2022**, *148*, 105858. [CrossRef]
41. Zamani, H.; Nadimi-Shahraki, M.H.; Gandomi, A.H. Starling murmuration optimizer: A novel bio-inspired algorithm for global and engineering optimization. *Comput. Methods Appl. Mech. Eng.* **2022**, *392*, 114616. [CrossRef]
42. Mirjalili, S.; Mirjalili, S.M.; Yang, X.-S. Binary bat algorithm. *Neural Comput. Appl.* **2014**, *25*, 663–681. [CrossRef]
43. Emary, E.; Zawbaa, H.M.; Hassanien, A.E. Binary grey wolf optimization approaches for feature selection. *Neurocomputing* **2016**, *172*, 371–381. [CrossRef]
44. Gandomi, A.H.; Deb, K.; Averill, R.C.; Rahnamayan, S.; Omidvar, M.N. Using semi-independent variables to enhance optimization search. *Expert Syst. Appl.* **2018**, *120*, 279–297. [CrossRef]
45. Kira, K.; Rendell, L.A. A practical approach to feature selection. In *Machine Learning Proceedings*; Elsevier: Amsterdam, The Netherlands, 1992; pp. 249–256.
46. Li, J.; Cheng, K.; Wang, S.; Morstatter, F.; Trevino, R.P.; Tang, J.; Liu, H. Feature selection: A data perspective. *ACM Comput. Surv. (CSUR)* **2017**, *50*, 1–45. [CrossRef]
47. Bolón-Canedo, V.; Sánchez-Maroño, N.; Alonso-Betanzos, A. A review of feature selection methods on synthetic data. *Knowl. Inf. Syst.* **2013**, *34*, 483–519. [CrossRef]
48. Brezočnik, L.; Fister Jr, I.; Podgorelec, V. Swarm intelligence algorithms for feature selection: A review. *Appl. Sci.* **2018**, *8*, 1521. [CrossRef]
49. Solorio-Fernández, S.; Carrasco-Ochoa, J.A.; Martínez-Trinidad, J.F. A review of unsupervised feature selection methods. *Artif. Intell. Rev.* **2020**, *53*, 907–948. [CrossRef]
50. Aljawarneh, S.; Aldwairi, M.; Yassein, M.B. Anomaly-based intrusion detection system through feature selection analysis and building hybrid efficient model. *J. Comput. Sci.* **2018**, *25*, 152–160. [CrossRef]
51. Ambusaidi, M.A.; He, X.; Nanda, P.; Tan, Z. Building an Intrusion Detection System Using a Filter-Based Feature Selection Algorithm. *IEEE Trans. Comput.* **2016**, *65*, 2986–2998. [CrossRef]
52. Khater, B.; Wahab, A.A.; Idris, M.; Hussain, M.; Ibrahim, A.; Amin, M.; Shehadeh, H. Classifier Performance Evaluation for Lightweight IDS Using Fog Computing in IoT Security. *Electronics* **2021**, *10*, 1633. [CrossRef]
53. Naseri, T.S.; Gharehchopogh, F.S. A Feature Selection Based on the Farmland Fertility Algorithm for Improved Intrusion Detection Systems. *J. Netw. Syst. Manag.* **2022**, *30*, 1–27. [CrossRef]
54. Mohammadzadeh, H.; Gharehchopogh, F.S. Feature Selection with Binary Symbiotic Organisms Search Algorithm for Email Spam Detection. *Int. J. Inf. Technol. Decis. Mak.* **2021**, *20*, 469–515. [CrossRef]
55. Zhang, Y.; Wang, S.; Phillips, P.; Ji, G. Binary PSO with mutation operator for feature selection using decision tree applied to spam detection. *Knowl.-Based Syst.* **2014**, *64*, 22–31. [CrossRef]
56. Lin, F.; Liang, D.; Yeh, C.-C.; Huang, J.-C. Novel feature selection methods to financial distress prediction. *Expert Syst. Appl.* **2014**, *41*, 2472–2483. [CrossRef]
57. Kwak, N.; Choi, C.-H. Input feature selection for classification problems. *IEEE Trans. Neural Networks* **2002**, *13*, 143–159. [CrossRef]
58. Sharda, S.; Srivastava, M.; Gusain, H.S.; Sharma, N.K.; Bhatia, K.S.; Bajaj, M.; Kaur, H.; Zawbaa, H.M.; Kamel, S. A hybrid machine learning technique for feature optimization in object-based classification of debris-covered glaciers. *Ain Shams Eng. J.* **2022**, *13*, 101809. [CrossRef]
59. Xue, B.; Zhang, M.; Browne, W.N. Particle Swarm Optimization for Feature Selection in Classification: A Multi-Objective Approach. *IEEE Trans. Cybern.* **2012**, *43*, 1656–1671. [CrossRef]
60. Akyol, S.; Alatas, B. Plant intelligence based metaheuristic optimization algorithms. *Artif. Intell. Rev.* **2016**, *47*, 417–462. [CrossRef]
61. Alatas, B. Chaotic bee colony algorithms for global numerical optimization. *Expert Syst. Appl.* **2010**, *37*, 5682–5687. [CrossRef]
62. Alatas, B.; Bingol, H. Comparative Assessment Of Light-based Intelligent Search And Optimization Algorithms. *Light Eng.* **2020**, *6*, 51–59. [CrossRef]
63. Mafarja, M.; Eleyan, D.; Abdullah, S.; Mirjalili, S. S-shaped vs. V-shaped transfer functions for ant lion optimization algorithm in feature selection problem. In Proceedings of the International Conference on Future Networks and Distributed Systems, Cambridge, UK, 19–20 July 2017; pp. 1–7.

64. Rashedi, E.; Nezamabadi-Pour, H.; Saryazdi, S. BGSA: Binary gravitational search algorithm. *Nat. Comput.* **2009**, *9*, 727–745. [CrossRef]
65. De Souza, R.C.T.; dos Santos Coelho, L.; De Macedo, C.A.; Pierezan, J. A V-shaped binary crow search algorithm for feature selection. In Proceedings of the 2018 IEEE Congress on Evolutionary Computation (CEC), Rio de Janeiro, Brazil, 8–13 July 2018; pp. 1–8.
66. Mafarja, M.; Aljarah, I.; Heidari, A.A.; Faris, H.; Fournier-Viger, P.; Li, X.; Mirjalili, S. Binary dragonfly optimization for feature selection using time-varying transfer functions. *Knowl.-Based Syst.* **2018**, *161*, 185–204. [CrossRef]
67. Mirjalili, S.; Zhang, H.; Mirjalili, S.; Chalup, S.; Noman, N. A Novel U-Shaped Transfer Function for Binary Particle Swarm Optimisation. In *Soft Computing for Problem Solving 2019*; Springer: Berlin/Heidelberg, Germany, 2020; pp. 241–259. [CrossRef]
68. Ahmed, S.; Ghosh, K.K.; Mirjalili, S.; Sarkar, R. AIEOU: Automata-based improved equilibrium optimizer with U-shaped transfer function for feature selection. *Knowl. Based. Syst.* **2021**, *228*, 107283. [CrossRef]
69. Ghosh, K.K.; Singh, P.K.; Hong, J.; Geem, Z.W.; Sarkar, R. Binary social mimic optimization algorithm with X-shaped transfer function for feature selection. *IEEE Access* **2020**, *8*, 97890–97906. [CrossRef]
70. Guo, S.-S.; Wang, J.-S.; Guo, M.-W. Z-Shaped Transfer Functions for Binary Particle Swarm Optimization Algorithm. *Comput. Intell. Neurosci.* **2020**, *2020*, 1–21. [CrossRef]
71. Ramasamy, A.; Mondry, A.; Holmes, C.; Altman, D.G. Key Issues in Conducting a Meta-Analysis of Gene Expression Microarray Datasets. *PLOS Med.* **2008**, *5*, e184. [CrossRef]
72. Alirezanejad, M.; Enayatifar, R.; Motameni, H.; Nematzadeh, H. Heuristic filter feature selection methods for medical datasets. *Genomics* **2019**, *112*, 1173–1181. [CrossRef]
73. Varzaneh, Z.A.; Orooji, A.; Erfannia, L.; Shanbehzadeh, M. A new COVID-19 intubation prediction strategy using an intelligent feature selection and K-NN method. *Informatics Med. Unlocked* **2021**, *28*, 100825. [CrossRef]
74. Pashaei, E.; Pashaei, E. An efficient binary chimp optimization algorithm for feature selection in biomedical data classification. *Neural Comput. Appl.* **2022**, *34*, 6427–6451. [CrossRef]
75. Nadimi-Shahraki, M.H.; Fatahi, A.; Zamani, H.; Mirjalili, S. Binary Approaches of Quantum-Based Avian Navigation Optimizer to Select Effective Features from High-Dimensional Medical Data. *Mathematics* **2022**, *10*, 2770. [CrossRef]
76. Alweshah, M.; Alkhalaileh, S.; Al-Betar, M.A.; Abu Bakar, A. Coronavirus herd immunity optimizer with greedy crossover for feature selection in medical diagnosis. *Knowl.-Based Syst.* **2021**, *235*, 107629. [CrossRef] [PubMed]
77. Anter, A.M.; Ali, M. Feature selection strategy based on hybrid crow search optimization algorithm integrated with chaos theory and fuzzy c-means algorithm for medical diagnosis problems. *Soft Comput.* **2019**, *24*, 1565–1584. [CrossRef]
78. Singh, N.; Singh, P. A hybrid ensemble-filter wrapper feature selection approach for medical data classification. *Chemom. Intell. Lab. Syst.* **2021**, *217*, 104396. [CrossRef]
79. Too, J.; Abdullah, A.R. Binary atom search optimisation approaches for feature selection. *Connect. Sci.* **2020**, *32*, 406–430. [CrossRef]
80. Elgamal, Z.; Sabri, A.Q.M.; Tubishat, M.; Tbaishat, D.; Makhadmeh, S.N.; Alomari, O.A. Improved Reptile Search Optimization Algorithm using Chaotic map and Simulated Annealing for Feature Selection in Medical Filed. *IEEE Access* **2022**, *10*, 51428–51446. [CrossRef]
81. Emary, E.; Zawbaa, H.M.; Hassanien, A.E. Binary ant lion approaches for feature selection. *Neurocomputing* **2016**, *213*, 54–65. [CrossRef]
82. Zhang, Y.; Song, X.-F.; Gong, D.-W. A return-cost-based binary firefly algorithm for feature selection. *Inf. Sci.* **2017**, *418*, 561–574. [CrossRef]
83. Sayed, G.I.; Tharwat, A.; Hassanien, A.E. Chaotic dragonfly algorithm: An improved metaheuristic algorithm for feature selection. *Appl. Intell.* **2018**, *49*, 188–205. [CrossRef]
84. Wang, J.; Khishe, M.; Kaveh, M.; Mohammadi, H. Binary Chimp Optimization Algorithm (BChOA): A New Binary Meta-heuristic for Solving Optimization Problems. *Cogn. Comput.* **2021**, *13*, 1297–1316. [CrossRef]
85. Kundu, R.; Chattopadhyay, S.; Cuevas, E.; Sarkar, R. AltWOA: Altruistic Whale Optimization Algorithm for feature selection on microarray datasets. *Comput. Biol. Med.* **2022**, *144*, 105349. [CrossRef]
86. Balakrishnan, K.; Dhanalakshmi, R.; Seetharaman, G. S-shaped and V-shaped binary African vulture optimization algorithm for feature selection. *Expert Syst.* **2022**, *10*, e13079. [CrossRef]
87. Akinola, O.A.; Ezugwu, A.E.; Oyelade, O.N.; Agushaka, J.O. A hybrid binary dwarf mongoose optimization algorithm with simulated annealing for feature selection on high dimensional multi-class datasets. *Sci. Rep.* **2022**, *12*, 1–22. [CrossRef]
88. Huang, C.-L.; Wang, C.-J. A GA-based feature selection and parameters optimizationfor support vector machines. *Expert. Syst. Appl.* **2006**, *31*, 231–240. [CrossRef]
89. Mirjalili, S.; Lewis, A. S-shaped versus V-shaped transfer functions for binary Particle Swarm Optimization. *Swarm Evol. Comput.* **2013**, *9*, 1–14. [CrossRef]
90. Kennedy, J.; Eberhart, R.C. A discrete binary version of the particle swarm algorithm. In Proceedings of the 1997 IEEE International Conference on Systems, man, and Cybernetics Computational Cybernetics and Simulation, Orlando, FL, USA, 12–15 October 1997; pp. 4104–4108.
91. Blake, C. UCI repository of machine learning databases. Available online: http://www.ics.uci.edu/~mlearn/MLRepository-html-1998 (accessed on 22 July 2021).

92. Iwendi, C.; Bashir, A.K.; Peshkar, A.; Sujatha, R.; Chatterjee, J.M.; Pasupuleti, S.; Mishra, R.; Pillai, S.; Jo, O. COVID-19 patient health prediction using boosted random forest algorithm. *Front. Public Health* **2020**, *8*, 357. [CrossRef] [PubMed]

93. Guo, G.; Wang, H.; Bell, D.; Bi, Y.; Greer, K. KNN model-based approach in classification. In Proceedings of the OTM Confederated International Conferences on the Move to Meaningful Internet Systems, Rhodes, Greece, 22–26 October 2003; pp. 986–996.

94. Zhang, S.; Li, X.; Zong, M.; Zhu, X.; Cheng, D. Learning k for knn classification. *ACM Trans. Intell. Syst. Technol. (TIST)* **2017**, *8*, 1–19. [CrossRef]

95. Zhang, S.; Li, X.; Zong, M.; Zhu, X.; Wang, R. Efficient kNN Classification With Different Numbers of Nearest Neighbors. *IEEE Trans. Neural Networks Learn. Syst.* **2017**, *29*, 1774–1785. [CrossRef] [PubMed]

96. Garcia, S.; Fernandez, A.; Luengo, J.; Herrera, F. A study of statistical techniques and performance measures for genetics-based machine learning: Accuracy and interpretability. *Soft Comput.* **2008**, *13*, 959–977. [CrossRef]

97. Sokolova, M.; Lapalme, G. A systematic analysis of performance measures for classification tasks. *Inf. Process. Manag.* **2009**, *45*, 427–437. [CrossRef]

98. Glas, A.S.; Lijmer, J.G.; Prins, M.H.; Bonsel, G.J.; Bossuyt, P.M.M. The diagnostic odds ratio: A single indicator of test performance. *J. Clin. Epidemiol.* **2003**, *56*, 1129–1135. [CrossRef]

99. Ciotti, M.; Ciccozzi, M.; Terrinoni, A.; Jiang, W.-C.; Wang, C.-B.; Bernardini, S. The COVID-19 pandemic. *Crit. Rev. Clin. Lab. Sci.* **2020**, *57*, 365–388. [CrossRef]

100. Chakraborty, I.; Maity, P. COVID-19 outbreak: Migration, effects on society, global environment and prevention. *Sci. Total Environ.* **2020**, *728*, 138882. [CrossRef]

101. Dokeroglu, T.; Sevinc, E.; Kucukyilmaz, T.; Cosar, A. A survey on new generation metaheuristic algorithms. *Comput. Ind. Eng.* **2019**, *137*, 106040. [CrossRef]

Article

Innovative Tool for Automatic Detection of Arterial Stenosis on Cone Beam Computed Tomography

Agnese Simoni [1,2,†], Eleonora Barcali [1,3,†], Cosimo Lorenzetto [2], Eleonora Tiribilli [1,2], Vieri Rastrelli [3], Leonardo Manetti [2], Cosimo Nardi [3,*], Ernesto Iadanza [4,‡] and Leonardo Bocchi [1,‡]

[1] Department of Information Engineering, University of Florence, v. Santa Marta, 50139 Florence, Italy
[2] Epica Imaginalis, Via Rodolfo Morandi 13/15, 50019 Sesto Fiorentino, Italy
[3] Department of Biomedical Experimental and Clinical Sciences "Mario Serio", University of Florence, 50139 Florence, Italy
[4] Department of Medical Biotechnologies, Via Aldo Moro 2, 53100 Siena, Italy
* Correspondence: cosimo.nardi@unifi.it
† These authors contributed equally to this work.
‡ These authors contributed equally to this work.

Abstract: Arterial stenosis is one of the main vascular diseases that are treated with minimally invasive surgery approaches. The aim of this study was to provide a tool to support the medical doctor in planning endovascular surgery, allowing the rapid detection of stenotic vessels and the quantification of the stenosis. Skeletonization was used to improve vessels' visualization. The distance transform was used to obtain a linear representation of the diameter of critical vessels selected by the user. The system also provides an estimate of the exact distance between landmarks on the vascular tree and the occlusion, important information that can be used in the planning of the surgery. The advantage of the proposed tool is to lead the examination on the linear representation of the chosen vessels that are free from tortuous vascular courses and from vessel crossings.

Keywords: cone beam computed tomography; segmentation; stenosis; software

Citation: Simoni, A.; Barcali, E.; Lorenzetto, C.; Tiribilli, E.; Rastrelli, V.; Manetti, L.; Nardi, C.; Iadanza, E.; Bocchi, L. Innovative Tool for Automatic Detection of Arterial Stenosis on Cone Beam Computed Tomography. *Appl. Sci.* **2023**, *13*, 805. https://doi.org/10.3390/app13020805

Academic Editor: Marco Giannelli

Received: 13 December 2022
Revised: 31 December 2022
Accepted: 1 January 2023
Published: 6 January 2023

1. Introduction

Image-guided surgical navigation is one of the main technologies used for minimally invasive surgery, which, in turn, is a technique that introduces several advantages over traditional open surgery. This reduces the size and number of incisions that need to be made on the body of the patients, inducing a faster recovery of the patient and a reduced risk of injury-related complications, as well as lower hospitalization costs [1]. On the other hand, minimally invasive surgery often requires computed tomography (CT) and/or magnetic resonance imaging (MRI) of the anatomical area to be combined with the operating scenario. In this context, surgical planning acquires a decisive role as the best outcomes are achieved when surgeons are preoperatively prepared with a deep understanding of the anatomy they will face. To obtain an accurate model of the anatomical part that needs to be studied, some steps need to be performed. First, high-quality images of the anatomical structures of interest must be acquired, followed by segmentation of the parts of interest and rendering the surface or volume using specialized software [2,3]. The resulting models allow the surgeon to observe the anatomy from different angles and define the best-possible route to reach the area of interest [4,5]. The realization of models before surgery is spreading more and more. It allows not only the surgeon to gain an idea and act in a more targeted way, but also students to practice and learn techniques of intervention [6]. Hoetznecker et al. in their work realized a color-coded 3D model of benign glotto-subglottic stenosis and a control airway using a commercial 3D printer starting from CT scans. They showed how the realization of a 3D model resulted in being the more accurate diagnostic strategy compared to the endoscopy and the CT scan usually used for these diagnoses [7]. Furthermore,

Shi et al. showed, as the application of a preoperative planning software to treat the lumbar foraminal stenosis, advantages such as the reduced puncture channel establishment time, operative time, and number of intraoperative fluoroscopic images taken without affecting the clinical outcomes [8]. Marragianis et al. showed the possibility of making patient-specific models of the anatomic and functional characteristics of severe aortic valve stenosis using 3D stereolithographic printing to convert high-resolution CT images into life-sized physical models. The CT digital data were processed with a computer-aided design (CAD) software and exported to a multi-material 3D printer to create dual-material fused 3D models of severe aortic stenosis. The realized model accurately reflected the anatomy with excellent visual correlation to the corresponding clinical CT. The possibility of obtaining specific tailored models can lead to further advances in surgery [9]. Our study focused on the applications of the 3D model in the field of endovascular surgery, a branch of minimally invasive surgery, with the aim of restoring natural blood flow in pathological conditions affecting the vascular system such as the occluded arteries. Numerous diseases, as well as aging induce pathological changes in vascular structures. One of the most-common alterations consists of a reduction in the diameter of the the vessel, resulting in stenosis of varying severity up to a complete occlusion; this type of alteration leads to changes in blood flow, thus reducing the supply of the normal amount of oxygen-rich blood to the tissues, eventually leading to serious complications [10,11]. For this reason, the study of the properties of the arterial walls is the subject of several papers, as well as providing important information that can increase the performance of doctors in both the diagnosis and therapy of vascular pathology [12–17].

The purpose of this study was the development of a new vessel analysis tool that supports the endovascular surgeon in preoperative planning; in particular, it performs an analysis of the vascular tree of the area of interest, enabling a rapid localization and quantification of vascular stenosis. The peculiarity of the suggested tool is to conduct the recognition of occlusion directly on the 3D rendering of the vessels, thus helping in the workflow of surgery.

2. Materials and Methods

We analyzed two cone beam computed tomography (CBCT) scans of the chest of a cat, pre- and post-contrast administration, acquired with a VIMAGO scanner (VIMAGO™ HU, Epica Medical) using the following parameters: 9 ms, 90 mA, 80 peak kilo voltage (KVP). Before the second acquisition, a bolus of Lopamiro, 600 mg/kg, was administered to the animal. Both acquisitions were performed by a professional veterinarian for diagnostic purposes. The process for obtaining the 3D rendering of the vessels, the analysis of their diameter, and the stenosis detection follows the steps described in the following paragraphs.

2.1. Segmentation

The first step is the segmentation. Accurate vascular segmentation is crucial as further analysis of vessels' properties depends on the accuracy of this procedure. In this study, we exploited a segmentation algorithm presented in a previous paper by Simoni et al. [18] and briefly summarized below. This algorithm consists of two consecutive steps: bone segmentation and vessel segmentation. The images are acquired with two CBCT scans, carried out before and after the administration of the contrast medium, respectively. The first acquisition is used to segment the bones. This step is necessary to obtain a binary mask to apply on the contrast image. This binary mask improves the segmentation of vessels in the contrast image, avoiding errors related to a similar gray level distribution of bones and vessels with contrast. Thus, segmented vessels are obtained first by subtracting the segmented bones from the image with contrast and, then, applying a second phase of segmentation on the resulting image. Both segmentations are based on a threshold approach (with an empirically selected threshold), which provides provisional segmentation, further refined by a region-growing method, as explained in the following. The region-growing method begins with a seed point, which is determined with the help of a function that is able to identify all spherical shapes in the frame by exploiting the circular Hough transform

(CHT). Then, the method iteratively groups voxels according to a proximity criterion and a predefined acceptance rule. This consists of comparing the voxel under consideration with the seed point: the difference in the gray levels between the seed point and voxels under test must be lower than a certain local threshold, determined empirically. The algorithms used to determine the segmentation of the bone and the vessel are shown, respectively, in Figures 1 and 2.

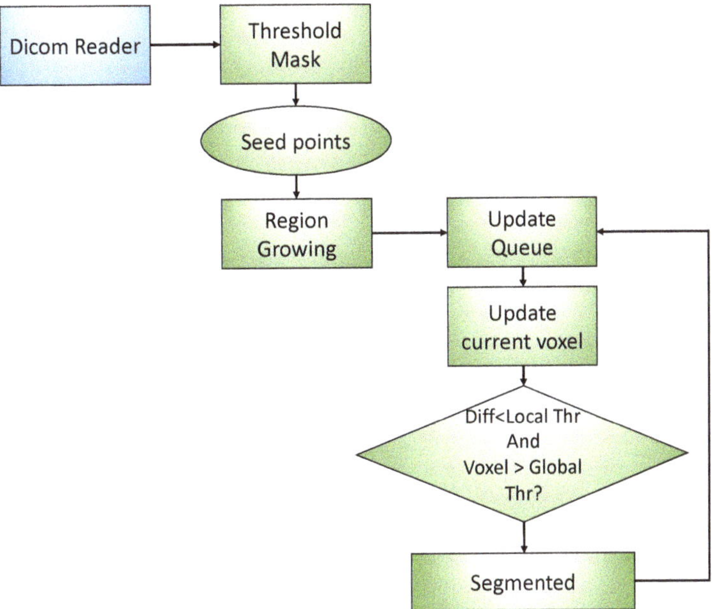

Figure 1. Overview of the bone segmentation algorithm.

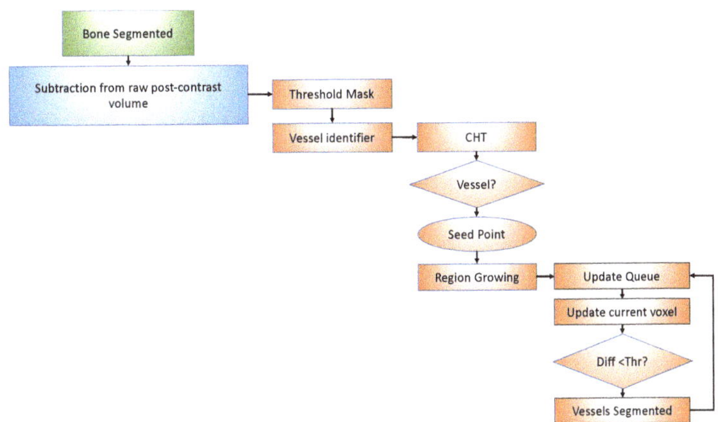

Figure 2. Overview of the blood vessel segmentation algorithm.

2.2. Skeletonization

Skeletonization consists of an iterative process of segmented image thinning, producing centerlines from binary or gray-scale images by extracting medial axes or ridges [19]. The skeleton is a basic representation of the shape of an object that has been reduced to its minimum level. In this case, it takes the binarized volume as the input and outputs a 3D binary image with the same measurements as the reference binarized volume. Vessel skeletonization is performed through an appropriate function present in MATLAB. The skeletonized vessel tree provides a simplified representation of the complex vascular ramifications and pixels that form it. By reducing the volume of blood vessels to their skeleton, the overlap between vascular segments is eliminated accordingly, increasing the understanding of the flow path followed by the blood. The structure of the skeletonized image is then analyzed with the algorithm developed by Kollmannsberger [20], which, starting from a 3D binary skeleton, produces a topological description of the tree providing the adjacency matrix of the graph and two structures containing information about the nodes (coordinates, list of connections for each node, etc.) and links, respectively. By the term "nodes", we refer to the points of bifurcation where a vessel splits into two smaller vessels, while by "link", the set of points that are on the branch.

2.3. Diameter Analysis

The skeleton is then associated with a local representation of the vessel diameters. The local diameter of the vessel can be estimated with the distance transform function. The distance transform operates on binary images and calculates the distance of each pixel in the foreground from its nearest point in the background. Specifically, it associates the intensity of each pixel with the value of the distance between that pixel under study and the nearest pixel in the background [21]. In our case, the input binary volume of vessels constitutes the foreground, so that each pixel of the resulting image, belonging to the vessels, represents its distance from the vascular lumen. Figure 3a shows an example of a frame of the output volume of the distance transform. We can observe that the intensity of the pixels is higher in the center of the vessels, where the distance from the vascular wall reaches its maximum value, while it gradually decreases near the edge of the vessels. The maximum of the distance transform in each vessel provides an accurate estimate of the vessel radius in the given section. Therefore, we associated each point of the skeleton with the value returned by the distance transform applied to the corresponding complementary of the binary volume of the vessel under examination. At the end of the procedure, each pixel k along each branch of the skeleton i is associated with the corresponding distance transform value $r_{i,k}$, providing information of the diameter of the vessel associated with the skeleton.

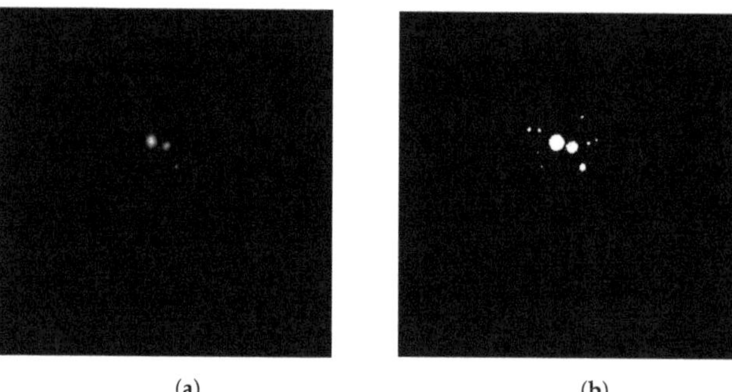

(a)   (b)

Figure 3. Distance transform. (a) Frame of the output image resulting from the application of the distance transform on the binary mask produced by the segmentation of the vessels. (b) The corresponding frame of vessels' binary mask.

2.4. User Interface

The proposed tool includes a graphical user interface (GUI) that allows the user to view the analysis results both in 3D and 2D representation. The 3D representation shows the vessel skeleton, color-coded according to the estimated diameter of the vessel, providing the user with explicit information about the thickness of each vessel directly on the 3D representation of the vessels. Each point of each vessel in the tree is painted using a color-coding scheme based on the classic "jet" colormap, where smaller vessels appear as dark blue; as the diameter grows, these appear lighter up to the larger ones, which appear as bright red. From this representation, the user can select a single vessel composed of several tree segments, for a detailed analysis of the diameter. The procedure begins by selecting the first segment of the vessel, which turns red, while all adjacent segments of the vascular tree are highlighted in green. At this point, the user may select multiple segments, which are consecutively added to the vessel under study. Once a new segment is selected, the representation is updated by drawing all selected segments in red, while the green color identifies the segments that may be further selected for extending the vessel, i.e., segments connected at one of the extremities of the vessel itself.

In order to provide a more detailed view of the variation of the diameter along the selected vessel, the tool also provides a 2D view mode. This two-dimensional plot provides a means to make the search for stenosis more intuitive for medical doctors, compared to the 3D visualization of the skeleton. When the user selects a vessel on the skeleton, the corresponding link and pixels that belong to it are identified and the corresponding linear representation appears, where the horizontal axis represents a linear coordinate L_t along the vessel axis, and the vertical axis indicates the corresponding vessel diameter.

The points on the x-axis are estimated as an integral of the Euclidean distance ΔL_i between each pair of consecutive points in the list describing the structure of the vessel tree. The measure of the diameter of the vessel is obtained by representing, on the y-axis, the value $r_{i,k}$, associated with each pixel of the branch k (or branches) chosen on the skeleton. The plot is also mirrored along the x-axis, to provide a better visual representation of the vessel structure.

The diameter and length of the vessel are scaled in real units (mm) referring to two DICOM tags present in the original pre-contrast CBCT image: the pixel spacing and the slice thickness. The first one is expressed by a pair of values $(\hat{x}, \hat{y})$ that correspond, respectively, to the spacing between the centers of adjacent rows (mm) and the spacing between the centers of adjacent columns (mm) of the image. The other one is represented by $\hat{z}$, which corresponds to the nominal slice thickness measured in mm.

Information about the length of the vessel is obtained as $L_t = \sum_0^t \Delta L_i$, where ΔL_i is computed using the following equation:

$$\Delta L_i = \sqrt{(x_i - x_{i-1})^2 \hat{x}^2 + (y_i - y_{i-1})^2 \hat{y}^2 + (z_i - z_{i-1})^2 \hat{z}^2} \tag{1}$$

where (x_i, y_i, z_i) are the coordinates of the i-th voxel on the axis of the vessel.

As for the vessel diameters, since CBCT scanners acquire isotropic volumes, both DICOM tags have the same value, v. Thus, diameter conversion from pixel units into millimeters is performed by multiplying $r_{i,k}$ by v. In the case of extension to non-isotropic acquisitions, a correct calculation of the diameter of the vessel will require either developing an interpolation strategy for transforming the volumetric image into an isotropic one or the development of a modified distance transform algorithm, taking into account the different scales.

Figure 4 shows the flow chart of the proposed image analysis tool, summarizing both the diameter calculation and visualization.

2.5. Detection of Suspected Stenosis

The system also includes a module for detecting and emphasizing diameter reductions that could be a manifestation of a pathological stenosis. The algorithm employs a moving average filter sliding its window along the segment under study. The filter computes the average radius associated with the pixels contained in each window, rejecting noise and artifacts possibly occurring during image segmentation and diameter estimation. If the difference between the radius of the current pixel and the output of the filter exceeds the fixed threshold, the point is considered as an occlusion of the vessel and brought to the attention of the clinicians. The length of the window was chosen in order to analyze tracts about 1.5 mm long. We chose this value arbitrarily in order to smooth out the noise that may appear in the estimation procedure, while not rejecting small stenoses. However, the operator can properly tune this value in order to obtain the optimal results. Furthermore, the plot highlights the variations in diameter that occur at vessel bifurcations, where a change in the diameter of the vessel is physiologically plausible.

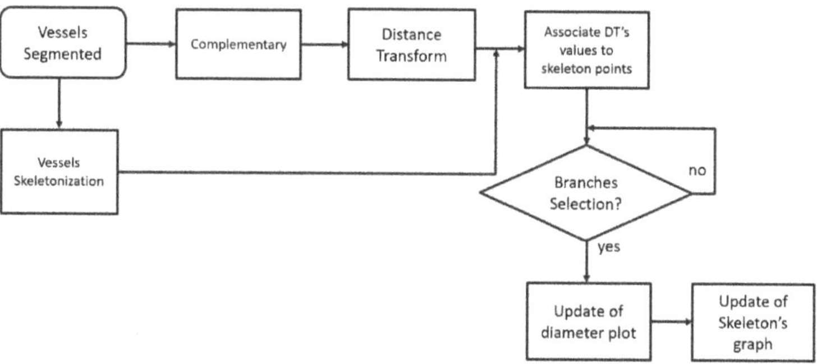

Figure 4. Flow chart of the proposed vessel analysis tool.

To validate the performance of our system, we compared the results detected by our tool with the measurements identified by expert radiologists on the post-contrast CBCT scan DICOM images using an open-source software, Horos™ v3.3.5 software (© 2019, HorosProject).

3. Results

This study provides a tool to facilitate surgeons in detecting arterial stenosis. Critical vessels are identified on the corresponding 3D skeleton of the analyzed CBCT volume, as shown in Figure 5a, where the thickness of each branch is specified. When a branch is selected by the user, the colored skeleton turns gray, while the chosen segment turns red and its consecutive branches turn green, as shown in Figure 5b. Extraction of further vessels follows the same steps as described above. All vessels that have been selected at least once remain red, allowing the user to follow the flow of the vessel under study. This representation is schematized in Figure 5c. Once the skeleton is selected, the two-dimensional plot corresponding to that tract of the segment is also realized, as seen in Figure 5d. The phase of segmentation was previously validated in the paper of Simoni et al., from which we resumed the process to perform our segmentation [18]. They compared the results with the reference images obtained performing a manual segmentation. This analysis showed a sensitivity of 0.748 and a specificity of 0.999.

Our system was able to identify the presence of stenosis in the upper mediastinum at the level of the epiaortic vessels. The value of the stenosis shown in the linear graph is 0.84 mm. This corresponds to a little more than a 50% stenosis of the left subclavian artery. This value was calculated from the y-coordinate measurement of 0.419 mm. Once the user selects the tract of interest, the corresponding linear representation appears. On the plot shown in Figure 6 is reported the two-dimensional representation of the segment and the values of the coordinates x and y. This graph shows contemporaneously the length of the tract along the x-axis and the value of the diameter along the y-axis.

The values on the y-axis are plotted with respect to each point of the skeleton of that branch starting from the values obtained with the distance transform. Specifically, the value of the diameter on the graph is obtained by taking these values and reversing them symmetrically with respect to the x-axis.

Thus, given that the coordinate y on the graph of Figure 6 has value $y = 0.419$ mm, on the y-axis, the diameter is equal to double that value, that is 0.84 mm.

This value was compared with the measurement performed by the radiologists, who identified, on the coronal plane, a diameter of about 1.06 mm, as shown in Figure 7. The relative error between the two estimates corresponds to approximately 20%. These measurements were performed manually using a digital ruler, and thus, as with any other type of manual measurement, they are subject to human mistakes and inter-individual variability. However, the fact that the values measured both manually and by our tool are approximately similar suggests, first of all, that the system has detected correctly the narrowing point of interest and, secondly, that our tool can improve the accuracy of the human measurements by reducing human subjectivity.

The system is able to detect even minimal narrowing, which might not be visible simply by just the CBCT volume, as shown in the example reported in Figure 8.

The tool has the same ability for both wide and narrow vessels segments, as the detection of structural variations is based on short vessels tracts by comparing the variation of the diameter between the current point and the mean diameter of the analyzed tract.

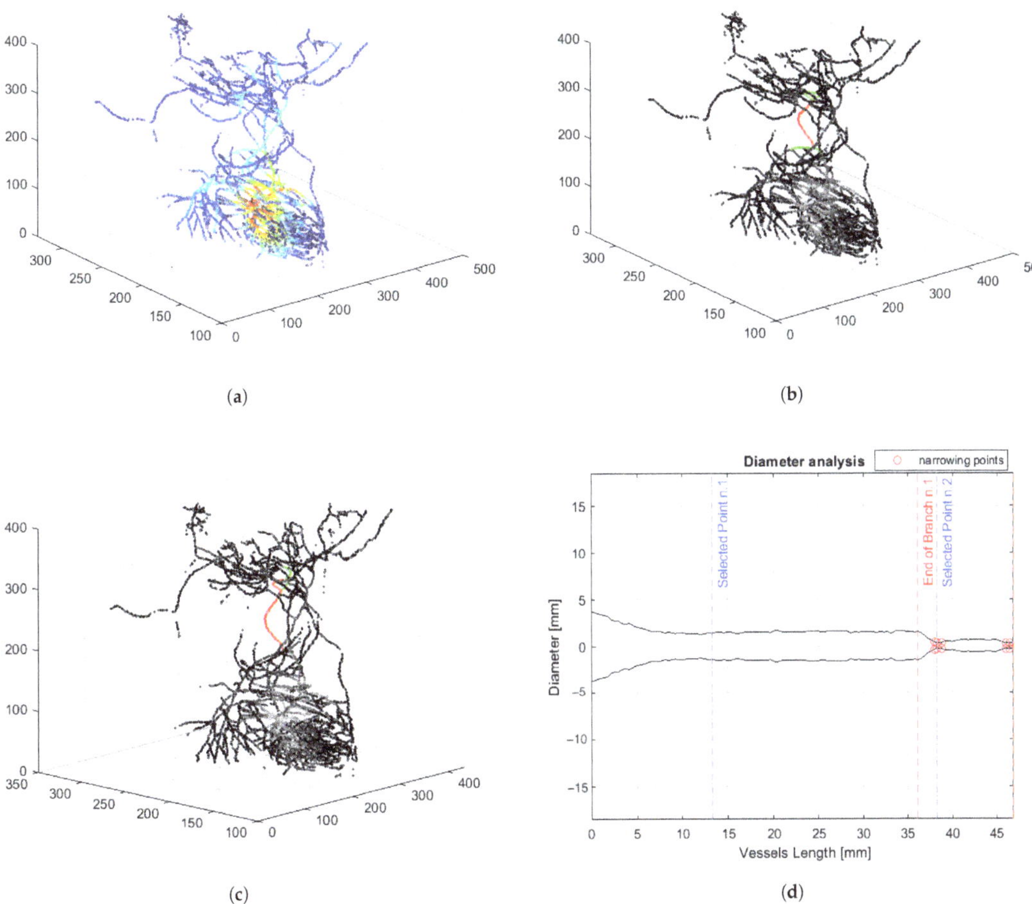

Figure 5. Overview of the vessel analysis tool functioning in response to the selection of two branches on the skeleton. (**a**) Skeleton of the vascular tree. (**b**) After selecting the first branch, the skeleton is shown in gray-scale skeleton, the selected branch is red, and its adjacent segments are green. (**c**) Once the second branch is selected, the GUI shows all selected branches in red and adjacent tracts of the last selected segment in green. (**d**) Linear representation of the vessels.

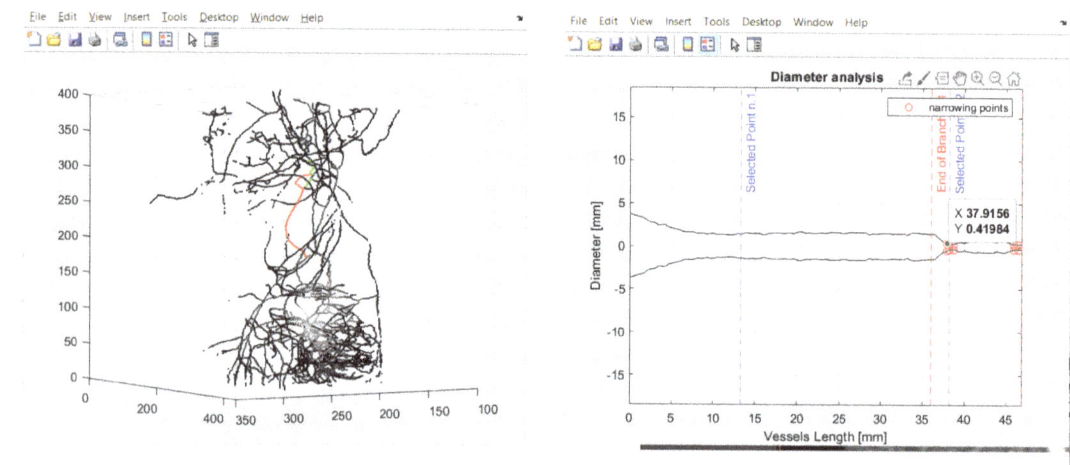

Figure 6. Skeletonization and linear representation of the tract in which the tool has identified the stenosis.

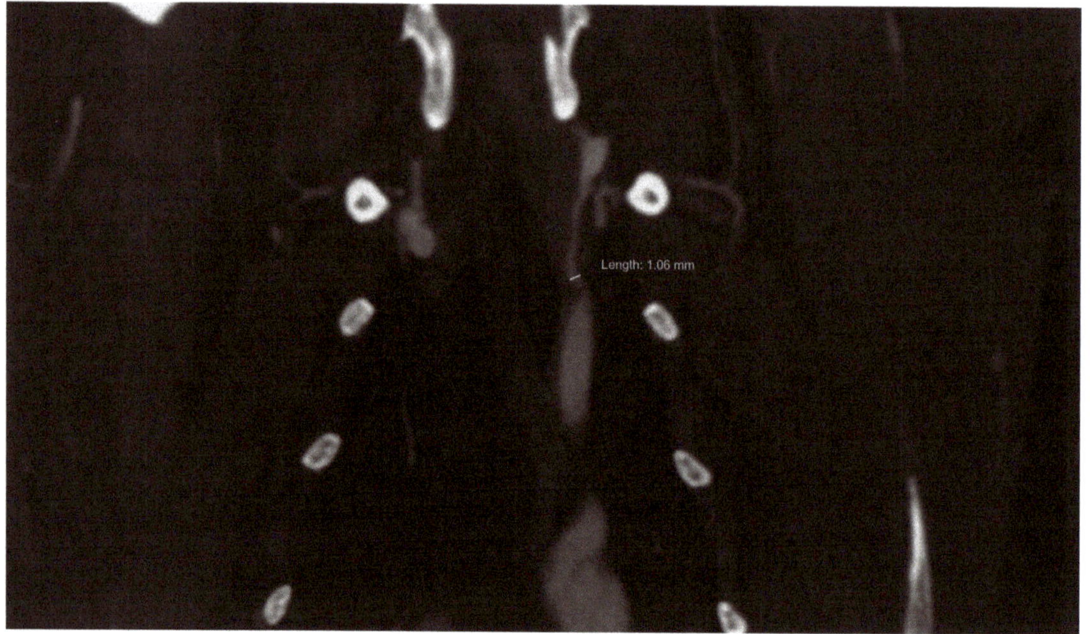

Figure 7. Measurement of the stenosis in the coronal view made by expert radiologists.

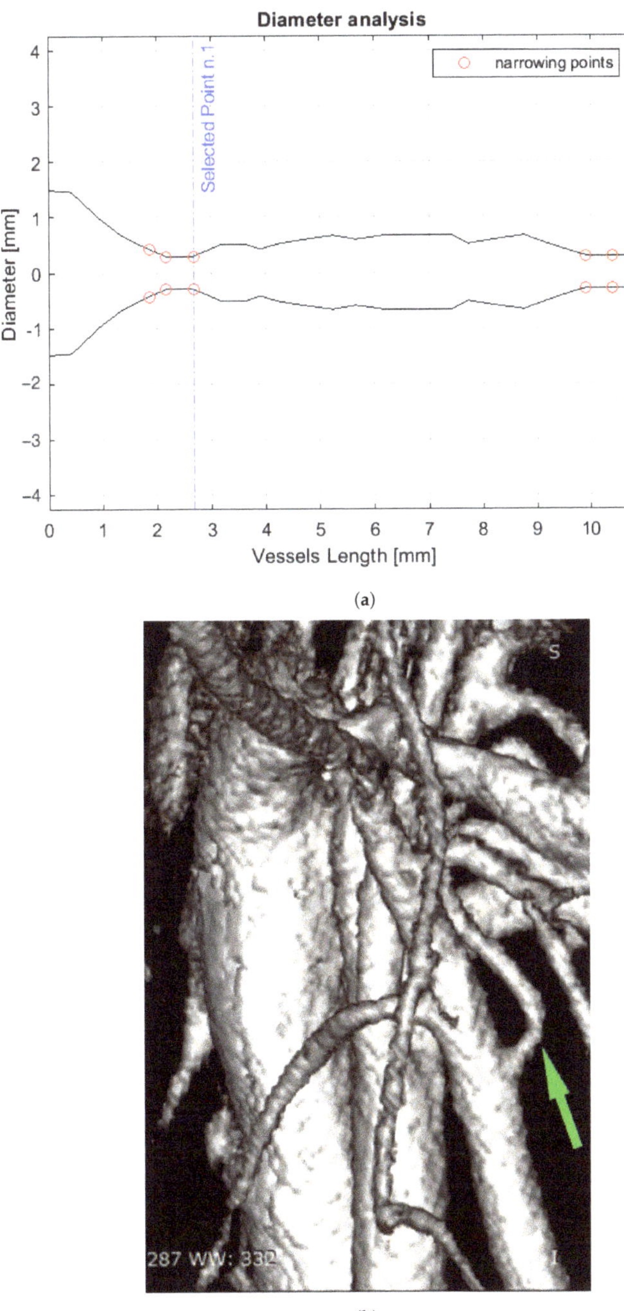

(a)

(b)

Figure 8. (**a**) Linear representation of a vessel presenting five narrowing points that are highlighted with red circles. (**b**) A view of the 3D image of the vascular tree. The green arrow indicates the analyzed vessel in (**a**).

4. Discussion

The proposed system aims to perform an analysis of the vascular tree of the area of interest, supporting the user in the rapid recognition of stenotic branches during preoperative planning. The system contributes to the identification of arterial stenosis, bringing the attention of the clinician to the tracts with suspicious alterations of the vascular lumen. These points are detected by comparing diameter changes along each segment and highlighting narrowing points that exceed the tolerance. The aim of minimally invasive approaches and related tools is to highlight the precision of surgery, improving outcomes such as reducing intraoperative complications and the operative time, and, overall, to improve the safety of the patients. In general, 3D models can be virtual, printed, or augmented reality [22] and are based on high-resolution imaging such as multi-parametric MRI or CT. Vilser et al. introduced a technique for studying the behavior of large retinal vessels using diameter measurements [23], while Heneghan et al. focused on calculating both the tortuosity and the width of the retinal vessels using a morphological processing carried out on segmented vessels [24]. Vascular structural parameters can be exploited in several analysis tools as in the case of Boskamp et al. [25], who developed a software that enriches the visualization of datasets from angiographic CT and MR imaging. The system supplements traditional 2D viewers of the original images with the 3D rendering of the vascular tree, in addition to quantitative morphometric information such as curvature and tortuosity measurements.

The system we realized implements an interactive three-dimensional view of blood vessels, which allows a deepened examination of these anatomical structures by the surgeon. The clinician identifies stenotic points by progressively selecting consecutive blood vessels on the vascular tree rendering. The tool analyzes the chosen segments and emphasizes the tracts that present suspected narrowing, basing the identification of stenotic points on vessels diameter measures obtained with the distance transform technique. It takes advantage of a graphical user interface, which improves the performance of the tool for the user (reported in Figure 9). The GUI presents a section that allows the user to interact with the skeleton of the vascular tree selecting the branches of interest on the skeleton itself. The chosen vascular segments are linearized and represented on a diagram, where the vessel diameter is drawn as a function of vessel length. Moreover, narrowing points, where the vessel diameter drops below a predefined threshold, are highlighted on the plot and brought to the attention of the clinician.

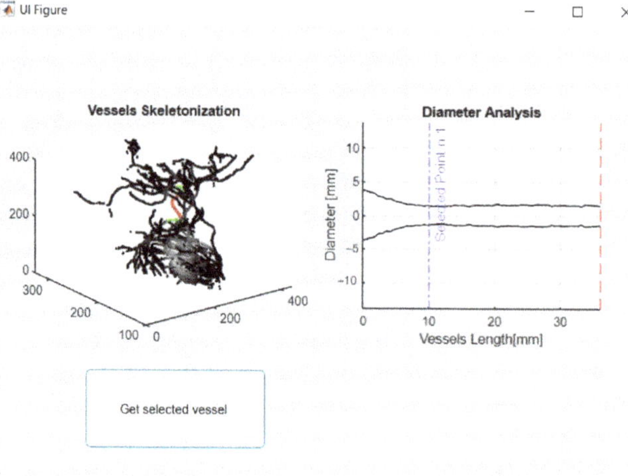

Figure 9. Main window of the system; on the (**left**), 3D plot of the skeletonized vessel tree; on the (**right**), plot of the vessel(s) diameter as a function of length.

The system replaces the complicated 3D rendering of vascular trees with a representation that follows the centerlines of vessels. This "skeletal" image has the advantage of enhancing the bifurcations points and tortuosity of vascular segments, easing the visualization of vascularization and the detection of critical vessels. Vessels are selected by the medical doctor directly on the skeleton. The tool offers a rendered image, which follows the natural course of the vessel under study as its constituent tracts are selected by the user. The skeleton adopts a color code that assists the user in the choice of the segments that are linked to the analyzed vessel, in order to reconstruct the right vascular course. Besides, the surgeon is supported in the identification of stenotic points, as they are marked on the linear representation of the vessel. A further functionality of the vessel analysis tool is allowing the computation of the distance between a reference point on the vascular tree and the point affected by the stenosis. This important clue can be used in the surgical planning phase as it determines the length of the path that will be covered by the catheter during the surgery. At the same time, this information might be used during the surgery, as it has the possibility of reducing the number of angiographic acquisitions that are necessary to control the advancement of the catheter inside the vessel.

We chose to analyze a CBCT volume of the thorax of a cat before and after contrast administration. The system successfully detected the presence of a stenosis of about 50% in the upper mediastinum at the level of the epiaortic vessels.

The advantage of the proposed tool is to lead the examination by the linear representation of the chosen vessels, as shown in Figure 5d, where the vessel is represented as if it were stretched along its length. In fact, the linear representation is free from tortuous vascular courses and vessel crossings, which often increase the effort of clinicians in detecting diameter variations. The limitations of this study include the need for more tests to validate the results, the arbitrariness in the choice of the threshold during the segmentation step, and the problem of vascular enhancement due to the imaging technique used. The problem with the low number of tests is related to the difficulty of recruiting vascular stenosis of animals performed on CBCT scans. Dedicated CBCT scanners are not very popular for studying animals, and it is even more difficult to spot animals with stenotic arteries as collateral findings in examinations carried out for other purposes. Exposing animals to X-rays without a real clinical question is not ethically justified [26]. We are planning to create a simulated test bench for an objective assessment of the proposed algorithm.

The contrast agent needs to be homogeneously distributed to give the correct information. Since CBCT exposure times are longer than traditional multislice CT, there is a risk that the enhancement is not distributed evenly during the whole acquisition and that this can therefore affect the correct display and subsequent segmentation. A solution could be the application of a programmable veterinary infusion pump to administer the contrast agent in order to generate a uniform and prolonged enhancement profile.

5. Conclusions and Future Developments

The results of this study indicate that surgical planning with preoperative 3D imaging of the vessels may potentially reduce the rate of complications during surgeries. Thanks to this system, medical doctors may visualize how the branch tree evolves and the measures by which it is featured. However, larger studies will be needed to confirm these results. In the future, apart from realizing models to test the performance of the algorithm, an extension of this paper could be to implement this system in the environments of virtual and augmented reality, which are experiencing great interest for preoperative planning together with intraoperative navigation, as these technologies can create completely artificial environments, enriching the surgical immersive experience of the team with more available information.

Author Contributions: A.S., E.B., L.M., E.T., C.L. and L.B. designed the study. A.S., L.B., E.T. and C.L. performed the tests. E.B., V.R. and C.N. performed the bibliographic research and organized the results. E.I., E.B., C.N. and L.M. aided in interpreting the results and wrote the final version of the manuscript with the support of all authors. All authors have read and agreed to the published version of the manuscript.

Funding: This work was supported by Fondazione Cassa di Risparmio di Firenze, Florence, Italy (Grant Number 2020.1515).

Institutional Review Board Statement: Ethical review and approval were waived for this study because it is a retrospective study on already available data.

Data Availability Statement: Not applicable.

Conflicts of Interest: A. Simoni, C. Lorenzetto, E. Tiribilli, and L. Manetti work for Epica Imaginalis, a company producing CT scanners.

References

1. Cleary, K.; Peters, T.M. Image-guided interventions: Technology review and clinical applications. *Annu. Rev. Biomed. Eng.* **2010**, *12*, 119–142. [CrossRef] [PubMed]
2. Sakamoto, T. Roles of universal three-dimensional image analysis devices that assist surgical operations. *J. Hepato-Biliary-Pancreat. Sci.* **2014**, *21*, 230–234. [CrossRef] [PubMed]
3. Stella, F.; Dolci, G.; Dell'Amore, A.; Badiali, G.; De Matteis, M.; Asadi, N.; Marchetti, C.; Bini, A. Three-dimensional surgical simulation-guided navigation in thoracic surgery: A new approach to improve results in chest wall resection and reconstruction for malignant diseases. *Interact. Cardiovasc. Thorac. Surg.* **2014**, *18*, 7–12. [CrossRef] [PubMed]
4. Tiribilli, E.; Iadanza, E.; Lorenzetto, C.; Manetti, L.; Bocchi, L. A Novel Implementation of Road Mapping from Digital Subtraction Angiography Images. In Proceedings of the CMBEBIH 2021, Mostar, Bosnia and Herzegovina, 21–24 April 2021.
5. Porumb, M.; Iadanza, E.; Massaro, S.; Pecchia, L. A convolutional neural network approach to detect congestive heart failure. *Biomed. Signal Process. Control.* **2020**, *55*, 101597. [CrossRef]
6. Neumuth, T. Surgical process modeling. *Innov Surg Sci* **2017**, *2*, 123–137. [CrossRef]
7. Hoetzenecker, K.; Chan, H.H.; Frommlet, F.; Schweiger, T.; Keshavjee, S.; Waddell, T.K.; Klepetko, W.; Irish, J.C.; Yasufuku, K. 3D Models in the Diagnosis of Subglottic Airway Stenosis. *Ann. Thorac. Surg.* **2019**, *107*, 1860–1865. [CrossRef]
8. Shi, C.; Sun, B.; Tang, G.; Xu, N.; He, H.; Ye, X.; Xu, G.; Gu, X. Clinical and radiological outcomes of endoscopic foraminoplasty and decompression assisted with preoperative planning software for lumbar foraminal stenosisSurgical process modeling. *Int. J. Comput. Assist. Radiol. Surg.* **2021**, *16*, 1829–1839. [CrossRef]
9. Maragiannis, D.; MS, J.; Igo, S.; Schutt, R.; Connell, P.; Grande-Allen, J.; Barker, C.; Chang, S.; Reardon, M.; Zoghbi, W.; et al. Replicating Patient-Specific Severe Aortic Valve Stenosis With Functional 3D Modeling. *Int. J. Comput. Assist. Radiol. Surg.* **2015**, *8*, e00362. [CrossRef]
10. Blankenhorn, D.H.; Kramsch, D.M. Reversal of atherosis and sclerosis. The two components of atherosclerosis. *Circulation* **1989**, *79*, 1–7. [CrossRef]
11. Sorelli, M.; Perrella, A.; Bocchi, L. Detecting vascular age using the analysis of peripheral pulse. *IEEE Trans. Biomed. Eng.* **2018**, *65*, 2742–2750. [CrossRef]
12. Selzer, R.H.; Mack, W.J.; Lee, P.L.; Kwong-Fu, H.; Hodis, H.N. Improved common carotid elasticity and intima-media thickness measurements from computer analysis of sequential ultrasound frames. *Atherosclerosis* **2001**, *154*, 185–193. [CrossRef] [PubMed]
13. Dobbe, J.G.; Streekstra, G.J.; Atasever, B.; Van Zijderveld, R.; Ince, C. Measurement of functional microcirculatory geometry and velocity distributions using automated image analysis. *Med. Biol. Eng. Comput.* **2008**, *46*, 659. [CrossRef] [PubMed]
14. Jamal, A.; Hazim Alkawaz, M.; Rehman, A.; Saba, T. Retinal imaging analysis based on vessel detection. *Microscopy research and technique* **2017**, *80*, 799–811. [CrossRef] [PubMed]
15. Kovács, G.; Hajdu, A. A self-calibrating approach for the segmentation of retinal vessels by template matching and contour reconstruction. *Med. Image Anal.* **2016**, *29*, 24–46. [CrossRef] [PubMed]
16. Sorelli, M.; Perrella, A.; Bocchi, L. Cardiac pulse waves modeling and analysis in laser doppler perfusion signals of the skin microcirculation. In Proceedings of the IFMBE Proceedings, Sarajevo, Bosnia and Herzegovina, 16–18 March 2017; Volume 62, pp. 20–25.
17. Rogai, F.; Manfredi, C.; Bocchi, L. Metaheuristics for specialization of a segmentation algorithm for ultrasound images. *IEEE Trans. Evol. Comput.* **2016**, *20*, 730–741. [CrossRef]
18. Simoni, A.; Tiribilli, E.; Lorenzetto, C.; Manetti, L.; Iadanza, E.; Bocchi, L. 3D Vessel Segmentation in CT for Augmented and Virtual Reality. In Proceedings of the Mediterranean Forum—Data Science Conference. Springer Computer Science, Sarajevo, Bosnia and Herzegovina, 24 October 2020. [CrossRef]
19. Babin, D.; Pižurica, A.; Velicki, L.; Matić, V.; Galić, I.; Leventić, H.; Zlokolica, V.; Philips, W. Skeletonization method for vessel delineation of arteriovenous malformation. *Comput. Biol. Med.* **2018**, *93*, 93–105. [CrossRef] [PubMed]

20. Kollmannsberger, P.; Kerschnitzki, M.; Repp, F.; Wagermaier, W.; Weinkamer, R.; Fratzl, P. The small world of osteocytes: Connectomics of the lacuno-canalicular network in bone. *New J. Phys.* **2017**, *19*, 073019. [CrossRef]
21. Felzenszwalb, P.F.; Huttenlocher, D.P. Distance transforms of sampled functions. *Theory Comput.* **2012**, *8*, 415–428. [CrossRef]
22. Barcali, E.; Iadanza, E.; Manetti, L.; Francia, P.; Nardi, C.; Bocchi, L. Augmented Reality in Surgery: A Scoping Review. *Appl. Sci.* **2022**, *12*, 6890. [CrossRef]
23. Vilser, W. Retinal vessel analyzer (RVA)-a new measuring system for examination of local and temporal vessel behaviour. *Investig. Ophthalmol. Vis. Sci.* **1997**, *38*, 678. [CrossRef]
24. Heneghan, C.; Flynn, J.; O'Keefe, M.; Cahill, M. Characterization of changes in blood vessel width and tortuosity in retinopathy of prematurity using image analysis. *Med. Image Anal.* **2002**, *6*, 407–429. [CrossRef] [PubMed]
25. Boskamp, T.; Rinck, D.; Link, F.; Kummerlen, B.; Stamm, G.; Mildenberger, P. New vessel analysis tool for morphometric quantification and visualization of vessels in CT and MR imaging data sets. *Radiographics* **2004**, *24*, 287–297. [CrossRef] [PubMed]
26. Nardi, C.; Salerno, S.; Molteni, R.; Occhipinti, M.; Grazzini, G.; Norberti, N.; Cordopatri, C.; Colagrande, S. Radiation dose in non-dental cone beam CT applications: A systematic review. *Radiol. Medica* **2018**, *123*, 765–777. [CrossRef] [PubMed]

applied sciences

MDPI

Article

Morphological, Functional and Texture Analysis Magnetic Resonance Imaging Features in the Assessment of Radiotherapy-Induced Xerostomia in Oropharyngeal Cancer

Leonardo Calamandrei [1], Luca Mariotti [1], Eleonora Bicci [1], Linda Calistri [1], Eleonora Barcali [2], Martina Orlandi [3], Nicholas Landini [4], Francesco Mungai [5], Luigi Bonasera [5], Pierluigi Bonomo [6], Isacco Desideri [7,*], Leonardo Bocchi [2] and Cosimo Nardi [1]

[1] Radiodiagnostic Unit n.2, Department of Experimental and Clinical Biomedical Sciences, University of Florence-Azienda Ospedaliero-Universitaria Careggi, Largo Brambilla 3, 50134 Florence, Italy
[2] Department of Information Engineering, University of Florence, Via S. Marta, 3, 50139 Florence, Italy
[3] Rheumatology Unit, Department of Clinical and Experimental Medicine, University of Florence, 50134 Florence, Italy
[4] Department of Radiological Sciences, Oncology and Pathology, Policlinico Umberto I, Sapienza University, 00185 Rome, Italy
[5] Department of Radiology, University of Florence-Azienda Ospedaliero-Universitaria Careggi, Largo Brambilla 3, 50134 Florence, Italy
[6] Radiation Oncology, University of Florence-Azienda Ospedaliero-Universitaria Careggi, 50134 Florence, Italy
[7] Radiotherapy Unit, Department of Experimental and Clinical Biomedical Sciences, University of Florence-Azienda Ospedaliero-Universitaria Careggi, Viale Morgagni 85, 50134 Florence, Italy
* Correspondence: isacco.desideri@unifi.it

Citation: Calamandrei, L.; Mariotti, L.; Bicci, E.; Calistri, L.; Barcali, E.; Orlandi, M.; Landini, N.; Mungai, F.; Bonasera, L.; Bonomo, P.; et al. Morphological, Functional and Texture Analysis Magnetic Resonance Imaging Features in the Assessment of Radiotherapy-Induced Xerostomia in Oropharyngeal Cancer. *Appl. Sci.* **2023**, *13*, 810. https://doi.org/10.3390/app13020810

Academic Editor: Marco Giannelli

Received: 14 December 2022
Revised: 2 January 2023
Accepted: 3 January 2023
Published: 6 January 2023

Abstract: The aim of this single-center, observational, retrospective study was to investigate magnetic resonance imaging (MRI) biomarkers for the assessment of radiotherapy (RT)-induced xerostomia. Twenty-seven patients who underwent radiation therapy for oropharyngeal cancer were divided into three groups according to the severity of their xerostomia—mild, moderate, and severe—clinically confirmed with the Common Terminology Criteria for Adverse Events (CTCAE). No severe xerostomia was found. Conventional and functional MRI (perfusion- and diffusion- weighted imaging) performed both pre- and post-RT were studied for signal intensity, mean apparent diffusion coefficient (ADC) values, k-trans, and area under the perfusion curves. Contrast-enhanced T1 images and ADC maps were imported into 3D slicer software, and salivary gland volumes were segmented. A total of 107 texture features were derived. T-Student and Wilcoxon signed-rank tests were performed on functional MRI parameters and texture analysis features to identify the differences between pre- and post-RT populations. A p-value < 0.01 was defined as acceptable. Receiver operating characteristic (ROC) curves were plotted for significant parameters to discriminate the severity of xerostomia in the pre-RT population. Conventional and functional MRI did not yield statistically significant results; on the contrary, five texture features showed significant variation between pre- and post-RT on the ADC maps, of which only informational measure of correlation 1 (IMC 1) was able to discriminate the severity of RT-induced xerostomia in the pre-RT population (area under the curve (AUC) > 0.7). Values lower than the cut-off of -1.473×10^{-11} were associated with moderate xerostomia, enabling the differentiation of mild xerostomia from moderate xerostomia with a 73% sensitivity, 75% specificity, and 75% diagnostic accuracy. Therefore, the texture feature IMC 1 on the ADC maps allowed the distinction between different degrees of severity of RT-induced xerostomia in the pre-RT population. Accordingly, texture analysis on ADC maps should be considered a useful tool to evaluate salivary gland radiosensitivity and help identify patients at risk of developing more serious xerostomia before radiation therapy is administered.

Keywords: xerostomia; magnetic resonance imaging; texture analysis; radiomics; head and neck

1. Introduction

Head and neck cancers represent the sixth most common form of cancer worldwide. More than 90% of head and neck cancers are squamous cell carcinomas of the oral cavity, oropharynx, and larynx mucosal tissues [1]. Radiotherapy (RT) is generally included in the primary oncologic treatment, as it improves clinical and functional outcomes for cancer patients, especially in the case of oropharyngeal cancers (OPC) [2]. In addition to its therapeutic effects, RT can cause injuries to the tissues inside and surrounding the irradiated area, and approximately 90% of patients undergoing radiotherapy for head and neck cancers suffer from clinically relevant xerostomia [3,4].

Xerostomia is defined as dryness of the oral cavity resulting from insufficient saliva secretion [5,6]. RT-induced xerostomia is caused by salivary gland dysfunction resulting from X-ray related tissue damage [5,7,8]. In regard to the diagnosis of xerostomia, the basic tests include the determination of stimulated and unstimulated salivary flow rate, palatal secretion, and parotid secretion [9–11]. These measurements constitute the simplest methods of assessing the secretory function of salivary glands. Very low unstimulated and stimulated salivary flow rates are defined as <0.1 mL/min and <0.7 mL/min, respectively [10–12]. Such values are confirmatory of xerostomia, whether or not they co-exist with specific symptoms for this condition, such as oral soreness, dry lips, halitosis, decreased or altered sense of taste, recurrent mouth infections, tooth decay and gum disease, and difficulty speaking, eating, or swallowing [13–15].

Because of the high-contrast resolution and the ability to study complex anatomical regions without the use of radiation, magnetic resonance imaging (MRI) is considered the most relevant imaging technique for the identification of head and neck lesions [16–19]. Diffusion-weighted imaging (DWI) is an established diagnostic tool that evaluates the tissue microanatomy by studying the spontaneous molecular diffusion of protons corresponding to the stochastic Brownian motion of water molecules [16,20,21]. The apparent diffusion coefficient (ADC) is calculated from DWI and consists of the quantitative assessment of the impedance of water molecule diffusion within tissues [20]. The ADC is typically reduced in hypercellular tissues and increased in situations where water molecules are free to move [16,17]. The ADC values are calculated automatically and integrated into a parametric map, upon which regions of interest (ROIs) can be traced at a workstation to determine the ADC values of specific portions of a tissue [16–18]. Dynamic contrast-enhanced, perfusion-weighted imaging (DCE-PWI) is a functional MRI technique that allows one to infer blood perfusion to specific tissues by measuring the changes in tissues over time after the intravenous administration of a contrast agent [22]. DCE-PWI is crucial in the detection of focal lesions since it mainly assesses the vascular permeability [23]. Changes in hemodynamic parameters can precede abnormalities on conventional MRI and can thus be used to help with the diagnosis [22–26].

Radiomics has been rapidly developing over the last few years; it is a hybrid analytical process aimed at determining the correlation between the characteristics of tissues and their corresponding digital images [17,27]. Texture analysis is a form of radiomics, in which macroscopic heterogeneities of tissues can be non-invasively studied to infer information about their microscopic architecture beyond the possibilities of the human eye as if it were a "virtual biopsy" [17,28,29]. Texture analysis is based on the extraction of parameters representing the distribution frequency, intensity, or direction of gray levels within the ROI in order to evaluate the single pixel, its interactions with adjacent pixels, and the distribution of pixels and voxels in the image [17,29].

Some previous studies evaluated RT-induced xerostomia with different MRI techniques, such as DWI [21,30–32] and DCE-PWI [33]. However, neither used both techniques on the same cohort of patients in pre- and post-RT MRI examinations. Furthermore, the texture analysis was mainly performed with CT imaging [34–36], and only one paper can be found on ultrasound [37] and conventional T1 MRI sequences [38].

The current study represents the first attempt to evaluate RT-induced xerostomia by using multiparametric MRI techniques, including DWI, DCE-PWI, and texture analysis, carried out in both pre- and post-RT imaging. The aim of this retrospective study was to correlate such MRI techniques with the severity of RT-induced xerostomia clinically confirmed with the Common Terminology Criteria for Adverse Events (CTCAE).

2. Materials and Methods

2.1. Patient Selection

From January 2014 to December 2021, 180 patients who underwent RT for OPC were assessed in the radiotherapy department of the University Hospital of Florence (Italy).

Inclusion criteria:

- Patients aged over 18 years.
- Histological diagnosis of OPC confirmed with biopsy.
- RT to defeat OPC.
- No disease of the salivary glands.
- MRI for both tumor staging and 4-month follow-up after ending RT.
- DWI and DCE-PWI MRI sequences.

Exclusion criteria:

- Previous head and neck radiation or surgical treatments.
- No MRI carried out in our institute.
- MRI not performed for both tumor staging and follow-up.
- No DWI and DCE-PWI sequences.
- No clinically confirmed xerostomia with CTCAE.
- No sialometric data available.

The initial population included 180 patients who underwent RT to treat oropharyngeal cancer (Figure 1). Among them, 128 did not have available MRI data. Of the remaining 52 patients who were studied with MRI, 21 patients had no DWI and DCE-PWI performed or MRI examinations both before and after RT were not carried out. In addition, 4 patients were not clinically assessed or did not develop clinically confirmed xerostomia. Therefore, the final number of patients included in the study was 27. They developed xerostomia after RT and were studied with MRI both before and after RT with DWI and DCE-PWI sequences.

2.2. Patients' Differentiations into Groups Based on Clinical Evaluation

The severity of dry mouth was clinically assessed 1 month after the end of RT by a radiotherapist with 10-years' experience using the U.S. National Cancer Institute's CTCAE v4. This score scale has three increasing levels of severity [39]. Accordingly, we decided to divide the patients into three groups:

1. Group 1 (mild xerostomia): Feeling of dry or thick saliva with no significant dietary alteration; unstimulated saliva flow > 0.2 mL/min;
2. Group 2 (moderate xerostomia): Moderate symptoms; oral intake alterations (e.g., copious water, other lubricants, diet limited to purees and/or soft, moist foods); unstimulated saliva flow 0.1 to 0.2 mL/min;
3. Group 3 (severe xerostomia): Inability to adequately aliment orally; tube feeding or total parenteral nutrition indicated; unstimulated saliva flow < 0.1 mL/min.

2.3. Image Acquisition and Analysis

MRI examinations were performed via 1.5 T Magnetom aera (Siemens Healthcare, Erlangen, Germany) with a devoted head and neck coil. The MRI acquisition protocol included pre- and post-contrast scans. An axial fat-saturated, echo-planar, imaging-based DWI with two different b-values (b 50–800 s/mm^2) was acquired. The apparent diffusion coefficient (ADC) values of the parotid and sub-mandibular glands before and after RT were calculated by positioning three regions of interest (ROI) within the salivary gland parenchyma in three contiguous axial sections (Figure 2). Time/intensity curve (I/t), area under the curve (AUC), and K(trans) values of the parotid and submandibular glands before and after RT were generated by using IntelliSpace software version 9.0 (Philips, Amsterdam,

The Netherlands) from the native DCE-PWI images by drawing a ROI including the largest gland section possible (Figure 2). Before the sampling, a ROI was automatically placed on the internal carotid artery to obtain an arterial input function curve, defined as the contrast concentration in the vessels feeding the tissue at each point in time during the contrast passage. The vessels, lymph nodes, and cystic areas within the salivary gland parenchyma were excluded on both DWI and DCE-PWI analyses. The ADC, I/t, AUC, and K(trans) values of the trapezius muscle were also obtained as control parameters.

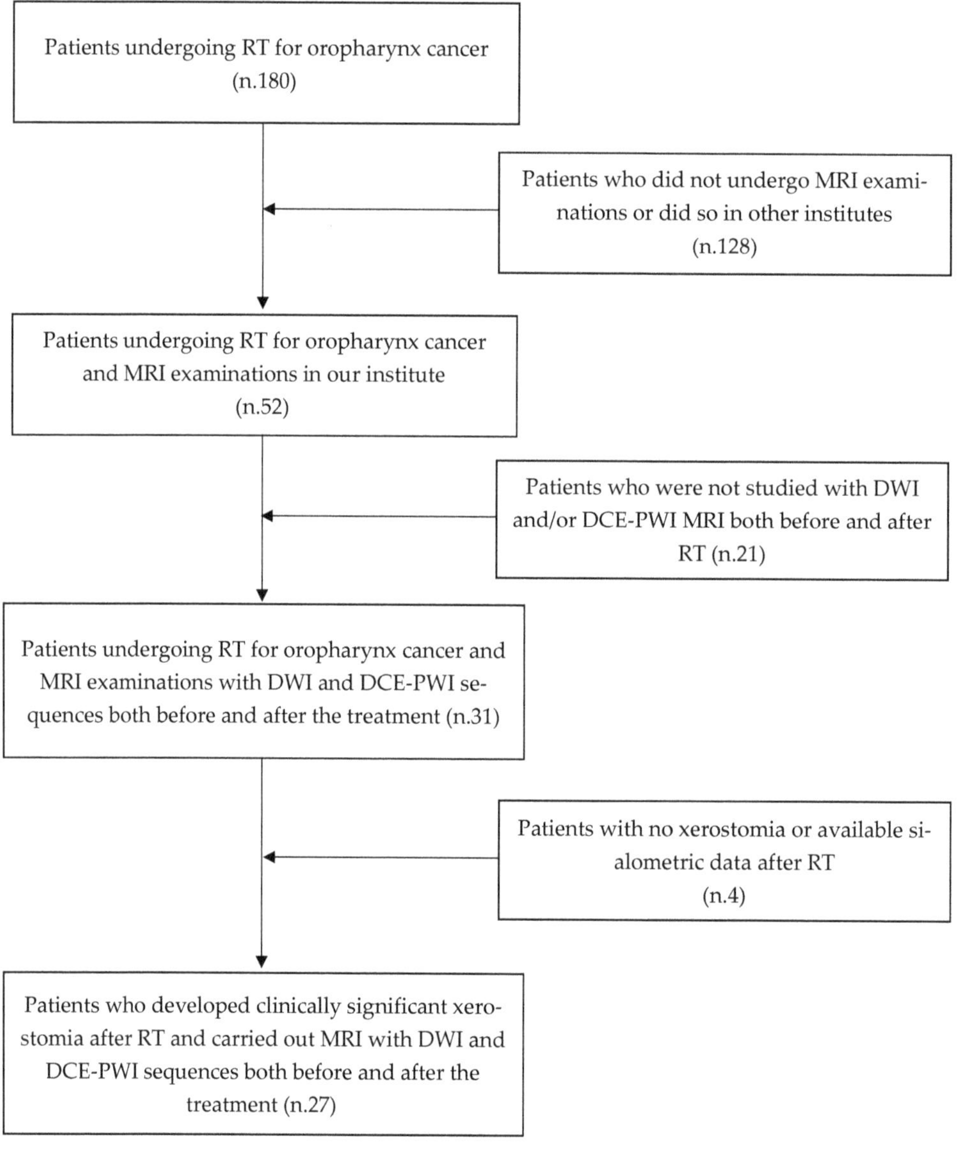

Figure 1. Patient inclusion/exclusion flowchart. RT = radiotherapy, MRI = magnetic resonance imaging, DWI = diffusion-weighted imaging, DCE-PWI = dynamic contrast-enhanced, perfusion-weighted imaging.

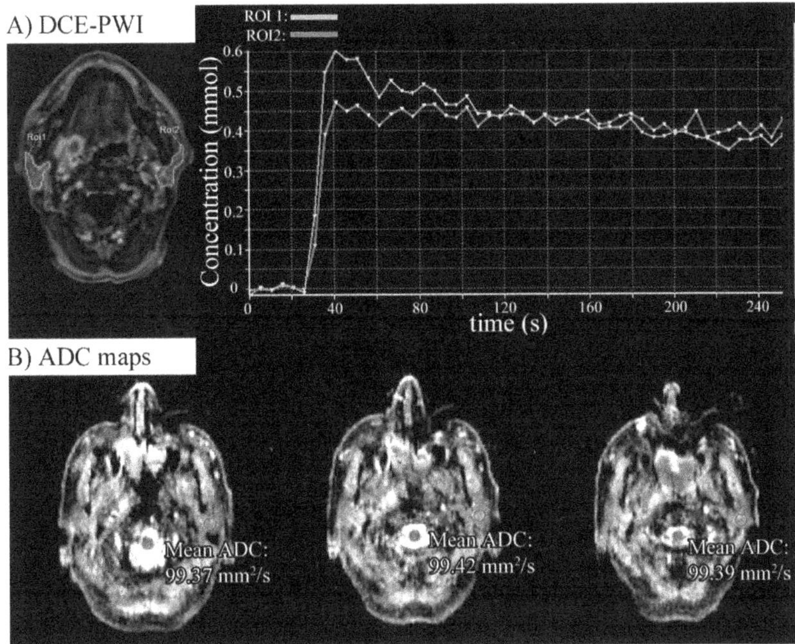

Figure 2. Region of interest (ROI) definition for functional MRI. (**A**) DCE-PWI ROI definition for kinetic parameters drawn on both the parotid glands. The graph shows the contrast agent concentration over time for the two ROIs (left and right parotid gland, respectively). (**B**) ADC map ROI definition for local ADC values (see purple circle ROIs on the left parotid gland) with mean ADC values.

The following morphologic, DWI, and DCE-PWI features were assessed:

1. T2 signal intensity (SI) hyper-, iso-, or hypointense with respect to the muscle signal of the parotid and submandibular glands before and after RT;
2. SI hyper- or hypointense of the parotid and submandibular glands before and after RT on DWIb800 images;
3. Mean ADC values of the parotid and submandibular glands before and after RT (ADC pre-post) on DWI sequences;
4. Mean AUC and K(trans) values of the parotid and submandibular glands before RT (AUCpre, K(trans)pre) and after RT (AUCpost, K(trans)post) on DCE-PWI sequences;
5. Ratio between AUC values of parotid and submandibular glands before and after RT (AUCpost/pre);
6. Ratio between K(trans) values of the parotid and submandibular glands before and after RT (K(trans)post/pre).

The MRI acquisition parameters are shown in Table A1 in Appendix A.

2.4. Texture Analysis

The MRI images obtained with T1 post-contrast sequences and the ADC maps before and after RT were imported into 3D slicer (www.3dslicer.org (accessed on 12 February 2022)) v10.4.2 software. The parotid and submandibular glands located on the same side of the oropharyngeal cancer, corresponding to the irradiated side, were segmented for the entirety of their volumes by a radiologist with 3-years' experience in head and neck cancer using the "segmentation wizard" extension for 3D slicer (Figure 3). Specifically, the segmentation of the gland located on the irradiated side was performed on CE-T1 sequences both before and after RT. The same method was carried out on the ADC maps thus resulting in a total of 4 different segmented volumes per gland. This process resulted in a total number of 216 volumes being segmented with 4 submandibular and 4 parotid gland volumes investigated for each of the 27 patients. Texture features were analyzed and extracted from such volumes using the extension "Pyradiomics" for 3D slicer.

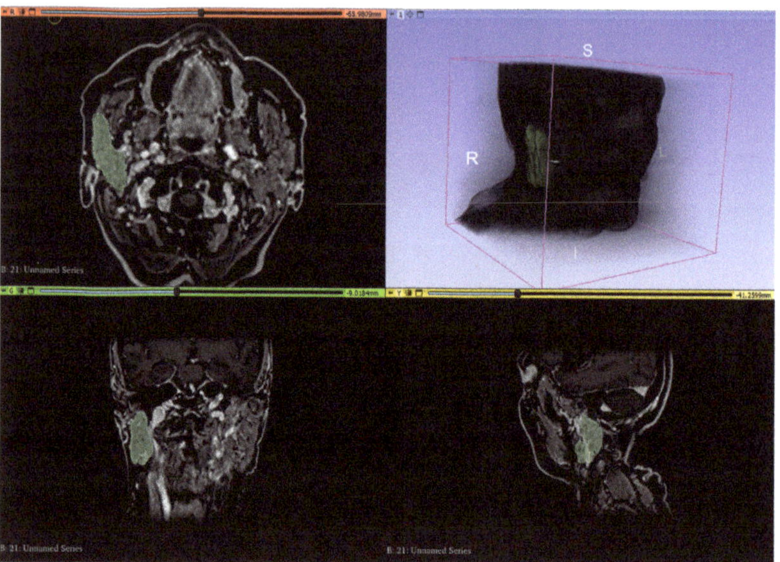

Figure 3. Whole volume segmentation of the right parotid gland using 3D slicer software on contrast-enhanced T1 sequences. The entire process was performed for submandibular and parotid glands of the irradiated side on both pre- and post-RT contrast-enhanced T1 sequences and ADC maps.

A total of 107 radiomics features were extracted, belonging to the following categories: First Order, Shape-based (2D and 3D), Gray Level Co-occurrence Matrix, Gray Level Size Zone Matrix, Gray Level Run Length Matrix, Gray Level Dependence Matrix, and Neighboring Gray Tone Difference Matrix. The detailed explanations of the texture subclasses can be found in Table A2 (Appendix A).

2.5. Statistical Analysis

Texture analysis and functional DWI/DCE-PWI MRI were assessed with the same statistical methods.

First, a Shapiro–Wilk test was performed to determine the nature of the distribution of the data obtained from both the functional MRI and texture analysis. After separating the subjects into three groups (group 1, 2, and 3) according to their CTCAE v4 score, each parameter from the functional MRI and texture analysis was evaluated for its variation between pre- and post-RT for each group separately in order to study the correlation between RT-induced variation and the severity of xerostomia. More specifically, the parameters were tested for variations for each gland separately.

The parameters that showed normal and non-normal distributions were analyzed with the parametric t-Student test and Wilcoxon signed-rank test, respectively. A p-value < 0.01 was defined as acceptable. Once p-values were obtained using both methods (t- Student and Wilcoxon), receiver operating characteristics (ROC) curves were produced by calculating and plotting true positive and false positive rates for each statistically significant parameter within the pre-RT population to determine the area under the curve (AUC) and cut-off values to identify the subjects who would develop more or less severe xerostomia after RT (group 1 vs. group 2 vs. group 3).

3. Results

Eighteen patients and nine patients belonged to group 1 (mild xerostomia) and group 2 (moderate xerostomia), respectively. No group 3 patients (severe xerostomia) were found among the 27 patients enrolled. T and N stages at baseline for each patient are shown in Table 1. The radiation dose data are provided in Table A3 (Appendix A).

Table 1. T and N stages at baseline for the patients with mild xerostomia (Group 1) and moderate xerostomia (Group 2).

Patients	Stage	
	T	N
Group 1		
Patient 1	3	2c
Patient 2	3	3
Patient 3	2	1
Patient 4	3	2b
Patient 5	4a	2b
Patient 6	3	0
Patient 7	2	0
Patient 8	4b	0
Patient 9	2	2b
Patient 10	4a	2b
Patient 11	2	2c
Patient 12	4a	2a
Patient 13	4a	1
Patient 14	3	0
Patient 15	2	0
Patient 16	2	2a
Patient 17	2	1
Patient 18	3	2c
Group 2		
Patient 19	3	2c
Patient 20	3	1
Patient 21	2	2b
Patient 22	4a	2c
Patient 23	1	2b
Patient 24	1	2b
Patient 25	3	2a
Patient 26	4a	2c
Patient 27	1	2b

3.1. Morphological and Functional MRI

The SI of the submandibular and parotid glands did not change before and after RT compared to the muscle tissue on both T2 and DWI b800 images since the SI was hyperintense in all cases. In addition, the mean ADC values, K(trans) parameters, and AUC parameters did not show significant variation between pre- and post-RT. The p-value, mean value, median value, and standard deviation for the functional MRI parameters are shown in Tables 2 and 3.

3.2. Texture Analysis

No texture features showed statistically significant variation between pre- and post-RT on both the ADC maps and CE-MRI T1w sequences in group 1 (p-value > 0.01). The same results were found for group 2 on the CE-MRI T1w sequences (p-value > 0.01). On the contrary, in group 2, the ADC map values of the parotid and submandibular glands observed before RT were significantly different than those observed after RT (p-value < 0.01) (Table 4).

Table 2. *p*-values for the parameters studied in group 1 (mild xerostomia) and group 2 (moderate xerostomia) on DWI and DCE-PWI sequences.

Parameter	*p*-Value	
	Group 1	Group 2
DWI MRI		
ADC Parotid	0.82	0.18
ADC Submandibular	0.60	0.54
DCE-PWI MRI		
AUC PAROTID	0.07	0.03
AUC SUBMANDIBULAR	0.26	0.21
KTRANS PAROTID	0.18	0.13
KTRANS SUBMANDIBULAR	0.65	0.82

Table 3. Mean, median, and standard deviation for the parameters studied in group 1 (mild xerostomia) and group 2 (moderate xerostomia) on DWI and DCE-PWI sequences. P = parotid gland. S = submandibular gland.

	ADC P Pre RT	ADC P Post RT	ADC S Pre RT	ADC S Post RT	AUC P Pre RT	AUC P Post RT	AUC S Pre RT	AUC S Post RT	Ktrans P Pre RT	Ktrans P Post RT	Ktrans S Pre RT	Ktrans S Post RT
Group 1 Mean	0.84	0.90	1.16	1.26	88.40	152.45	12. 62	176.65	123.80	229.31	169.44	233.20
Standard deviation	0.08	0.25	0.28	0.23	30.83	70.80	55.19	91.40	46.81	131.29	71.71	133.59
Group 2 Mean	0.80	0.90	1.19	1.32	94.28	146.35	88.77	154.92	158.76	203.08	149.22	211.09
Standard deviation	0.13	0.51	0.16	0.24	22.22	82.03	26.01	89.34	61.82	87.80	68.78	87.75

Table 4. *p*-values for statistically significant texture features. See Table A1 (Appendix A) for the definition of First Order and Gray Level Run Length Matrix.

Feature Name	*p*-Value
Parotid	
Informational measure of correlation 1 (Gray Level Run Length Matrix)	0.002
Informational measure of correlation 2 (Gray Level Run Length Matrix)	0.003
Submandibular	
Gray Level Non-Uniformity Normalized (First Order)	0.002
Informational measure of correlation 2 (First Order)	0.006
Gray Level Non-Uniformity Normalized (Gray Level Run Length Matrix)	0.006

Two features were statistically significant for the parotid glands:

- Informational measure of correlation 1 (IMC 1)—Gray level co-occurrence matrix class.
- Informational measure of correlation 2 (IMC 2)—Gray level co-occurrence matrix class.

Three features were statistically significant for the submandibular glands:

- Informational measure of correlation 2 (IMC 2)—First-order class.
- Gray Level Non-uniformity Normalized (GLNN)—First-order class.
- Gray Level Non-uniformity Normalized (GLNN)—Gray Level Run Length Matrix class.

After producing ROC curves on the pre-RT values, only the feature IMC 1 for the parotid glands showed an acceptable level of diagnostic accuracy (AUC = 0.727) (Figure 4).

The cut-off value defined by the ROC curve for IMC 1 in the pre-RT population was -1.473×10^{-11}, which allowed the distinction between group 1 and group 2 with 73% sensitivity, 75% specificity, and 75% diagnostic accuracy. Values lower than -1.473×10^{-11} corresponded to moderate xerostomia (group 2), and values higher than -1.473×10^{-11} corresponded to mild xerostomia (group 1).

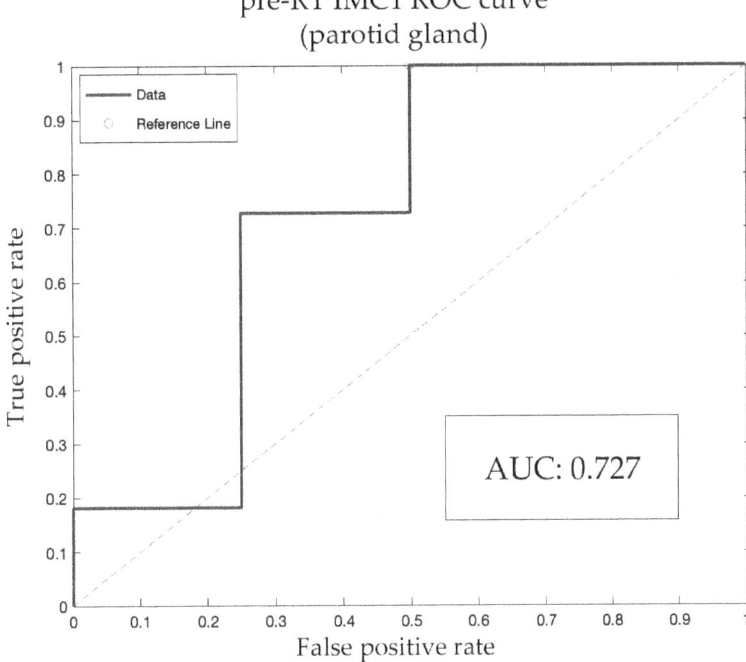

Figure 4. Receiver operating characteristics (ROC) curve for IMC 1 in the parotid gland (pre-RT population). Diagnostic accuracy was considered acceptable when the AUC values were higher than 0.7.

4. Discussion

Xerostomia is a common complication in patients receiving radiation therapy as a consequence of damage to the salivary glands. Morphological and structural changes in the irradiated glands can be non-invasively evaluated with MRI [32,40,41]. In addition, the indirect assessment of the microarchitecture of the salivary glands with texture analysis has been recently hypothesized to be a useful tool in the identification of the severity of RT-induced xerostomia [21,34,37,42]. Overall, predictive models employing texture analysis alongside imaging techniques have shown very promising results in the assessment of head and neck disease [17,18,34,43,44]. Therefore, interest towards artificial intelligence and its applications to imaging is steadily growing. In the context of the ever-growing academic importance of texture analysis [17,28,29,34], the present study was an effort to investigate its role in the assessment of salivary gland alterations and, specifically, RT-induced xerostomia.

While it is known that the development of RT-induced xerostomia correlates with the dose distribution to the salivary glands [45,46], the correlation between imaging techniques and specific alterations in the gland microarchitectural structure is still unclear [34,47]. The current study represented the first attempt to assess RT-induced xerostomia by taking advantage of different MRI techniques, including functional imaging—DCE-PWI and DWI—and texture analysis performed on CE-T1 sequences and ADC maps in both pre-and post-RT imaging.

No parameter for neither morphological (T2 sequences) nor functional (DWI and DCE-PWI) MRI yielded statistically significant results. Such findings were different from previous studies conducted by Juan et al. [33] on DCE-PWI and Zhang et al. [21,30,31] on DWI since they suggested a possible role for parameters obtained from these imaging techniques as biomarkers in the evaluation of RT-induced xerostomia. Possible explanations for the disagreement between our findings and those studies [21,30,31,33] could be that, in those last studies, a larger sample was enrolled and acid stimulation was used to assess the ADC values while also taking into account time-related parameters, such as time-to-peak ADC. In addition to that, the parameters obtained from DWI and DCE-PWI were correlated with the different amount of radiation dose emitted, whereas in our case, all patients enrolled received the same treatment.

Texture analysis techniques on CT images were used in previous studies to investigate RT-induced xerostomia [34–36,42], but they have not yet been employed on CE-MRI and ADC maps. In the present study, the CE-T1 MRI texture features did not show statistically significant correlation with the development of RT-induced xerostomia. On the contrary, the texture analysis carried out on the ADC maps yielded significant results. More specifically, three texture features for the submandibular glands (GLNN—First Order, IMC 2—First Order, and GLNN—Gray Level Run Length Matrix) and two texture features for the parotid glands (IMC 1 and IMC 2, both belonging to the Gray Level Run Length Matrix) showed significant differences between pre- and post-RT imaging. Among the aforementioned texture features, only IMC 1 showed acceptable levels of diagnostic accuracy (AUC = 0.727) for the development of moderate xerostomia when applied on the pre-RT population where, in fact, was found a significant decrease in its values in patients with moderate xerostomia as opposed to patients with mild xerostomia. While no feature except for IMC 1 yielded an acceptable diagnostic accuracy on its own, a more complex model taking all of them into account might better discriminate different degrees of severity of RT-induced xerostomia. In this same framework, the poor results obtained in both morphological and functional MRI would be useful as part of a wider analysis.

IMC 1 is a second-order feature belonging to the gray level co-occurrence matrix group—values ranging from 0 to $-\infty$ (values ≤ 0)—that represents a measure of the level of the textural complexity and tissue heterogeneity. In the current study, low values of IMC 1 in the pre-RT population, more precisely, values lower than the optimal cut-off of -1473×10^{-11}, were associated with the development of moderate xerostomia (group 2). This finding seemed to be in line with other similar studies carried out on CT imaging, in which features relating to higher textural complexity and more heterogeneous distribution of grays correlated with xerostomia of greater severity [36,38]. A study by Nardone et al. [34] performed on planning CT postulated that more heterogeneous textures might be indicative of a higher salivary gland radiosensitivity. Irregular microarchitectural structure on histopathology—altered vascularization or loss of acinar cells replaced by adipose tissue—has an impact on the development of RT-induced xerostomia as stated by Teshima et al. [47]. In this context, the hypothesized higher textural complexity of the salivary glands with normal acinar tissue replaced by a variable amount of adipose tissue might explain the correlation between the severity of xerostomia and greatest values in features related to the textural heterogeneity, as suggested in previous studies on CT examinations by van Dijk et al. [36,38].

In addition, performing texture analysis on ADC maps and CE-T1 MRI sequences, as was done in our study, might prove extremely advantageous. The possibility of using MRI tools with the help of artificial intelligence to differentiate functional gland tissue from adipose tissue [17,19,48] intuitively suggested the possible benefit of these advanced techniques in the characterization of the gland radiosensitivity, albeit at the cost of a more complex standardization. It is well known that the voxel intensity on CT images relates to the intrinsic physical properties of a tissue; on the contrary, the voxel intensity on MRI acquisition techniques is highly dependent on machine-specific characteristics [44]. This makes quantitative assessments with radiomics more prone to variation based on hardware-specific settings [44,49,50] and standardization as a whole more difficult. However, this is only partially true for quantitative functional MRI analyses, as those carried out in the present study on ADC and DCE-PWI; the hardware specifics are less impactful [44,50], which suggests other possible benefits to their implementation on texture analysis.

The main limitation of our study was the small sample size and, especially, the relatively smaller cohort of patients with moderate RT-induced xerostomia than patients with mild RT-induced xerostomia. It is reasonable to assume that all the features resulting as statistically significant in the current study may also yield acceptable results in terms of diagnostic accuracy when larger samples are selected. However, the limited number of cases enrolled (27 patients with oropharyngeal cancer) has to be connected to the originality of our study design. The assessment of both DCE-PWI and DWI in pre- and post-RT MRI and the process of texture analysis on both ADC maps and CE-T1 sequences required a wide variety of different examination and investigation techniques to be performed, thus sensibly reducing the number of eligible patients.

The authors are fully aware that a more elaborate model that combines all functional and textural parameters examined should be used to investigate such a complex phenomenon as RT-induced xerostomia. However, the necessity of a much larger sample required to design a performing classifier has made it impossible for authors to do that. This shortcoming was another limitation of the present study. The definition of a more complex classifier in the future would likely be better able to assess RT-induced xerostomia.

Finally, the results attained by our investigation were entirely related to the pre-RT cohort and, therefore, represented a cautionary glimpse into the possibilities of employing texture analysis techniques to predict salivary gland radiosensitivity before radiation therapy is administered. This remains a topic of discussion for further investigation in order to stratify patients according to the risk of developing xerostomia of different severity.

5. Conclusions

In our series, the texture analysis performed on the ADC maps of the parotid glands showed good accuracy in the assessment of the severity of RT-induced xerostomia in the pre-RT population (AUC = 0.727). The differentiation between mild and moderate RT-induced xerostomia was achieved with IMC1 (cut-off -1.473×10^{-11}) with 73% sensitivity, 75% specificity, and 75% diagnostic accuracy. Therefore, texture analysis should be considered a useful tool to estimate salivary gland radiosensitivity and help identify patients more prone to develop serious xerostomia before radiation therapy is administered.

No statistically significant parameter was found for both morphological and functional MRI (DCE-PWI and DWI) or for texture analysis on CE-T1.

Author Contributions: Conceptualization, L.C. (Leonardo Calamandrei), I.D. and C.N.; methodology, I.D., F.M., L.B. (Leonardo Bocchi), E.B. (Eleonora Barcali) and C.N.; validation, F.M., L.C. (Linda Calistri) and C.N.; investigation, L.M., E.B. (Eleonora Bicci), N.L., M.O. and L.C. (Leonardo Calamandrei); resources, I.D., L.C. (Linda Calistri), F.M., L.B. (Luigi Bonasera), E.B. (Eleonora Barcali) and C.N.; data curation, L.M., E.B. (Eleonora Barcali), P.B., L.C. (Leonardo Calamandrei) and F.M.; writing—original draft preparation, L.C. (Leonardo Calamandrei), E.B. (Eleonora Bicci), L.M. and C.N.; writing—review and editing, F.M., L.C. (Linda Calistri), I.D., P.B. and C.N.; visualization, L.M., N.L., I.D., M.O. and L.C. (Leonardo Calamandrei); supervision, L.B. (Luigi Bonasera)., M.O., N.L., L.C. (Linda Calistri) and C.N.; project administration, C.N. All authors have read and agreed to the published version of the manuscript.

Funding: This research received no external funding.

Institutional Review Board Statement: The study was conducted according to the guidelines of the Declaration of Helsinki. This is a monocentric, retrospective, comparative study approved by the Ethical Review Board of the AOU Careggi (# 21800).

Informed Consent Statement: Written informed consent was obtained from all patients involved in the study.

Data Availability Statement: Not applicable.

Conflicts of Interest: The authors declare no conflict of interest.

Appendix A

Table A1. MRI acquisition parameters. Unenhanced scans included T1 and T2 sampling perfection with application-optimized contrasts using different flip angle evolution (SPACE) sequences with axial, coronal, and sagittal multiplanar reconstructions; axial T2 turbo spin echo; axial fat-saturated, echo-planar DWI spectral attenuated inversion recovery (SPAIR) with two b-values (b 50–800 s/mm^2) and ADC maps. Enhanced scans carried out after intravenous gadolinium contrast agent (gadobutrol, 1 mL/10 kg, flow 3 mL/s followed by 20 mL saline flush) consisted of an axial T1 turbo spin echo and axial T1 volumetric interpolated breath-hold examination (VIBE) Dixon.

Sequence	Contrast Agent	Repetition Time (ms)	Echo Time (ms)	Slice Thickness (mm)	Interslice Gap (mm)	Field of View (mm)	Matrix	Acceleration Factor	Number of Signal Averaged	Band Width (Hz/Px)	Acquisition Time (min:s)	Voxel Size
SPACE T1-w Sagittal	pre	500	7.2	0.9	-	229 × 229	230 × 256	2	1.4	630	5:47	0.9 × 0.9 × 0.9
SPACE T2-w Sagittal Fat-Sat	pre	3000	380	0.9	-	229 × 229	230 × 256	2	1.4	698	5:56	0.9 × 0.9 × 0.9
TSE T2-w Axial	pre	5050	117	3	0.9	210 × 190	261 × 484	2	3	191	2:23	0.5 × 0.5 × 3.0
SPAIR EPI-DWI Axial (b 50/800 s/mm^2)	pre	4100	55	3	0.9	240 × 240	102 × 128	3	1	1608	3:09	1.6 × 1.6 × 3.0
VIBE T1-w DCE-PWI Axial; FA 5°, 15°	pre	4.65	1.66	3.5	0.7	250 × 226	139 × 132	3	1	390	1:04	1.3 × 1.3 × 3.5
TSE T1-w Axial	post	440	17	3	0.9	200 × 181	384 × 384	3	3	200	2:31	0.5 × 0.5 × 3.0
VIBE Dixon Axial	post	10	2.4	0.9	0.18	225 × 225	212 × 256	-	1	340	4:37	0.9 × 0.9 × 0.9
VIBE T1-w DCE-PWI Axial; FA 30°	post	4.65	1.66	3.5	0.7	250 × 226	139 × 132	3	1	300	4:17	1.3 × 1.3 × 3.5

Table A2. Texture features subclasses.

First-Order Statistics	Describes the distribution of voxel intensities within the image region defined by the mask through commonly used and basic metrics.
Gray Level Co-occurrence Matrix (GLCM)	Describes the second-order joint probability function of an image region constrained by the mask.
Gray Level Dependence Matrix (GLDM)	Quantifies gray level dependencies in an image. A gray level dependency is defined as the number of connected voxels within distance δ that are dependent on the center voxel.
Gray Level Size Zone (GLSZM)	Quantifies gray level zones in an image. A gray level zone is defined as the number of connected voxels that share the same gray level intensity.
Gray Level Run Length Matrix (GLRLM)	Quantifies gray level runs, which are defined as the length in number of pixels of consecutive pixels that have the same gray level value.
Neighboring Gray Tone Difference Matrix (NGTDM)	Quantifies the difference between a gray value and the average gray value of its neighbors within distance δ.

Table A3. Radiation dose data for the patients with mild xerostomia (Group 1) and moderate xerostomia (Group 2). Volumes have been stratified for oncologic risk (likelihood of persistence/recurrence of disease in the selected volume).

Patients	Dose to High Risk Volume (Gy)	Dose to Intermediate Risk Volume (Gy)	Dose to Low Risk Volume (Gy)	Overall RT Treatment Time(Days)
Group 1				
Patient 1	69.9	60	54	49
Patient 2	69.9	60	54	53
Patient 3	70	50	50	86
Patient 4	69.9	60	54	46
Patient 5	70	60	50	55
Patient 6	69.9	60	54	50
Patient 7	69.9	60	54	56
Patient 8	699	60	54	49
Patient 9	69.9	60	54	55
Patient 10	69.9	60	54	46
Patient 11	69.9	59.4	54	47
Patient 12	69.9	59.4	54	52
Patient 13	69.9	60	54	67
Patient 14	69.9	60	54	53
Patient 15	69.9	60	54	46
Patient 16	69.9	59.4	52.8	47
Patient 17	66	60	54	46
Patient 18	70	60	52.8	48
Group 2				
Patient 19	69.9	60	54	58
Patient 20	69.9	60	54	43
Patient 21	69.9	60	54	51
Patient 22	69.9	59.4	54	54
Patient 23	70.5	60	54	42
Patient 24	69.9	59.4	52.8	44
Patient 25	69.9	59.4	52.8	47
Patient 26	69.9	59.4	54	45
Patient 27	69.9	59.4	52.8	58

References

1. Vigneswaran, N.; Williams, M.D. Epidemiologic trends in head and neck cancer and aids in diagnosis. Oral Maxillofac. *Surg. Clin.* **2014**, *26*, 123–141.
2. Virnig, B.A.; Warren, J.L.; Cooper, G.S.; Klabunde, C.N.; Schussler, N.; Freeman, J. Studying radiation therapy using SEER-Medicare-linked data. *Med. Care* **2002**, *40* (Suppl. S8), IV-49–IV-54. [CrossRef]
3. Vissink, A.; Jansma, J.; Spijkervet, F.K.L.; Burlage, F.R.; Coppes, R.P. Oral sequelae of head and neck radiotherapy. *Crit. Rev. Oral Biol. Med.* **2003**, *14*, 199–212. [CrossRef]
4. Trotti, A.; A Bellm, L.; Epstein, J.B.; Frame, D.; Fuchs, H.J.; Gwede, C.K.; Komaroff, E.; Nalysnyk, L.; Zilberberg, M.D. Mucositis incidence, severity and associated outcomes in patients with head and neck cancer receiving radiotherapy with or without chemotherapy: A systematic literature review. *Radiother. Oncol.* **2003**, *66*, 253–262. [CrossRef]
5. Tanasiewicz, M.; Hildebrandt, T.; Obersztyn, I. Xerostomia of Various Etiologies: A Review of the Literature. *Adv. Clin. Exp. Med.* **2016**, *25*, 199–206. [CrossRef]
6. Wiener, R.C.; Wu, B.; Crout, R.; Wiener, M.; Plassman, B.; Kao, E.; McNeil, D. Hyposalivation and xerostomia in dentate older adults. *J. Am. Dent. Assoc.* **2010**, *141*, 279–284. [CrossRef] [PubMed]
7. Vissink, A.; Burlage, F.; Spijkervet, F.; Jansma, J.; Coppes, R.P. Prevention and treatment of the consequences of head and neck radiotherapy. *Crit. Rev. Oral Biol. Med.* **2003**, *14*, 213–225. [CrossRef] [PubMed]
8. Valdez, I.H. Radiation-induced salivary dysfunction: Clinical course and significance. *Spec. Care Dent.* **1991**, *11*, 252–255. [CrossRef] [PubMed]
9. Sroussi, H.Y.; Epstein, J.B.; Bensadoun, R.-J.; Saunders, D.P.; Lalla, R.V.; Migliorati, C.A.; Heaivilin, N.; Zumsteg, Z.S. Common oral complications of head and neck cancer radiation therapy: Mucositis, infections, saliva change, fibrosis, sensory dysfunctions, dental caries, periodontal disease, and osteoradionecrosis. *Cancer Med.* **2017**, *6*, 2918–2931. [CrossRef]
10. Turner, M.; Jahangiri, L.; Ship, J. Hyposalivation, xerostomia and the complete denture: A systematic review. *J. Am. Dent. Assoc.* **2008**, *139*, 146–150. [CrossRef]

11. Dawes, C. Physiological factors affecting salivary flow rate, oral sugar clearance, and the sensation of dry mouth in man. *J. Dent. Res.* **1987**, *66*, 648–653. [CrossRef] [PubMed]
12. Valdez, I.H.; Fox, P.C. Diagnosis and management of salivary dysfunction. *Crit. Rev. Oral Biol. Med.* **1993**, *4*, 271–277. [CrossRef] [PubMed]
13. Al-Dwairi, Z.; Lynch, E. Xerostomia in complete denture wearers: Prevalence, clinical findings and impact on oral functions. *Gerodontology* **2014**, *31*, 49–55. [CrossRef] [PubMed]
14. Berti-Couto, S.D.A.; Couto-Souza, P.H.; Jacobs, R.; Nackaerts, O.; Rubira-Bullen, I.R.F.; Westphalen, F.H.; Moysés, S.J.; Ignácio, S.A.; Costa, M.B.D.; Tolazzi, A.L. Clinical diagnosis of hyposalivation in hospitalized patients. *J. Appl. Oral Sci.* **2012**, *20*, 157–161. [CrossRef]
15. Stephen, T.S.; Linda, S.E.; Dorothy, K.; Douglas, E.P.; Mark, S.; Martin, H.J. Perspectives on cancer therapy-induced mucosal injury: Pathogenesis, measurement, epidemiology, and consequences for patients. *Cancer* **2004**, *100* (Suppl. S9), 1995–2025.
16. Widmann, G.; Henninger, B.; Kremser, C.; Jaschke, W. MRI Sequences in Head & Neck Radiology—State of the Art. *Rofo* **2017**, *189*, 413–422.
17. Bicci, E.; Nardi, C.; Calamandrei, L.; Pietragalla, M.; Cavigli, E.; Mungai, F.; Bonasera, L.; Miele, V. Role of Texture Analysis in Oropharyngeal Carcinoma: A Systematic Review of the Literature. *Cancers* **2022**, *14*, 2445. [CrossRef]
18. Nardi, C.; Tomei, M.; Pietragalla, M.; Calistri, L.; Landini, N.; Bonomo, P.; Mannelli, G.; Mungai, F.; Bonasera, L.; Colagrande, S. Texture analysis in the characterization of parotid salivary gland lesions: A study on MR diffusion weighted imaging. *Eur. J. Radiol.* **2021**, *136*, 109529. [CrossRef]
19. Maraghelli, D.; Pietragalla, M.; Cordopatri, C.; Nardi, C.; Peired, A.J.; Maggiore, G.; Colagrande, S. Magnetic resonance imaging of salivary gland tumours: Key findings for imaging characterisation. *Eur. J. Radiol.* **2021**, *139*, 109716. [CrossRef]
20. Khoo, M.M.Y.; Tyler, P.A.; Saifuddin, A.; Padhani, A. Diffusion-weighted imaging (DWI) in musculoskeletal MRI: A critical review. *Skelet. Radiol.* **2011**, *40*, 665–681. [CrossRef]
21. Zhang, Y.; Ou, D.; Gu, Y.; He, X.; Peng, W. Evaluation of Salivary Gland Function Using Diffusion-Weighted Magnetic Resonance Imaging for Follow-Up of Radiation-Induced Xerostomia. *Korean J. Radiol.* **2018**, *19*, 758–766. [CrossRef] [PubMed]
22. Patel, P.; Baradaran, H.; Delgado, D.; Askin, G.; Christos, P.; Tsiouris, A.J.; Gupta, A. MR perfusion-weighted imaging in the evaluation of high-grade gliomas after treatment: A systematic review and meta-analysis. *Neuro-Oncology* **2017**, *19*, 118–127. [CrossRef]
23. Wang, L.; Wei, L.; Wang, J.; Li, N.; Gao, Y.; Ma, H.; Qu, X.; Zhang, M. Evaluation of perfusion MRI value for tumor progression assessment after glioma radiotherapy: A systematic review and meta-analysis. *Medicine* **2020**, *99*, e23766. [CrossRef] [PubMed]
24. Sade, R.; Kantarci, M.; Karaca, L.; Okur, A.; Ogul, H.; Keles, M.; Çankaya, E.; Ayan, A.K. Value of dynamic MRI using the Ktrans technique for assessment of native kidneys in pre-emptive renal transplantation. *Acta Radiol.* **2017**, *58*, 1005–1011. [CrossRef] [PubMed]
25. Mungai, F.; Verrone, G.B.; Pietragalla, M.; Berti, V.; Addeo, G.; Desideri, I.; Bonasera, L.; Miele, V. CT assessment of tumor heterogeneity and the potential for the prediction of human papillomavirus status in oropharyngeal squamous cell carcinoma. *Radiol. Med.* **2019**, *124*, 804–811. [CrossRef]
26. Maraghelli, D.; Pietragalla, M.; Calistri, L.; Barbato, L.; Locatello, L.G.; Orlandi, M.; Landini, N.; Casto, A.L.; Nardi, C. Techniques, Tricks, and Stratagems of Oral Cavity Computed Tomography and Magnetic Resonance Imaging. *Appl. Sci.* **2022**, *12*, 1473. [CrossRef]
27. Gardin, I.; Grégoire, V.; Gibon, D.; Kirisli, H.; Pasquier, D.; Thariat, J.; Vera, P. Radiomics: Principles and radiotherapy applications. *Crit. Rev. Oncol. Hematol.* **2019**, *138*, 44–50. [CrossRef]
28. Soni, N.; Priya, S.; Bathla, G. Texture Analysis in Cerebral Gliomas: A Review of the Literature. *AJNR Am. J. Neuroradiol.* **2019**, *40*, 928–934. [CrossRef]
29. Cozzi, D.; Bicci, E.; Cavigli, E.; Danti, G.; Bettarini, S.; Tortoli, P.; Mazzoni, L.N.; Busoni, S.; Pradella, S.; Miele, V. Radiomics in pulmonary neuroendocrine tumours (NETs). *Radiol. Med.* **2022**, *127*, 609–615. [CrossRef]
30. Zhang, Q.; Wei, Y.-M.; Qi, Y.-G.; Li, B.-S. Early Changes in Apparent Diffusion Coefficient for Salivary Glands during Radiotherapy for Nasopharyngeal Carcinoma Associated with Xerostomia. *Korean J. Radiol.* **2018**, *19*, 328–333. [CrossRef]
31. Zhang, Y.; Ou, D.; Gu, Y.; He, X.; Peng, W.; Mao, J.; Yue, L.; Shen, X. Diffusion-weighted MR imaging of salivary glands with gustatory stimulation: Comparison before and after radiotherapy. *Acta Radiol.* **2013**, *54*, 928–933. [CrossRef] [PubMed]
32. Zhou, N.; Chu, C.; Dou, X.; Li, M.; Liu, S.; Zhu, L.; Liu, B.; Guo, T.; Chen, W.; He, J.; et al. Early evaluation of irradiated parotid glands with intravoxel incoherent motion MR imaging: Correlation with dynamic contrast-enhanced MR imaging. *BMC Cancer* **2016**, *16*, 865. [CrossRef] [PubMed]
33. Juan, C.J.; Chen, C.-Y.; Jen, Y.-M.; Liu, H.-S.; Liu, Y.-J.; Hsueh, C.-J.; Wang, C.-Y.; Chou, Y.-C.; Chai, Y.-T.; Huang, G.-S.; et al. Perfusion characteristics of late radiation injury of parotid glands: Quantitative evaluation with dynamic contrast-enhanced MRI. *Eur. Radiol.* **2009**, *19*, 94–102. [CrossRef] [PubMed]
34. Nardone, V.; Tini, P.; Nioche, C.; Mazzei, M.A.; Carfagno, T.; Battaglia, G.; Pastina, P.; Grassi, R.; Sebaste, L.; Pirtoli, L. Texture analysis as a predictor of radiation-induced xerostomia in head and neck patients undergoing IMRT. *Radiol. Med.* **2018**, *123*, 415–423. [CrossRef]
35. van Dijk, L.V.; Langendijk, J.A.; Sijtsema, N.M.; Steenbakkers, R.J. Reply letter to "Texture analysis of parotid gland as a predictive factor of radiation induced xerostomia: A subset analysis". *Radiother. Oncol.* **2017**, *122*, 322. [CrossRef]

36. van Dijk, L.V.; Brouwer, C.L.; Van Der Schaaf, A.; Burgerhof, J.G.; Beukinga, R.J.; Langendijk, J.A.; Sijtsema, N.M.; Steenbakkers, R.J. CT image biomarkers to improve patient-specific prediction of radiation-induced xerostomia and sticky saliva. *Radiother. Oncol.* **2017**, *122*, 185–191. [CrossRef]

37. Yang, X.; Tridandapani, S.; Beitler, J.J.; Yu, D.S.; Yoshida, E.J.; Curran, W.J.; Liu, T. Ultrasound GLCM texture analysis of radiation-induced parotid-gland injury in head-and-neck cancer radiotherapy: An in vivo study of late toxicity. *Med. Phys.* **2012**, *39*, 5732–5739. [CrossRef]

38. van Dijk, L.V.; Thor, M.; Steenbakkers, R.J.; Apte, A.; Zhai, T.T.; Borra, R.; Noordzij, W.; Estilo, C.; Lee, N.; Langendijk, J.A.; et al. Parotid gland fat related Magnetic Resonance image biomarkers improve prediction of late radiation-induced xerostomia. *Radiother. Oncol.* **2018**, *128*, 459–466. [CrossRef]

39. Berger, T.; Noble, D.J.; Shelley, L.E.; McMullan, T.; Bates, A.; Thomas, S.; Carruthers, L.J.; Beckett, G.; Duffton, A.; Paterson, C.; et al. Predicting radiotherapy-induced xerostomia in head and neck cancer patients using day-to-day kinetics of radiomics features. *Phys. Imaging Radiat. Oncol.* **2022**, *24*, 95–101. [CrossRef]

40. Shi, D.; Qian, J.-J.; Fan, G.-H.; Shen, J.-K.; Tian, Y.; Xu, L. Salivary gland function in nasopharyngeal carcinoma before and late after intensity-modulated radiotherapy evaluated by dynamic diffusion-weighted MR imaging with gustatory stimulation. *BMC Oral Health* **2019**, *19*, 288. [CrossRef]

41. Zhou, N.; Chu, C.; Dou, X.; Li, M.; Liu, S.; Guo, T.; Zhu, L.; Liu, B.; Chen, W.; He, J.; et al. Early Changes of Irradiated Parotid Glands Evaluated by T1rho-Weighted Imaging: A Pilot Study. *J. Comput. Assist. Tomogr.* **2017**, *41*, 472–476. [CrossRef] [PubMed]

42. Scalco, E.; Moriconi, S.; Rizzo, G. Texture analysis to assess structural modifications induced by radiotherapy. In Proceedings of the 2015 37th Annual International Conference of the IEEE Engineering in Medicine and Biology Society (EMBC), Milan, Italy, 25–29 August 2015; 2015; 2015, pp. 5219–5222.

43. Litvin, A.; Burkin, D.; Kropinov, A.; Paramzin, F. Radiomics and Digital Image Texture Analysis in Oncology (Review). *Sovrem. Technol. Med.* **2021**, *13*, 97–104. [CrossRef] [PubMed]

44. Wahid, K.A.; He, R.; McDonald, B.A.; Anderson, B.M.; Salzillo, T.; Mulder, S.; Wang, J.; Sharafi, C.S.; McCoy, L.A.; Naser, M.A.; et al. Intensity standardization methods in magnetic resonance imaging of head and neck cancer. *Phys. Imaging Radiat. Oncol.* **2021**, *20*, 88–93. [CrossRef] [PubMed]

45. Beetz, I.; Steenbakkers, R.J.H.M.; Chouvalova, O.; Leemans, C.R.; Doornaert, P.; van der Laan, B.F.A.M.; Christianen, M.E.M.C.; Vissink, A.; Bijl, H.P.; van Luijk, P.; et al. The QUANTEC criteria for parotid gland dose and their efficacy to prevent moderate to severe patient-rated xerostomia. *Acta Oncol.* **2014**, *53*, 597–604. [CrossRef]

46. Tribius, S.; Prosch, C.; Tennstedt, P.; Bajrovic, A.; Kruell, A.; Petersen, C.; Rapp, W.; Muenscher, A.; Goy, Y. Xerostomia after radiotherapy: What matters—Mean total dose or dose to each parotid gland? *Strahlenther. Onkol.* **2013**, *189*, 216–222. [CrossRef]

47. Teshima, K.; Murakami, R.; Yoshida, R.; Nakayama, H.; Hiraki, A.; Hirai, T.; Nakaguchi, Y.; Tsujita, N.; Tomitaka, E.; Furusawa, M.; et al. Histopathological changes in parotid and submandibular glands of patients treated with preoperative chemoradiation therapy for oral cancer. *J. Radiat. Res.* **2012**, *53*, 492–496. [CrossRef]

48. Burke, C.J.; Thomas, R.; Howlett, D. Imaging the major salivary glands. *Br. J. Oral Maxillofac. Surg.* **2011**, *49*, 261–269. [CrossRef]

49. Bloem, J.L.; Reijnierse, M.; Huizinga, T.W.J.; van der Helm-van Mil, A.H.M. MR signal intensity: Staying on the bright side in MR image interpretation. *RMD Open* **2018**, *4*, e000728. [CrossRef]

50. van Timmeren, J.E.; Cester, D.; Tanadini-Lang, S.; Alkadhi, H.; Baessler, B. Radiomics in medical imaging-"how-to" guide and critical reflection. *Insights Imaging* **2020**, *11*, 91. [CrossRef]

Article

Skin Cancer Classification Framework Using Enhanced Super Resolution Generative Adversarial Network and Custom Convolutional Neural Network

Sufiyan Bashir Mukadam and Hemprasad Yashwant Patil *

School of Electronics Engineering, Vellore Institute of Technology, Vellore 632014, Tamil Nadu, India
* Correspondence: hemprasad.patil@vit.ac.in

Simple Summary: Skin cancer is one of the most fatal diseases for mankind. The early detection of skin cancer will facilitate its overall treatment and contribute towards lowering the mortalities. This paper presents the deep learning-based algorithm along with pre-processing for the classification of skin cancer images. The image resolution of publicly available HAM10000 data after resizing is low and hence, when we pre-process the data to enhance the image resolution and then subject it to the deep neural network, overall performance metrics namely accuracy, is typically competitive.

Abstract: Melanin skin lesions are most commonly spotted as small patches on the skin. It is nothing but overgrowth caused by melanocyte cells. Skin melanoma is caused due to the abnormal surge of melanocytes. The number of patients suffering from skin cancer is observably rising globally. Timely and precise identification of skin cancer is crucial for lowering mortality rates. An expert dermatologist is required to handle the cases of skin cancer using dermoscopy images. Improper diagnosis can cause fatality to the patient if it is not detected accurately. Some of the classes come under the category of benign while the rest are malignant, causing severe issues if not diagnosed at an early stage. To overcome these issues, Computer-Aided Design (CAD) systems are proposed which help to reduce the burden on the dermatologist by giving them accurate and precise diagnosis of skin images. There are several deep learning techniques that are implemented for cancer classification. In this experimental study, we have implemented a custom Convolution Neural Network (CNN) on a Human-against-Machine (HAM10000) database which is publicly accessible through the Kaggle website. The designed CNN model classifies the seven different classes present in HAM10000 database. The proposed experimental model achieves an accuracy metric of 98.77%, 98.36%, and 98.89% for protocol-I, protocol-II, and protocol-III, respectively, for skin cancer classification. Results of our proposed models are also assimilated with several different models in the literature and were found to be superior than most of them. To enhance the performance metrics, the database is initially pre-processed using an Enhanced Super Resolution Generative Adversarial Network (ESRGAN) which gives a better image resolution for images of smaller size.

Keywords: benign; malignant; skin cancer; ESRGAN; CAD

Citation: Mukadam, S.B.; Patil, H.Y. Skin Cancer Classification Framework Using Enhanced Super Resolution Generative Adversarial Network and Custom Convolutional Neural Network. *Appl. Sci.* **2023**, *13*, 1210. https://doi.org/10.3390/app13021210

Academic Editor: Cosimo Nardi

Received: 6 December 2022
Revised: 22 December 2022
Accepted: 28 December 2022
Published: 16 January 2023

1. Introduction

Skin melanoma occurs due to fast procreation of aberrant skin cells in human anatomy. The count of skin malignancy cases has significantly increased over the past years [1]. As the skin is comprised of three lamina, the topmost lamina is the Epidermis, the middle lamina is the Dermis, and the deepest lamina is the Hypodermis, which is for the formation of fat and fibrous connective tissue. As skin is the outer most organ of human anatomy, it is most likely to be affected by fungal growth and bacteria which can be identified under microscopic examination. It results in varying textures and colours of the skin [2]. Skin cancer is classified under two sub-classifications, namely non-melanoma and malignant melanoma cancer. Non-melanoma cancer is less hazardous and occurs due to repeated exposure to UV radiation. The most common reason for skin cancer-related mortality is malignant melanoma. According to the survey by WHO, one out of three patients who are diagnosed with cancer have a skin cancer specifically. There are nearly 2–3 million non-malignant patients and 1.32 lakh malignant melanoma patients [3]. Melanoma is caused due to an imbalance of melanocytes in skin cells. The diagnosis of skin lesions are difficult due to the lack of standard guidelines for the detection of skin cancer. In addition to this, skin lesion classification is more challenging due to obscure boundaries, and the involvement of obstacles like veins, hairs, and moles [4]. The dermatologists who work on different skin diseases face limitations in visualising the dermoscopic images manually. Due to the similarity in skin lesions (inter-class similarity of skin diseases) leads to a degree of subjectivity and thus, human error [5]. There are further issues presented by clinical examinations: they are costlier and require highly skilled medical experts to operate the specialized medical diagnostic tools [6]. In recent years, researchers have developed various techniques, namely via a computer-aided diagnosis (CAD) system in an effort to lessen the workload of medical professionals by supporting them in providing an accurate diagnosis of cancer [7]. The CAD systems can categorize the lesion images into the melanoma and non-melanoma cancer [8]. In this proposed work, we implemented a Custom Convolution Neural Network (CCNN) which helps us to categorize the seven distinct classes of skin cancer stated in the database Human Against Machine (HAM10000) [9]. The HAM10000 database, consisting of 10,015 images of dermoscopic skin lesions, is used in this proposed work. The pre-processing of the HAM10000 database is carried out using an Enhanced Super Resolution Generative Adversarial Network (ESRGAN) which enhanced the quality of dermoscopic images to acquire better results compared to existing models. The proposed model was implemented on the HAM10000 dataset which is split into two subsets stated as the training and testing datasets in an 80:20 ratio (protocol-I), as well as the train:val:test split as per protocol-II and protocol-III. The paper is organized as follows. Section 2 presents the related published works on skin cancer classification. The description of the HAM10000 dataset is mentioned in Section 3. The proposed methodology including preprocessing techniques, and the design and building of the custom CNN model is indicated in Section 4. The results of the proposed framework are presented in Section 5. Section 6 concludes the work and discusses the future scope for further enhancement of performance metrics.

2. Related Work

The majority of research on the classification of melanomas focuses on the use of the dermoscopic data, which provides more visual information and is frequently employed by professional dermatologists. Recent research on the CAD system for skin lesion categorization employs deep learning-based approaches. In most of the approaches, it is seen that the model requires more training time due to larger image size. Furthermore, the presently available public databases for skin lesion classification are mostly imbalanced, which hinders the performance of the model. To classify skin lesions, a study was performed by Aladhadh et al. [7] in which they employed a deep learning method based on vision transformers. A two-layer architecture is used in this work to accurately classify skin cancer. The transformer splits the augmented data into different patches and feeds the input to a multi-layer perceptron classifier to define its class with an accuracy of 96.14%. The study carried out by Bansal et al. [10] have pre-processed the HAM10000 database using different morphological operations. The handcrafted method for feature extraction is used to retrieve the features. The two-transfer learning models named EfficientNet-B0 and ResNet50V2 are used for skin lesion classification and obtained an accuracy of 94.9%. A research study by Basak et al. [11] worked on the HAM10000 database by employing a multi-focus segmentation network (MFS-Net) based on a deep learning algorithm. The retrieval of deep features is performed using the parallel partial decoder technique to produce a segmentation map. Finally, two different attention modules are implemented to obtain a segmentation output. The authors achieved a dice score of 90.6% using the prescribed algorithm. Nakai et al. [12] used a transformer model based on a deep bottleneck. This model integrates the self-attention block to form a model with deep extracted features. It helped them to enhance the performance of overall categorization. The model accomplished a total accuracy of 96.1%. In the research work by Popescu et al. [13], a collective intelligence-based transfer learning system was presented. This system comprises of nine different transfer learning models. This individual model is trained using the HAM10000 database and the outputs of the individual network are combined using a decision-level fusion module. It helped them to boost their overall performance by 3%. This approach yields an accuracy of 86.71%. Qian et al. [14] used multi-scale attention blocks which are a deep learning-based approach. This technique was implemented on the HAM10000 database to retrieve special features which will focus on skin lesion area. It has also adopted a loss weighting which helped to solve the issue of imbalanced data per class. The performance of this model gains an accuracy of 91.6%. The study stated in [15] is utilized for the categorization of skin lesions using the HAM10000 dataset. Multi-Scale Multi-CNN (MSM-CNN), a DL model built on a three-tier ensemble approach, was employed in this work. The proposed model results are then compared to the pre-trained CNN models such as EfficientNetB0, SeResNeXt-50, and EfficientNetB1. The MSM-CNN achieves the highest accuracy of 96.3% compared to other models. Panthakkan et al. [16] used a concatenation of Xception and ResNet50 models on the HAM10000 database. A sliding window method is implemented for the purpose of training as well as testing the system. The presented approach yields a good accuracy of 97.8% on testing data. In the study article [17], a classification of skin lesions is performed using the fusion of handcrafted and DL-based features and is further classified using ML classifiers to achieve an accuracy of 92.4%.

Through these related work studies we have identified some shortcomings which have been overcome by our proposed model. We can clearly see that there is a scope to improvise the performance metrics in terms of accuracy. Furthermore, it is identified that complexion in the model leads to the maximum execution time in training the model.

3. Materials and Methods

This section presents a detailed description of the HAM10000 dataset and also explains the seven different classes present in it.

3.1. HAM10000 Dataset

To train any neural network for obtaining good classification results, a huge dataset is required. The datasets used for the classification of skin pigmented lesions were small and inadequate for training. To overcome this issue, Tschandl and his team released the Human against Machine (HAM10000) dataset [9]. The dataset consists of 10,000 skin pigmented lesions of seven different important classes that can be used for the diagnosis of skin cancer. Due to the diverse population of dermoscopic images, data organization, cleaning, and defining a workflow to train a neural network is required. The final database version consists of 10,015 images and was released for academic research purpose and is made available on ISIC archive [9]. The ground truth of the database was confirmed by the expert pathologist in the field of dermoscopy. The seven important diagnosis classes are the following.

3.1.1. Actinic Keratosis (akiec)

Actinic Keratosis is the most common and non-obtrusive carcinoma. It is a sub-variant of squamous cell carcinoma which is cured locally without any surgical operation. It is said that akiec is an early sign of cell carcinoma and not a real carcinoma. This akiec lesion may grow into an intrusive squamous cell carcinoma [18]. Actinic Keratosis mostly appears on the face of the human body and is induced due to excessive exposure to UV light [9].

3.1.2. Basal Cell Carcinoma (bcc)

Basal cell carcinoma is a specific class of melanoma which arises in basal melanocytes that make new cells rather than shedding old ones. It is the most prevalent kind of melanoma [19]. It is more likely to appear in areas that are susceptible to direct sun light, such as the neck and head of human body [20]. It generally occurs in the form of pink growths, recurrent sores, and red patches on the skin. These lesions develop gradually and hardly disseminate [19].

3.1.3. Benign Keratosis-Like Lesions (bkl)

The bkl category in the database has three distinct classes of lesions that lacked cancerous traits. These sorts of lesions include Lichenoid Keratosis, Solar Lentigo, and Seborrheic Keratosis [19]. A benign skin condition known as lichenoid keratosis often manifests as a tiny, single, grey-brown lesion on the chest and upper limbs [21]. Solar Lentigo is a kind of macular hyper-pigmented infection that may differ in size, ranging from a few millimetres to more than one centimetre [22]. Seborrheic Keratosis is a benign condition that does not necessitate in-depth treatment. It is reddish-brown or greyish brown in color and often appears on the back, collar, scalp, and chest [23].

3.1.4. Dermatofibroma (df)

Dermatofibroma is a relatively common dermatological condition that mostly impacts adolescent or elderly humans, with little women preponderance [24]. Clinically speaking, dermatofibroma presents as stiff soles, or many hard pustules, patches, or lumps, with a soft surface and a color that may range from pale brown to darkly brown, purplish-red, or yellow [24]. These benign skin lesions often appear on the upper arm, upper back, and lower leg [19].

3.1.5. Melanocytic Nevi (nv)

The list of seven subclasses includes all of the innocuous melanocyte malignancies known as melanocytic nevi, which may have numerous variations [9]. They are skin tumours brought on by the expansion of melanocytes (the skin's pigment-producing cells). It is mainly induced due to UV rays emitted from the sun at the early childhood age [19].

3.1.6. Vascular Lesions (vasc)

The majority of vasc are inherited; however, they may arise later in life and are seldom malignant. They are sores of various appearances that form on the epidermis and surrounding tissues and are often referred to as birthmarks [19].

3.1.7. Melanoma (mel)

Malignant melanocytes give rise to melanoma, a cancer that may manifest in many different forms. If removed at a preliminary phase, it is curable with simple surgical intervention. Melanomas may be either intrusive or harmless [9]. It is particularly apparent on sun-exposed body parts that include the face, trunk, hands, collar, and legs. Melanoma may be identified by patches that have an irregular shape, uneven borders, and distinct colours, are larger than 6 mm, and tend to expand. It might disseminate to different organs of the body and can cause fatality if it remains untreated [19].

The HAM10000 dataset comprises seven different classes as described above and the class-wise categorization for the number of images is stated in Table 1. The distribution of images seems to be imbalanced. To make it balanced, data is augmented which is elaborated in the pre-processing section.

Table 1. Class-wise images present in HAM10000 dataset.

Class	akiec	bcc	bkl	df	nv	vasc	mel
Images	327	514	1099	115	6705	142	1113

4. Proposed Methodology

This section elaborates on two different pre-processing techniques implemented in this proposed work. In addition to pre-processing, we have also discussed the custom convolutional neural network and we built a CNN model from scratch.

4.1. Pre-Processing

It is one of the most important steps while working with clinical image data [25]. It is primarily applied on the raw database, before feeding them for training the Convolutional Neural Network based system [15]. The pre-processing algorithm provides enhancement in images, which helps to boost the inclusive performance metrics pertaining to the model. One of the substantive contributions of the proposed research work is to enhance the quality of HAM10000 data using the ESRGAN algorithm which indeed leads to better extraction of features from the clinical image by the model. There are two different pre-processing techniques that are implemented in this study, namely ESRGAN and data augmentation which are discussed in detail in the following sub-sections.

4.1.1. Enhanced Super-Resolution Generative Adversarial Network (ESRGAN)

Pre-processing is a crucial step for the enhancement of images which helps in achieving superior performance metrics [26]. The Super-Resolution Generative Adversarial Network (SRGAN) is a foundational technique which can generate photorealistic patterns while super-resolving a single picture. The reckoning of a high resolved image from a low-resolution image is termed a super-resolution. The major optimization focus of super-resolution is to cut back the mean square error from the obtained highly resolved image and original image. GANs offer a potent framework for creating realistic pictures that seem believable and have excellent perceptual quality [27]. The visual hallucinated features, though, are very often associated with undesirable effects [28]. The Enhanced SRGAN is the adaptive technique which mainly addresses the three shortcomings of SRGAN that are Adversarial loss, Network design, and Perceptual loss. It also facilitates maintenance of a better ocular peculiarity with more pragmatic and natural-looking colors than SRGAN. To achieve the Enhanced Super-Resolution Generative Adversarial Network (ESRGAN), Wang et al. introduced an additional Residual layer in Residual Dense Block (RDB) in [28] by removing the Batch Normalization (BN) layer. The Residual in Residual Dense Network (RRDN) comprises four different blocks, namely Dense Feature Fusion, Residual Dense Blocks, Shallow Feature Extraction, and up-sampling net [29]. The Local Feature Fusion layer and the Local Residual Learning layer are the two dense layers that form the RRDB.

a.   Local Feature Fusion (LFF): It is an adaptive state derived from RRDB and a convolution layer in a new RRDB and is given by Equation (1).

$$f_{d,LF} = h_{LFF}^{D}\left(\left\{f_{d-1},\, f_{d,1},\ldots\ldots,\, f_{d,c},\ldots,f_{d,C}\right\}\right) \tag{1}$$

where h_{LFF}^{D} indicates the convolution layer of size 1×1 in the dth RRDB block and f_{d-1}, $f_{d,1}$, etc., are the input and output of dth RRDB correspondingly.

b.   Local Residual Learning (LRL): It is implemented for the improvement of overall information flow. It also helps to get the final output of dth RRDB as shown in Equation (2).

$$f_d = \left(\left\{f_{d-1} + f_{d,LF}\right\}\right) \tag{2}$$

Other than the improvement of visual qualities using RRDB, Wang et al. also calculated different loss functions which gave the overall performance of the generator. The different loss functions are stated as [29].

(i)   Discriminator loss: It is the loss calculated during misclassification of real and fake instances. Some of the fake instances are obtained from the generator by expanding the equation given in Equation (3).

$$l_D^{Ra} = -E_{x_r}\left(log\left\{D_{Ra}\left\{x_r, x_f\right\}\right\}\right) - E_{x_f}\left(log\left\{1 - D_{Ra}\left\{x_f, x_r\right\}\right\}\right) \tag{3}$$

where $log\left\{D_{Ra}\left\{x_r, x_f\right\}\right\}$ is the probability of classification by generator correctly and $log\left\{1 - D_{Ra}\left\{x_f, x_r\right\}\right\}$ helps to accurately label the fake images from the generator.

(ii)   Generator loss: The generator loss is calculated if the discriminator misclassifies the fake images which helps the discriminator to improvise. It is given by Equation (4)

$$l_G^{Ra} = -E_{x_r}\left(log\left\{1 - D_{Ra}\left\{x_r, x_f\right\}\right\}\right) - E_{x_f}\left(log\left\{D_{Ra}\left\{x_f, x_r\right\}\right\}\right) \tag{4}$$

It is observed that the generator can achieve better results from both real and generated data in adversarial training.

(iii) Perpetual Loss: In ESRGAN, the perpetual loss is also improved by confining the features prior to activation, as compared to features after activation in SRGAN. The perpetual loss function is given by Equation (5).

$$l_G = l_{percep} + \lambda l_G^{Ra} + \eta l_1 \tag{5}$$

where the terms λ and η are the factors to equalize various loss functions and l_G^{Ra} is the generator loss function.

(iv) Content Loss: The element wise Mean Square Error (MSE). It is most broadly used in targeting the super resolved image and is given by Equation (6)

$$L_{MSE}^{SR} = \frac{1}{r^2 wh} \sum_{x=1}^{rw} \sum_{y=1}^{rh} \left(i_{x,y}^{HR} - g_{\theta_g}\left(i^{LR}\right)_{x,y}\right)^2 \tag{6}$$

where $g_{\theta_g}\left(i^{LR}\right)$ is the reformed image and $i_{x,y}^{HR}$ is the down sampled operation with a factor r [29].

Figure 1 presents the comparison of sample images with their respective ESRGAN-enhanced images.

4.1.2. Data Augmentation

In order to train the CNN model with multiple variations of the dermoscopic images, a data augmentation method is included in our research work. Minority oversampling is the most widely implemented method in restoring the model's robustness and reducing the dataset's bias when there is a significant imbalance in classes [30]. The deep learning model performs well when it is feed with a huge training dataset. The HAM10000 dataset used for our proposed work is imbalanced, as seen in Table 1. Data augmentation helps the network from overfitting issues caused due to imbalanced data. The main reason for augmenting the data is that there are only 8012 images in the training dataset. The different augmentation methods are implemented such as rescaling, rotating the image, zooming with factor of 0.1, and height and width shift with range factor 0.1. It makes the dataset more balanced and improves overall performance of the model.

4.2. Custom Convolutional Neural Network

CNN is a category of deep-learning system which detects and extracts features from images automatically [31]. It has acquired significance in medical image analysis, as it has in many other fields as a result of its higher performance. The layers of a standard CNN include convolution layer, dropout layer, activation function, fully connected layer, and pooling layer [32]. The image pixels need to be processed and are given as an input to the

CNN. The original input pixels are subjected to detecting feature vectors, also termed as filters, in the convolution layer in order to extract a collection of features [33]. CNN's primary function, convolution, allows automated feature extraction [34]. During the step of pooling, a dimensionality reduction process is conducted by applying filters to an input vector [33]. The reduction technique is carried out by taking the minimum, maximum, or median of the values in the filtering window, which is strung across the initial input vector [19].

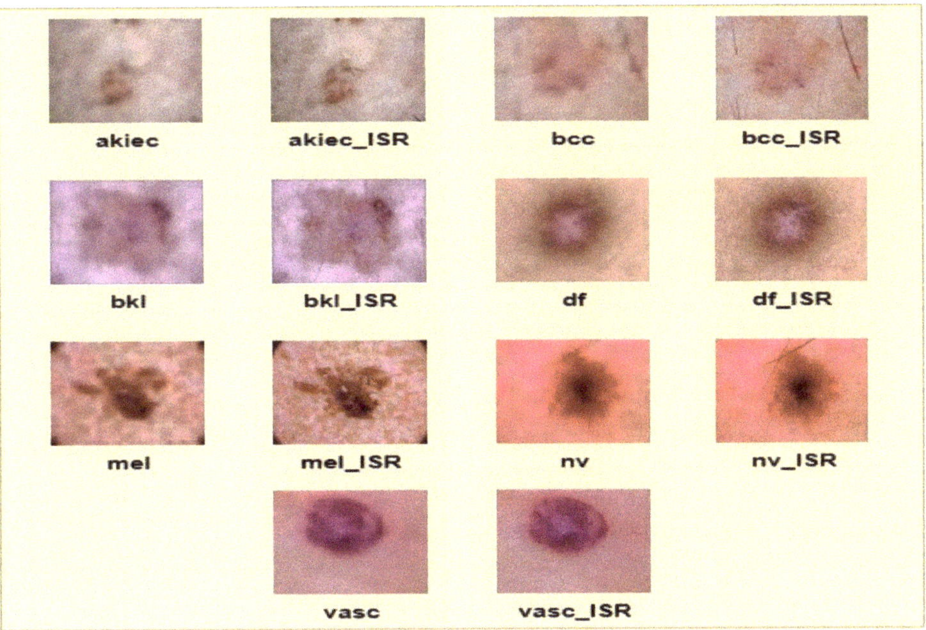

Figure 1. Comparison of sample images with their respective ESRGAN enhanced images.

In neural network models, overfitting problems can arise, especially when the manifold training samples is insufficient. With a view to address this issue, a dropout operation was used, which increased the network's capacity to alter distinct environments by arbitrarily deactivating a fraction of its neurons during training. The fully connected layer helps the process go on to the categorization stage. The output matrix is flattened before being sent on to the classifier after the feature extraction and pooling procedures. The proposed algorithm is shown in Figure 2.

The dataset has two fundamental aspects. The first component aspect is a metadata file that contains specific data for cancer lesion images. The skin lesion's location, the patient's age and gender, the lesion's diagnosis, and the skin lesion directory are all included in the metadata file. The second and primary section of the collection is comprised of visual files.

The objective of this study is to categorize skin lesions only based on digital images. Thus, the data file was reorganized to simply include the lesion type and the image file directory. Each lesion's textual labelling was transformed into digital values between 0 and 6. Each subtype labelling codes are shown in Table 2.

The original dermoscopic images are of 600×400 pixels resolution and are saved in the RGB format. It was observed that the processing burden increases proportionally with picture size. Hence, image size reduction increases processing speed. Therefore, all samples in the collection are downsized to 24×24 pixels. Since the colour is a distinguishing factor in diagnosing the kind of lesion, the original colours of the photographs were maintained. The sharpening filters are implemented to enhance the contrast of every applied image.

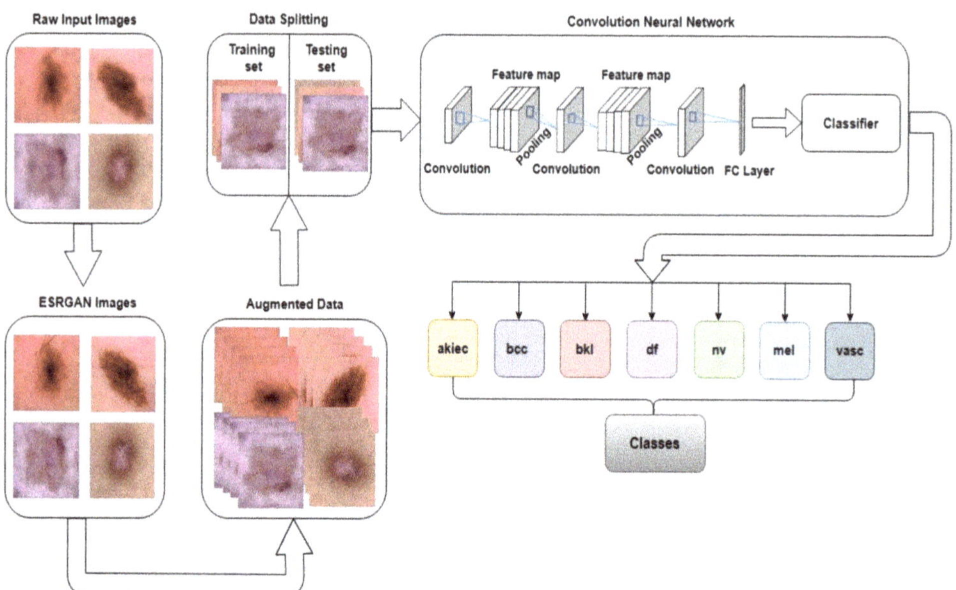

Figure 2. Overview of the proposed ESRGAN-based CNN algorithm.

Table 2. Notations for each class in the HAM10000 database.

Class	akiec	bcc	bkl	df	nv	vasc	mel
Label	0	1	2	3	4	5	6

4.3. Building a Custom CNN Model

When dealing with a large dataset, deep learning is typically regarded as an effective algorithm [35]. Conventionally, deep learning techniques demand a significant amount of computing time and large storage space [25]. Figure 3 depicts the customized CNN network model for classifying skin lesions. The custom CNN model is comprised of 4×2 layers. RGB input image of size 28×28 was utilized. The convolution operation is performed on the first two layers in each of these layers and 3×3 sized 32 filters are applied with a ReLu activation function. It is followed by the implementation of max pooling 2D layer with pool size of 2×2 and a batch normalization layer. In the second layer, the same convolution operation is performed with change in the parameters. In this layer 3×3 sized 64 filters are used with a ReLu activation function. After ReLu function, a max pooling 2D layer of 2×2 size and a batch normalization layer are employed. As a part of the third layer, the same convolution operation is performed with alteration in the parameters. In this layer, 3×3 sized 128 neurons are implemented with a ReLu activation function. It is then followed by the max pooling 2D layer of 2×2 size and a batch normalization layer. The fourth layer contains the similar convolution operation is performed with another set of parameters. In this layer, 3×3 sized 256 filters are used which are then followed by max 2D pooling of size 2×2, batch normalization layer, a dropout layer of 20%, and a flattening layer. In the final stage, the classifier receives the output of the flattening layer. Tables 3 and 4 show the summary and hyper parameters used for designing the model, respectively. Proposed method for classification of skin lesions is illustrated in Algorithm 1.

Algorithm 1: Proposed algorithm for classification of skin lesions

---

Step 1: Pre-processing

    a.   Raw input images are first pre-processed using the ESRGAN generator model.
    b.   The images are then resized to 28×28 for faster classification using the CNN model.
    c.   The imbalanced dataset is balanced using the data augmentation processes.
    d.   The augmented data is first split up into training data and testing data.

Step 2: Training custom CNN model

    a.   Feature map *Fmap* are extracted from the input images
    b.   Set *Fc* = 2D Conv (*Fmap, size(32)*);
    c.   Set *Fr* = ReLu (*Fc*);
    d.   Set *Fp* = MaxPooling2D (*Fr*);
    e.   Set *Fb* = BatchNormalization (*Fp*);
    f. $size_1$ = [64,128,256]

        for i = 0 to 2:

Set Fc_1 = 2D Conv (*Fmap, $size_1(i)$*);
Set Fr_1 = ReLu (*Fc_1*);
Set Fc_2 = 2D Conv (*Fmap, $size_1(i)$*);
Set Fr_2 = ReLu (*Fc_2*);
Set Fp_1 = MaxPooling2D (*Fr_2*);
Set Fb_1 = BatchNormalization (*Fp_1*);

end for

    g.   Set *Ff* = Flattening (*Fb_1*);
    h.   Set *F∂* = Dropout (*Ff*);
    i. $size_2$ = [256,128,64,32]

        for j = 0 to 3
        Set *Fd* = Dense (*F∂,size2(j)*);
        Set *Fb* = BatchNormalization (*Fd*);
        end j
    j.   Set *Foc* = OutputClassifier (*Fb*);

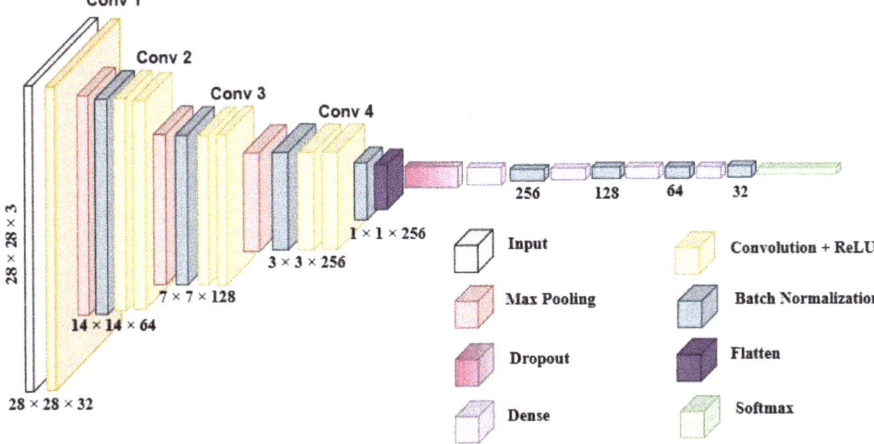

Figure 3. Layered architecture of proposed CNN model.

Table 3. CNN Model Summary.

Layer	Output Shape	Parameters
Input Layer	[(None, 28, 28, 3)]	0
Convolution 2D_1	(None, 28, 28, 32)	896
MaxPooling2D_1	(None, 14, 14, 32)	0
Batch Normalization_1	(None, 14, 14, 32)	128
Convolution 2D_2	(None, 14, 14, 64)	18,496
Convolution 2D_3	(None, 14, 14, 64)	36,928
MaxPooling2D_2	(None, 7, 7, 64)	0
Batch Normalization_2	(None, 7, 7, 64)	256
Convolution 2D_4	(None, 7, 7, 128)	73,856
Convolution 2D_5	(None, 7, 7, 128)	147,584
MaxPooling2D_3	(None, 3, 3, 128)	0
Batch Normalization_3	(None, 3, 3, 128)	512
Convolution 2D_6	(None, 3, 3, 256)	295,168
Convolution 2D_7	(None, 3, 3, 256)	590,080
Batch Normalization_4	(None, 1, 1, 256)	0
Flatten	(None, 256)	0
Dropout	(None, 256)	0
Dense_1	(None, 256)	65,792
Batch Normalization_5	(None, 256)	1024
Dense_2	(None, 128)	32,896
Batch Normalization_6	(None, 128)	512
Dense_3	(None, 64)	8256
Batch Normalization_7	(None, 64)	256
Dense_4	(None, 32)	2080
Batch Normalization_8	(None, 32)	128
Classifier	(None, 7)	231

Table 4. Hyper parameters for training the model.

Parameter	Value
Batch size	128
Number of epochs	25
Number of iterations	294
Optimizer	Adam
Optimizer parameters	Lr = 0.00001

5. Results and Discussion

In this section, we discuss the model's performance over a range of metrics and present a comparative study that illustrates how the suggested technique outperforms the current melanoma detection algorithms.

5.1. Performance Metrics

To assess the efficiency of the presented model, we used performance metrics such as Accuracy, F1-Score, Recall, and Precision. Performance metrics shown in Table 5 are calculated from a confusion matrix and are given by Equations (7), (8), (9) and (10), respectively. Performance measurement of the deep learning model comprises the following terms: (a) True Positive (Tp), (b) True Negative (Tn), (c) False Positive (Fp), and (d) False Negative (Fn) [36].

Table 5. Performance metrics and their formulas.

Performance Metrics	Formula	Equation
Accuracy	$\frac{(T_n+T_p)}{(T_p+F_p+F_n+T_n)}$	(7)
F1-Score	$\frac{(2*Precision*Recall)}{(Precision+Recall)}$	(8)
Recall	$\frac{T_p}{(T_p+F_n)}$	(9)
Precision	$\frac{T_p}{T_p+F_p}$	(10)

5.2. Protocol-I (Train:Test = 80:20 Ratio)

After performing data augmentation, the entire dataset is split into two partitions, namely, train and test with the ratio of 80:20. For protocol-I, for this augmented data, the total number of training images is 37,548 and of test images is 9387. The details of class-wise training and test images are depicted in Table 6.

Table 6. Class-wise images present in HAM10000 dataset with protocol-I.

Class	akiec	bcc	bkl	df	nv	vasc	mel
Training Samples	5383	5352	5408	5417	5325	5341	5322
Testing Samples	1322	1353	1297	1288	1380	1364	1383

The model was trained for 25 epochs on the Google Colaboratory Pro platform with 12 GB RAM and Python 3 Google compute backend engine GPU Accelerator. We interrupt the model's continuing execution using the early stopping method and record the model's best-performing parameters, such as its maximum accuracy and minimum cross-entropy loss. Every time the model fails to reach an accuracy greater than those acquired in the previous two epochs, we decreased the learning rate of the model to prevent additional stalling in the learning phase. The training and testing accuracies and loss graphs are displayed in Figures 4 and 5, respectively. The highest testing accuracy was obtained as 98.77% on the 25th epoch. Accuracy is one of the important metrics to characterize the achievement of the model if the dataset is proportionate. To get the different evaluation scores, we have used the confusion matrix which gives the exact classifications as shown in Figure 6. In this experiment, a confusion matrix is incurred for seven classes as mentioned in the dataset. The confusion matrix scores are computed to examine the performance of the model for different classes. From Table 7, it can be observed that the model works very well in classifying class 0, class 3, and class 5. The scores obtained for class 4 are slightly low.

Various Approaches That Follow Protocol-I

Table 8 specifies accuracies in the context of current research performed on the HAM10000 database. In the Agyenta et al. [37], the authors carried out research work on the HAM10000 database. Transfer learning techniques like InceptionV3, ResNet50, DenseNet201, and comparative study is accomplished on the HAM1000 database and achieved accuracies of 85.80%, 86.69%, and 86.91%, respectively. The authors have reached the highest accuracy for the DenseNet201 model. In another work by Onur et al. in [19], the presented approach included a custom CNN model and experiments with an image size of 75×100. An accuracy of 91.51% was achieved in this study. Qian et al. [14] presented an experimental study using the CNN model concatenated with the Grouping Of Multi-Scale Attention Blocks (GMAB) technique. This study achieved an accuracy of 91.6 %. Shetty et al. [8] developed a CNN model along with a k-fold cross-validation method. The accuracy of this model was 95.18%. The study carried out by Panthakkan et al. [16], and was based on the Concatenated Xception-ResNet50 model for the diagnosis of skin cancer. This model yields competitive results with an accuracy of 97.8%. The proposed work presents a custom CNN model and implements it on pre-processed data using the ESRGAN algorithm to achieve an accuracy of 98.77% which is much higher when compared to other literature studies carried out on the HAM10000 database.

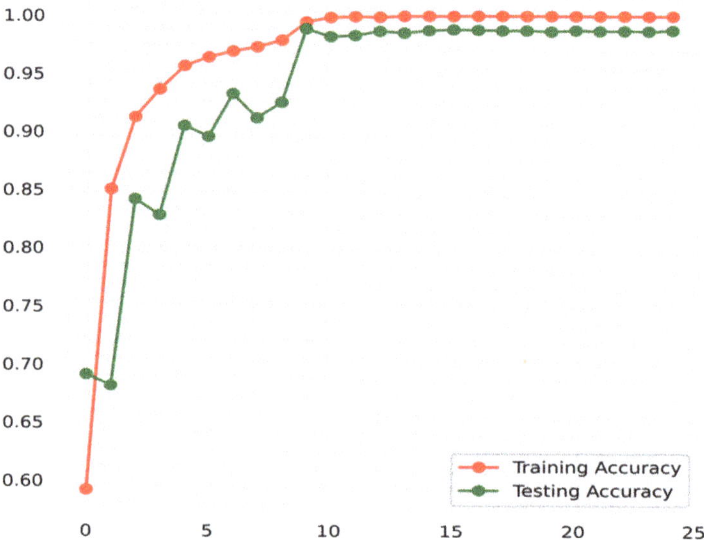

Figure 4. Accuracy graph for training and testing for protocol-I.

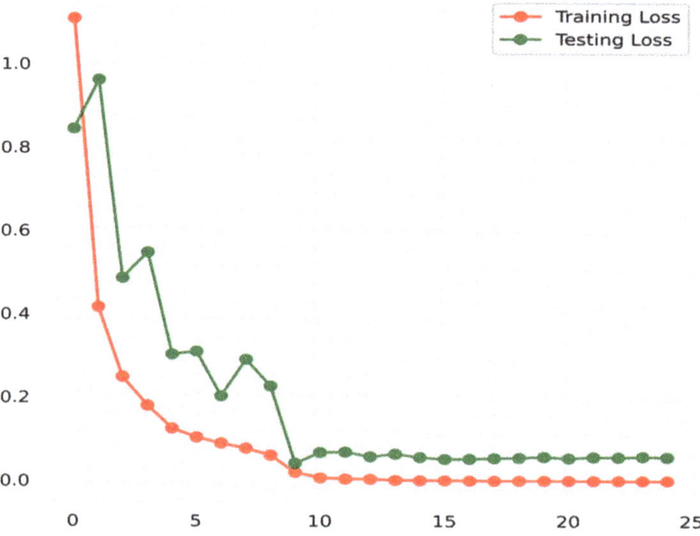

Figure 5. Training and testing losses for protocol-I.

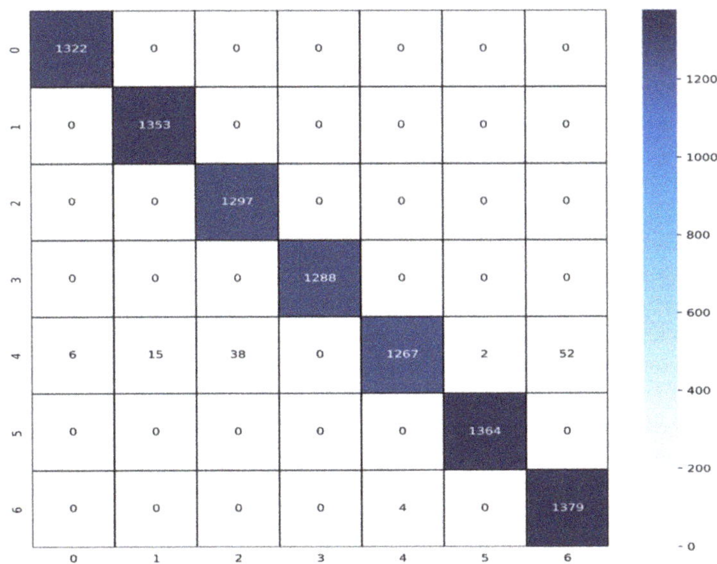

Figure 6. Confusion matrix for protocol-I (80:20 Train:Test split).

Table 7. Class-wise performance measures of the model.

Lesion Class	Precision	Recall	F1-Score
0-akiec	1.00	1.00	1.00
1-bcc	0.99	1.00	0.99
2-bkl	0.97	1.00	0.99
3-df	1.00	1.00	1.00
4-nv	1.00	0.92	0.96
5-vasc	1.00	1.00	1.00
6-mel	0.96	1.00	0.98

Table 8. Highest Accuracy for protocol-I.

Sr. No.	Work	Data Augmentation/Balancing? (Yes/No). Total Number of Images after Data Augmentation/Balancing	Methodology	Accuracy (%)
1	Agyenta et al. [37]	Yes, 7283	InceptionV3	85.80%
			ResNet50	86.69%
			DenseNet201	86.91%
2	Qian et al. [14]	Yes, Not mentioned	Grouping of Multi-scale Attention Blocks (GMAB)	91.6%
3	Shetty et al. [8]	Yes, 1400	Convolutional neural network (CNN)	95.18%
4	Panthakkan et al. [16]	No	Concatenated Xception-ResNet50 -	97.8%
5	Proposed algorithm	Yes, 46,935	ESRGAN-CNN	98.77%

5.3. Protocol II

For the purpose of parameter tuning with more test images, the following protocol-II is chosen where the dataset is split into the following ratio ((Train + Val):Test) = ((90 + 10):20).

It indicates that at first, the dataset is divided into 80: 20 Train: Test split. Subsequently, the training set is subdivided into 90% for training and 10% for validation. For this augmented data the number of training images is 33,793, validation images is 3755 as well as 9387 test images. Class-wise samples for this experimentation are depicted in Table 9. The model was trained using a machine with 12 GB RAM and GPU attached to it. Training and validation accuracies and losses are indicated in Figures 7 and 8, respectively. The confusion matrix for protocol-II is indicated in Figure 9.

Table 9. Class-wise images present in the HAM10000 dataset with protocol-II.

Class	akiec	bcc	bkl	df	nv	vasc	mel
Training Samples	4845	4817	4867	4875	4792	4807	4790
Validation Samples	538	535	541	542	533	534	532
Testing Samples	1322	1353	1297	1288	1380	1364	1383

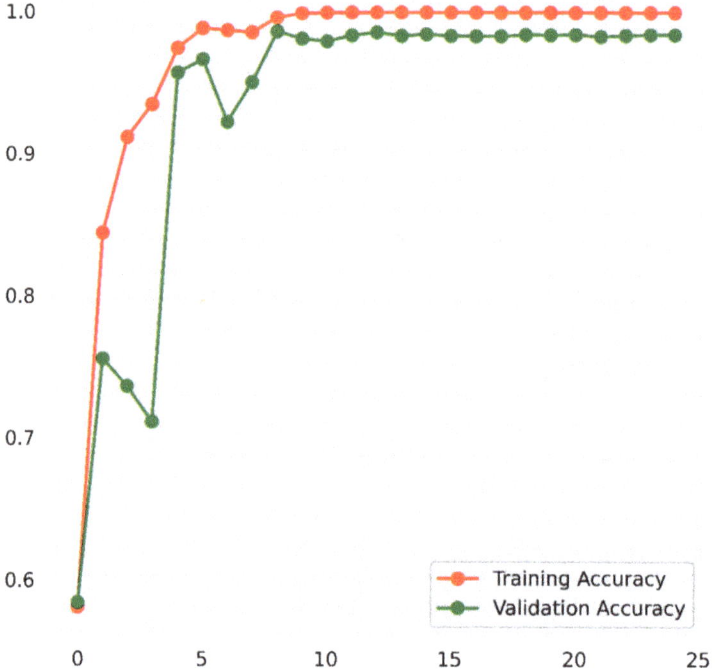

Figure 7. Accuracy graph for training and testing for protocol-II.

Various Approaches That Follow Protocol-II

Table 10 specifies accuracies in the context of current research performed on the HAM10000 database. In Sevli et al. [19], a deep convolutional neural network was implemented for the classification of skin lesions. This study accomplished an accuracy of 91.51%. Saarela et al. [38] worked on the HAM10000 dataset for skin-lesion classification. In this study, the robustness, stability, and fidelity studies of the deep convolutional neural network are carried out. Their model gives a classification accuracy of 80%. The proposed method for protocol-II gives a better accuracy of 98.36% as indicated in Table 10.

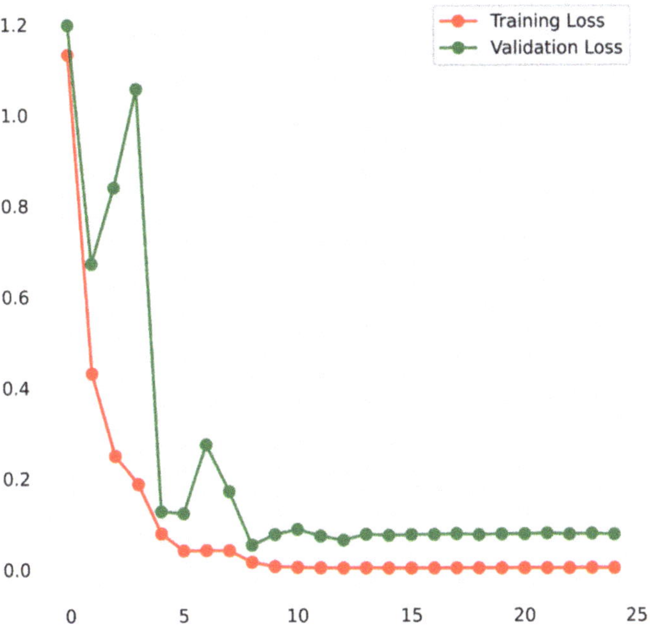

Figure 8. Training and testing losses for protocol-II.

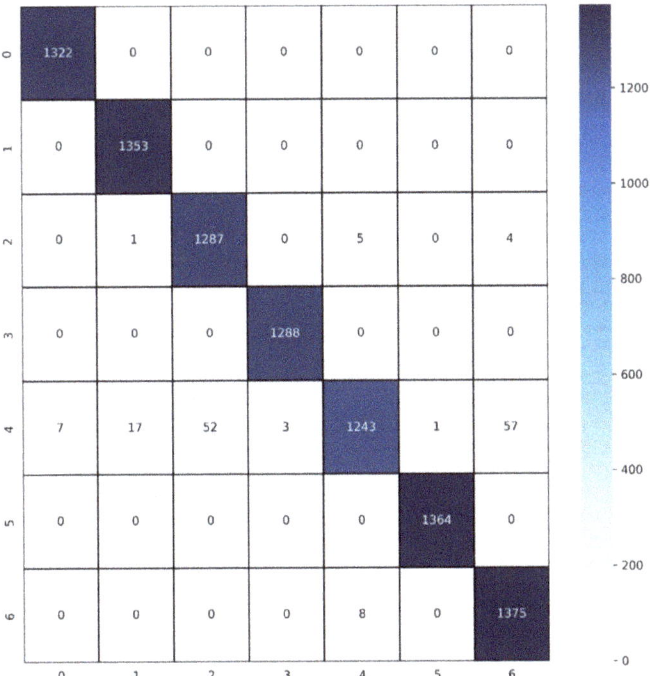

Figure 9. Confusion matrix for protocol-II.

Table 10. Highest accuracy for protocol-II.

Sr. No.	Work	Data Augmentation/Balancing? (Yes/No). Total Number of Images after Data Augmentation/Balancing	Methodology	Accuracy (%)
1	Onur et al. [19]	Yes, Not Mentioned	Convolutional neural network (CNN)	91.51%
2	Saarela et al. [38]	No	Deep Convolutional neural network (CNN)	80%
3	Proposed algorithm	Yes, 46,935	ESRGAN-CNN	98.36%

5.4. Protocol III

For the purpose of parameter tuning with fewer numbers of images for testing, the following protocol-III is implemented where the dataset is split into the following ratio ((Train + Val):Test) = ((90 + 10):10). It indicates that at first, the dataset is divided into 90: 10 Train: Test split. Subsequently, the training set is subdivided into 90% for training and 10% for validation. For this augmented data, the number of training images is 38,017, the number of validation images is 4224, and the number of test images is 4694. Class-wise samples for this experimentation are depicted in Table 11.

Table 11. Class-wise images present in HAM10000 dataset with protocol-III.

Class	akiec	bcc	bkl	df	nv	vasc	mel
Training Samples	5346	5557	5338	5184	5500	5513	5579
Validation Samples	594	617	593	576	611	613	620
Testing Samples	660	686	659	640	679	681	689

The CNN model was trained using a machine with 12 GB RAM and GPU attached to it. Training and validation accuracies and losses are indicated in Figures 10 and 11, respectively. The confusion matrix for protocol-III is indicated in Figure 12.

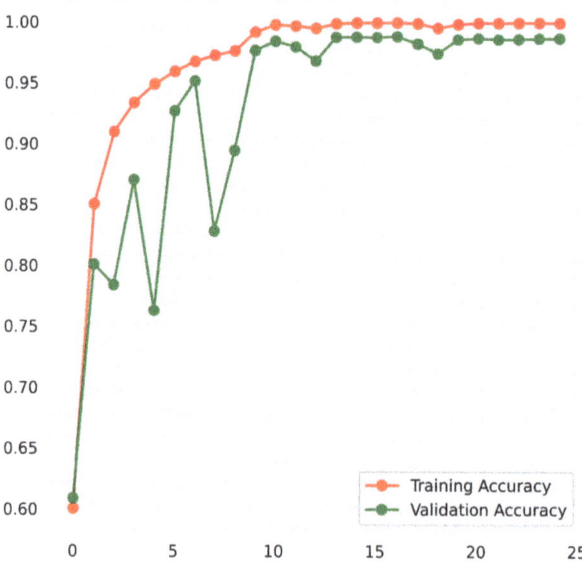

Figure 10. Accuracy graph for training and testing for protocol-III.

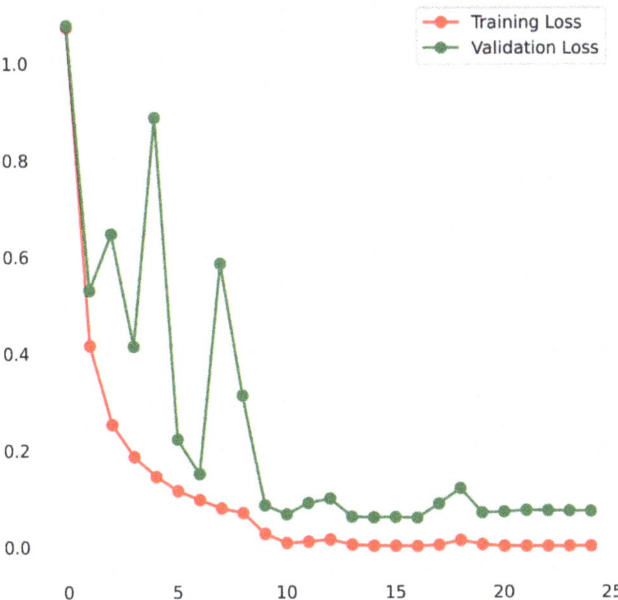

Figure 11. Training and testing losses for Protocol-III.

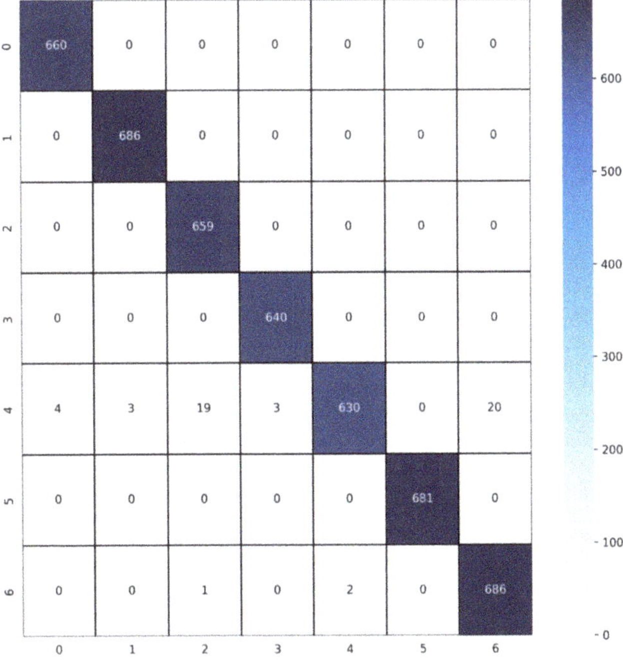

Figure 12. Confusion matrix for Protocol-III.

Various Approaches That Follow Protocol-III

Table 12 specifies accuracies, in the context of current research performed on the HAM10000 database. The research article presented by Aldhyani et al. [2] focused on kernel-based CNN. In this study, a lightweight dynamic kernel deep-learning-based convolutional neural network is implemented. This algorithm achieved an accuracy of 97.8% when the model was tested on the HAM10000 database. Alam et al. [39] presented an approach that works on the segmentation-based sub-network. In this study, the S2C-DeLeNet algorithm was applied on skin cancer data. This algorithm has obtained an accuracy of 90.58%. The proposed Custom CNN using ESRGAN technique achieved an accuracy of 98.89%.

Table 12. Highest accuracy for Protocol-III.

Sr. No.	Work	Data Augmentation/Balancing? (Yes/No). Total Number of Images after Data Augmentation/Balancing	Methodology	Accuracy (%)
1	Aldhyani et al. [2]	Yes, 54,907	Lightweight Dynamic Kernel Deep-Learning-Based Convolutional Neural Network	97.8%
2	Alam et al. [39]	No	S2C-DeLeNet	90.58%
3	Proposed algorithm	Yes, 46,935	ESRGAN-CNN	98.89%

6. Conclusions and Future Scope

Melanocyte cells are responsible for the formation of pigmented lesions. Skin malignancies such as melanoma are caused by the unregulated division of melanocyte cells, which may have a damaging effect on the human body. The dermatologists with extensive training interpret dermoscopic images. Due to the insufficiency of educated specialists and the need to minimize human-induced mistakes, the use of computer-assisted systems is emphasized. Convolution neural network, a method for deep learning that retrieves features from images, achieved huge success in the domain of computer vision. The preprocessing with ESRGAN helps us to reduce the size of images with better resolution and overall execution time for the experiment. The complexion in the model in terms of image shape that it considers as an input leads to maximum execution time in training the model. Hence, in this work, we have used images with a resolution of 28×28 pixels. Before resampling the images, the original images were enhanced using the ESRGAN dataset which helps to preserve the eminent features in the input images after down sampling. In this experimental analysis, we have implemented a HAM10000 dataset having 10,015 images of skin lesions and are categorized into seven different classes using a custom CNN model. The experimental model achieved accuracies of 98.77%, 98.36%, and 98.89% for protocol-I, protocol-II, and protocol-III, respectively, and it is seen to be competitively high as compared to the pretrained models presented by different researchers.

In the future, our aim is to work on the diagnosis of real-time skin lesions with improvement in the testing accuracy. We also hope to implement our proposed model to work on larger datasets if available for skin-cancer image categorization. It will in turn help us to enhance the performance metric scores. It is anticipated that the proposed work will help the dermatologist to examine and classify the class of skin cancer in lesser time duration and with more precision. Additionally, it will assist in reducing the total costs associated with skin cancer diagnosis. There is a scope for further enhancement in performance metrics such as accuracy, precision, and recall.

Author Contributions: Conceptualization, S.B.M. and H.Y.P.; methodology, S.B.M.; software, S.B.M.; validation, S.B.M.; formal analysis, S.B.M.; investigation, S.B.M.; resources, S.B.M.; data curation, S.B.M.; writing—original draft preparation, S.B.M.; writing—review and editing, S.B.M.; visualization, S.B.M.; supervision, H.Y.P.; project administration, H.Y.P. All authors have read and agreed to the published version of the manuscript.

Funding: This research received no external funding.

Institutional Review Board Statement: Not applicable.

Informed Consent Statement: Not applicable.

Data Availability Statement: Data is publicly available.

Conflicts of Interest: The author declares no conflict of interest.

References

1. Afza, F.; Sharif, M.; Khan, M.A.; Tariq, U.; Yong, H.S.; Cha, J. Multiclass Skin Lesion Classification Using Hybrid Deep Features Selection and Extreme Learning Machine. *Sensors* **2022**, *22*, 799. [CrossRef] [PubMed]
2. Aldhyani, T.H.H.; Verma, A.; Al-Adhaileh, M.H.; Koundal, D. Multi-Class Skin Lesion Classification Using a Lightweight Dynamic Kernel Deep-Learning-Based Convolutional Neural Network. *Diagnostics* **2022**, *12*, 2048. [CrossRef] [PubMed]
3. World Health Organization. Radiation: Ultraviolet (UV) Radiation and Skin Cancer—How Common Is Skin Cancer. Available online: https://www.who.int/news-room/questions-and-answers/item/radiation-ultraviolet-(uv)-radiation-and-skin-cancer (accessed on 12 October 2022).
4. Jeyakumar, J.P.; Jude, A.; Priya Henry, A.G.; Hemanth, J. Comparative Analysis of Melanoma Classification Using Deep Learning Techniques on Dermoscopy Images. *Electronics* **2022**, *11*, 2918. [CrossRef]
5. Ali, K.; Shaikh, Z.A.; Khan, A.A.; Laghari, A.A. Multiclass Skin Cancer Classification Using EfficientNets—A First Step towards Preventing Skin Cancer. *Neurosci. Inform.* **2022**, *2*, 100034. [CrossRef]
6. Hebbar, N.; Patil, H.Y.; Agarwal, K. Web Powered CT Scan Diagnosis for Brain Hemorrhage Using Deep Learning. In Proceedings of the 2020 IEEE 4th Conference on Information & Communication Technology (CICT), Chennai, India, 3 December 2020; IEEE: Piscataway, NJ, USA, 2020; pp. 1–5.
7. Aladhadh, S.; Alsanea, M.; Aloraini, M.; Khan, T.; Habib, S.; Islam, M. An Effective Skin Cancer Classification Mechanism via Medical Vision Transformer. *Sensors* **2022**, *22*, 4008. [CrossRef]
8. Shetty, B.; Fernandes, R.; Rodrigues, A.P. Skin Lesion Classiication of Dermoscopic Images Using Machine Learning and Convolutional Neural Network. *Sci. Rep.* **2022**, *12*, 18134. [CrossRef]
9. Tschandl, P.; Rosendahl, C.; Kittler, H. The HAM10000 Dataset, a Large Collection of Multi-Source Dermatoscopic Images of Common Pigmented Skin Lesions. *Sci. Data* **2018**, *5*, 180161. [CrossRef] [PubMed]
10. Bansal, P.; Garg, R.; Soni, P. Detection of Melanoma in Dermoscopic Images by Integrating Features Extracted Using Handcrafted and Deep Learning Models. *Comput. Ind. Eng.* **2022**, *168*, 108060. [CrossRef]
11. Basak, H.; Kundu, R.; Sarkar, R. MFSNet: A Multi Focus Segmentation Network for Skin Lesion Segmentation. *Pattern Recognit.* **2022**, *128*, 108673. [CrossRef]
12. Nakai, K.; Chen, Y.W.; Han, X.H. Enhanced Deep Bottleneck Transformer Model for Skin Lesion Classification. *Biomed. Signal Process. Control* **2022**, *78*, 103997. [CrossRef]
13. Popescu, D.; El-Khatib, M.; Ichim, L. Skin Lesion Classification Using Collective Intelligence of Multiple Neural Networks. *Sensors* **2022**, *22*, 4399. [CrossRef] [PubMed]
14. Qian, S.; Ren, K.; Zhang, W.; Ning, H. Skin Lesion Classification Using CNNs with Grouping of Multi-Scale Attention and Class-Specific Loss Weighting. *Comput. Methods Programs Biomed.* **2022**, *226*, 107166. [CrossRef] [PubMed]
15. Mahbod, A.; Schaefer, G.; Wang, C.; Dorffner, G.; Ecker, R.; Ellinger, I. Transfer Learning Using a Multi-Scale and Multi-Network Ensemble for Skin Lesion Classification. *Comput. Methods Programs Biomed.* **2020**, *193*, 105475. [CrossRef]
16. Panthakkan, A.; Anzar, S.M.; Jamal, S.; Mansoor, W. Concatenated Xception-ResNet50—A Novel Hybrid Approach for Accurate Skin Cancer Prediction. *Comput. Biol. Med.* **2022**, *150*, 106170. [CrossRef]
17. Almaraz-Damian, J.A.; Ponomaryov, V.; Sadovnychiy, S.; Castillejos-Fernandez, H. Melanoma and Nevus Skin Lesion Classification Using Handcraft and Deep Learning Feature Fusion via Mutual Information Measures. *Entropy* **2020**, *22*, 484. [CrossRef]
18. Zalaudek, I.; Giacomel, J.; Schmid, K.; Bondino, S.; Rosendahl, C.; Cavicchini, S.; Tourlaki, A.; Gasparini, S.; Bourne, P.; Keir, J.; et al. Dermatoscopy of Facial Actinic Keratosis, Intraepidermal Carcinoma, and Invasive Squamous Cell Carcinoma: A Progression Model. *J. Am. Acad. Dermatol.* **2012**, *66*, 589–597. [CrossRef] [PubMed]
19. Sevli, O. A Deep Convolutional Neural Network-Based Pigmented Skin Lesion Classification Application and Experts Evaluation. *Neural Comput. Appl.* **2021**, *33*, 12039–12050. [CrossRef]
20. Lallas, A.; Apalla, Z.; Argenziano, G.; Longo, C.; Moscarella, E.; Specchio, F.; Raucci, M.; Zalaudek, I. The Dermatoscopic Universe of Basal Cell Carcinoma. *Dermatol. Pract. Concept.* **2014**, *4*, 11–24. [CrossRef]
21. BinJadeed, H.; Aljomah, N.; Alsubait, N.; Alsaif, F.; AlHumidi, A. Lichenoid Keratosis Successfully Treated with Topical Imiquimod. *JAAD Case Rep.* **2020**, *6*, 1353–1355. [CrossRef]
22. Ortonne, J.P.; Pandya, A.G.; Lui, H.; Hexsel, D. Treatment of Solar Lentigines. *J. Am. Acad. Dermatol.* **2006**, *54*, 262–271. [CrossRef] [PubMed]
23. Zaballos, P.; Salsench, E.; Serrano, P.; Cuellar, F.; Puig, S.; Malvehy, J. Studying Regression of Seborrheic Keratosis in Lichenoid Keratosis with Sequential Dermoscopy Imaging. *Dermatology* **2010**, *220*, 103–109. [CrossRef]
24. Zaballos, P.; Puig, S.; Llambrich, A.; Malvehy, J. Dermoscopy of Dermatofibromas. *Arch. Dermatol.* **2008**, *144*, 75–83. [CrossRef] [PubMed]
25. Sarkar, R.; Chatterjee, C.C.; Hazra, A. Diagnosis of Melanoma from Dermoscopic Images Using a Deep Depthwise Separable Residual Convolutional Network. *IET Image Process* **2019**, *13*, 2130–2142. [CrossRef]

26. Teja, K.U.V.R.; Reddy, B.P.V.; Likith Preetham, A.; Patil, H.Y.; Poorna Chandra, T. Prediction of Diabetes at Early Stage with Supplementary Polynomial Features. In Proceedings of the 2021 Smart Technologies, Communication and Robotics (STCR)STCR, Sathyamangalam, India, 9–10 October 2021; pp. 7–11. [CrossRef]

27. Ledig, C.; Theis, L.; Huszár, F.; Caballero, J.; Cunningham, A.; Acosta, A.; Aitken, A.P.; Tejani, A.; Totz, J.; Wang, Z.; et al. Photo-Realistic Single Image Super-Resolution Using a Generative Adversarial Network. *Cvpr* **2017**, *2*, 4.

28. Wang, X.; Yu, K.; Wu, S.; Gu, J.; Liu, Y.; Dong, C.; Qiao, Y.; Loy, C.C. ESRGAN: Enhanced Super-Resolution Generative Adversarial Networks. In *Computer Vision – ECCV 2018 Workshops*; Lecture Notes in Computer Science (LNCS), including its subseries Lecture Notes in Artificial Intelligence (LNAI) and Lecture Notes in Bioinformatics (LNBI); Springer: Cham, Switzerland, 2019; Volume 11133, pp. 63–79. [CrossRef]

29. Le-Tien, T.; Nguyen-Thanh, T.; Xuan, H.P.; Nguyen-Truong, G.; Ta-Quoc, V. Deep Learning Based Approach Implemented to Image Super-Resolution. *J. Adv. Inf. Technol.* **2020**, *11*, 209–216. [CrossRef]

30. Milton, M.A.A. Automated Skin Lesion Classification Using Ensemble of Deep Neural Networks in ISIC 2018: Skin Lesion Analysis Towards Melanoma Detection Challenge. *arXiv* **2019**, arXiv:1901.10802.

31. Naeem, A.; Farooq, M.S.; Khelifi, A.; Abid, A. Malignant Melanoma Classification Using Deep Learning: Datasets, Performance Measurements, Challenges and Opportunities. *IEEE Access* **2020**, *8*, 110575–110597. [CrossRef]

32. Hu, Z.; Tang, J.; Wang, Z.; Zhang, K.; Zhang, L.; Sun, Q. Deep Learning for Image-Based Cancer Detection and diagnosis – A Survey. *Pattern Recognit.* **2018**, *83*, 134–149. [CrossRef]

33. Srivastava, V.; Kumar, D.; Roy, S. A Median Based Quadrilateral Local Quantized Ternary Pattern Technique for the Classification of Dermatoscopic Images of Skin Cancer. *Comput. Electr. Eng.* **2022**, *102*, 108259. [CrossRef]

34. Patil, P.; Ranganathan, M.; Patil, H. *Ship Image Classification Using Deep Learning Method BT—Applied Computer Vision and Image Processing*; Iyer, B., Rajurkar, A.M., Gudivada, V., Eds.; Springer: Singapore, 2020; pp. 220–227.

35. Barua, S.; Patil, H.; Desai, P.; Manoharan, A. *Deep Learning-Based Smart Colored Fabric Defect Detection System*; Springer: Berlin/Heidelberg, Germany, 2020; pp. 212–219. ISBN 978-981-15-4028-8.

36. Sarkar, A.; Maniruzzaman, M.; Ahsan, M.S.; Ahmad, M.; Kadir, M.I.; Taohidul Islam, S.M. Identification and Classification of Brain Tumor from MRI with Feature Extraction by Support Vector Machine. In Proceedings of the 2020 International Conference for Emerging Technology (INCET), Belgaum, India, 5–7 June 2020; Volume 2, pp. 9–12. [CrossRef]

37. Agyenta, C.; Akanzawon, M. Skin Lesion Classification Based on Convolutional Neural Network. *J. Appl. Sci. Technol. Trends* **2022**, *3*, 14–19. [CrossRef]

38. Saarela, M.; Geogieva, L. Robustness, Stability, and Fidelity of Explanations for a Deep Skin Cancer Classification Model. *Appl. Sci.* **2022**, *12*, 9545. [CrossRef]

39. Alam, M.J.; Mohammad, M.S.; Hossain, M.A.F.; Showmik, I.A.; Raihan, M.S.; Ahmed, S.; Mahmud, T.I. S2C-DeLeNet: A Parameter Transfer Based Segmentation-Classification Integration for Detecting Skin Cancer Lesions from Dermoscopic Images. *Comput. Biol. Med.* **2022**, *150*, 106148. [CrossRef] [PubMed]

Systematic Review

Applications of Deep Learning to Neurodevelopment in Pediatric Imaging: Achievements and Challenges

Mengjiao Hu [1], Cosimo Nardi [2], Haihong Zhang [1] and Kai-Keng Ang [1,3,*]

[1] Institute for Infocomm Research (I2R), Agency for Science, Technology and Research (A*STAR), Singapore 138632, Singapore
[2] Department of Experimental and Clinical Biomedical Sciences, University of Florence—Azienda Ospedaliero-Universitaria Careggi, 50134 Florence, Italy
[3] School of Computer Science and Engineering, Nanyang Technological University, Singapore 639798, Singapore
* Correspondence: kkang@i2r.a-star.edu.sg; Tel.: +65-64082678

Abstract: Deep learning has achieved remarkable progress, particularly in neuroimaging analysis. Deep learning applications have also been extended from adult to pediatric medical images, and thus, this paper aims to present a systematic review of this recent research. We first introduce the commonly used deep learning methods and architectures in neuroimaging, such as convolutional neural networks, auto-encoders, and generative adversarial networks. A non-exhaustive list of commonly used publicly available pediatric neuroimaging datasets and repositories are included, followed by a categorical review of recent works in pediatric MRI-based deep learning studies in the past five years. These works are categorized into recognizing neurodevelopmental disorders, identifying brain and tissue structures, estimating brain age/maturity, predicting neurodevelopment outcomes, and optimizing MRI brain imaging and analysis. Finally, we also discuss the recent achievements and challenges on these applications of deep learning to pediatric neuroimaging.

Keywords: pediatric; magnetic resonance imaging; neurodevelopment; deep learning

Citation: Hu, M.; Nardi, C.; Zhang, H.; Ang, K.-K. Applications of Deep Learning to Neurodevelopment in Pediatric Imaging: Achievements and Challenges. *Appl. Sci.* **2023**, *13*, 2302. https://doi.org/10.3390/app13042302

Academic Editor: Qi-Huang Zheng

Received: 15 December 2022
Revised: 3 February 2023
Accepted: 8 February 2023
Published: 10 February 2023

1. Introduction

Machine learning has achieved extraordinary achievements during the past decades. Conventional machine learning algorithms such as support vector machine and logistic regression have been widely applied to image analysis for pattern recognition and identification [1]. Yet applications of such approaches are limited by the reliance on feature extraction procedure and restrictions on high dimensionality of data. Feature extraction requires high expertise in domain knowledge to transform raw data into a different representation. Further dimension reduction techniques are required to fit the high-dimensional features to the machine learning algorithms [2]. Evolution of deep learning algorithms such as convolutional neural networks has advanced the development of machine learning to another triumph. The end-to-end framework of deep learning allows automatic feature learning of the complicated data patterns which migrates the subjectivity in feature extraction procedure. The deep architecture and nonlinear processing units empower the deep learning algorithm to deal with a vast amount of data [3,4]. Successful applications of conventional machine learning and deep learning to medical imaging have been widely reported [5,6]. Specifically, neuroimaging studies based on magnetic resonance imaging (MRI) have applied machine learning to the study of the brain in many aspects [7,8].

MRI has become a crucial diagnostic imaging technique for the study of the brain for its advantage of non-ionic and high-contrast resolution [9]. MRI relies on the nuclear magnetic resonance phenomenon, in which atomic nuclei will re-emit radio signals when placed in a magnetic field and stimulated by oscillating radio waves. Human body contains rich hydrogen nuclei and the nuclei align to the magnetic field generated by the MRI scanner. Then, an oscillating radio frequency deviates the magnetic momentum of the nuclei from the field. When the oscillating radio pulse is removed, signals generated by the realignment of hydrogen nuclei can be detected by a reciever coil [10,11]. The most common MRI modality is the structural MRI (sMRI) which provides morphostructural information based on the concentration of hydrogen protons. sMRI measures the signals produced by aligned hydrogen protons in water molecules in the body and creates excellent contrast among different tissues. Functional MRI (fMRI) quantifies the blood oxygenation level-dependent (BOLD) signals based on the blood flow and blood oxygen changes around cells and reflects the brain activity information [12]. Resting-state fMRI (rs-fMRI) is measured when the subject is at rest while task fMRI monitors the brain function during an assigned task. Diffusion tensor imaging (DTI) estimates the motion of water molecules in the brain. The water molecules' diffusion speed and directions are restricted by tissue types and fiber architectures. DTI therefore provides information based on the quantitative anisotropy and orientation [13]. Deep learning methods have been widely applied to neuroimaging studies in adult for neuropsychiatric disorder recognition, brain tissues and structures segmentation, and clinical outcome prediction [8,14,15]. In comparison, relatively few deep learning studies have been conducted in pediatric MRI. Most previous reviews on pediatric MRI involved a large number of studies using conventional machine learning approaches instead of deep learning algorithms and some reviews focused on specific topics such as Autism [7,16,17]. To illustrate the most recent achievements of deep learning in pediatric MRI, this systematic review summarized the advanced deep learning approaches applied to multiple neurodevelopmental topics in MRI-based research in the past five years. Section 2 introduces the most commonly utilized deep learning algorithms as well as a list of available public datasets for neurodevelopment. Section 3 categorizes the recent studies into five main topics: recognizing neurodevelopmental disorders, identifying brain and tissue structures, estimating brain age/maturity, predicting neurodevelopment outcomes, and optimizing MRI brain imaging and analysis. The challenges and insights of applying deep learning to pediatric MRI are discussed in Section 4. We conclude in Section 5.

2. Methods

2.1. Deep Learning Model Architectures

Multi-layer perceptron (MLP) has the most basic architecture of deep neural networks, which is composed a stack of processing layers: an input layer, several hidden layers, and an output layer (Figure 1) [18]. The neurons in the processing layers allow nonlinear computation and empower the model to learn different representations of the training data at multiple levels of abstraction [3].

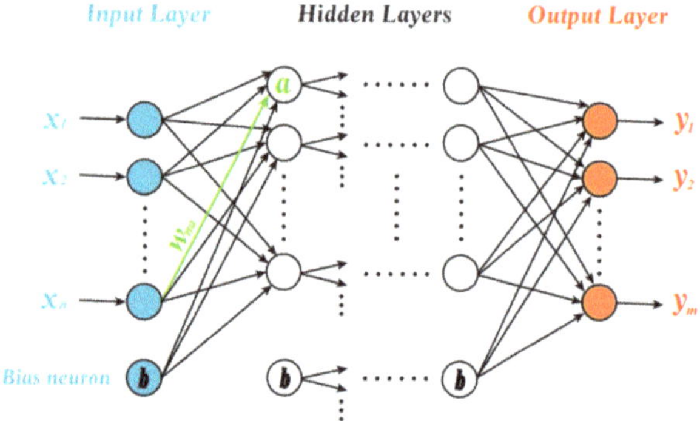

Figure 1. Architecture of multi-layer perceptron (MLP) [18].

Convolutional neural network (CNN) is the most widely applied deep learning algorithm for medical imaging studies. A typical CNN consists of convolutional layers with activation functions, pooling layers, and fully connected layers (Figure 2) [2]. Convolutional layers convolve an image with different types of kernel functions to extract image features. The kernels are applied to the entire image, thus greatly reducing the number of weights to be trained compared to fully connected neural networks. Activation functions such as sigmoid and ReLu (Rectified Linear Unit) serve as nonlinear feature detectors to introduce nonlinearities to CNN. Pooling layers reduce feature map resolution with translational invariance. The combination of convolutional and pooling layers enables CNN to learn spatial hierarchies among feature patterns. Fully connected layers function as a classifier or regressor to predict the desired outcomes [2]. The weight sharing and translational invariance properties facilitate CNN the efficient and precise power on image processing tasks. Depending on the input data dimensionality, 1D, 2D, and 3D convolutional kernels can be employed. Besides the basic stacking of convolutional layers, pooling layers and fully connected layers, models with complex architectures have been developed to further improve the performance of CNN. AlexNet was the first big CNN model which showed the great potential of CNN on image recognition tasks [19]. Inception blocks utilize convolution kernels of different sizes at the same level to optimize the accuracy and computation time of the model [20]. Residual connection from a previous layer to a later layer without extra parameters solves the vanishing gradients issues and thereby make the CNN model with many layers [21]. Dense blocks formed by many convolution operations and a final pooling and connecting the input and output of each convolution are proposed to train even deeper models [22]. Many other CNN models with different architectures have been proposed. A detailed summary can be found in the review paper by Celard et al. [2].

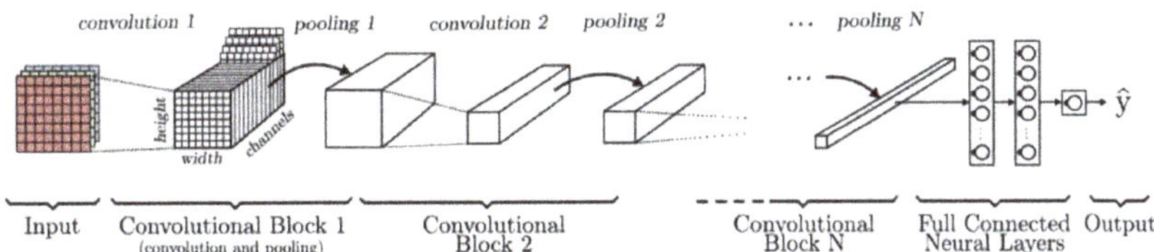

Figure 2. Architecture of convolutional neural networks [2].

U-net was proposed for semantic segmentation in 2015 and is still one of the most used CNN architectures for medical image segmentation. The typical U-net is composed of symmetrical encoder and decoder paths connected by skip connections (Figure 3) [23]. The model first performs a set of convolutions at the encoder side to extract features from the input data and then reconstructs the input image while including new information by transposed convolutions at the decoder side. Skip connections connect the encoder and decoder at each level. Complex architectures have also been applied to U-net to further improve its performance, for example, the Res-U-net and U-net with attention mechanism [24,25].

Auto-encoder plays a pivotal role in unsupervised deep learning. Auto-encoder follows the encoder and decoder architecture (Figure 4). The encoder aims at learning a latent representation with low dimensionality which retains only the significant information while ignoring the noise. The decoder utilizes the latent representation to reconstruct the input data. Auto-encoder provides an effective approach for feature learning in recognition tasks with unlabeled data. Variational auto-encoders are applied as generative models which randomly generate new data that are similar to the input data [2].

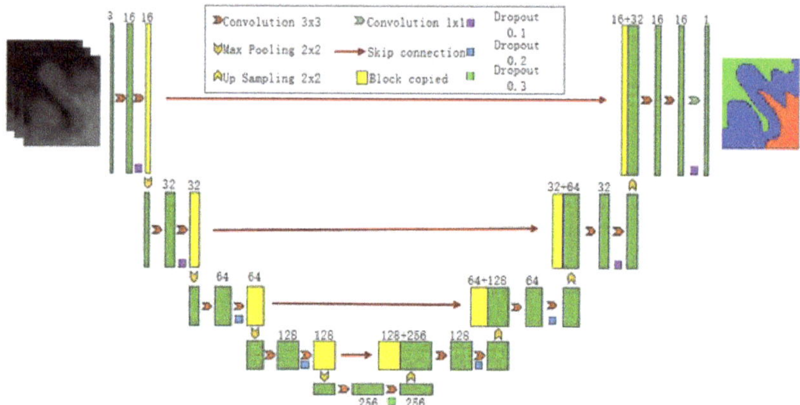

Figure 3. Architecture of U-net [26].

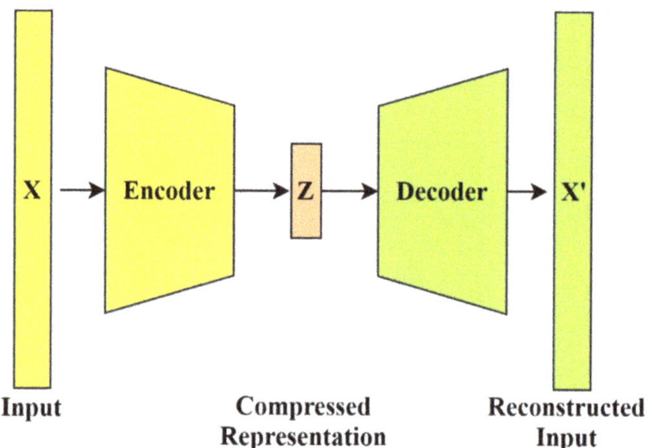

Figure 4. Architecture of auto-encoder [27].

Generative adversarial network (GAN) has attracted attention with its ability to model data distributions and generate realistic data since proposed in 2014 [28]. GAN consists of one generator network which captures the data distribution in real images and generates a fake image and one discriminator which classifies the generated fake images and real images (Figure 5). Two networks are trained alternatively in a competitive manner. A large number of variations of GAN have been proposed and applied to object detection, localization, segmentation, data augmentation, and image quality improvement tasks [29]. A review paper [30] introduced various architectures of GAN and their applications in medical imaging.

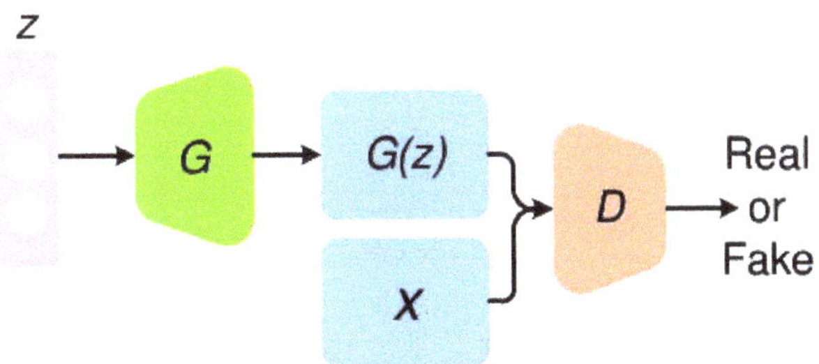

Figure 5. Architecture of generative adversarial networks (GAN) [29].

2.2. Public Datasets and Repositories

Sample size is one of the most critical issues for training a deep learning algorithm as the number of trainable parameters grows exponentially with deep architectures. However, data collection is expensive and time-consuming for medical images. Fortunately, more and more data repositories and data-sharing platforms are available recently, making it possible to conduct medical imaging studies on a large scale. Table 1 lists the available public datasets and repositories involved in the studies reviewed in this manuscript. Some repositories collect data from multiple independent sites and provide a large number of subjects. The Autism Brain Imaging Data Exchange (ABIDE) dataset and IMaging-PsychiAtry Challenge (IMPAC) dataset focus on autism spectrum disorder (ASD) recognition and provide data of subjects with ASD and healthy controls. The ADHD-200 consortium collects data for attention deficit hyperactivity disorder (ADHD) patients and healthy controls. The Healthy Brain Network (HBN) dataset and Human Connectome Project Development (dHCP) project are data collections for typically developed individuals. The UNC/UMN Baby Connectome Project (BCP) collects data of infants and pre-school age children. Other datasets including a large number of participants such as UK Biobank and International Consortium for Brain Mapping (ICBM) involve healthy controls as well as patients with various neurodevelopmental disorders at all ages.

Table 1. Public datasets.

Dataset	No. of Sites/Projects	Population	Technique	Citation
Autism Brain Imaging Data Exchange I (ABIDE I)	17 independent imaging sites	539 subjects with ASD and 573 healthy controls (age 7–64 years)	sMRI, rs-fMRI	[31]
Autism Brain Imaging Data Exchange II (ABIDE II)	19 independent imaging sites	521 subjects with ASD and 593 healthy controls (age 5–64 years)	sMRI, rs-fMRI, DTI	[32]
IMaging-PsychiAtry Challenge (IMPAC)	-	549 subjects with ASD 601 healthy controls (age 0–80 years)	sMRI, rs-fMRI	[33]
ADHD-200 Consortium	8 independent imaging sites	285 subjects with ADHD 491 healthy controls (age 7–21 years)	sMRI, rs-fMRI	[34]
UK Biobank	-	500,000 subjects (age 40–69 years)	sMRI, rs-fMRI, DTI	[35]

Table 1. *Cont.*

Dataset	No. of Sites/Projects	Population	Technique	Citation
National Database for Autism Research (NDAR)	hundreds of research projects	117,573 subjects by age (57,510 affected subjects and 59,763 control subjects)	sMRI, rs-fMRI, DTI	[36]
Open fMRI	95 datasets	3375 subjects across all datasets	sMRI, rs-fMRI, task fMRI	[37]
International Consortium for Brain Mapping (ICBM)	-	853 subjects (age 18–89 years)	sMRI, rs-fMRI, DTI	[38]
1000 funtional connectome	33 independent imaging sites	1355 subjects (age 13–80 years)	rs-fMRI	[39]
The Adolescent Brain Cognitive Development (ABCD) Study	-	12,000 subjects (age 9–10 years)	sMRI, rs-fMRI, task fMRI	[40]
ENIGMA ADHD working group	34 cohorts	over 4000 subjects	sMRI, rs-fMRI, DTI	[41]
Philadelphia Neurodevelopmental Cohort (PNC)	-	9500 subjects (age 8–21 years)	sMRI, rs-fMRI, task fMRI, DTI	[42]
Healthy Brain Network (HBN)	-	10,000 subjects (age 5–21 years)	sMRI, rs-fMRI, task fMRI, DTI	[43]
Human Connectome Project Development (dHCP)	-	1350 subjects (age 5–21 years)	sMRI, rs-fMRI, task fMRI	[44]
The UNC/UMN Baby Connectome Project (BCP)	2 sites	500 subjects (age 0–5 years)	sMRI, rs-fMRI, DTI	[45]

Abbreviations: sMRI—structural MRI, rs-fMRI—resting-state functional MRI, DTI—Diffusion Tensor Imaging.

2.3. Review Parameters

The paper selection and review procedure in this study follows the preferred reporting items for systematic reviews and meta-analysis (PRISMA) guidelines [46,47]. The search terms employed were <deep learning brain MRI neurodevelopment> or <deep learning pediatric brain MRI> or <deep learning child brain MRI> or <deep learning adolescent brain MRI> to include the deep learning studies based on MRI for pediatric neurodevelopment studies. The initial search was performed on PubMed and Web of Science databases on 26 October 2022. Search engines ScienceDirect and Google Scholar were excluded due to the large number of search results returned (thousands of results).

The initial search yielded 412 papers from PubMed and 252 papers from Web of science. Following the PRISMA protocols, we performed selection and review steps in Figure 6. A total of 304 duplicate records was removed in the first step. Secondly, we examined the keywords, titles, and abstracts of the remaining 360 papers and excluded review papers, case reports, papers with foreign language (French), and animal studies. Furthermore, we identified studies with topics on adult population, genetics, maternity, and non-deep learning approaches as irrelevant and excluded them. We retrieved the full paper for 184 out of the remaining 185 studies. The full papers were further examined for eligibility and 67 studies with non-pediatric population, non-MRI modality or non-deep learning methods were removed. Then, 120 Studies were carefully reviewed and 113 of them are categorized and reported in the next chapter. The remaining 7 studies on gender prediction, functional connectivity estimation, and fascicles detection are not reported.

Three researchers independently examined the eligibility of the studies and conflict decisions were resolved by discussion. Data extracted from selected studies include but are not limited to the year of the study, clinical questions, study population, imaging techniques, preprocessing protocols and tools, deep learning approach, training and validation settings, results, results interpretation, and limitations. Extracted information is presented and discussed in the following chapters. Specifically, risk of bias analysis was performed following the Risk Of Bias In Non-randomized Studies of Interventions [48] for (1) risk of bias due to confounding; (2) risk of bias in selection of participants into the study; (3) risk of bias in classification of interventions; (4) risk of bias due to deviations from intended interventions; (5) risk of bias due to missing data; (6) risk of bias arising from measurement of outcomes; (7) risk of bias in selection of reported results. Risk of bias analysis is presented in Appendix A (Table A1).

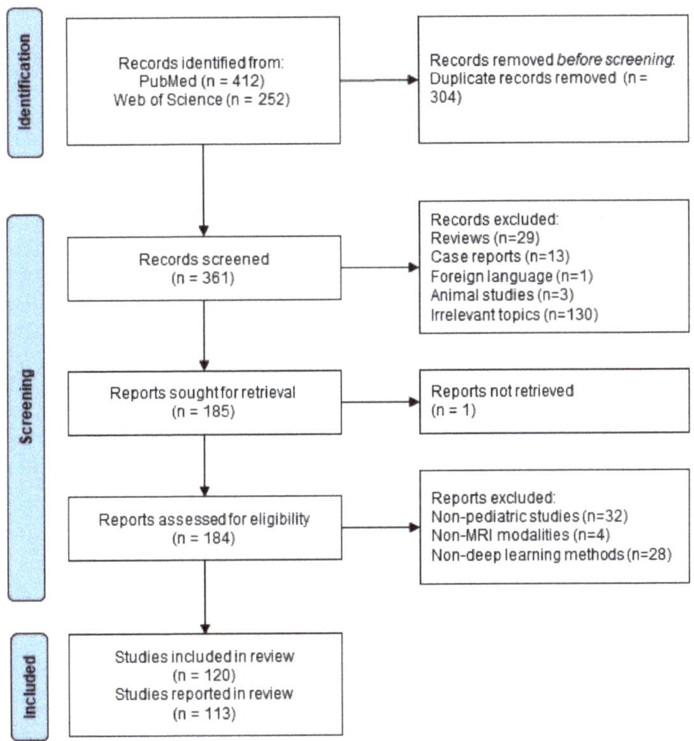

Figure 6. Study selection procedure.

3. Results

3.1. Recognizing Neurodevelopmental Disorders

Neurodevelopmental disorders are common brain disorders in children, bringing a variety of challenges to the affected patients and causing great burdens to their families. Various genetic and environmental factors may perturb the developmental process and result in neurodevelopmental disorders [49].

Autism spectrum disorder (ASD) is one of the most common neurodevelopmental disorders [50]. ASD is characterized by early deficits in social interactions and communication accompanied by restricted and repetitive behaviors [49]. Review papers [7,17] summarized a selected number of studies using artificial intelligence approaches to classify ASD patients and healthy controls including both conventional machine learning methods and deep learning methods. This review listed the recent deep learning advancements using MLP, CNN, RNN, and auto-encoder models (Table 2). Rs-fMRI is widely utilized for ASD recognition. Connectomes derived from fMRI were used as inputs to MLP, CNN, and RNN for classification [51–54]. A multimodal study [55] combined sMRI, rs-fMRI, and task fMRI.

Attention deficit hyperactivity disorder (ADHD) is another common neurodevelopmental disorder [50]. ADHD patients often suffer from hyperactivity, impulsivity, and inattention, and ADHD often continues to adulthood [56]. Previous ADHD recognition studies were summarized in the review paper [7] in conventional machine learning category and deep learning category. This review paper focuses on more recent studies utilizing deep learning approaches for ADHD detection (Table 2). Both rs-fMRI and sMRI are employed as inputs for deep learning networks.

Neurodevelopmental disorders which are less common such as cerebellar dysplasia [57], dyslexic [58], epilepsy [59,60], conduct disorder [61], disruptive behavior disorder [62], and post-traumatic stress disorder [63] are also reviewed in this study. We also include three studies for detection of posterior fossa tumors and tubers in tuberous sclerosis complex [64–66], and two studies for white matter pathway classification [67,68]. This review aims to investigate the deep learning methods utilized in various pediatric topics in an overall manner and therefore includes multiple disorders. Structural imaging techniques such as sMRI and DTI are more commonly utilized in these studies.

Overall, the selected studies are summarized in Table 2. Most studies conducted baseline comparisons using conventional machine learning approaches and reported the superior performance of deep learning approaches [53,69]. CNN dominates in the image recognition tasks. A total of 41 out of 48 neurodevelopmental disorder classification studies in this review utilized CNN approaches. Advanced CNN architectures such as inception and residual modules were employed in 2D CNN models [70–72]. Several studies trained 3D CNN with a limited number of sample size [61,69,73,74], bringing concerns on overfitting. Large-scale studies which involve thousands of training data were conducted using public datasets and repositories [55,75–78]. Multimodal studies combined features from multiple MRI modalities showed better performance than single modality [62,76].

Table 2. Recognizing neurodevelopmental disorders.

Study	Year	Disorder	Population	Technique	Preprocessing	Method	Results
[79]	2017	Autism	ABIDE I dataset 55 ASD (age 14.2 ± 3.2 years) 55 HC (age 12.7 ± 2.4 years)	rs-fMRI	Preprocessed Connectomes Project	MLP	Accuracy 86.36%
[80]	2018	Autism	62 ASD 48 HC	task fMRI	FSL	MLP	Accuracy 87.1%
[51]	2018	Autism	ABIDE I dataset 529 ASD 571 HC	rs-fMRI	In-house pipeline	RNN	Accuracy 70.1%
[81]	2018	Autism	ABIDE I & II dataset 116 ASD 69 HC (age 5–10 years)	sMRI, rs-fMRI	SPM8	Deep Belief Network	Accuracy 65.56%
[53]	2019	Autism	ABIDE I & II dataset 210 ASD 249 HC (age 5–10 years)	rs-fMRI	SPM8	CNN	Accuracy 72.73%
[52]	2019	Autism	ABIDE II dataset 117 ASD 81 HC (age 5–12 years)	rs-fMRI	FSL	Auto-encoder	Accuracy 96.26%

Table 2. *Cont.*

Study	Year	Disorder	Population	Technique	Preprocessing	Method	Results
[55]	2020	Autism	multi datasets: ABCD, ABIDE I, II, BioBank, NDAR, ICBM, Open fMRI, 1000 Functional Connectomes 43,838 total connectomes 1711 ASD (age 0.42–78 years)	rs-fMRI, task-fMRI	SPT, AFNI, SpeddyPP	CNN	AUROC 0.6774
[82]	2020	Autism	YUM dataset 40 ASD (age 29.4 ± 11.6 years) 33 HC (age 30.1 ± 5.3 years) ABIDE I dataset 521 ASD (age 29.4 ± 11.6 years) 593 HC (age 30.1 ± 5.3 years)	sMRI	SPM8	3D CNN	Accuracy 88% (YUM) 64% (ABIDE)
[69]	2021	Autism	ABIDE I dataset 55 ASD (age 14.52 ± 6.97 years) 55 HC (age 15.81 ± 6.25 years)	rs-fMRI	Configurable Pipeline for the Analysis of Connectomes	3D CNN	Accuracy 77.74%
[74]	2021	Autism	50 ASD 50 HC (age 12–40 months)	task-fMRI	FSL, FEAT	3D CNN	Accuracy 80%
[83]	2021	Autism	ABIDE I & II dataset 1060 ASD 1146 HC (age 5–64 years)	rs-fMRI	In-house pipeline	CNN	Accuracy 89.5%
[84]	2021	Autism	ABIDE I dataset 506 ASD 532 HC (age 10–28 years)	rs-fMRI	DPABI	MLP	Accuracy 78.07 ± 4.38%
[85]	2021	Autism	52 ASD 195 HC infants (age 24 months)	MRI	iBEAT	CNN	Accuracy 92%
[76]	2021	Autism	multi datasets: ABCD, ABIDE I, II, BioBank, NDAR, Open fMRI 29,288 total connectomes 1555 ASD (age 0.42–78 years)	sMRI, rs-fMRI, task-fMRI	AFNI, SpeddyPP	CNN	AUROC 0.7354
[54]	2022	Autism	ABIDE & UM dataset 411 HC for offline learning 48 ASD 65 HC for testing (age 13.8 ± 2 years)	rs-fMRI	Connectome Computation System	Auto-encoder	Accuracy 67.2%
[73]	2022	Autism	Preschool dataset 110 subjects ABIDE I dataset 1099 subjects	sMRI	SPM8	CNN	AUROC 0.787 (preschool) 0.856 (ABIDE)
[86]	2022	Autism	151 ASD 151 HC (age 1–6 years)	sMRI	In-house pipeline	3D CNN	Accuracy 84.4%
[75]	2022	Autism	IMPAC dataset 418 ASD 497 hc (age 17 ± 9.6 years)	sMRI, rs-fMRI	In-house pipeline	MLP	AUROC 0.79 ± 0.01

Table 2. *Cont.*

Study	Year	Disorder	Population	Technique	Preprocessing	Method	Results
[87]	2019	ADHD	ADHD-200 consortium 776 subjects	rs-fMRI	In-house pipeline	3D CNN	Accuracy 69.01%
[88]	2020	ADHD	ADHD-200 consortium 262 subjects	rs-fMRI	AFNI, FSL	CNN	Accuracy 73.1%
[78]	2021	ADHD	ENIGMA-ADHD Working Group 2192 ADHD 1850 HC (age 4–63 years)	sMRI	FreeSurfer	MLP	Testing AUROC 0.60
[89]	2022	ADHD	ADHD-200 consortium NI site25 ADHD 23 HC (age 11–22 years) NYU site: 118 ADHD 98 HC (age 7–18 years) KKI site: 22 ADHD 61 HC (age 8–13 years) PU site: 78 ADHD 116 HC (age 8–17 years) PU-1 site: 24 ADHD 62 HC (age 8–17 years)	rs-fMRI	Preprocessed Connectomes Project	Auto-encoder	Accuracy >99%
[90]	2022	ADHD	ADHD-200 consortium NI site: 28 ADHD-I 37 HC NYU site: 72 ADHD-I, 42 ADHD-C, 96 HC OHSU site: 27 ADHD-I, 13 ADHD-C, 70 HC KKI site: 16 ADHD-I, 5 ADHD-C 60 HC PU-1 site: 16 ADHD-I, 26 ADHD-C, 88 HC PU-2 site: 15 ADHD-I, 20 ADHD-C, 31 HC PU-3 site: 7 ADHD-I, 12 ADHD-C, 23 HC	rs-fMRI	DPABI	CNN	Accuracy >99%
[91]	2022	ADHD	ADHD-200 consortium Training: 69 ADHD 99HC Testing: 24 ADHD 27 HC (age 7–21 years	rs-fMRI	Athena pipeline	CNN	Testing accuracy 67%
[77]	2022	ADHD	ADHD-200 consortium 325 ADHD 547 HC (age 12 ± 3.0 years)	rs-fMRI	Athena pipeline	CNN	Accuracy 78.7 ± 4.3%
[92]	2022	ADHD	19 ADHD (age 10.25 ± 1.94 years) 20 HC (age 10.15 ± 2.13 years)	sMRI	SPM	CNN	Accuracy 93.45 ± 1.18%
[93]	2022	ADHD	ABCD Dataset 127 ADHD 127 HC (age 9–10 years)	sMRI	ANTs	CNN	Accuracy 71.1%
[57]	2018	Cerebellar Dysplasia	90 patients, 40 HC	sMRI	FSL, ANTs	3D CNN	Accuracy 98.5 ± 2.41%
[61]	2020	Conduct Disorder	60 patients (age 15.3 ± 1.0 years) 60 HC (age 15.5 ± 0.7 years)	sMRI	-	3D CNN	Accuracy 85%

Table 2. *Cont.*

Study	Year	Disorder	Population	Technique	Preprocessing	Method	Results
[62]	2021	Disruptive Behavior Disorder	ABCD Study: 550 patients, 550 HC (age 9–11 years)	sMRI, rs-fMRI, DTI	FSL	3D CNN	Accuracy 72%
[58]	2020	Dyslexic	36 patients, 19 HC (age 9–12 years)	task fMRI	SPM	3D CNN	Accuracy 72.73%
[94]	2020	Embryonic Neurodevelopmental Disorders	114 patients, 113 HC (age 16–39 weeks)	sMRI	—	CNN	Accuracy 87.7%
[59]	2020	Epilepsy	30 patients, 13 HC	sMRI	BET	CNN	Accuracy 66–73%
[60]	2020	Epilepsy	59 patients, 70 HC (age 7–18 years)	DTI	SPM	CNN	Accuracy 90.75%
[70]	2021	Neonatal Hyperbilirubinemia	47 patients, 32 HC (age 1–18 days)	sMRI		CNN	Accuracy 72.15%
[63]	2021	PTSD	33 patients (age 14.3 ± 3.3 years) 53 HC (age 15.0 ± 2.3 years)	rs-fMRI	SPM12	MLP	Accuracy 72%
[64]	2020	Tuber	260 patients, 260 HC	sMRI	FSL	3D CNN	Accuracy 97.1%
[65]	2022	Tuber	296 patients, 245 HC (age 0–8 years)	sMRI	-	3D CNN	Accuracy 86%
[71]	2020	Tuber	114 patients (age 5–15.3 years), 114 HC (age 6.9–15.7 years)	sMRI	In-house pipeline	CNN	Accuracy 95%
[95]	2021	Tumor	136 patients, 22 HC (age 0–11 years)	sMRI	SPM	CNN	Accuracy 87 ± 2%
[72]	2020	Tumor	617 patients with tumor (age 0.2–34 years)	sMRI	Pydicom	CNN	Accuracy 72%
[66]	2018	Tumor	233 subjects	sMRI	-	Capsule Network	Accuracy 86.56%
[96]	2020	Tumor	39 pediatric patients	sMRI	-	CNN	Accuracy 87.8%
[67]	2020	White Matter Pathways	89 patients with focal epilepsy (age 9.95 ± 5.41 years)	DTI	FreeSurfer	CNN	Accuracy 98%
[68]	2019	White Matter Pathways	70 HC (age 12.01 ± 4.80 years), 70 patients with focal epilepsy (age 11.60 ± 4.80 years)	DTI	FreeSurfer, FSL, NIH TORTOISE	CNN	F1 score 0.9525 ± 0.0053

Abbreviations: ASD—Autism spectrum disorder, HC—healthy control, ADHD—Attention deficit hyperactivity disorder, sMRI—structural MRI, rs-fMRI—resting-state functional MRI, DTI—Diffusion Tensor Imaging, MLP—Multi-layer perceptron, CNN—Convolutional neural network.

3.2. Identifying Brain and Tissue Structures

Identifying brain and tissue structures is of great importance in facilitating studies investigating changes in a specific region of interest. Accurate segmentation of brain tissues and structures lays the foundation for volumetric and morphologic analysis. Volumetric analysis of gray matter, white matter, cerebrospinal fluid, and specific brain structure such as amygdala assist in computer-aided diagnosis of neurodevelopmental disorders. Localization and segmentation of brain tumor is essential for assessment of the tumor burden as well as treatment response and tumor progression [97]. Brain masking isolates the brain from surrounding tissues across non-stationary 3D brain volumes in fMRI, which is important and challenging, especially for fetal imaging [98]. Specific challenges for pediatric brain segmentation exist due to the variations in head size and shape in children compared to adults. Rapid changes in tissue contrast and low contrast to noise ratio in fetal and newborn MRIs lead to further demanding techniques [99]. This study reviews segmentation of pediatric brain tissues, structures, tumors, and masking of fetal brain (Table 3).

Most of the studies employed U-net for segmentation. Dice scores vary across studies. 3D U-net models were implemented for brain tissue and volume segmentation [25,100–102]. Transfer learning and active learning greatly reduced the number of samples that need to be labeled for training a high-quality patch-wise segmentation method [99]. FetalGAN was proposed to segment a fetal functional brain MRI using a segmentor as the generator in GAN architecture and achieved better performance than 3D U-net [98]. Adversarial domain adaptation was used to adapt a pre-trained U-net to another segmentation task in an unsupervised learning manner [103]. Transfer learning and GAN stand for the opportunity of training segmentation algorithms with weakly labeled or unlabeled data, which may greatly reduce the tedious and time-consuming process of creating groundtruth for segmentation tasks.

Table 3. Identifying brain and tissue structures.

Study	Year	Structure	Population	Technique	Preprocessing	Method	Results
[104]	2020	Amygdala	171 infants (age 6 months) 204 infants (age 12 months) 201 infants (age 24 months)	sMRI	-	U-net	Dice score 0.882 (6-month) 0.882 (12-month) 0.903 (24-month)
[105]	2020	Anterior Visual Pathway	18 subjects	sMRI	-	GAN	Dice score 0.602 ± 0.201
[106]	2018	Brain Mask	10 adolescent subjects (age 10–15 years), 25 newborn subjects from dHCP dataset	sMRI	-	CNN	F1 score 95.21 ± 0.94 (adolescent) 90.24 ± 1.84 (newborns)
[99]	2019	Brain Mask	10 adolescent subjects, 26 newborn subjects from dHCP dataset, 25 other subjects (age 0.2–2.5 years)	sMRI	-	CNN	Improve dice score after labeling a very small portion of target dataset (<0.25%)

Table 3. *Cont.*

Study	Year	Structure	Population	Technique	Preprocessing	Method	Results
[107]	2020	Brain Mask	197 fetuses (gestation age 24–39 weeks)	rs-fMRI	FSL	U-net	Dice score 0.94
[98]	2020	Brain Mask	71 scans of fetuses	rs-fMRI	AFNI	GAN	Dice score 0.973 ± 0.013
[108]	2020	Brain Mask	37 healthy fetuses (gestation age 27.3 ± 4.11 weeks) 32 fetuses with spina bifida pre-surgery (gestation age 23.06 ± 1.64 weeks) 16 fetuses post-surgery (gestation age 25.69 ± 1.21 weeks)	sMRI	-N4ITK	U-net	Dice score 0.9321 (healthy), 0.9387 (pre-surgery), 0.9294 (post-surgery)
[101]	2021	Brain Mask	214 fetuses (gestation age 22–38 weeks)	sMRI	-	3D U-net	Testing dice score 0.944
[109]	2021	Brain Mask	30 subjects (ages 2.34–4.31 years)	sMRI	-	CNN	Dice score 0.90 ± 0.14
[110]	2019	Brain Tissue	29 subjects (age 9.96 ± 7.16 years)	sMRI	-	3D CNN	Dice score 0.888 (gray matter), 0.863 (white matter), 0.937 (CSF)
[111]	2019	Brain Tissue	12 fetuses (gestation age 22.9–34.6 weeks)	sMRI	-	CNN	Dice score 0.88
[112]	2019	Brain Tissue	95 very pre-term infants (gestation age 28.5 ± 2.5 weeks, scan at term age), 28 very pre-term infants (gestation age 26.8 ± 2.1 weeks, scan at term age)	sMRI	-	CNN	Dice score 0.895 ± 0.098 testing dice score 0.845 ± 0.079
[113]	2020	Brain Tissue	47 patients with pediatric hydrocephalus (age 5.8 ± 5.4 years)	sMRI	-	CNN	Dice score 0.86
[114]	2021	Brain Tissue	35 subjects (age 4.2 ± 0.7 years)	sMRI	-	3D CNN	JS = 0.83 for gray matter JS = 0.92 for white matter
[25]	2021	Brain Tissue	98 preterm infants (gestation age ≤ 32 weeks)	DTI	In-house pipeline	3D U-net	Dice score 0.907 ± 0.041
[102]	2022	Brain Tissue	106 fetuses (gestation age 23–39 weeks)	sMRI	FSL	3D U-net	Dice score 0.897
[115]	2022	Brain Tissue	dHCP datast: 150 term (gestation age 37–44 weeks) 50 preterm (gestation age ≤ 32 weeks, scan at term-equivalent age)	sMRI	-	CNN	Dice score 0.88
[116]	2022	Brain Tissue	23 infants (age 6 ± 0.5 months)	sMRI	In-house pipeline	U-net	Dice score 0.92 (gray matter), 0.901 (white matter), 0.955 (CSF)

Table 3. *Cont.*

Study	Year	Structure	Population	Technique	Preprocessing	Method	Results
[117]	2020	Cerebral Arteries	48 subjects (age 0.8–22 years)	sMRI	In-house pipeline	U-net	Testing dice score 0.75
[118]	2021	Cerebral Ventricle	200 patients with obstructive hydrocephalus (age 0–22 years) 199 HC (age 0–19 years)	sMRI	In-house pipeline	U-net	Dice score 0.901
[103]	2021	Cortical Parcellation Network	dHCP datast: 403 infants, ePRIME dataset: 486 infants (gestation age 23–42 weeks, scanned at term-equivalent age)	sMRI	-MRITK	GAN	Dice score 0.96–0.99
[119]	2020	Cortical Plate	52 fetuses (gestation age 22.9–31.4 weeks)	sMRI	In-house pipeline	CNN	Testing dice score 0.907 ± 0.027
[120]	2021	Cortical Plate	12 fetuses (gestation age 16–39 weeks)	sMRI	-AutoNet, ITK-SNAP	CNN	Dice score 0.87
[121]	2019	Intracranial Volume	80 scans of fetuses (gestation age 22.9–34.6 weeks) 101 scans of infants (age 30–44 weeks)	sMRI	-	U-net	Dice score 0.976
[122]	2022	Limbic Structure	dHCPdataset: 473 subjects (40.65 ± 2.19)	sMRI	-	CNN	Dice score 0.87
[123]	2022	Posterior Limb of Internal Capsule	450 preterm infants (gestation age ≤ 32 weeks, scan at term-equivalent age)	sMRI	In-house pipeline	U-net	Dice score 0.690
[124]	2022	Tuber	29 subjects (age 9.96 ± 7.16 years)	sMRI	-	U-net	Testing dice score 0.59 ± 0.23
[125]	2022	Tumor	311 pediatric subjects	sMRI	-	U-net	Dice score 0.773
[126]	2022	Tumor	177 patients (age 0.27–17.87 years)	sMRI	CaPTk software	CNN	Dice score 0.910
[100]	2022	Tumor	122 patients (age 0.2–17.9 years)	sMRI	ANTs	3D U-net	Dice score 0.724
[97]	2022	Tumor	BraTS 2020 Dataset: 369 patients local dataset: 22 patients (average age 7.5–9 years)	sMRI	In-house pipeline	U-net	Dice score 0.896

Abbreviations: sMRI—structural MRI, rs-fMRI—resting-state functional MRI, DTI—Diffusion Tensor Imaging, CNN—Convolutional neural network, GAN—Generative adversarial network.

3.3. Predicting Brain Age

The brain development of children experiences a rapid and complex stage, especially for children younger than two years. Early brain development is critical for cognitive, sensory, and motor ability. Delayed brain development can lead to many neurodevelopmental disorders in children and affect their quality of life [127]. Accurate evaluation of brain development via brain age estimation based on neuroimaging is of clinical importance to understand healthy brain development and study the brain maturity deviation caused by neurodevelopmental disorders [128].

We summarized age prediction studies involved both infants and young children (Table 4). Structural MRI techniques are commonly utilized in 2D and 3D CNN models.

Study [128] using 2D CNN on DTI achieved comparison results with human experts. Study [127] demonstrated superior performance of 3D CNN compared to conventional machine learning approaches and 2D CNN. Multimodal study [129] combined sMRI, rs-fMRI, and DTI features and yielded a mean absolute error of 0.381 years for children and adolescents aged 8–21 years old. The age difference for the study population varies and thus reporting of the relative error rate is necessary for comparing different methods in different studies.

Table 4. Predicting brain age.

Study	Year	Population	Technique	Preprocessing	Method	Results
[84]	2017	115 infants (gestation age 24–32 weeks)	DTI	In-house pipeline	CNN	MAE 2.17 weeks
[130]	2019	317 MRI images of 112 infants age 2 weeks (8 to 35 days); 12 months (each ±2-weeks) and 3 years (each ±4-weeks).	sMRI	In-house pipeline	3D CNN	Accuracy 98.4% classifying three age groups
[131]	2019	PNC Dataset: 857 subject (age 8–22 years) 20% as children 20% as young adult	rs-fMRII	SPM12	MLP	Accuracy 96.64% predicting children and young adult
[132]	2020	ABIDE II dataset 382 subjects ADHD200 consortium 378 subjects	sMRI	SPM12	3D CNN	MAE 1.11 years (ABIDE II dataset) 1.16 years (ADHD200 consortium)
[127]	2020	220 subjects (age 0–5 years)	sMRI	In-house pipeline	CNN	MAE 2.26 months
[129]	2020	PNC Dataset: 839 subject (age 8–21 years)	sMRI, rs-fMRI, DTI	SPM12, DPARSF, PANDA	MLP	MAE 0.381 ± 0.119 years
[128]	2021	161 subjects (age 0–2 years)	sMRI	In-house pipeline	CNN	MAE 8.2 weeks
[133]	2021	84 infants (age 8 days–3 years)	sMRI	In-house pipeline	CNN	Accuracy 90%
[134]	2021	119 subjects (age 0–2 years)	sMRI	In-house pipeline	CNN	MAE 0.98 months
[135]	2021	220 fetuses (gestation age 15.9–38.7 weeks)	sMRI	In-house pipeline	CNN	MAE 0.125 weeks
[136]	2021	167 patients with Rolandic epilepsy (age 9.81 ± 2.55 years), 107 HC (age 9.43 ± 2.57 years)	sMRI	CAT12, SPM12	CNN	MAE 1.05 years for HC 1.21 years for patients
[137]	2022	524 infants (gestation age 23–42 weeks)	sMRI, DTI	Neonatal specific segmentation pipeline	CNN	MAE 0.72 weeks (term-born) 2.21 weeks (preterm)

Abbreviations: sMRI—structural MRI, rs-fMRI—resting-state functional MRI, DTI—Diffusion Tensor Imaging, CNN—Convolutional neural network, GAN—Generative adversarial network, MAE—mean absolute error.

3.4. Predicting Neurodevelopment Outcomes

The relationship between brain structure and cognitive function is complex. Research on brain activity and connectivity builds the network theory to capture the brain trajectories. It remains a challenge in the field of neuroscience to relate basic structural properties of brain to complex cognitive functions [138]. This study reviewed research on correlating brain structure and measurable neurodevelopment outcomes such as fluid intelligence, language function, and motor function (Table 5).

The ABCD dataset provides neuroimaging data including sMRI, rsfMRI, and DTI as well as cognitive assessments such as fluid intelligence and oral reading scores. Large-scale studies based on the ABCD dataset involve thousands of data and a variety of modalities to predict neurodevelopment outcomes [138–142]. CNN models were also employed to predict motor function and cognitive deficits in very preterm infants [143,144].

Table 5. Predicting neurodevelopment outcomes.

Study	Year	Score	Population	Technique	Preprocessing	Method	Results
[143]	2021	Cognitive Deficits	261 very preterm infants (gestation age ≤32 weeks , scan at 39–44 weeks postmenstrual age)	DTI, rs-fMRI	FSL	CNN	Accuracy 88.4%
[145]	2020	Fluid Intelligence	ABCD Study 8333 subjects (age 9–10 years)	sMRI	-	3D CNN	MSE 0.75626
[141]	2021	Fluid Intelligence	ABCD Dataset 7709 subjects (age 9–10 years)	sMRI	FSL, ANFI, FreeSuerfer	CNN	Pearson's correlation coefficient r = 0.18
[138]	2022	Fluid Intelligence	ABCD Dataset 8070 subjects (age 9–11 years) HCP Dataset 1079 subjects (age 22–35 years)	sMRI	FreeSurfer	CNN	MSE 0.919 (ABCD Dataset) 0.834 (HCP dataset)
[140]	2022	Fluid Intelligence	ABCD Dataset 7693 subjects (age 9–11 years)	rs-fMRI	FreeSurfer	CNN	MAE 5.582 ± 0.012
[142]	2022	Fluid Intelligence	ABCD Dataset Training: 3739 subjects, Validation 415 subjects, Testing 4515 subjects (age 9–11 years)	sMRI	FSL, ANFI, FreeSuerfer	CNN	MSE 82.56 for testing
[146]	2021	Language Scores	31 subjects with persistent language concerns (age 4.25 ± 2.38years)	DTI	In-house pipeline	CNN	MAE 0.28
[147]	2021	Language Scores	37 subjects with epilepsy (age 11.8 ± 3.1years)	DTI	FSL	CNN	MAE 7.77

Table 5. *Cont.*

Study	Year	Score	Population	Technique	Preprocessing	Method	Results
[144]	2020	Motor	77 very pre-term infants (gestation age <31 weeks)	DTI	ANTS	CNN	Accuracy 73%
[139]	2021	Oral Reading	ABCD Study 5252 subjects (age 9–10 years)	sMRI, DTI	-	Auto-encoder	MSE 206.5

Abbreviations: sMRI—structural MRI, rs-fMRI—resting-state functional MRI, DTI—Diffusion Tensor Imaging, CNN—Convolutional neural network, MAE—mean absolute error, MSE—mean squared error.

3.5. Optimizing MRI Brain Imaging and Analysis

Assessing imaging quality and optimizing image acquisition are significant for medical imaging analysis. Reconstruction techniques adjust the scanning parameters to maximize the image quality and control the scanning time, which is of great benefit for pediatric imaging in which many subjects cannot stay still for a long time [148]. Furthermore, some scans may be missing or with low quality due to inadequate scanning time or fail completion by the participants. Image generation algorithms synthesize pseudo-images from low-resolution image or latent space, which provide a solution to recapture missing data or rectify scans with low quality [149]. Here, we review the deep learning algorithms for image quality assessment, reconstruction, and synthesis (Table 6).

Image quality assessment tools were constructed with 2D CNN for structural MRI and DTI [150–152]. Study [153] utilized a two-stage transfer learning strategy which showed near-perfect accuracy in evaluating image quality and is capable of real-time large-scale assessment. GANs are widely applied in image generation tasks [149,154–157]. GANs showed great capability in generating synthetic images to implement missing data or improve the signal-to-noise ratio of poor quality images [24,149]. Study [148] proposed CNN models for reconstruction which reduced the scan time by 42% while maintaining image quality and lesion detectability. CNN combined with RNN also showed superior performance in improving the signal-to-noise ratio [24].

Table 6. Optimizing MRI brain imaging and analysis.

Study	Year	Task	Population	Technique	Preprocessing	Method	Results
[158]	2020	Image Enhancement	131 neuro-oncology patients (age 0.4–17.1 years)	ASL	-	Auto-encoder	SNR Gain 62%
[159]	2018	Image Generation	28 infants (scan at birth, 3 months, and 6 months)	DTI	FSL	CNN	MAE 44.4 ± 17.5 (3-month-old from neonates) 40.1 ± 10.6 (6-month-old from 3-month-old)
[154]	2019	Image Generation	16 subjects (age 1.1–21.3 years)	sMRI	-	GAN	MAE 52.4 ± 17.6
[155]	2020	Image Generation	60 subjects (age 2.6–19 years)	sMRI	In-house pipeline	GAN	MAE 61.0 ± 14.1

Table 6. *Cont.*

Study	Year	Task	Population	Technique	Preprocessing	Method	Results
[156]	2022	Image Generation	ABCD Dataset: 1517 subjects (age 9–10 years)	sMRI	-	GAN	PSNR 31.371 ± 1.813
[149]	2022	Image Generation	127 neonates (postmenstrual age = 41.1 ± 1.5 weeks)	sMRI	ANTs	3D GAN	RMAE 5.6 ± 1.1%
[157]	2022	Image Generation	125 subjects (age 1–20 years)	sMRI	FSL	GAN	PSNR 28.5 ± 2.2
[150]	2019	Image Quality Evaluation	ABIDE Dataset: 1112 subjects (age 7–64 years)	sMRI	SPM12	CNN	Accuracy 84%
[153]	2020	Image Quality Evaluation	BCP dataset: 534 images (age 0–6 years)	sMRI	-	CNN	capable of real-time large-scale assessment with near-perfect accuracy.
[151]	2021	Image Quality Evaluation	211 fetuses (gestation age 30.9 ± 5.5 weeks)	sMRI	In-house pipeline	CNN	Accuracy 85 ± 1%
[152]	2022	Image Quality Evaluation	ABCD Dataset: 2494 subjects (age 9–10 years) HBN Dataset: 4226 subjects (age 5–21 years)	DTI	MATRIX, FSL	CNN	Accuracy 96.61% (ABCD Dataset) 97.52% (HBN Dataset)
[160]	2021	Image Reconstruction	20 fetuses (gestation age 23.4–38 weeks)	DTI	SVR pipeline	CNN	RMSE 0.0379 ± 0.0030
[24]	2021	Image Reconstruction	305 subjects (age 0–15 years)	sMRI	In-house pipeline	CNN+RNN	PSNR 27.85+/−2.12
[161]	2022	Image Reconstruction	107 subjects (age 0.2–18 years)	sMRI	-	CNN	image quality improved significantly by qualitative assessment
[148]	2022	Image Reconstruction	47 subjects (age 2.3–14.7 years)	sMRI	-	CNN	Reduce scan time by 42%

Abbreviations: sMRI—structural MRI, ASL—Arterial spin labeling, DTI—Diffusion Tensor Imaging, CNN—Convolutional neural network, GAN—Generative adversarial network, MAE—mean absolute error, PSNR—Peak signal-to-noise ratio.

4. Discussion

4.1. Advancements in Deep Learning Applied to Pediatric MRI

This study reviews pediatric MRI studies for recognition, segmentation, and prediction tasks in neurodevelopment. Throughout the review, CNN is the most commonly utilized model. Variations and advancement based on the basic architecture have been proposed to improve the performance in multi-tasks. Multi-view 2D CNN and 3D CNN have been proposed to deal with the 3D volumes in neuroimaging [57,82,84]. The multi-view 2D CNN processes 3D volumes with slices generated from sagittal, axial, and coronal sections while 3D CNN utilizes 3D kernels in the networks. Multi-branch CNN models also utilize multimodal imaging to study the brain from different perspectives. Structural connectomes and functional connectomes were combined for age prediction in study [129] and cognitive function prediction in study [139]. Multimodal studies classified children with ASD from healthy controls using combinations of sMRI and rs-fMRI [75,76,81]. sMRI provides structural information, fMRI provides information based on brain activity, and DTI provides information regarding quantitative anisotropy and orientation. Multimodal neuroimaging allows researchers to understand the brain from different perspectives and plays an essential role in investigating the brain functional and structural changes in pediatric neurodevelopment. Variations of U-net dominate in the segmentation tasks. Dilated-Dense U-Net and U-net with attention mechanism achieved great performance in brain structure segmentation [104,120]. Meanwhile, semi-supervised learning and transfer learning initiated studies with a small number of training data [103,122]. GAN shows its superiority in image generation tasks. Variations of GANs have been proposed to synthesize pseudo-images from low-resolution images or latent space [149,155,156]. Overall, the development of computational powers has enabled deep learning models to have more complex structures and greater ability to process 3D volumes for a variety of tasks.

4.2. Challenges and Future Directions

4.2.1. Overfitting Caused by Small Sample Size

Overfitting remains a major concern for deep learning models with deep and complex architectures, especially the models with 3D structures as the number of training parameters grows exponentially with an extra dimension [2]. The sample size should also increase to train models with many parameters to avoid overfitting. Otherwise the model might be overfitted to the training data and fail to predict new data accurately. However, neuroimaging acquisition via MRI is expensive and time-consuming. Many studies are limited to a small number of training data, experiencing the risk of overfitting [162]. In our review, some studies use cross-validation to report results while some others also report results on an independent testing dataset. The testing results are important indicators of the capability to apply the trained model on unseen new data.

Data-sharing projects and platforms provide a vast amount of neuroimaging data, facilitating large-scale studies to train deep and complex models. We share a non-exhaustive list of available public datasets and repositories in Section 2. In common practice, supervised learning, in which the deep learning model is trained with labeled data is the most widely applied learning process [15,163]. Open datasets and repositories prepared data and labels in pairs where labels can be disease diagnosis, clinical outcomes, and semantic segmentation ground truth. Other than labeled data, there are tons of neuroimaging data without labels or with a limited number of labels. Unsupervised learning and semi-supervised learning show great potential in dealing with such data. Unsupervised learning utilizes training data without any labels by separating the data into different categories with automatically learned patterns during training [15,163]. Semi-supervised learning utilizes the unlabeled data to learn the feature patterns and use the labeled data to update model weights, which has yielded superior performance with a limited number of training samples in both classification and segmentation tasks [70,110]. Transfer learning accommodates another possibility for developing deep learning algorithms with a limited number of training data. Transfer learning takes advantage of models pre-trained on large datasets and fine-tunes the system with a small number of data, providing an applicable solution for neuroimaging studies with a small sample size [60,94,97].

4.2.2. Inconsistent Preprocessing Pipelines

Preprocessing is another challenge in pediatric neuroimaging studies. It is necessary to remove the non-brain tissue and noise in many tasks, especially for neuroimaging data of children with significant motion artifacts. However, replication and validation of results are often thus challenged by the variations in data inclusion criteria and preprocessing pipelines. The common preprocessing steps for sMRI include brain extraction, normalization to standard templates, brain tissue segmentation, and brain surface reconstruction [93].

The fMRI preprocessing steps include brain extraction, motion correction, slice time correction, distortion correction, alignment to structural images, and confounds regression [52,90]. The DTI preprocessing steps include distortion correction, Eddy current correction, brain extraction, alignment to structural images, and tensor fitting [60]. The mentioned preprocessing steps may involve multiple preprocessing softwares and adjustments may be applied to different pipelines in different studies. We listed the specified softwares and pipelines in our results. Common preprocessing softwares include SPM [164], AFNI [165], ANTs [166], FSL [167], Dpabi [168], and FreeSurfer [169]. Some studies use in-house preprocessing pipelines or did not specify the preprocessing steps. Preprocessing in single research projects may be time- and effort-consuming while variations of preprocessing pipelines restrict the replication of research results.

Standardization in data preparation and preprocessing is an urgent need for conducting large-scale neuroimaging studies. Fortunately, efforts towards standardization have been contributed by different organizations. Many data-sharing platforms employ the Brain Imaging Data Structure (BIDS) format to adopt a standardized way of organizing neuroimaging and behavioral data [170]. Furthermore, the ABIDE dataset and ADHD200 consortium release both raw and preprocessed data with shared preprocessing pipelines [31,34]. Standardization of preprocessing pipelines will greatly improve the efficacy of neuroimaging studies in the future.

4.2.3. Difficulty in Interpreting Deep Learning Results

Deep learning has been criticized for its "black-box nature" which poses challenges for the interpretability and explainability of trained models, and thus brings concerns to medical decision-making. The deep learning system must provide the rationale behind the decision-making process to make trustworthy predictions [171]. Various approaches have been proposed to interpret deep learning algorithms. One of the common methods is the graph-based visualization approach, which identifies the critical regions for predicting results based on activation maps derived from model weights [172,173]. Study [92] applied such an approach to identify the brain regions where children with ADHD differed from controls. The attention mechanism which focuses selectively on information of interest also plays a vital role in the interpretability of deep learning [174]. Functional connectivity differences between ADHD patients and healthy controls were identified using deep self-attention factorization in the study [90]. There are some other techniques for interpretation such as feature importance and analyzing trends and outliers in predictions. However, studies in this review have not utilized such techniques. Deep model interpretation provides crucial information for understanding brain functions and neurodevelopment, which is of great importance for pediatric neuroimaging studies. Interpretability should be one of the research focuses in future neuroimaging studies.

4.3. Limitations

Although some of the studies did not specify the limitations, there are some common limitations shared across individual studies. Firstly, many studies trained with a limited number of training samples, risking the bias of overfitting. The lack of independent testing results greatly restrains the generalizability of trained models to unseen data. Secondly, architectures of deep neural networks in many studies are trained in a non-exhausted exploration manner that is restricted by computational power. Thirdly, interpretation of the results is lacking in many studies and thus inhibits the interpretability and explainability of trained models. Lastly, for multi-site data which have different scanning protocols, confounding factors might cause risks of bias in the results.

This review systematically organized the most recent research on deep learning applied to pediatric MRI. However, we are unable to include the thousands of results returned by databases GoogleScholar and ScienceDirect, which remains a limitation of the study. Further investigations on unlisted studies may be applied with automatic review tools for paper selection. Keywords selected for the review are not disorder-specific and hence may neglect some studies optimal for the inclusion criteria but not included in the initial research. Future studies on specific disorders may accommodate the limitations.

5. Conclusions

Deep learning plays an essential role in recent neuroimaging studies. Advancements in applications of deep learning to pediatric neuroimaging have been illustrated in this review. Complex deep learning models such as CNN and GAN have shown superior performance in neuroimaging recognition, prediction, segmentation, and generation tasks. Semi-supervised learning demonstrated great potential in the utilization of weakly labeled or unlabeled data. Challenges such as overfitting, preprocessing variations, and interpretation issues remain in many neuroimaging studies, but data-sharing platforms, standardization of preprocessing protocols, and advanced interpretation approaches have been proposed to tackle such difficulties. Future neuroimaging research on large scales will not only achieve high accuracy but also benefit the understanding of the brain functions and neurodevelopment.

Author Contributions: Writing—original draft preparation, M.H.; writing—review and editing, K.-K.A., H.Z. and C.N. All authors have read and agreed to the published version of the manuscript.

Funding: The research is supported by Institute for Infocomm Research (I2R), Agency for Science, Technology and Research (A*STAR), Singapore, and also by the A*STAR Strategic Programme Funds Project No. C211817001 Brain Body Initiative.

Institutional Review Board Statement: Not applicable.

Informed Consent Statement: Not applicable.

Data Availability Statement: Not applicable.

Conflicts of Interest: The authors declare no conflict of interest.

Abbreviations

The following abbreviations are used in this manuscript:

ABCD	The Adolescent Brain Cognitive Development
ABIDE	Autism Brain Imaging Data Exchange
ADHD	Attention deficit hyperactivity disorder
ASD	Autism spectrum disorder
ASL	Arterial spin labeling
CNN	Convolutional neural network
dHCP	Human Connectomme Project Development
DTI	Diffusion tensor imaging
fMRI	functional MRI
GAN	Generative adversarial network
HBN	Human Brain Network
HC	Healthy control
ICBM	International Consortium for Brain Mapping
IMPAC	Imaging Psychiatry Challenge
MAE	mean absolute error
MLP	Multi-layer perceptron
MRI	Magnetic resonance imaging
MSE	mean squared error
NDAR	National Dtabase for Autism Research
PNC	Philadelphia Neurodevelopmental Cohort
PRISMA	preferred reporting items for systematic reviews and meta-analysis
PSNR	Peak signal-to-noise ratio
rs-fMRI	resting-state fMRI
sMRI	structural MRI

Appendix A. Risk of Bias Analysis

Risk of bias analysis were performed following the Risk Of Bias In Non-randomized Studies of Interventions [48] for (1) risk of bias due to confounding (age, gender, scanning parameters); (2) risk of bias in selection of participants into the study (population, sample size); (3) risk of bias in classification of interventions; (4) risk of bias due to deviations from intended interventions (unexpected results); (5) risk of bias due to missing data; (6) risk of bias arising from measurement of outcomes (assessment parameters, validation protocol, independent testing protocols); (7) risk of bias in selection of reported results.

Each risk of bias is rated with "N"—No, "PN"—Probably No, "PY"—Probably Yes, and "Y"—Yes. Most studies are well-designed and have low risks in most criteria while some studies with small sample sizes have the risk of bias due to confounding, selection of participants, and measurement of outcomes. Studies with at least two "PY"s are rated "Moderate" in the summary. Ratings of individual studies are listed in Table A1.

Table A1. Risk of bias analysis.

Study	Confounding	Selection of Participants	Classification of Interventions	Deviations from Intended Interventions	Missing Data	Measurement of Outcomes	Selection of Reported Results	Summary
[79]	PN	PY	N	N	N	PY	N	Moderate
[80]	N	PY	N	N	N	PY	N	Moderate
[51]	PN	N	N	N	N	PY	N	Low
[81]	PN	PY	N	N	N	PY	N	Moderate
[53]	PN	PN	N	N	N	PY	N	Low
[52]	PN	PY	N	N	N	PY	N	Moderate
[55]	PN	N	N	N	N	PY	N	Low
[82]	PN	N	N	N	N	PY	N	Low
[69]	PN	PY	N	N	N	PY	N	Moderate
[74]	N	PY	N	N	N	PY	N	Moderate
[83]	PN	N	N	N	N	PY	N	Low
[84]	PN	N	N	N	N	PY	N	Low
[85]	N	PY	N	N	N	PY	N	Moderate
[76]	PN	N	N	N	N	PY	N	Low
[54]	PN	N	N	N	N	N	N	Low
[73]	PN	N	N	N	N	PY	N	Low
[86]	N	PN	N	N	N	PY	N	Low
[75]	PN	PN	N	N	N	PY	N	Low
[87]	PN	N	N	PY	N	PY	N	Moderate
[88]	PN	PN	N	N	N	PY	N	Low
[78]	PN	N	N	N	N	N	N	Low
[89]	PN	N	N	N	N	PY	N	Low
[90]	PN	N	N	N	N	PY	N	Low
[91]	PN	PY	N	N	N	N	N	Low
[77]	PN	N	N	N	N	PY	N	Low
[92]	N	PY	N	N	N	PY	N	Moderate
[93]	N	PN	N	N	N	PY	N	Low
[57]	N	PY	N	N	N	PY	N	Moderate
[61]	N	PY	N	N	N	PY	N	Moderate

Table A1. *Cont.*

Study	Confounding	Selection of Participants	Classification of Interventions	Deviations from Intended Interventions	Missing Data	Measurement of Outcomes	Selection of Reported Results	Summary
[62]	PN	N	N	N	N	PY	N	Low
[58]	N	PY	N	N	N	PY	N	Moderate
[70]	N	PN	N	N	N	PY	N	Low
[59]	N	PY	N	N	N	PY	N	Moderate
[60]	N	PY	N	N	N	PY	N	Moderate
[94]	N	PY	N	N	N	PY	N	Moderate
[63]	N	PY	N	N	N	PY	N	Moderate
[64]	N	PN	N	N	N	PY	N	Low
[65]	N	PN	N	N	N	PY	N	Low
[71]	N	PN	N	N	N	PY	N	Low
[95]	PY	PY	N	N	N	PY	N	Moderate
[72]	N	N	N	N	N	PY	N	Low
[66]	N	PN	N	N	N	PY	N	Low
[96]	N	PY	N	N	N	PY	N	Moderate
[67]	N	PY	N	N	N	PY	N	Moderate
[68]	N	PY	N	N	N	PY	N	Moderate
[104]	N	PN	N	N	N	PY	N	Low
[105]	N	PY	N	N	N	PY	N	Moderate
[106]	N	PY	N	N	N	PY	N	Moderate
[99]	N	PY	N	N	N	PY	N	Moderate
[107]	N	PN	N	N	N	PY	N	Low
[98]	N	PY	N	N	N	PY	N	Moderate
[108]	N	PY	N	N	N	PY	N	Moderate
[101]	N	PN	N	N	N	PY	N	Low
[109]	N	PY	N	N	N	PY	N	Moderate
[110]	N	PY	N	N	N	PY	N	Moderate
[111]	N	PY	N	N	N	PY	N	Moderate
[112]	N	PY	N	N	N	PN	N	Low
[113]	N	PY	N	N	N	PY	N	Moderate
[114]	N	PY	N	N	N	PY	N	Moderate
[25]	N	PY	N	N	N	PY	N	Moderate
[102]	N	PN	N	N	N	PY	N	Low
[115]	PN	PN	N	N	N	PY	N	Low
[116]	N	PY	N	N	N	PY	N	Moderate
[117]	N	PY	N	N	N	PN	N	Low
[118]	N	PN	N	N	N	PY	N	Low
[103]	PN	PN	N	N	N	PY	N	Low
[119]	N	PY	N	N	N	PN	N	Low
[120]	N	PY	N	N	N	PY	N	Moderate
[121]	N	PY	N	N	N	PY	N	Moderate
[122]	PN	PN	N	N	N	PY	N	Low
[123]	N	PN	N	N	N	PY	N	Low
[124]	N	PY	N	N	N	PN	N	Low

Table A1. Cont.

Study	Confounding	Selection of Participants	Classification of Interventions	Deviations from Intended Interventions	Missing Data	Measurement of Outcomes	Selection of Reported Results	Summary
[125]	N	PN	N	N	N	PY	N	Low
[126]	N	PN	N	N	N	PY	N	Low
[100]	N	PN	N	N	N	PY	N	Low
[97]	N	PN	N	N	N	PY	N	Low
[84]	N	PN	N	N	N	PY	N	Low
[130]	N	N	N	N	N	PY	N	Low
[131]	N	N	N	N	N	PY	N	Low
[132]	PN	N	N	N	N	N	N	Low
[127]	N	PN	N	N	N	PY	N	Low
[129]	N	N	N	N	N	PY	N	Low
[128]	N	PY	N	N	N	PY	N	Moderate
[133]	N	PY	N	N	N	PY	N	Moderate
[134]	N	PN	N	N	N	PY	N	Low
[135]	N	PN	N	N	N	PY	N	Low
[136]	N	PN	N	N	N	PY	N	Low
[137]	N	N	N	N	N	PY	N	Low
[143]	N	PN	N	N	N	PY	N	Low
[145]	PN	N	N	N	N	PY	N	Low
[141]	PN	N	N	N	N	PY	N	Low
[138]	PN	N	N	N	N	PY	N	Low
[140]	PN	N	N	N	N	PY	N	Low
[142]	PN	N	N	N	N	N	N	Low
[146]	N	PY	N	N	N	PY	N	Moderate
[147]	N	PY	N	N	N	PY	N	Moderate
[144]	N	PY	N	N	N	PY	N	Moderate
[139]	PN	N	N	N	N	PY	N	Low
[158]	N	PN	N	N	N	PY	N	Low
[159]	N	PY	N	N	N	PY	N	Moderate
[154]	N	PY	N	N	N	PY	N	Moderate
[155]	N	PY	N	N	N	PY	N	Moderate
[156]	N	N	N	N	N	PY	N	Low
[149]	N	PN	N	N	N	PY	N	Low
[157]	N	PN	N	N	N	PY	N	Low
[150]	PN	N	N	N	N	PY	N	Low
[153]	PN	N	N	N	N	PY	N	Low
[151]	N	PN	N	N	N	PY	N	Low
[152]	PN	N	N	N	N	PY	N	Low
[160]	N	PY	N	N	N	PY	N	Moderate
[24]	N	PN	N	N	N	PY	N	Low
[161]	N	PN	N	N	N	PN	N	Low
[148]	N	PN	N	N	N	PY	N	Low

Abbreviations: N—No, PN—Probably No, PY—Probably Yes.

References

1. Jordan, M.I.; Mitchell, T.M. Machine learning: Trends, perspectives, and prospects. *Science* **2015**, *349*, 255–260. [CrossRef] [PubMed]
2. Celard, P.; Iglesias, E.; Sorribes-Fdez, J.; Romero, R.; Vieira, A.S.; Borrajo, L. A survey on deep learning applied to medical images: From simple artificial neural networks to generative models. *Neural Comput. Appl.* **2022**, *35*, 2291–2323. [CrossRef] [PubMed]
3. LeCun, Y.; Bengio, Y.; Hinton, G. Deep learning. *Nature* **2015**, *521*, 436–444. [CrossRef] [PubMed]
4. Gu, J.; Wang, Z.; Kuen, J.; Ma, L.; Shahroudy, A.; Shuai, B.; Liu, T.; Wang, X.; Wang, G.; Cai, J.; et al. Recent advances in convolutional neural networks. *Pattern Recognit.* **2018**, *77*, 354–377. [CrossRef]
5. Hosny, A.; Parmar, C.; Quackenbush, J.; Schwartz, L.H.; Aerts, H.J.W.L. Artificial intelligence in radiology. *Nat. Reviews. Cancer* **2018**, *18*, 500–510. [CrossRef]
6. Reig, B.; Heacock, L.; Geras, K.J.; Moy, L. Machine learning in breast MRI. *J. Magn. Reson. Imaging JMRI* **2020**, *52*, 998–1018. [CrossRef]
7. Eslami, T.; Almuqhim, F.; Raiker, J.S.; Saeed, F. Machine learning methods for diagnosing autism spectrum disorder and attention-deficit/hyperactivity disorder using functional and structural MRI: A survey. *Front. Neuroinformatics* **2021**, *14*, 575999. [CrossRef]
8. Zhang, Z.; Li, G.; Xu, Y.; Tang, X. Application of artificial intelligence in the MRI classification task of human brain neurological and psychiatric diseases: A scoping review. *Diagnostics* **2021**, *11*, 1402. [CrossRef]
9. Yousaf, T.; Dervenoulas, G.; Politis, M. Advances in MRI methodology. *Int. Rev. Neurobiol.* **2018**, *141*, 31–76. [CrossRef]
10. Pykett, I.L.; Newhouse, J.H.; Buonanno, F.S.; Brady, T.J.; Goldman, M.R.; Kistler, J.P.; Pohost, G.M. Principles of nuclear magnetic resonance imaging. *Radiology* **1982**, *143*, 157–168. [CrossRef]
11. Van Geuns, R.J.M.; Wielopolski, P.A.; de Bruin, H.G.; Rensing, B.J.; van Ooijen, P.M.; Hulshoff, M.; Oudkerk, M.; de Feyter, P.J. Basic principles of magnetic resonance imaging. *Prog. Cardiovasc. Dis.* **1999**, *42*, 149–156. [CrossRef]
12. Huettel, S.A.; Song, A.W.; McCarthy, G. *Functional Magnetic Resonance Imaging*; Sinauer Associates Sunderland: Sunderland, MA, USA, 2004; Volume 1.
13. Mori, S.; Zhang, J. Principles of diffusion tensor imaging and its applications to basic neuroscience research. *Neuron* **2006**, *51*, 527–539. [CrossRef] [PubMed]
14. Colombo, E.; Fick, T.; Esposito, G.; Germans, M.; Regli, L.; van Doormaal, T. Segmentation techniques of brain arteriovenous malformations for 3D visualization: A systematic review. *Radiol. Medica* **2022**, *127*, 1333–1341. [CrossRef] [PubMed]
15. Castiglioni, I.; Rundo, L.; Codari, M.; Di Leo, G.; Salvatore, C.; Interlenghi, M.; Gallivanone, F.; Cozzi, A.; D'Amico, N.C.; Sardanelli, F. AI applications to medical images: From machine learning to deep learning. *Phys. Medica* **2021**, *83*, 9–24. [CrossRef] [PubMed]
16. Khodatars, M.; Shoeibi, A.; Sadeghi, D.; Ghaasemi, N.; Jafari, M.; Moridian, P.; Khadem, A.; Alizadehsani, R.; Zare, A.; Kong, Y.; et al. Deep learning for neuroimaging-based diagnosis and rehabilitation of autism spectrum disorder: A review. *Comput. Biol. Med.* **2021**, *139*, 104949. [CrossRef] [PubMed]
17. Bahathiq, R.A.; Banjar, H.; Bamaga, A.K.; Jarraya, S.K. Machine learning for autism spectrum disorder diagnosis using structural magnetic resonance imaging: Promising but challenging. *Front. Neuroinform.* **2022**, *16*, 949926. [CrossRef] [PubMed]
18. Wang, S.; Di, J.; Wang, D.; Dai, X.; Hua, Y.; Gao, X.; Zheng, A.; Gao, J. State-of-the-Art Review of Artificial Neural Networks to Predict, Characterize and Optimize Pharmaceutical Formulation. *Pharmaceutics* **2022**, *14*, 183. [CrossRef] [PubMed]
19. Krizhevsky, A.; Sutskever, I.; Hinton, G.E. Imagenet classification with deep convolutional neural networks. *Commun. ACM* **2017**, *60*, 84–90. [CrossRef]
20. Szegedy, C.; Liu, W.; Jia, Y.; Sermanet, P.; Reed, S.; Anguelov, D.; Erhan, D.; Vanhoucke, V.; Rabinovich, A. Going deeper with convolutions. In Proceedings of the IEEE Conference on Computer Vision and Pattern Recognition, Boston, MA, USA, 7–12 June 2015; pp. 1–9. [CrossRef]
21. He, K.; Zhang, X.; Ren, S.; Sun, J. Deep residual learning for image recognition. In Proceedings of the IEEE Conference on Computer Vision and Pattern Recognition, Las Vegas, NV, USA, 27–30 June 2016; pp. 770–778. [CrossRef]
22. Huang, G.; Liu, Z.; Van Der Maaten, L.; Weinberger, K.Q. Densely connected convolutional networks. In Proceedings of the IEEE Conference on Computer Vision and Pattern Recognition, Honolulu, HI, USA, 21–26 July 2017; pp. 4700–4708. [CrossRef]
23. Ronneberger, O.; Fischer, P.; Brox, T. U-net: Convolutional networks for biomedical image segmentation. In Proceedings of the International Conference on Medical image Computing and Computer-Assisted Intervention, Munich, Germany, 5–9 October 2015; Springer: Berlin/Heidelberg, Germany, 2015; pp. 234–241. [CrossRef]
24. Li, Z.; Yu, J.; Wang, Y.; Zhou, H.; Yang, H.; Qiao, Z. Deepvolume: Brain structure and spatial connection-aware network for brain mri super-resolution. *IEEE Trans. Cybern.* **2019**, *51*, 3441–3454. [CrossRef]
25. Li, H.; Chen, M.; Wang, J.; Illapani, V.S.P.; Parikh, N.A.; He, L. Automatic Segmentation of Diffuse White Matter Abnormality on T2-weighted Brain MR Images Using Deep Learning in Very Preterm Infants. *Radiol. Artif. Intell.* **2021**, *3*, e200166. [CrossRef]
26. Yuan, J.; Ran, X.; Liu, K.; Yao, C.; Yao, Y.; Wu, H.; Liu, Q. Machine learning applications on neuroimaging for diagnosis and prognosis of epilepsy: A review. *J. Neurosci. Methods* **2021**, *368*, 109441. [CrossRef] [PubMed]
27. Elbattah, M.; Loughnane, C.; Guérin, J.L.; Carette, R.; Cilia, F.; Dequen, G. Variational Autoencoder for Image-Based Augmentation of Eye-Tracking Data. *J. Imaging* **2021**, *7*, 83. [CrossRef] [PubMed]

28. Goodfellow, I.; Pouget-Abadie, J.; Mirza, M.; Xu, B.; Warde-Farley, D.; Ozair, S.; Courville, A.; Bengio, Y. Generative Adversarial Nets. In *Proceedings of the Advances in Neural Information Processing Systems*; Ghahramani, Z., Welling, M., Cortes, C., Lawrence, N., Weinberger, K.Q., Eds.; Curran Associates, Inc.: New York, NY, USA, 2014; Volume 27.

29. Pan, Z.; Yu, W.; Yi, X.; Khan, A.; Yuan, F.; Zheng, Y. Recent progress on generative adversarial networks (GANs): A survey. *IEEE Access* **2019**, *7*, 36322–36333. [CrossRef]

30. Yi, X.; Walia, E.; Babyn, P. Generative adversarial network in medical imaging: A review. *Med. Image Anal.* **2019**, *58*, 101552. [CrossRef]

31. Di Martino, A.; Yan, C.G.; Li, Q.; Denio, E.; Castellanos, F.X.; Alaerts, K.; Anderson, J.S.; Assaf, M.; Bookheimer, S.Y.; Dapretto, M.; et al. The autism brain imaging data exchange: Towards a large-scale evaluation of the intrinsic brain architecture in autism. *Mol. Psychiatry* **2014**, *19*, 659–667. [CrossRef]

32. Di Martino, A.; O'Connor, D.; Chen, B.; Alaerts, K.; Anderson, J.S.; Assaf, M.; Balsters, J.H.; Baxter, L.; Beggiato, A.; Bernaerts, S.; et al. Enhancing studies of the connectome in autism using the autism brain imaging data exchange II. *Sci. Data* **2017**, *4*, 170010. [CrossRef]

33. IMPAC—Imaging-Psychiatry Challenge: Predicting Autism. Available online: https://paris-saclay-cds.github.io/autism_challenge/ (accessed on 15 December 2022).

34. Consortium, T.A. The ADHD-200 consortium: A model to advance the translational potential of neuroimaging in clinical neuroscience. *Front. Syst. Neurosci.* **2012**, *6*, 62. [CrossRef]

35. Sudlow, C.; Gallacher, J.; Allen, N.; Beral, V.; Burton, P.; Danesh, J.; Downey, P.; Elliott, P.; Green, J.; Landray, M.; et al. UK biobank: An open access resource for identifying the causes of a wide range of complex diseases of middle and old age. *PLoS Med.* **2015**, *12*, e1001779. [CrossRef]

36. Payakachat, N.; Tilford, J.M.; Ungar, W.J. National Database for Autism Research (NDAR): Big Data Opportunities for Health Services Research and Health Technology Assessment. *PharmacoEconomics* **2016**, *34*, 127–138. [CrossRef]

37. Poldrack, R.A.; Barch, D.M.; Mitchell, J.P.; Wager, T.D.; Wagner, A.D.; Devlin, J.T.; Cumba, C.; Koyejo, O.; Milham, M.P. Toward open sharing of task-based fMRI data: The OpenfMRI project. *Front. Neuroinform.* **2013**, *7*, 12. [CrossRef]

38. Mazziotta, J.; Toga, A.; Evans, A.; Fox, P.; Lancaster, J.; Zilles, K.; Woods, R.; Paus, T.; Simpson, G.; Pike, B.; et al. A probabilistic atlas and reference system for the human brain: International Consortium for Brain Mapping (ICBM). *Philos. Trans. R. Soc. London. Ser. B Biol. Sci.* **2001**, *356*, 1293–1322. [CrossRef] [PubMed]

39. Yan, C.G.; Craddock, R.C.; Zuo, X.N.; Zang, Y.F.; Milham, M.P. Standardizing the intrinsic brain: Towards robust measurement of inter-individual variation in 1000 functional connectomes. *Neuroimage* **2013**, *80*, 246–262. [CrossRef] [PubMed]

40. Casey, B.J.; Cannonier, T.; Conley, M.I.; Cohen, A.O.; Barch, D.M.; Heitzeg, M.M.; Soules, M.E.; Teslovich, T.; Dellarco, D.V.; Garavan, H. The adolescent brain cognitive development (ABCD) study: Imaging acquisition across 21 sites. *Dev. Cogn. Neurosci.* **2018**, *32*, 43–54. [CrossRef]

41. Thompson, P.M.; Stein, J.L.; Medland, S.E.; Hibar, D.P.; Vasquez, A.A.; Renteria, M.E.; Toro, R.; Jahanshad, N.; Schumann, G.; Franke, B. The ENIGMA Consortium: Large-scale collaborative analyses of neuroimaging and genetic data. *Brain Imaging Behav.* **2014**, *8*, 153–182. [CrossRef]

42. Satterthwaite, T.D.; Elliott, M.A.; Ruparel, K.; Loughead, J.; Prabhakaran, K.; Calkins, M.E.; Hopson, R.; Jackson, C.; Keefe, J.; Riley, M. Neuroimaging of the Philadelphia neurodevelopmental cohort. *Neuroimage* **2014**, *86*, 544–553. [CrossRef]

43. Alexander, L.M.; Escalera, J.; Ai, L.; Andreotti, C.; Febre, K.; Mangone, A.; Vega-Potler, N.; Langer, N.; Alexander, A.; Kovacs, M. An open resource for transdiagnostic research in pediatric mental health and learning disorders. *Sci. Data* **2017**, *4*, 1–26. [CrossRef]

44. Van Essen, D.C.; Ugurbil, K.; Auerbach, E.; Barch, D.; Behrens, T.E.J.; Bucholz, R.; Chang, A.; Chen, L.; Corbetta, M.; Curtiss, S.W. The Human Connectome Project: A data acquisition perspective. *Neuroimage* **2012**, *62*, 2222–2231. [CrossRef]

45. Howell, B.R.; Styner, M.A.; Gao, W.; Yap, P.T.; Wang, L.; Baluyot, K.; Yacoub, E.; Chen, G.; Potts, T.; Salzwedel, A.; et al. The UNC/UMN Baby Connectome Project (BCP): An overview of the study design and protocol development. *NeuroImage* **2019**, *185*, 891–905. [CrossRef]

46. Moher, D.; Liberati, A.; Tetzlaff, J.; Altman, D.G.; Group, P. Preferred reporting items for systematic reviews and meta-analyses: the PRISMA statement. *Ann. Intern. Med.* **2009**, *151*, 264–269. [CrossRef]

47. Tricco, A.C.; Lillie, E.; Zarin, W.; O'Brien, K.K.; Colquhoun, H.; Levac, D.; Moher, D.; Peters, M.D.J.; Horsley, T.; Weeks, L. PRISMA extension for scoping reviews (PRISMA-ScR): Checklist and explanation. *Ann. Intern. Med.* **2018**, *169*, 467–473. [CrossRef] [PubMed]

48. Sterne, J.A.; Hernán, M.A.; Reeves, B.C.; Savović, J.; Berkman, N.D.; Viswanathan, M.; Henry, D.; Altman, D.G.; Ansari, M.T.; Boutron, I.; et al. ROBINS-I: A tool for assessing risk of bias in non-randomised studies of interventions. *BMJ* **2016**, *355*. [CrossRef]

49. Edition, F. Diagnostic and statistical manual of mental disorders. *Am. Psychiatr. Assoc.* **2013**, *21*, 591–643.

50. Morris-Rosendahl, D.J.; Crocq, M.A. Neurodevelopmental disorders-the history and future of a diagnostic concept. *Dialogues Clin. Neurosci.* **2020**, *22*, 65–72. [CrossRef] [PubMed]

51. Dvornek, N.C.; Ventola, P.; Duncan, J.S. Combining phenotypic and resting-state fMRI data for autism classification with recurrent neural networks. In Proceedings of the 2018 IEEE 15th International Symposium on Biomedical Imaging (ISBI 2018), Washington, DC, USA, 4–7 April 2018; IEEE: Piscataway, NJ, USA, 2018; pp. 725–728.

52. Xiao, Z.; Wu, J.; Wang, C.; Jia, N.; Yang, X. Computer-aided diagnosis of school-aged children with ASD using full frequency bands and enhanced SAE: A multi-institution study. *Exp. Ther. Med.* **2019**, *17*, 4055–4063. [CrossRef] [PubMed]

53. Aghdam, M.A.; Sharifi, A.; Pedram, M.M. Diagnosis of autism spectrum disorders in young children based on resting-state functional magnetic resonance imaging data using convolutional neural networks. *J. Digit. Imaging* **2019**, *32*, 899–918. [CrossRef]

54. Li, H.; Parikh, N.A.; He, L. A novel transfer learning approach to enhance deep neural network classification of brain functional connectomes. *Front. Neurosci.* **2018**, *12*, 491. [CrossRef]

55. Leming, M.; Górriz, J.M.; Suckling, J. Ensemble deep learning on large, mixed-site fMRI datasets in autism and other tasks. *Int. J. Neural Syst.* **2020**, *30*, 2050012. [CrossRef] [PubMed]

56. Sibley, M.H.; Swanson, J.M.; Arnold, L.E.; Hechtman, L.T.; Owens, E.B.; Stehli, A.; Abikoff, H.; Hinshaw, S.P.; Molina, B.S.; Mitchell, J.T.; et al. Defining ADHD symptom persistence in adulthood: Optimizing sensitivity and specificity. *J. Child Psychol. Psychiatry* **2017**, *58*, 655–662. [CrossRef]

57. Ceschin, R.; Zahner, A.; Reynolds, W.; Gaesser, J.; Zuccoli, G.; Lo, C.W.; Gopalakrishnan, V.; Panigrahy, A. A computational framework for the detection of subcortical brain dysmaturation in neonatal MRI using 3D Convolutional Neural Networks. *NeuroImage* **2018**, *178*, 183–197. [CrossRef]

58. Zahia, S.; Garcia-Zapirain, B.; Saralegui, I.; Fernandez-Ruanova, B. Dyslexia detection using 3D convolutional neural networks and functional magnetic resonance imaging. *Comput. Methods Programs Biomed.* **2020**, *197*, 105726. [CrossRef]

59. Aminpour, A.; Ebrahimi, M.; Widjaja, E. Deep learning-based lesion segmentation in paediatric epilepsy. In Proceedings of the Medical Imaging 2021: Computer-Aided Diagnosis, Online, 15–19 February 2021; SPIE: Washington, DC, USA, 2021; Volume 11597, pp. 635–641. [CrossRef]

60. Huang, J.; Xu, J.; Kang, L.; Zhang, T. Identifying epilepsy based on deep learning using DKI images. *Front. Hum. Neurosci.* **2020**, *14*, 590815. [CrossRef] [PubMed]

61. Zhang, J.; Li, X.; Li, Y.; Wang, M.; Huang, B.; Yao, S.; Shen, L. Three dimensional convolutional neural network-based classification of conduct disorder with structural MRI. *Brain Imaging Behav.* **2020**, *14*, 2333–2340. [CrossRef] [PubMed]

62. Menon, S.S.; Krishnamurthy, K. Multimodal Ensemble Deep Learning to Predict Disruptive Behavior Disorders in Children. *Front. Neuroinform.* **2021**, *15*, 742807. [CrossRef] [PubMed]

63. Yang, J.; Lei, D.; Qin, K.; Pinaya, W.H.; Suo, X.; Li, W.; Li, L.; Kemp, G.J.; Gong, Q. Using deep learning to classify pediatric posttraumatic stress disorder at the individual level. *BMC Psychiatry* **2021**, *21*, 535. [CrossRef] [PubMed]

64. Jiang, D.; Hu, Z.; Zhao, C.; Zhao, X.; Yang, J.; Zhu, Y.; Liao, J.; Liang, D.; Wang, H. Identification of Children's Tuberous Sclerosis Complex with Multiple-contrast MRI and 3D Convolutional Network. In Proceedings of the 2022 44th Annual International Conference of the IEEE Engineering in Medicine & Biology Society (EMBC), Scotland, UK, 11–15 July 2022; IEEE: Piscataway, NJ, USA, 2022; pp. 2924–2927.

65. Shabanian, M.; Imran, A.A.Z.; Siddiqui, A.; Davis, R.L.; Bissler, J.J. 3D deep neural network to automatically identify TSC structural brain pathology based on MRI. In Proceedings of the Medical Imaging 2022: Image Processing, San Diego, CA, USA, 20–24 February 2022; SPIE: Washington, DC, USA, 2022; Volume 12032, pp. 613–619.

66. Afshar, P.; Mohammadi, A.; Plataniotis, K.N. Brain tumor type classification via capsule networks. In Proceedings of the 2018 25th IEEE international conference on image processing (ICIP), Athens, Greece, 7–10 October 2018; IEEE: Piscataway, NJ, USA, 2018; pp. 3129–3133. [CrossRef]

67. Lee, M.H.; O'Hara, N.; Sonoda, M.; Kuroda, N.; Juhasz, C.; Asano, E.; Dong, M.; Jeong, J.W. Novel deep learning network analysis of electrical stimulation mapping-driven diffusion MRI tractography to improve preoperative evaluation of pediatric epilepsy. *IEEE Trans. Biomed. Eng.* **2020**, *67*, 3151–3162. [CrossRef] [PubMed]

68. Xu, H.; Dong, M.; Lee, M.H.; O'Hara, N.; Asano, E.; Jeong, J.W. Objective detection of eloquent axonal pathways to minimize postoperative deficits in pediatric epilepsy surgery using diffusion tractography and convolutional neural networks. *IEEE Trans. Med. Imaging* **2019**, *38*, 1910–1922. [CrossRef]

69. Yang, M.; Cao, M.; Chen, Y.; Chen, Y.; Fan, G.; Li, C.; Wang, J.; Liu, T. Large-scale brain functional network integration for discrimination of autism using a 3-D deep learning model. *Front. Hum. Neurosci.* **2021**, *15*, 277. [CrossRef]

70. Wu, M.; Shen, X.; Lai, C.; Zheng, W.; Li, Y.; Shangguan, Z.; Yan, C.; Liu, T.; Wu, D. Detecting neonatal acute bilirubin encephalopathy based on T1-weighted MRI images and learning-based approaches. *BMC Med. Imaging* **2021**, *21*, 103. [CrossRef]

71. Sánchez Fernández, I.; Yang, E.; Calvachi, P.; Amengual-Gual, M.; Wu, J.Y.; Krueger, D.; Northrup, H.; Bebin, M.E.; Sahin, M.; Yu, K.H.; et al. Deep learning in rare disease. Detection of tubers in tuberous sclerosis complex. *PLoS ONE* **2020**, *15*, e0232376. [CrossRef]

72. Quon, J.; Bala, W.; Chen, L.; Wright, J.; Kim, L.; Han, M.; Shpanskaya, K.; Lee, E.; Tong, E.; Iv, M.; et al. Deep learning for pediatric posterior fossa tumor detection and classification: A multi-institutional study. *Am. J. Neuroradiol.* **2020**, *41*, 1718–1725. [CrossRef]

73. Li, S.; Tang, Z.; Jin, N.; Yang, Q.; Liu, G.; Liu, T.; Hu, J.; Liu, S.; Wang, P.; Hao, J.; et al. Uncovering Brain Differences in Preschoolers and Young Adolescents with Autism Spectrum Disorder Using Deep Learning. *Int. J. Neural Syst.* **2022**, *32*, 2250044. [CrossRef] [PubMed]

74. Haweel, R.; Shalaby, A.; Mahmoud, A.; Seada, N.; Ghoniemy, S.; Ghazal, M.; Casanova, M.F.; Barnes, G.N.; El-Baz, A. A robust DWT–CNN-based CAD system for early diagnosis of autism using task-based fMRI. *Med. Phys.* **2021**, *48*, 2315–2326. [CrossRef] [PubMed]

75. Mellema, C.J.; Nguyen, K.P.; Treacher, A.; Montillo, A. Reproducible neuroimaging features for diagnosis of autism spectrum disorder with machine learning. *Sci. Rep.* **2022**, *12*, 3057. [CrossRef] [PubMed]

76. Leming, M.J.; Baron-Cohen, S.; Suckling, J. Single-participant structural similarity matrices lead to greater accuracy in classification of participants than function in autism in MRI. *Mol. Autism* **2021**, *12*, 34. [CrossRef] [PubMed]

77. Chen, M.; Li, H.; Fan, H.; Dillman, J.R.; Wang, H.; Altaye, M.; Zhang, B.; Parikh, N.A.; He, L. ConCeptCNN: A novel multi-filter convolutional neural network for the prediction of neurodevelopmental disorders using brain connectome. *Med. Phys.* **2022**, *49*, 3171–3184. [CrossRef]

78. Zhang-James, Y.; Helminen, E.C.; Liu, J.; Franke, B.; Hoogman, M.; Faraone, S.V. Evidence for similar structural brain anomalies in youth and adult attention-deficit/hyperactivity disorder: A machine learning analysis. *Transl. Psychiatry* **2021**, *11*, 82. [CrossRef]

79. Guo, X.; Dominick, K.C.; Minai, A.A.; Li, H.; Erickson, C.A.; Lu, L.J. Diagnosing autism spectrum disorder from brain resting-state functional connectivity patterns using a deep neural network with a novel feature selection method. *Front. Neurosci.* **2017**, *11*, 460. [CrossRef]

80. Li, X.; Dvornek, N.C.; Zhuang, J.; Ventola, P.; Duncan, J.S. Brain biomarker interpretation in ASD using deep learning and fMRI. In Proceedings of the International Conference on Medical Image Computing and Computer-Assisted Intervention, Granada, Spain, 16–20 September 2018; Springer: Berlin/Heidelberg, Germany, 2018; pp. 206–214.

81. Akhavan Aghdam, M.; Sharifi, A.; Pedram, M.M. Combination of rs-fMRI and sMRI data to discriminate autism spectrum disorders in young children using deep belief network. *J. Digit. Imaging* **2018**, *31*, 895–903. [CrossRef]

82. Ke, F.; Choi, S.; Kang, Y.H.; Cheon, K.A.; Lee, S.W. Exploring the structural and strategic bases of autism spectrum disorders with deep learning. *IEEE Access* **2020**, *8*, 153341–153352. [CrossRef]

83. Husna, R.N.S.; Syafeeza, A.; Hamid, N.A.; Wong, Y.; Raihan, R.A. Functional magnetic resonance imaging for autism spectrum disorder detection using deep learning. *J. Teknol.* **2021**, *83*, 45–52. [CrossRef]

84. Kawahara, J.; Brown, C.J.; Miller, S.P.; Booth, B.G.; Chau, V.; Grunau, R.E.; Zwicker, J.G.; Hamarneh, G. BrainNetCNN: Convolutional neural networks for brain networks; towards predicting neurodevelopment. *NeuroImage* **2017**, *146*, 1038–1049. [CrossRef] [PubMed]

85. Gao, K.; Sun, Y.; Niu, S.; Wang, L. Unified framework for early stage status prediction of autism based on infant structural magnetic resonance imaging. *Autism Res.* **2021**, *14*, 2512–2523. [CrossRef] [PubMed]

86. Guo, X.; Wang, J.; Wang, X.; Liu, W.; Yu, H.; Xu, L.; Li, H.; Wu, J.; Dong, M.; Tan, W.; et al. Diagnosing autism spectrum disorder in children using conventional MRI and apparent diffusion coefficient based deep learning algorithms. *Eur. Radiol.* **2022**, *32*, 761–770. [CrossRef]

87. Wang, T.; Kamata, S.I. Classification of structural MRI images in Adhd using 3D fractal dimension complexity map. In Proceedings of the 2019 IEEE International Conference on Image Processing (ICIP), Taipei, Taiwan, 22–25 September 2018; IEEE: Piscataway, NJ, USA, 2019; pp. 215–219.

88. Riaz, A.; Asad, M.; Alonso, E.; Slabaugh, G. DeepFMRI: End-to-end deep learning for functional connectivity and classification of ADHD using fMRI. *J. Neurosci. Methods* **2020**, *335*, 108506. [CrossRef]

89. Tang, Y.; Sun, J.; Wang, C.; Zhong, Y.; Jiang, A.; Liu, G.; Liu, X. ADHD classification using auto-encoding neural network and binary hypothesis testing. *Artif. Intell. Med.* **2022**, *123*, 102209. [CrossRef]

90. Ke, H.; Wang, F.; Ma, H.; He, Z. ADHD identification and its interpretation of functional connectivity using deep self-attention factorization. *Knowl.-Based Syst.* **2022**, *250*, 109082. [CrossRef]

91. Wang, D.; Hong, D.; Wu, Q. Attention Deficit Hyperactivity Disorder Classification Based on Deep Learning. *IEEE/Acm Trans. Comput. Biol. Bioinform.* **2022**, 1. [CrossRef]

92. Uyulan, C.; Erguzel, T.T.; Turk, O.; Farhad, S.; Metin, B.; Tarhan, N. A Class Activation Map-Based Interpretable Transfer Learning Model for Automated Detection of ADHD from fMRI Data. *Clin. EEG Neurosci.* **2022**, *54*, 15500594221122699. [CrossRef]

93. Stanley, E.A.M.; Rajashekar, D.; Mouches, P.; Wilms, M.; Plettl, K.; Forkert, N.D. A fully convolutional neural network for explainable classification of attention deficit hyperactivity disorder. In Proceedings of the Medical Imaging 2022: Computer-Aided Diagnosis, Leicester, UK, 20-21 November 2022; SPIE: Washington, DC, USA, 2022; Volume 12033, pp. 296–301.

94. Attallah, O.; Sharkas, M.A.; Gadelkarim, H. Deep learning techniques for automatic detection of embryonic neurodevelopmental disorders. *Diagnostics* **2020**, *10*, 27. [CrossRef]

95. Artzi, M.; Redmard, E.; Tzemach, O.; Zeltser, J.; Gropper, O.; Roth, J.; Shofty, B.; Kozyrev, D.A.; Constantini, S.; Ben-Sira, L. Classification of pediatric posterior fossa tumors using convolutional neural network and tabular data. *IEEE Access* **2021**, *9*, 91966–91973. [CrossRef]

96. Prince, E.W.; Whelan, R.; Mirsky, D.M.; Stence, N.; Staulcup, S.; Klimo, P.; Anderson, R.C.; Niazi, T.N.; Grant, G.; Souweidane, M.; et al. Robust deep learning classification of adamantinomatous craniopharyngioma from limited preoperative radiographic images. *Sci. Rep.* **2020**, *10*, 1–13. [CrossRef] [PubMed]

97. Nalepa, J.; Adamski, S.; Kotowski, K.; Chelstowska, S.; Machnikowska-Sokolowska, M.; Bozek, O.; Wisz, A.; Jurkiewicz, E. Segmenting pediatric optic pathway gliomas from MRI using deep learning. *Comput. Biol. Med.* **2022**, *142*, 105237. [CrossRef] [PubMed]

98. Asis-Cruz, D.; Krishnamurthy, D.; Jose, C.; Cook, K.M.; Limperopoulos, C. FetalGAN: Automated Segmentation of Fetal Functional Brain MRI Using Deep Generative Adversarial Learning and Multi-Scale 3D U-Net. *Front. Neurosci.* **2022**, *16*, 852. [CrossRef]

99. Sourati, J.; Gholipour, A.; Dy, J.G.; Tomas-Fernandez, X.; Kurugol, S.; Warfield, S.K. Intelligent labeling based on fisher information for medical image segmentation using deep learning. *IEEE Trans. Med. Imaging* **2019**, *38*, 2642–2653. [CrossRef] [PubMed]

100. Peng, J.; Kim, D.D.; Patel, J.B.; Zeng, X.; Huang, J.; Chang, K.; Xun, X.; Zhang, C.; Sollee, J.; Wu, J.; et al. Deep learning-based automatic tumor burden assessment of pediatric high-grade gliomas, medulloblastomas, and other leptomeningeal seeding tumors. *Neuro-oncology* **2022**, *24*, 289–299. [CrossRef] [PubMed]

101. Avisdris, N.; Yehuda, B.; Ben-Zvi, O.; Link-Sourani, D.; Ben-Sira, L.; Miller, E.; Zharkov, E.; Ben Bashat, D.; Joskowicz, L. Automatic linear measurements of the fetal brain on MRI with deep neural networks. *Int. J. Comput. Assist. Radiol. Surg.* **2021**, *16*, 1481–1492. [CrossRef]

102. Zhao, L.; Asis-Cruz, J.; Feng, X.; Wu, Y.; Kapse, K.; Largent, A.; Quistorff, J.; Lopez, C.; Wu, D.; Qing, K.; et al. Automated 3D Fetal Brain Segmentation Using an Optimized Deep Learning Approach. *Am. J. Neuroradiol.* **2022**, *43*, 448–454. [CrossRef]

103. Grigorescu, I.; Vanes, L.; Uus, A.; Batalle, D.; Cordero-Grande, L.; Nosarti, C.; Edwards, A.D.; Hajnal, J.V.; Modat, M.; Deprez, M. Harmonized segmentation of neonatal brain MRI. *Front. Neurosci.* **2021**, *15*, 662005. [CrossRef]

104. Li, G.; Chen, M.H.; Li, G.; Wu, D.; Lian, C.; Sun, Q.; Rushmore, R.J.; Wang, L. Volumetric Analysis of Amygdala and Hippocampal Subfields for Infants with Autism. *J. Autism Dev. Disord.* **2022**, 1–15. [CrossRef]

105. Tor-Diez, C.; Porras, A.R.; Packer, R.J.; Avery, R.A.; Linguraru, M.G. Unsupervised MRI homogenization: Application to pediatric anterior visual pathway segmentation. In Proceedings of the International Workshop on Machine Learning in Medical Imaging 2020, Lima, Peru, 4 October 2020; pp. 180–188. [CrossRef]

106. Sourati, J.; Gholipour, A.; Dy, J.G.; Kurugol, S.; Warfield, S.K. Active deep learning with fisher information for patch-wise semantic segmentation. In Proceedings of the Deep Learning in Medical Image Analysis and Multimodal Learning for Clinical Decision Support: 4th International Workshop, DLMIA 2018, and 8th International Workshop, ML-CDS 2018, Held in Conjunction with MICCAI 2018, Granada, Spain, 20 September 2018; Springer: Berlin/Heidelberg, Germany, 2018; pp. 83–91. [CrossRef]

107. Rutherford, S.; Sturmfels, P.; Angstadt, M.; Hect, J.; Wiens, J.; van den Heuvel, M.I.; Scheinost, D.; Sripada, C.; Thomason, M. Automated brain masking of fetal functional MRI with open data. *Neuroinformatics* **2022**, *20*, 173–185. [CrossRef]

108. Ebner, M.; Wang, G.; Li, W.; Aertsen, M.; Patel, P.A.; Aughwane, R.; Melbourne, A.; Doel, T.; Dymarkowski, S.; De Coppi, P.; et al. An automated framework for localization, segmentation and super-resolution reconstruction of fetal brain MRI. *NeuroImage* **2020**, *206*, 116324. [CrossRef]

109. Bermudez, C.; Blaber, J.; Remedios, S.W.; Reynolds, J.E.; Lebel, C.; McHugo, M.; Heckers, S.; Huo, Y.; Landman, B.A. Generalizing deep whole brain segmentation for pediatric and post-contrast MRI with augmented transfer learning. In Proceedings of the Medical Imaging 2020: Image Processing, Houston, TX, USA, 17–20 February 2020; SPIE: Washington, DC, USA, 2020; Volume 11313, pp. 111–118.

110. Enguehard, J.; O'Halloran, P.; Gholipour, A. Semi-supervised learning with deep embedded clustering for image classification and segmentation. *IEEE Access* **2019**, *7*, 11093–11104. [CrossRef] [PubMed]

111. Khalili, N.; Lessmann, N.; Turk, E.; Claessens, N.; de Heus, R.; Kolk, T.; Viergever, M.A.; Benders, M.J.; Išgum, I. Automatic brain tissue segmentation in fetal MRI using convolutional neural networks. *Magn. Reson. Imaging* **2019**, *64*, 77–89. [CrossRef] [PubMed]

112. Li, H.; Parikh, N.A.; Wang, J.; Merhar, S.; Chen, M.; Parikh, M.; Holland, S.; He, L. Objective and automated detection of diffuse white matter abnormality in preterm infants using deep convolutional neural networks. *Front. Neurosci.* **2019**, *13*, 610. [CrossRef]

113. Grimm, F.; Edl, F.; Kerscher, S.R.; Nieselt, K.; Gugel, I.; Schuhmann, M.U. Semantic segmentation of cerebrospinal fluid and brain volume with a convolutional neural network in pediatric hydrocephalus—transfer learning from existing algorithms. *Acta Neurochir.* **2020**, *162*, 2463–2474. [CrossRef] [PubMed]

114. Yang, R.; Zuo, H.; Han, S.; Zhang, X.; Zhang, Q. Computer-Aided Diagnosis of Children with Cerebral Palsy under Deep Learning Convolutional Neural Network Image Segmentation Model Combined with Three-Dimensional Cranial Magnetic Resonance Imaging. *J. Healthc. Eng.* **2021**, *2021*, 1822776. [CrossRef]

115. Uus, A.U.; Ayub, M.U.; Gartner, A.; Kyriakopoulou, V.; Pietsch, M.; Grigorescu, I.; Christiaens, D.; Hutter, J.; Grande, L.C.; Price, A. Segmentation of Periventricular White Matter in Neonatal Brain MRI: Analysis of Brain Maturation in Term and Preterm Cohorts. In Proceedings of the International Workshop on Preterm, Perinatal and Paediatric Image Analysis, Messina, Italy, 13–15 July 2022; Springer: Berlin/Heidelberg, Germany, 2022; pp. 94–104.

116. Luan, X.; Li, W.; Liu, L.; Shu, Y.; Guo, Y. Rubik-Net: Learning Spatial Information via Rotation-Driven Convolutions for Brain Segmentation. *IEEE J. Biomed. Health Inform.* **2021**, *26*, 289–300. [CrossRef]

117. Quon, J.L.; Chen, L.C.; Kim, L.; Grant, G.A.; Edwards, M.S.; Cheshier, S.H.; Yeom, K.W. Deep learning for automated delineation of pediatric cerebral arteries on pre-operative brain magnetic resonance imaging. *Front. Surg.* **2020**, *7*, 89. [CrossRef]

118. Quon, J.L.; Han, M.; Kim, L.H.; Koran, M.E.; Chen, L.C.; Lee, E.H.; Wright, J.; Ramaswamy, V.; Lober, R.M.; Taylor, M.D.; et al. Artificial intelligence for automatic cerebral ventricle segmentation and volume calculation: A clinical tool for the evaluation of pediatric hydrocephalus. *J. Neurosurg. Pediatr.* **2020**, *27*, 131–138. [CrossRef]

119. Hong, J.; Yun, H.J.; Park, G.; Kim, S.; Laurentys, C.T.; Siqueira, L.C.; Tarui, T.; Rollins, C.K.; Ortinau, C.M.; Grant, P.E.; et al. Fetal cortical plate segmentation using fully convolutional networks with multiple plane aggregation. *Front. Neurosci.* **2020**, *14*, 591683. [CrossRef]

120. Dou, H.; Karimi, D.; Rollins, C.K.; Ortinau, C.M.; Vasung, L.; Velasco-Annis, C.; Ouaalam, A.; Yang, X.; Ni, D.; Gholipour, A. A deep attentive convolutional neural network for automatic cortical plate segmentation in fetal MRI. *IEEE Trans. Med. Imaging* **2020**, *40*, 1123–1133. [CrossRef] [PubMed]

121. Khalili, N.; Turk, E.; Benders, M.; Moeskops, P.; Claessens, N.; de Heus, R.; Franx, A.; Wagenaar, N.; Breur, J.; Viergever, M.; et al. Automatic extraction of the intracranial volume in fetal and neonatal MR scans using convolutional neural networks. *Neuroimage Clin.* **2019**, *24*, 102061. [CrossRef] [PubMed]

122. Wang, Y.; Haghpanah, F.S.; Zhang, X.; Santamaria, K.; da Costa Aguiar Alves, G.K.; Bruno, E.; Aw, N.; Maddocks, A.; Duarte, C.S.; Monk, C.; et al. ID-Seg: An infant deep learning-based segmentation framework to improve limbic structure estimates. *Brain Inform.* **2022**, *9*, 12. [CrossRef]

123. Gruber, N.; Galijasevic, M.; Regodic, M.; Grams, A.E.; Siedentopf, C.; Steiger, R.; Hammerl, M.; Haltmeier, M.; Gizewski, E.R.; Janjic, T. A deep learning pipeline for the automated segmentation of posterior limb of internal capsule in preterm neonates. *Artif. Intell. Med.* **2022**, *132*, 102384. [CrossRef] [PubMed]

124. Park, D.K.; Kim, W.; Thornburg, O.S.; McBrian, D.K.; McKhann, G.M.; Feldstein, N.A.; Maddocks, A.B.; Gonzalez, E.; Shen, M.Y.; Akman, C.; et al. Convolutional neural network-aided tuber segmentation in tuberous sclerosis complex patients correlates with electroencephalogram. *Epilepsia* **2022**, *63*, 1530–1541. [CrossRef] [PubMed]

125. Vafaeikia, P.; Wagner, M.W.; Hawkins, C.; Tabori, U.; Ertl-Wagner, B.B.; Khalvati, F. Improving the segmentation of pediatric low-grade gliomas through multitask learning. In Proceedings of the 2022 44th Annual International Conference of the IEEE Engineering in Medicine & Biology Society (EMBC), Scotland, UK, 11–15 July 2022; IEEE: Piscataway, NJ, USA, 2022; pp. 2119–2122.

126. Madhogarhia, R.; Kazerooni, A.F.; Arif, S.; Ware, J.B.; Familiar, A.M.; Vidal, L.; Bagheri, S.; Anderson, H.; Haldar, D.; Yagoda, S. Automated segmentation of pediatric brain tumors based on multi-parametric MRI and deep learning. In Proceedings of the Medical Imaging 2022: Computer-Aided Diagnosis, Leicester, UK, 10 August 2022; SPIE: Washington, DC, USA, 2022; Volume 12033, pp. 723–731.

127. Hong, J.; Feng, Z.; Wang, S.H.; Peet, A.; Zhang, Y.D.; Sun, Y.; Yang, M. Brain age prediction of children using routine brain MR images via deep learning. *Front. Neurol.* **2020**, *11*, 584682. [CrossRef]

128. Kawaguchi, M.; Kidokoro, H.; Ito, R.; Shiraki, A.; Suzuki, T.; Maki, Y.; Tanaka, M.; Sakaguchi, Y.; Yamamoto, H.; Takahashi, Y.; et al. Age estimates from brain magnetic resonance images of children younger than two years of age using deep learning. *Magn. Reson. Imaging* **2021**, *79*, 38–44. [CrossRef]

129. Niu, X.; Zhang, F.; Kounios, J.; Liang, H. Improved prediction of brain age using multimodal neuroimaging data. *Hum. Brain Mapp.* **2020**, *41*, 1626–1643. [CrossRef]

130. Shabanian, M.; Eckstein, E.C.; Chen, H.; DeVincenzo, J.P. Classification of neurodevelopmental age in normal infants using 3D-CNN based on brain MRI. In Proceedings of the 2019 IEEE International Conference on Bioinformatics and Biomedicine (BIBM), San Diego, CA, USA, 18–21 November 2019; IEEE: Piscataway, NJ, USA, 2019; pp. 2373–2378.

131. Hu, W.; Cai, B.; Zhang, A.; Calhoun, V.D.; Wang, Y.P. Deep collaborative learning with application to the study of multimodal brain development. *IEEE Trans. Biomed. Eng.* **2019**, *66*, 3346–3359. [CrossRef]

132. Qu, T.; Yue, Y.; Zhang, Q.; Wang, C.; Zhang, Z.; Lu, G.; Du, W.; Li, X. Baenet: A brain age estimation network with 3d skipping and outlier constraint loss. In Proceedings of the 2020 IEEE 17th International Symposium on Biomedical Imaging (ISBI), Iowa City, IA, USA, 3–7 April 2020; IEEE: Piscataway, NJ, USA, 2020; pp. 399–403.

133. Shabanian, M.; Wenzel, M.; DeVincenzo, J.P. Infant brain age classification: 2D CNN outperforms 3D CNN in small dataset. In Proceedings of the Medical Imaging 2022: Image Processing, San Diego, CA, USA, 20–24 February 2022; SPIE: Washington, DC, USA, 2022; Volume 12032, pp. 626–633.

134. Wada, A.; Saito, Y.; Fujita, S.; Irie, R.; Akashi, T.; Sano, K.; Kato, S.; Ikenouchi, Y.; Hagiwara, A.; Sato, K.; et al. Automation of a Rule-based Workflow to Estimate Age from Brain MR Imaging of Infants and Children Up to 2 Years Old Using Stacked Deep Learning. *Magn. Reson. Med. Sci.* **2023**, *22*, 57–66. [CrossRef]

135. Hong, J.; Yun, H.J.; Park, G.; Kim, S.; Ou, Y.; Vasung, L.; Rollins, C.K.; Ortinau, C.M.; Takeoka, E.; Akiyama, S.; et al. Optimal method for fetal brain age prediction using multiplanar slices from structural magnetic resonance imaging. *Front. Neurosci.* **2021**, *15*, 1284. [CrossRef]

136. Zhang, Q.; He, Y.; Qu, T.; Yang, F.; Lin, Y.; Hu, Z.; Li, X.; Xu, Q.; Xing, W.; Gumenyuk, V.; et al. Delayed brain development of Rolandic epilepsy profiled by deep learning–based neuroanatomic imaging. *Eur. Radiol.* **2021**, *31*, 9628–9637. [CrossRef]

137. Taoudi-Benchekroun, Y.; Christiaens, D.; Grigorescu, I.; Gale-Grant, O.; Schuh, A.; Pietsch, M.; Chew, A.; Harper, N.; Falconer, S.; Poppe, T.; et al. Predicting age and clinical risk from the neonatal connectome. *NeuroImage* **2022**, *257*, 119319. [CrossRef]

138. Wu, Y.; Besson, P.; Azcona, E.A.; Bandt, S.K.; Parrish, T.B.; Breiter, H.C.; Katsaggelos, A.K. A multicohort geometric deep learning study of age dependent cortical and subcortical morphologic interactions for fluid intelligence prediction. *Sci. Rep.* **2022**, *12*, 17760. [CrossRef]

139. Liu, M.; Zhang, Z.; Dunson, D.B. Graph auto-encoding brain networks with applications to analyzing large-scale brain imaging datasets. *Neuroimage* **2021**, *245*, 118750. [CrossRef]

140. Huang, S.G.; Xia, J.; Xu, L.; Qiu, A. Spatio-temporal directed acyclic graph learning with attention mechanisms on brain functional time series and connectivity. *Med. Image Anal.* **2022**, *77*, 102370. [CrossRef]

141. Saha, S.; Pagnozzi, A.; Bradford, D.; Fripp, J. Predicting fluid intelligence in adolescence from structural MRI with deep learning methods. *Intelligence* **2021**, *88*, 101568. [CrossRef]
142. Li, M.; Jiang, M.; Zhang, G.; Liu, Y.; Zhou, X. Prediction of fluid intelligence from T1-w MRI images: A precise two-step deep learning framework. *PLoS ONE* **2022**, *17*, e0268707. [CrossRef] [PubMed]
143. He, L.; Li, H.; Chen, M.; Wang, J.; Altaye, M.; Dillman, J.R.; Parikh, N.A. Deep multimodal learning from MRI and clinical data for early prediction of neurodevelopmental deficits in very preterm infants. *Front. Neurosci.* **2021**, *15*, 753033. [CrossRef] [PubMed]
144. Saha, S.; Pagnozzi, A.; Bourgeat, P.; George, J.M.; Bradford, D.; Colditz, P.B.; Boyd, R.N.; Rose, S.E.; Fripp, J.; Pannek, K. Predicting motor outcome in preterm infants from very early brain diffusion MRI using a deep learning convolutional neural network (CNN) model. *Neuroimage* **2020**, *215*, 116807. [CrossRef] [PubMed]
145. Han, S.; Zhang, Y.; Ren, Y.; Posner, J.; Yoo, S.; Cha, J. 3D distributed deep learning framework for prediction of human intelligence from brain MRI. In Proceedings of the Medical Imaging 2020: Biomedical Applications in Molecular, Structural, and Functional Imaging, Houston, TX, USA, 18–20 February 2020; SPIE: Washington, DC, USA, 2020; Volume 11317, pp. 484–490.
146. Jeong, J.W.; Banerjee, S.; Lee, M.H.; O'Hara, N.; Behen, M.; Juhasz, C.; Dong, M. Deep reasoning neural network analysis to predict language deficits from psychometry-driven DWI connectome of young children with persistent language concerns. *Hum. Brain Mapp.* **2021**, *42*, 3326–3338. [CrossRef] [PubMed]
147. Jeong, J.W.; Lee, M.H.; O'Hara, N.; Juhász, C.; Asano, E. Prediction of baseline expressive and receptive language function in children with focal epilepsy using diffusion tractography-based deep learning network. *Epilepsy Behav.* **2021**, *117*, 107909. [CrossRef] [PubMed]
148. Kim, E.; Cho, H.H.; Cho, S.; Park, B.; Hong, J.; Shin, K.; Hwang, M.; You, S.; Lee, S. Accelerated Synthetic MRI with Deep Learning–Based Reconstruction for Pediatric Neuroimaging. *Am. J. Neuroradiol.* **2022**, *43*, 1653–1659. [CrossRef] [PubMed]
149. Kaplan, S.; Perrone, A.; Alexopoulos, D.; Kenley, J.K.; Barch, D.M.; Buss, C.; Elison, J.T.; Graham, A.M.; Neil, J.J.; O'Connor, T.G.; et al. Synthesizing pseudo-T2w images to recapture missing data in neonatal neuroimaging with applications in rs-fMRI. *NeuroImage* **2022**, *253*, 119091. [CrossRef]
150. Sujit, S.J.; Coronado, I.; Kamali, A.; Narayana, P.A.; Gabr, R.E. Automated image quality evaluation of structural brain MRI using an ensemble of deep learning networks. *J. Magn. Reson. Imaging* **2019**, *50*, 1260–1267. [CrossRef]
151. Largent, A.; Kapse, K.; Barnett, S.D.; De Asis-Cruz, J.; Whitehead, M.; Murnick, J.; Zhao, L.; Andersen, N.; Quistorff, J.; Lopez, C.; et al. Image quality assessment of fetal brain MRI using multi-instance deep learning methods. *J. Magn. Reson. Imaging* **2021**, *54*, 818–829. [CrossRef]
152. Ettehadi, N.; Kashyap, P.; Zhang, X.; Wang, Y.; Semanek, D.; Desai, K.; Guo, J.; Posner, J.; Laine, A.F. Automated Multiclass Artifact Detection in Diffusion MRI Volumes via 3D Residual Squeeze-and-Excitation Convolutional Neural Networks. *Front. Hum. Neurosci.* **2022**, *16*, 877326. [CrossRef]
153. Liu, S.; Thung, K.H.; Lin, W.; Yap, P.T.; Shen, D. Real-time quality assessment of pediatric MRI via semi-supervised deep nonlocal residual neural networks. *IEEE Trans. Image Process.* **2020**, *29*, 7697–7706. [CrossRef]
154. Wang, C.; Uh, J.; He, X.; Hua, C.h.; Acharya, S. Transfer learning-based synthetic CT generation for MR-only proton therapy planning in children with pelvic sarcomas. In Proceedings of the Medical Imaging 2021: Physics of Medical Imaging, Online, 5–19 February 2021; SPIE: Washington, DC, USA, 2021; Volume 11595, pp. 1112–1118.
155. Maspero, M.; Bentvelzen, L.G.; Savenije, M.H.; Guerreiro, F.; Seravalli, E.; Janssens, G.O.; van den Berg, C.A.; Philippens, M.E. Deep learning-based synthetic CT generation for paediatric brain MR-only photon and proton radiotherapy. *Radiother. Oncol.* **2020**, *153*, 197–204. [CrossRef]
156. Zhang, H.; Li, H.; Dillman, J.R.; Parikh, N.A.; He, L. Multi-Contrast MRI Image Synthesis Using Switchable Cycle-Consistent Generative Adversarial Networks. *Diagnostics* **2022**, *12*, 816. [CrossRef]
157. Wang, C.; Uh, J.; Merchant, T.E.; Hua, C.h.; Acharya, S. Facilitating MR-Guided Adaptive Proton Therapy in Children Using Deep Learning-Based Synthetic CT. *Int. J. Part. Ther.* **2022**, *8*, 11–20. [CrossRef] [PubMed]
158. Hales, P.W.; Pfeuffer, J.; A Clark, C. Combined denoising and suppression of transient artifacts in arterial spin labeling MRI using deep learning. *J. Magn. Reson. Imaging* **2020**, *52*, 1413–1426. [CrossRef]
159. Kim, J.; Hong, Y.; Chen, G.; Lin, W.; Yap, P.T.; Shen, D. Graph-based deep learning for prediction of longitudinal infant diffusion MRI data. In Proceedings of the Computational Diffusion MRI: International MICCAI Workshop, Granada, Spain, 22 September 2018; Springer: Berlin/Heidelberg, Germany, 2019; pp. 133–141. [CrossRef]
160. Karimi, D.; Jaimes, C.; Machado-Rivas, F.; Vasung, L.; Khan, S.; Warfield, S.K.; Gholipour, A. Deep learning-based parameter estimation in fetal diffusion-weighted MRI. *Neuroimage* **2021**, *243*, 118482. [CrossRef] [PubMed]
161. Kim, S.H.; Choi, Y.H.; Lee, J.S.; Lee, S.B.; Cho, Y.J.; Lee, S.H.; Shin, S.M.; Cheon, J.E. Deep learning reconstruction in pediatric brain MRI: Comparison of image quality with conventional T2-weighted MRI. *Neuroradiology* **2022**, *65*, 1–8. [CrossRef] [PubMed]
162. Winterburn, J.L.; Voineskos, A.N.; Devenyi, G.A.; Plitman, E.; de la Fuente-Sandoval, C.; Bhagwat, N.; Graff-Guerrero, A.; Knight, J.; Chakravarty, M.M. Can we accurately classify schizophrenia patients from healthy controls using magnetic resonance imaging and machine learning? A multi-method and multi-dataset study. *Schizophr. Res.* **2019**, *214*, 3–10. [CrossRef]
163. Bishop, C.M.; Nasrabadi, N.M. *Pattern Recognition and Machine Learning*; Springer: Berlin/Heidelberg, Germany, 2006; Volume 4.
164. Tzourio-Mazoyer, N.; Landeau, B.; Papathanassiou, D.; Crivello, F.; Etard, O.; Delcroix, N.; Mazoyer, B.; Joliot, M. Automated anatomical labeling of activations in SPM using a macroscopic anatomical parcellation of the MNI MRI single-subject brain. *Neuroimage* **2002**, *15*, 273–289. [CrossRef]

165. Cox, R.W. AFNI: Software for analysis and visualization of functional magnetic resonance neuroimages. *Comput. Biomed. Res.* **1996**, *29*, 162–173. [CrossRef]

166. Avants, B.B.; Tustison, N.; Song, G. Advanced normalization tools (ANTS). *Insight J.* **2009**, *2*, 1–35.

167. Smith, S.M.; Jenkinson, M.; Woolrich, M.W.; Beckmann, C.F.; Behrens, T.E.J.; Johansen-Berg, H.; Bannister, P.R.; De Luca, M.; Drobnjak, I.; Flitney, D.E. Advances in functional and structural MR image analysis and implementation as FSL. *Neuroimage* **2004**, *23*, S208–S219. [CrossRef]

168. Yan, C.G.; Wang, X.D.; Zuo, X.N.; Zang, Y.F. DPABI: Data processing & analysis for (resting-state) brain imaging. *Neuroinformatics* **2016**, *14*, 339–351. [CrossRef]

169. Fischl, B. FreeSurfer. *Neuroimage* **2012**, *62*, 774–781. [CrossRef] [PubMed]

170. Gorgolewski, K.J.; Auer, T.; Calhoun, V.D.; Craddock, R.C.; Das, S.; Duff, E.P.; Flandin, G.; Ghosh, S.S.; Glatard, T.; Halchenko, Y.O.; et al. The brain imaging data structure, a format for organizing and describing outputs of neuroimaging experiments. *Sci. Data* **2016**, *3*, 160044. [CrossRef] [PubMed]

171. Yang, G.; Ye, Q.; Xia, J. Unbox the black-box for the medical explainable AI via multi-modal and multi-centre data fusion: A mini-review, two showcases and beyond. *Inf. Fusion* **2022**, *77*, 29–52. [CrossRef] [PubMed]

172. Zhou, B.; Khosla, A.; Lapedriza, A.; Oliva, A.; Torralba, A. Learning deep features for discriminative localization. In Proceedings of the IEEE Conference on Computer Vision and Pattern Recognition, Las Vegas, NV, USA, 27–30 June 2016; pp. 2921–2929. [CrossRef]

173. Selvaraju, R.R.; Cogswell, M.; Das, A.; Vedantam, R.; Parikh, D.; Batra, D. Grad-cam: Visual explanations from deep networks via gradient-based localization. In Proceedings of the IEEE International Conference on Computer Vision, Venice, Italy, 22–29 October 2017; pp. 618–626. [CrossRef]

174. Guidotti, R.; Monreale, A.; Ruggieri, S.; Turini, F.; Giannotti, F.; Pedreschi, D. A survey of methods for explaining black box models. *ACM Comput. Surv. (CSUR)* **2018**, *51*, 1–42. [CrossRef]

applied sciences

Article

Hybrid System Mixed Reality and Marker-Less Motion Tracking for Sports Rehabilitation of Martial Arts Athletes

Michela Franzò [1], Andrada Pica [1], Simona Pascucci [1,2], Franco Marinozzi [1,*] and Fabiano Bini [1]

[1] Department of Mechanical and Aerospace Engineering, "Sapienza" University of Rome, 00184 Rome, Italy
[2] National Centre for Clinical Excellence, Healthcare Quality and Safety, Italian National Institute of Health, 00161 Rome, Italy
* Correspondence: franco.marinozzi@uniroma1.it

Abstract: Rehabilitation is a vast field of research. Virtual and Augmented Reality represent rapidly emerging technologies that have the potential to support physicians in several medical activities, e.g., diagnosis, surgical training, and rehabilitation, and can also help sports experts analyze athlete movements and performance. In this study, we present the implementation of a hybrid system for the real-time visualization of 3D virtual models of bone segments and other anatomical components on a subject performing critical karate shots and stances. The project is composed of an economic markerless motion tracking device, Microsoft Kinect Azure, that recognizes the subject movements and the position of anatomical joints; an augmented reality headset, Microsoft HoloLens 2, on which the user can visualize the 3D reconstruction of bones and anatomical information; and a terminal computer with a code implemented in Unity Platform. The 3D reconstructed bones are overlapped with the athlete, tracked by the Kinect in real-time, and correctly displayed on the headset. The findings suggest that this system could be a promising technology to monitor martial arts athletes after injuries to support the restoration of their movements and position to rejoin official competitions.

Keywords: mixed reality; sport biomechanics; rehabilitation engineering; martial arts; posture

Citation: Franzò, M.; Pica, A.; Pascucci, S.; Marinozzi, F.; Bini, F. Hybrid System Mixed Reality and Marker-Less Motion Tracking for Sports Rehabilitation of Martial Arts Athletes. *Appl. Sci.* **2023**, *13*, 2587. https://doi.org/10.3390/app13042587

Academic Editor: Cosimo Nardi

Received: 31 January 2023
Revised: 14 February 2023
Accepted: 15 February 2023
Published: 17 February 2023

1. Introduction

Innovative technologies contribute to the growth of the rehabilitation sector by providing effective and safe solutions [1–4]. The technology involved in rehabilitation protocols is varied. Examples of applied technologies are assistive devices and robotics [5], optoelectronic systems [6], inertial measurement units [7], and virtual and augmented reality environments [8–11]. Virtual reality (VR) and augmented reality (AR) have aroused particular interest in the creation of customized rehabilitation protocols in the study of body mechanics. Biomechanical outcomes, such as joint movement analysis, are effective not only in diagnosing but also in understanding the mechanism of symptom progression. Most experts' clinical assessments of these situations are based on observing a given movement performed by the subject or the manual measurement of angles on clinical images [12]. Several studies have proposed different innovative systems based on markerless motion tracking devices [13–19] and virtual or augmented reality systems [20,21] to support diagnosis and biomechanical measurements. Furthermore, a significant aspect of rehabilitation in which these new cutting-edge technologies are applied is postural analysis, a fundamental factor regarding personal well-being. Various pathologies or bad habits can influence posture and lead to the deformation of the bone structure. Consequently, malfunctioning of the musculoskeletal, respiratory, and nervous systems might also occur. The study of posture is based on measurements of anatomical angles and alignment of bone and joint components [22,23]. In many studies, practicing sports, such as martial arts, is recommended for correcting postural problems. In fact, in martial arts, posture and balance are crucial to the correct performance of exercises [24,25]. For example, in karate, repetition of simple positions, such as the Juntzuki (lunge punch) or the Zenkutsu Dachi stance (forward leaning stance), is fundamental to learning the basics of karate. Positions are also subjected to evaluation in competition kata (i.e., a set of shots and stances combined to perform an imaginary fight), kumite (i.e., fights with other athletes) [26], or to pass the exam to proceed to the next belt level. To be able to perform the positions correctly, control of coordination and total body balance, perception of the surrounding space, and knowledge of the anatomical angles to be achieved are necessary [22,25–27]. In these disciplines, experienced instructors can guide their disciples on how to perform the movement and maintain the correct posture. Still, studies have highlighted the contribution that technological systems could make to support instructors and disciples who want to improve and correct their mistakes [28,29].

The present study proposes a project implemented to visualize in real-time bones and other anatomical components of a subject performing critical karate shots and stances. The system comprises an economic motion tracking device that recognizes the subject movements and the position of anatomical joints and an AR headset, with which the physicians can observe bones and anatomical information fidelity overlapped to the subject in real-time. This hybrid system has the potential to contribute to the monitoring of martial arts athletes after injuries to support the restoration of their movements and position to rejoin the official competitions by taking advantage of new innovative technologies.

2. Related Works

In the medical field, VR and mixed reality (MR) are mainly applied in surgical planning and training. Several reviews [1–4] highlight the current application of VR and MR in surgical training for orthopedic procedures. According to these studies, numerous randomized clinical trials (RCTs) demonstrate the proficiency of innovative virtual techniques in teaching orthopedic surgical skills. In this framework, pilot studies [11,30] and clinical trials [31] evaluate whether VR or MR improve learning effectiveness for surgical trainees compared to traditional preparatory methods in orthopedic surgery. Innovative surgical simulators are presented in [8–10,32], proposing new approaches in surgical navigation, training preparations, and patient-specific modeling.

The efficiency of the HoloLens as a suitable device for such applications is highlighted in the previously mentioned studies [4,8–11,20,32]. For instance, ref. [33] specifically analyzes the use of the HoloLens 2 (HL2) in orthopedic surgery and compares it with the previous version HoloLens 1. Moreover, several studies [33,34] evaluate and quantify errors committed by the device in positioning and overlapping the virtual object with the real object reproduction. The results of [33] show that the newest model improved the AR projection accuracy by almost 25 percent, while both HoloLens versions yielded a root mean square error (RMSE) below 3 mm. In addition, El-Hariri and colleagues [34] evaluate possible new orthopedic surgical guidelines.

In addition, the authors in [6,20,21] propose the applications of AR, VR, or MR in pose or posture evaluation and correction. In these studies, tracking algorithms and systems, such as OpenPose and Vuforia, are used to recognize the position of the subject and to identify the posture accordingly. The aim is to provide support in the sport and physiotherapy fields and the diagnosis of orthopedic disease.

Hämäläinen [35] was the first to introduce martial arts in an AR game where the player has to fight virtual enemies. In [29], Wu et al. composed an AR martial arts system using deep learning based on real-time human pose forecasting. An external RGB camera was used to capture the motion of the trainer. The student wore a VR-Head Mounted Display (HMD) and could see the results directly on the screen. Moreover, Shen et al. [36] focused their work on the construction and visualization of the posture-based graph that focuses on the standard postures for launching and ending actions. They propose two numerical indices, the Connectivity Index and the Action Strategy Index, to measure skill level and the strengths and weaknesses of the boxers.

In a physiotherapeutic application, Debarba and co-workers [6] developed an AR tool to accurately overlap anatomical structures on the subject in motion on the HoloLens device using the external tracking system VICON. In [6], they present the first real-time bone mapping system of a moving subject. The VICON represents the gold standard of tracking and gait analysis systems but realistically is not usable in the everyday medical field. A markerless and dynamic system, such as the Microsoft Kinect, although less accurate, can be considered a viable technology to introduce to clinics and hospitals [37–41]. Certain precautions are necessary, such as designing a suitable joint model to correct device error.

The last version of the Microsoft Kinect devices, the Azure Kinect, has been validated by several studies [42–44]. In [42], the Azure Kinect showed a significantly higher accuracy of the spatial gait parameters than the previous version. Results provided by [43] confirm the officially stated values of standard deviation and distance error, i.e., std dev ≤ 17 mm and distance error > 11 mm $+ 0.1\%$ of distance without multi-path interference. However, this study suggests a warmup of the device of 40 min before acquisition to obtain stable results. In [44], Antico and colleagues calculated an RMSE value of 0.47 between the marker-less tracking systems and the VICON, considering the average results among all joints. In contrast, the range of value of the angular mean absolute error is 5–15 degrees for all the upper joints [44].

Azure Kinect and its previous versions were applied in various circumstances in the medical field. For instance, [15,16] applied Kinect in evaluating patients with hip disorders. The inclination angles of the trunk and the pelvis were similar to the outcomes from the VICON system. Ref. [17] presents a tool for deducting forearm and wrist range of motion. In this study, results are obtained by a reliability test performed by a healthy group. In [18], evaluating the Global Gait Asymmetry index (GGA index) after knee joint surgery is accomplished using a set of Kinects. Moreover, in [19], the device was used to monitor the dynamic valgus of the knee. The Kinect measurements were compared with OptiTrack, and the absolute average difference for the pelvis was 1.3 ± 0.7 cm and for the knee in lateral-medial movement 0.7 ± 0.3 cm. Moreover, the Azure Kinect is also useful in a telemedicine system to teleport the knowledge and skills of doctors [45].

Many studies evaluate AR-based applications highlighting challenges that still need to be addressed. Ref. [46] reassumes and sifts through all the technical challenges of AR and MR: tracking, rendering, processing speed, and ergonomics. The new Microsoft headset for MR still needs hardware improvements to overcome these issues and to allow the real-time use of MR in daily life applications. Indeed, ref. [47] shows all limitations of AR in sports and training fields. The most impacting challenge is the tracking accuracy which depends on the speed of motion, distance, noise, and hardware performances, followed by the Field of View limited by the headset. The Kinect can be considered the pioneer among marker-less tracking systems [47]. However, skeleton tracking and motion reconstructions must still be monitored and filtered. Ref. [48] reports factors that can influence the results of Kinect performance, such as the absence of silhouette visual changes and the changeable hands and foot joint estimations. Newer Kinect versions have reduced some issues; however, other improvements or algorithm corrections may still be needed.

3. Materials and Methods

3.1. Materials

The device selected to implement the MR application of this study correctly is the HMD HL2, the second version of the Microsoft device.

This device is a stand-alone holographic computer composed of a pair of see-through transparent lenses (also called waveguides) with a holographic resolution of 2 k 3:2 light engine and holographic density major of 2.5 k light points per radiant. The waveguides are flat optical fibers in which the light can be projected by the specific projectors in each lens. The light bounces between the interior surfaces of the display to be directly sent to the user's pupils to display the holograms directly in front of the user's eyes. HL2 is also equipped with an IR camera for eye tracking, RGB cameras, a Depth camera, and an

IMU sensor that includes an accelerometer, gyroscope, and magnetometer. The device has a resolution of 2048 × 1080 for each eye and an FOV of 52 degrees (information on the Microsoft official site).

A high-performance computer with appropriate technical characteristics that allowed programming in MR was used to implement the application and to manage the distribution process on the device.

Azure Kinect is the new version of the camera system developed by Microsoft. The Azure Kinect has an RGB (red, green, and blue) camera, a depth camera, IR emitters, and IMU sensors LSM6DSMUS (gyroscope and accelerometer) are simultaneously sampled at 1.6 kHz. The samples are reported to the host at 208 Hz. Therefore, due to the presence of IMU sensors, it can measure and track the entire body in real-time and estimate 3 coordinates of every major joint of the human body in 3 planes without requiring any marker or other supplemental equipment (information available from the official site).

To implement the 3D models of anatomical districts, DICOM computed tomography (CT) scans were used. The scans were achieved from different free and open-access databases, in particular:

- Tibia and Fibula scans are from a subject in the National Cancer Institute's Clinical Proteomic Tumor Analysis Consortium Sarcomas (CPTAC-SAR) cohort;
- Femur scans are from the Cancer Imaging Archive;
- Humerus scans are from the image datasets of the Laboratory of Human Anatomy and Embryology, University of Brussels (ULB), Belgium.

Humerus scans were acquired at 120 kVp, exposure of 200 mAs, and an X-ray current of 200 mA. Ulna and radius scans were acquired using 130 kVp, time of exposure of 1000 ms, current of X-ray of 70 mA, and a generator power of 10 kW. Femur scans were acquired at 80 kVp, an X-ray current of 20 mA, and a generator power of 1600 kW.

Table 1 reports other key information about the CT scans grouped by the anatomical areas.

Table 1. Technical features of CT scans of the bones segmented.

Anatomical Part	Size	Slice Thickness (mm)	Pixel Spacing (mm)	No of Slices
Humerus (Distal, left)	512 × 512	1.1	0.352	102
Humerus (Proximal, left)	512 × 512	1.1	0.352	120
Humerus (Diaphysis, left)	512 × 512	1.1	0.352	199
Radius (left)	512 × 512	3	0.473	475
Ulna (left)	512 × 512	3	0.473	475
Femur (left)	1101 × 888	600.545	0.545	221
Tibia (left)	559 × 1889	0.977	0.416	975
Fibula (left)	559 × 1889	0.977	0.416	975

3D Slicer version 4.11.20210226, a free and open-source software for clinical and biomedical research applications, was used to develop the corresponding 3D models of the bones through segmenting the CT Scans. Subsequently, the 3D models are used in the MR application.

Blender is a free and open-source software for 3D manipulation, and it was used to convert the model into a readable format with Unity 3D.

Unity 3D version 2020.3.30f1 is a game engine that was combined with the Mixed Reality Toolkit package (vs. 2.7.0) (MRTK) to develop the MR application. The MRTK contains a set of basic features that added to a Unity project can implement MR behavior.

Visual Studio 2019 is a free integrated development environment (IDE) that allows C# scripts to be written.

3.2. Method

3.2.1. 3D Models Reconstruction

To build the 3D models of the bones, the corresponding CT scans were imported into the 3D Slicer software. The segmentation was performed using the manual segmentation algorithms of the threshold and smoothing effect. All the objects were exported as obj files to have a Unity-readable 3D object.

After the segmentation, Blender software was used to modify the system of references of each 3D model and to scale it. These actions are required to obtain models congruent with real anatomical dimensions and to allow correct object manipulation in the Unity application.

The final 3D models (Figure 1) are exported in an obj format file.

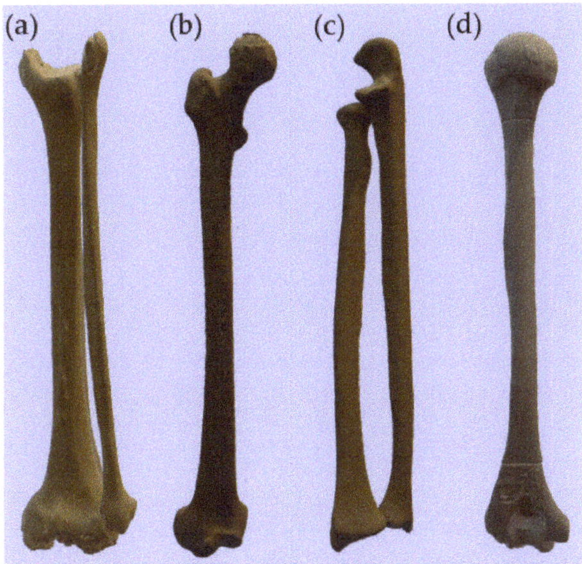

Figure 1. 3D models of bones. From the left: (**a**) Tibia and Fibula, (**b**) Femur, (**c**) Ulna and Radius, (**d**) Humerus.

3.2.2. Mixed Reality Behaviour and Unity Editor Settings

Since the 3D models were imported in Unity, it was necessary to deselect the conversion of measurement units to maintain the correct real proportions of 3D objects.

To implement skeleton tracking with Azure Kinect and to allow for the possibility of using the HL2, including the MR behavior in the project by adding the MRTK was required.

A skeleton was implemented to map all the bones and joints of the human body that the Azure Kinect can track (Figure 2a). In the first version, the bones were represented by a red cylinder and the joints as a grey ball. Then, each cylinder that corresponds to the anatomical part acquired was replaced with the segmented 3D object.

MRTK provides the elements to track both hands of the users correctly. Thus, it is possible to build a personalized prefab (3D object) that can reproduce the movement of all body parts in real-time. Through a C# code, it was possible to correctly assign each 3D object to the mapped joint through the Kinect (Figure 2b).

The script in Unity scales the 3D bones according to the distance between the centers of the Kinect mapped joints. For instance, the distance between the shoulder and elbow is considered by the algorithm in order to scale the humerus dimensions appropriately and show an adequate holographic overlay of the subject. Besides, the proposed 3D bone models can be substituted with a 3D reconstruction of the anatomical segments from DICOM images of the user, and, in this case, the scaling action will not be necessary.

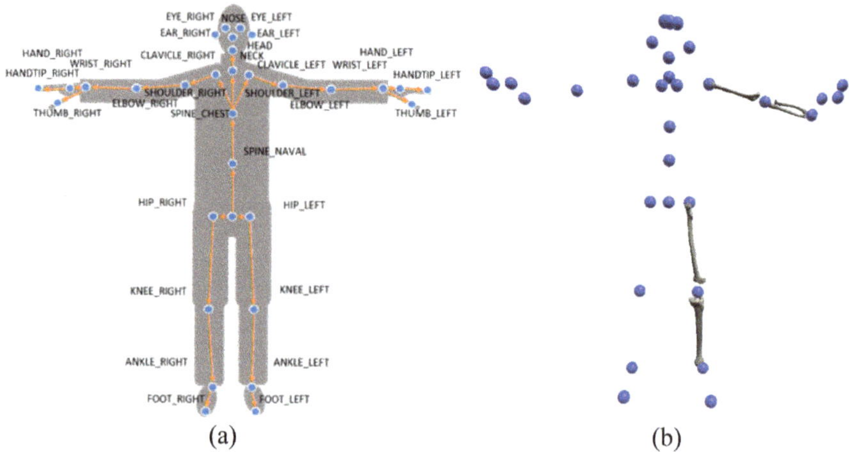

Figure 2. (**a**) Map of the joints tracked by the Azure Kinect; (**b**) Map of the 3D joint in Unity.

3.2.3. Distribution of Application

To correctly distribute the application on the HL2, the Holographic Remoting Player was used. Holographic Remoting is a complementary application that can be connected to the HL2 to display the game without deploying the application. In this manner, it is possible to modify the real-time application.

To link the HL2 and the Unity Editor, it was necessary to connect the computer and the HL2 to the same internet connection or connect them using a USB-C cable. After the pairing, it was possible to insert the IP address of the HL2 displayed on the home screen of the Holographic Remoting Player directly on the Unity Editor (Figure 3). Subsequently, the user could start the session.

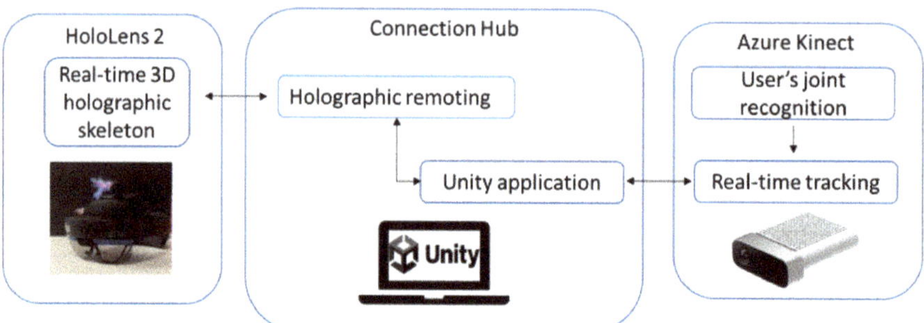

Figure 3. Block diagram of the system architecture and the functioning of each module.

Figure 3 shows the system architecture and the interaction among the hardware.

3.2.4. Experimental Protocol

Before launching the application, a subject person must be positioned in front of the Kinect Azure in its functioning area (within 1.5–2 m) to allow the mapping of the subject's joints.

The users analyzing the joint movement must wear the HL2 and activate the Holographic remoting player to allow the application to run on HL2.

When the application is running, the Azure Kinect starts to recognize all of the body segments, and the Unity application starts to associate each anatomical part with its own 3D object. To correctly verify the association between the bones-3D object and joint-3D object, it should be noted that the virtual skeleton reproduces exactly the same movement as the tracked person. The real-time movement of the 3D skeleton is visualized directly on the HL2 glasses.

During the streaming session, the user can walk around the holograms to better analyze the anatomical movement.

To evaluate the performance of the system concerning tracking karate positions of the Wado Ryu traditional Japanese style, a 28-year-old brown-belt karateka reproduced a set of karate shots and stances used during the training: Heiko Dachi (parallel stance) stance (Figure 4a); Juntzuki (lunge punch) on Zenkutsu Dachi (forward leaning stance) stance (Figure 4b); Shuto Uke (knife hand block) on Shomen Neko Aishi Dachi (front facing cat leg stance) stance (Figure 4c); Sokuto Geri (lateral kick) (Figure 4d).

Figure 4. Examples of karate shots and stances evaluated in the study: (**a**) Heiko Dachi (parallel stance) stance; (**b**) Juntzuki (lunge punch) on Zenkutsu Dachi (forward leaning stance) stance; (**c**) Shuto Uke (knife hand block) on Shomen Neko Aishi Dachi (front facing cat leg stance) stance; (**d**) Sokuto Geri (lateral kick).

The application saved the positions in the joint space of the subject during his movements. During the session, 1 crucial karate position was analyzed. The position analyzed was Shuto Uke on Shomen Neko Aishi Dachi, and data from left and right hip, left and right knee, left and right ankle, left foot, right foot, pelvis, left elbow, and wrist were measured.

Following a post-processing analysis of the data, the anatomical measurements necessary to evaluate the individual positions from a postural and competitive point of view of the discipline were derived.

4. Results

Figure 5 shows the HL2 view screenshots of the karate stances acquired by the system.

Figure 5. Karateka doing shots and stance and 3D holographic skeleton reproduced in HL2 glasses to evaluate the system performance: (**a**) Heiko Dachi (parallel stance) stance; (**b**) Juntzuki (lunge punch) on Zenkutsu Dachi (forward leaning stance) stance; (**c**) Shuto Uke (knife hand block) on Shomen Neko Aishi Dachi (front facing cat leg stance) stance; (**d**) Sokuto Geri (lateral kick) chudan.

Table 2 reports the mean value in real-time of the three coordinates for the selected joints during the Shuto Uke on Shomen Neko Aishi Dachi position.

Table 2. The table reports the mean value of the joints selected to analyze the Shuto Uke on Shomen Neko Aishi Dachi position.

Joints	Mean x (m)	Mean y (m)	Mean z (m)
right foot	1.72	0.24	−0.68
left foot	1.80	0.01	−0.68
right ankle	1.82	0.25	−0.57
left ankle	1.90	0.06	−0.56
right knee	1.74	0.29	−0.22
left knee	1.83	0.03	−0.22

Table 2. *Cont.*

Joints	Mean x (m)	Mean y (m)	Mean z (m)
right hip	1.81	0.24	0.14
left hip	1.85	0.10	0.14
left shoulder	1.88	0.04	0.57
left elbow	1.93	−0.05	0.38
left wrist	1.86	−0.04	0.30
pelvis	1.83	0.17	0.14

To evaluate karateka performance, 3D joints were used to compute the angles between the joints and were then compared with the standards (Table 3). The angles evaluated for the Shuto Uke on Neko Aishi Dachi are the angle of rotation between the right foot and the left foot and the angle of rotation between the wrist, the elbow, and the shoulder.

Table 3. The table compares the angles computed from the data output of the system and the karate standards for the Shuto Uke on Neko Aishi Dachi.

Condition	Computed	Standard
Right hip—right ankle joints	0.059 m	Aligned along z-axis
Left knee—Left ankle joints	0.063 m	Aligned along z-axis
Right foot/ankle axis—Left and Right feet axis	68.83°	Angle < 90°
Left shoulder/elbow axis—Left elbow/wrist axis	99.23°	Perpendicular (90°)

5. Discussion

The proposed system has the potential to provide support in assessing posture after sports injuries, particularly in martial arts, such as karate, where posture is fundamental to performing the sport correctly [24,25,36], and to monitor martial arts athletes after injuries to support the restoration of their movements and position. The superposition of 3D bone models reconstructed from medical imaging develops a more physiologically relevant environment. A more meaningful and detailed visualization of the body structures might be beneficial for experts to improve their assessment. Moreover, the 3D bone models overlayed on the subject allow observing how the bone segment is positioned during the athlete's performance without adding markers that could be affected by soft tissue artifacts.

The Unity application is not yet complete but shows adaptability to be used in sport application. Indeed, it emerges that the devices are adequate as a starting point in applying this type. HL2 and Azure Kinect represent valid substitutes for the gold standard systems of their categories, although not as much accurate.

The long-term purpose of this hybrid system composed of HL2 and Azure Kinect is to support athletes in restoring their abilities after an injury, but it still needs some improvements. Currently, MR-based systems that support athletes' recovery are not available. The systems presented in the literature are focused on the improvement of an athlete's performance. In Table 4, we compare the system illustrated in this paper with existing ones in the literature.

Studies proposing MR in orthopedics stop at the 3D reproduction of bones from DICOM and the possibility of interaction as an inanimate object in order to support experts in surgical planning [1–4]. Several studies recognize the contribution of such innovative technologies in reducing errors in surgery [8–11,30–32]. To our knowledge, a similar approach to the one presented in this work has not been suggested, except from [6]. It is worth noticing that in the work of [6], Vicon was the proposed device, which although allowing for the best possible accuracy in bone positioning and articulation is not applicable in clinical reality. Conversely, this project provides a system that is sufficiently accurate without the need for specific knowledge, given the absence of markers [42–44]. Furthermore, the system can also be considered low-cost if the HL2 is replaced by a cheaper VR visor, even though the AR or MR is more beneficial for this type of application for the overlap of the skeleton on the subject and lesser side effects, such as motion sickness. The system could also be considered applicable in a remote setting where the trainer is not present. In this situation, the overlay of bones on the subject is not applicable, which can occur if the examination is in the present. In this context, the video recorded by the HL2 camera might be visible to the trainer in real-time with the skeleton reconstruction.

Table 4. Comparison of the systems used for supporting martial arts athletes.

	Reference [35]	**Reference [29]**	**Reference [36]**	**This Study**
System proposed	2 screens Laptop (2.8 GHz Pentium 4) USB Cam OpenGL OpenCV	2D Pose estimation 3D pose forecasting 3D recovery HTC Vive Sony Camera DSC-QX10 Laptop	Optica Motion Capture system	HoloLens 2 Laptop Azure Kinect Unity 3D
Markless	yes	yes	yes	yes
HDM	no	yes	no	yes
technology	Virtual Reality	Virtual Reality	Virtual Reality	Mixed Reality
IMU tracking	no	yes	no	yes
Motion Sickness	no	no	no	no
Data acquisition	no	yes	yes	yes
Feedback	Audio-visual	visual	visual	visual
Personalization	Not available	Not available	Not available	yes
External control	yes	yes	yes	yes
DICOM	no	no	no	yes

Unfortunately, the use of remoting and real-time data saving introduces a delay between the movement of the subject and the movement repeated by the skeleton. The movement captured by the Kinect is correct and can follow even a fast movement, such as a kick or a punch, but it is reproduced on Unity with a delay of about half a second.

Furthermore, the system finds its greatest application when the subject has a fracture in the spine or a long bone. In these cases, the 3D reconstructions of the bones are directly built from their DICOM scans, and the fracture behavior can be studied during the sports movements. The percentage of fractures and dislocations to which karatekas are subject should not be underestimated [49,50]. In many cases, these are due to the incorrect execution of the basic position and stance assumed during a kick or a punch [51]. The adequate rotation of the foot, knee, and hips in a kick are essential to give more force and efficacy to the blow without suffering damage to the joint and bone to cushion the reaction force suffered on impact elegantly and correctly. Also, punches could be affected by the joint's wrong position. To better perform the punch, the wrist should be straight and parallel to the floor, the fingers might be correctly closed, and the punch's force is associated with the perfect rotation of the hips during the stance. The correct performance does not have a negative impact on the shoulder.

Anatomical fidelity is important for this type of application, so caution and improvements are required. As documented in other studies [42–44], Kinect-based systems can have poor joint tracking when a body part is not visible to the camera and during unusual poses or interactions with objects. It is worth pointing out that in this work, the assessment is characterized by the subject sited frontally to the Kinect Azure camera, and the athlete did not interact with any object. Complex poses are behind the scope of this preliminary study. Future implementations can be carried out to correct or minimize the 3D reconstruction errors of the devices.

Firstly, to effectively and immediately achieve the overlapping of the 3D anatomical components on the patient through the HoloLens viewer, the Kinect must be positioned as close as possible to the camera of the HoloLens device. In this way, the position detected by the Kinect could be used to locate the 3D reconstruction in MR correctly. In addition, other application components can be implemented depending on the purpose of use, such as the viewing of medical images or the possibility of remote sharing. Finally, the Kinect can determine the position of the body joints, albeit with a certain margin of error. If the subject is stationary and with arms outstretched, the joints are correctly recognized, and the skeletal overlap is coherent. When the subject bends the elbow, bringing the hand towards the shoulder, the joint remains in the correct position. Conversely, when the algorithm connects the elbow joint with the hand joint, it positions the forearm bones in an anatomically incorrect way. This situation is justified by the 3D prefab of the skeleton in which the bones are separated from each other and managed separately by the tracking algorithm. Therefore, the integration of algorithms for simulating the behavior of joints and bones is necessary to enable a more anatomically correct realization of the positions assumed by the skeleton in each situation [52]. In this context, an improvement of the system might take into account the integration of artificial intelligence and deep learning algorithms that can identify the position of the athlete and correct the position detected by the Azure Kinect.

Even more fundamental is this integration when considering patients with certain bone diseases or implanted prostheses that affect bone movement and joint function [53,54]. In these cases, patient-specific simulation studies are crucial to be considered. Thus, a user-friendly application that allows an in-depth analysis of a pathological joint in real-time represents a clinical need to improve the accuracy of the diagnosis or the surgical planning. For example, for joint-related pathologies, experts are interested in the range of motion and its value changing over time. For patients who underwent joint replacement surgery, the prostheses may affect posture or walking, and their effect should be examined [55]. Eventually, the system might be useful in assessing what would happen to the patient's movement in the case of an incorrect joint replacement, thanks to a properly trained and implemented artificial intelligence algorithm [56,57].

6. Conclusions

The combined system of HL2 and Azure Kinect shows the possibility of monitoring movement in certain conditions for athletes playing martial arts, such as karate. Due to its adaptability, this system could also be used to evaluate athletes after injuries and has shown high potential to support sports rehabilitation. However, the system needs to be tested with the engagements of professional athletes after injuries that need to restore their initial condition. In the future, the proposed system might also be used to train orthopedic clinicians. In fact, orthopedics students may interact with the virtual anatomical segments and may observe how bones could be affected by a pathology progression, such as valgus legs, an implanted prosthesis, or back sciatica. It could also be used in sports halls where the trainer can provide students with innovative technologies to objectively assess and correct their posture or for beginners' learning. We do not exclude the possibility of also using this system for boxing or other martial arts, such as kung-fu or jujutsu.

Furthermore, in official competitions and graduating exams, correct posture and execution of movements are the evaluated components [25–27]. The karateka must repeat the positions many times in training to reach perfection, and, with the help of this technology, a video can be recorded of his performances in conjunction with real-time observation.

Author Contributions: Conceptualization, F.B.; methodology, M.F. and S.P.; software, M.F. and S.P.; validation, M.F., S.P. and A.P.; formal analysis, M.F.; investigation, M.F.; resources, F.B. and F.M.; data curation, M.F. and S.P.; writing—original draft preparation, M.F. and S.P.; writing—review and editing, F.B. and A.P.; visualization, M.F., S.P. and A.P.; supervision, F.B. and F.M.; project administration, F.B.; funding acquisition, F.B. and F.M. All authors have read and agreed to the published version of the manuscript.

Funding: This research received no external funding.

Institutional Review Board Statement: Not applicable.

Informed Consent Statement: Not applicable.

Data Availability Statement: Not applicable.

Conflicts of Interest: The authors declare no conflict of interest.

References

1. Berthold, D.P.; Muench, L.N.; Rupp, M.C.; Siebenlist, S.; Cote, M.P.; Mazzocca, A.D.; Quindlen, K. Head-Mounted Display Virtual Reality Is Effective in Orthopaedic Training: A Systematic Review. *Arthrosc. Sport. Med. Rehabil.* **2022**, *4*, e1843–e1849. [CrossRef] [PubMed]
2. Clarke, E. Virtual Reality Simulation—The Future of Orthopaedic Training? A Systematic Review and Narrative Analysis. *Adv. Simul.* **2021**, *6*, 2. [CrossRef]
3. Hasan, L.K.; Haratian, A.; Kim, M.; Bolia, I.K.; Weber, A.E.; Petrigliano, F.A. Virtual Reality in Orthopedic Surgery Training. *Adv. Med. Educ. Pract.* **2021**, *12*, 1295–1301. [CrossRef] [PubMed]
4. Barcali, E.; Iadanza, E.; Manetti, L.; Francia, P.; Nardi, C.; Bocchi, L. Augmented Reality in Surgery: A Scoping Review. *Appl. Sci.* **2022**, *12*, 6890. [CrossRef]
5. Son, S.; Lim, K.B.; Kim, J.; Lee, C.; Cho, S.I.I.; Yoo, J. Comparing the Effects of Exoskeletal-Type Robot-Assisted Gait Training on Patients with Ataxic or Hemiplegic Stroke. *Brain Sci.* **2022**, *12*, 1261. [CrossRef]
6. Debarba, H.G.; De Oliveira, M.E.; Ladermann, A.; Chague, S.; Charbonnier, C. Augmented Reality Visualization of Joint Movements for Rehabilitation and Sports Medicine. In Proceedings of the 2018 20th Symposium on Virtual and Augmented Reality (SVR), Foz do Iguacu, Brazil, 28–30 October 2018; pp. 114–121. [CrossRef]
7. Bertoli, M.; Cereatti, A.; Croce, U.D.; Pica, A.; Bini, F. Can MIMUs Positioned on the Ankles Provide a Reliable Detection and Characterization of U-Turns in Gait? In Proceedings of the 2018 IEEE International Symposium on Medical Measurements and Applications (MeMeA), Rome, Italy, 11–13 June 2018; pp. 1–6. [CrossRef]
8. Condino, S.; Turini, G.; Parchi, P.D.; Viglialoro, R.M.; Piolanti, N.; Gesi, M.; Ferrari, M.; Ferrari, V. How to Build a Patient-Specific Hybrid Simulator for Orthopaedic Open Surgery: Benefits and Limits of Mixed-Reality Using the Microsoft Hololens. *J. Healthc. Eng.* **2018**, *2018*, 5435097. [CrossRef]
9. Turini, G.; Condino, S.; Parchi, P.D.; Viglialoro, R.M.; Piolanti, N.; Gesi, M.; Ferrari, M.; Ferrari, V. A Microsoft HoloLens Mixed Reality Surgical Simulator for Patient-Specific Hip Arthroplasty Training. In *Augmented Reality, Virtual Reality, and Computer Graphics, Proceedings of the 5th International Conference, AVR 2018, Otranto, Italy, 24–27 June 2018*; De Paolis, L., Bourdot, P., Eds.; Springer: Cham, Switzerland, 2018; pp. 201–210. [CrossRef]
10. Liebmann, F.; Roner, S.; von Atzigen, M.; Wanivenhaus, F.; Neuhaus, C.; Spirig, J.; Scaramuzza, D.; Sutter, R.; Snedeker, J.; Farshad, M.; et al. Registration Made Easy—Standalone Orthopedic Navigation with HoloLens. *arXiv* **2020**, arXiv:2001.06209. [CrossRef]
11. Cevallos, N.; Zukotynski, B.; Greig, D.; Silva, M.; Thompson, R.M. The Utility of Virtual Reality in Orthopedic Surgical Training. *J. Surg. Educ.* **2022**, *79*, 1516–1525. [CrossRef]
12. Vanicek, N.; Strike, S.; McNaughton, L.; Polman, R. Gait Patterns in Transtibial Amputee Fallers vs. Non-Fallers: Biomechanical Differences during Level Walking. *Gait Posture* **2009**, *29*, 415–420. [CrossRef]
13. Lau, I.Y.S.; Chua, T.T.; Lee, W.X.P.; Wong, C.W.; Toh, T.H.; Ting, H.Y. Kinect-Based Knee Osteoarthritis Gait Analysis System. In Proceedings of the 2020 IEEE 2nd International Conference on Artificial Intelligence in Engineering and Technology (IICAIET), Kota Kinabalu, Malaysia, 26–27 September 2020; pp. 1–6. [CrossRef]
14. Yoshimoto, K.; Shinya, M. Use of the Azure Kinect to Measure Foot Clearance during Obstacle Crossing: A Validation Study. *PLoS ONE* **2022**, *17*, e0265215. [CrossRef]
15. Lahner, M.; Mußhoff, D.; Von Schulze Pellengahr, C.; Willburger, R.; Hagen, M.; Ficklscherer, A.; Von Engelhardt, L.V.; Ackermann, O.; Lahner, N.; Vetter, G. Is the Kinect System Suitable for Evaluation of the Hip Joint Range of Motion and as a Screening Tool for Femoroacetabular Impingement (FAI)? *Technol. Heal. Care* **2015**, *23*, 75–82. [CrossRef] [PubMed]
16. Asaeda, M.; Kuwahara, W.; Fujita, N.; Yamasaki, T.; Adachi, N. Validity of Motion Analysis Using the Kinect System to Evaluate Single Leg Stance in Patients with Hip Disorders. *Gait Posture* **2018**, *62*, 458–462. [CrossRef] [PubMed]
17. Aleksandra, K.; Maj, A.; Dejnek, M.; Prill, R.; Skotowska-Machaj, A.; Kołcz, A. Wrist Motion Assessment Using Microsoft Azure Kinect DK: A Reliability Study in Healthy Individuals. *Adv. Clin. Exp. Med.* **2022**, *32*. [CrossRef] [PubMed]
18. Cho, H.M.; Seon, J.; Park, J.Y.; Ahn, J.; Lee, Y. Usefulness of the Kinect-V2 System for Determining the Global Gait Index to Assess Functional Recovery after Total Knee Arthroplasty. *Orthop. Surg.* **2022**, *14*, 3216–3224. [CrossRef]
19. Uhlár, Á.; Ambrus, M.; Kékesi, M.; Fodor, E.; Grand, L.; Szathmáry, G.; Rácz, K.; Lacza, Z. Kinect Azure–Based Accurate Measurement of Dynamic Valgus Position of the Knee—A Corrigible Predisposing Factor of Osteoarthritis. *Appl. Sci.* **2021**, *11*, 5536. [CrossRef]
20. Johnson, P.B.; Jackson, A.; Saki, M.; Feldman, E.; Bradley, J. Patient Posture Correction and Alignment Using Mixed Reality Visualization and the HoloLens 2. *Med. Phys.* **2022**, *49*, 15–22. [CrossRef]
21. Jan, Y.F.; Tseng, K.W.; Kao, P.Y.; Hung, Y.P. Augmented Tai-Chi Chuan Practice Tool with Pose Evaluation. In Proceedings of the 2021 IEEE 4th International Conference on Multimedia Information Processing and Retrieval (MIPR), Tokyo, Japan, 8–10 September 2021; pp. 35–41. [CrossRef]
22. Singla, D.; Veqar, Z.; Hussain, M.E. Photogrammetric Assessment of Upper Body Posture Using Postural Angles: A Literature Review. *J. Chiropr. Med.* **2017**, *16*, 131–138. [CrossRef]

23. Do Rosário, J.L.P. Photographic Analysis of Human Posture: A Literature Review. *J. Bodyw. Mov. Ther.* **2014**, *18*, 56–61. [CrossRef]
24. Byun, S.; An, C.; Kim, M.; Han, D. The Effects of an Exercise Program Consisting of Taekwondo Basic Movements on Posture Correction. *J. Phys. Ther. Sci.* **2014**, *26*, 1585–1588. [CrossRef]
25. Cherepov, E.A.; Eganov, A.V.; Bakushin, A.A.; Platunova, N.Y.; Sevostyanov, D.Y. Maintaining Postural Balance in Martial Arts Athletes Depending on Coordination Abilities. *J. Phys. Educ. Sport* **2021**, *21*, 3427–3432. [CrossRef]
26. Gauchard, G.C.; Lion, A.; Bento, L.; Perrin, P.P.; Ceyte, H. Postural Control in High-Level Kata and Kumite Karatekas. *Mov. Sport. Sci.-Sci. Mot.* **2017**, *100*, 21–26. [CrossRef]
27. Güler, M.; Ramazanoglu, N. Evaluation of Physiological Performance Parameters of Elite Karate-Kumite Athletes by the Simulated Karate Performance Test. *Univers. J. Educ. Res.* **2018**, *6*, 2238–2243. [CrossRef]
28. Petri, K.; Emmermacher, P.; Danneberg, M.; Masik, S.; Eckardt, F.; Weichelt, S.; Bandow, N.; Witte, K. Training Using Virtual Reality Improves Response Behavior in Karate Kumite. *Sport. Eng.* **2019**, *22*, 2. [CrossRef]
29. Wu, E.; Koike, H. FuturePose—Mixed Reality Martial Arts Training Using Real-Time 3D Human Pose Forecasting with a RGB Camera. In Proceedings of the 2019 IEEE Winter Conference on Applications of Computer Vision (WACV), Waikoloa, HI, USA, 7–11 January 2019; pp. 1384–1392. [CrossRef]
30. Lu, L.; Wang, H.; Liu, P.; Liu, R.; Zhang, J.; Xie, Y.; Liu, S.; Huo, T.; Xie, M.; Wu, X.; et al. Applications of Mixed Reality Technology in Orthopedics Surgery: A Pilot Study. *Front. Bioeng. Biotechnol.* **2022**, *10*, 740507. [CrossRef] [PubMed]
31. Lohre, R.; Bois, A.J.; Pollock, J.W.; Lapner, P.; McIlquham, K.; Athwal, G.S.; Goel, D.P. Effectiveness of Immersive Virtual Reality on Orthopedic Surgical Skills and Knowledge Acquisition among Senior Surgical Residents: A Randomized Clinical Trial. *JAMA Netw. Open* **2020**, *3*, e2031217. [CrossRef] [PubMed]
32. Gregory, T.M.; Gregory, J.; Sledge, J.; Allard, R.; Mir, O. Surgery Guided by Mixed Reality: Presentation of a Proof of Concept. *Acta Orthop.* **2018**, *89*, 480–483. [CrossRef]
33. Pose-Díez-De-la-lastra, A.; Moreta-Martinez, R.; García-Sevilla, M.; García-Mato, D.; Calvo-Haro, J.A.; Mediavilla-Santos, L.; Pérez-Mañanes, R.; von Haxthausen, F.; Pascau, J. HoloLens 1 vs. HoloLens 2: Improvements in the New Model for Orthopedic Oncological Interventions. *Sensors* **2022**, *22*, 4915. [CrossRef] [PubMed]
34. El-Hariri, H.; Pandey, P.; Hodgson, A.J.; Garbi, R. Augmented Reality Visualisation for Orthopaedic Surgical Guidance with Pre- and Intra-Operative Multimodal Image Data Fusion. *Healthc. Technol. Lett.* **2018**, *5*, 189–193. [CrossRef]
35. Hämäläinen, P.; Ilmonen, T.; Höysniemi, J.; Lindholm, M.; Nykänen, A. Martial Arts in Artificial Reality. In Proceedings of the CHI05: CHI 2005 Conference on Human Factors in Computing Systems, Portland, OR, USA, 2–7 April 2005; pp. 781–790. [CrossRef]
36. Shen, Y.; Wang, H.; Ho, E.S.L.; Yang, L.; Shum, H.P.H. Posture-Based and Action-Based Graphs for Boxing Skill Visualization. *Comput. Graph.* **2017**, *69*, 104–115. [CrossRef]
37. Franzo', M.; Pascucci, S.; Serrao, M.; Marinozzi, F.; Bini, F. Kinect-Based Wearable Prototype System for Ataxic Patients Neurorehabilitation: Software Update for Exergaming and Rehabilitation. In Proceedings of the 2021 IEEE International Symposium on Medical Measurements and Applications (MeMeA), Lausanne, Switzerland, 23–25 June 2021. [CrossRef]
38. Franzo', M.; Pascucci, S.; Serrao, M.; Marinozzi, F.; Bini, F. Kinect-Based Wearable Prototype System for Ataxic Patients Neurorehabilitation: Control Group Preliminary Results. In Proceedings of the 2020 IEEE International Symposium on Medical Measurements and Applications (MeMeA), Bari, Italy, 1–3 June 2020. [CrossRef]
39. Franzo', M.; Pascucci, S.; Serrao, M.; Marinozzi, F.; Bini, F. Exergaming in Mixed Reality for the Rehabilitation of Ataxic Patients. In Proceedings of the 2022 IEEE International Symposium on Medical Measurements and Applications (MeMeA), Messina, Italy, 22–24 June 2022. [CrossRef]
40. Yeung, L.F.; Cheng, K.C.; Fong, C.H.; Lee, W.C.C.; Tong, K.Y. Evaluation of the Microsoft Kinect as a Clinical Assessment Tool of Body Sway. *Gait Posture* **2014**, *40*, 532–538. [CrossRef]
41. Otte, K.; Kayser, B.; Mansow-Model, S.; Verrel, J.; Paul, F.; Brandt, A.U.; Schmitz-Hübsch, T. Accuracy and Reliability of the Kinect Version 2 for Clinical Measurement of Motor Function. *PLoS ONE* **2016**, *11*, e0166532. [CrossRef] [PubMed]
42. Albert, J.A.; Owolabi, V.; Gebel, A.; Brahms, C.M.; Granacher, U.; Arnrich, B. Evaluation of the Pose Tracking Performance of the Azure Kinect and Kinect v2 for Gait Analysis in Comparison with a Gold Standard: A Pilot Study. *Sensors* **2020**, *20*, 5104. [CrossRef] [PubMed]
43. Tölgyessy, M.; Dekan, M.; Chovanec, L. Skeleton Tracking Accuracy and Precision Evaluation of Kinect V1, Kinect V2, and the Azure Kinect. *Appl. Sci.* **2021**, *11*, 5756. [CrossRef]
44. Antico, M.; Balletti, N.; Laudato, G.; Lazich, A.; Notarantonio, M.; Oliveto, R.; Ricciardi, S.; Scalabrino, S.; Simeone, J. Postural Control Assessment via Microsoft Azure Kinect DK: An Evaluation Study. *Comput. Methods Programs Biomed.* **2021**, *209*, 106324. [CrossRef]
45. Bailey, J.L.; Jensen, B.K. Telementoring: Using the Kinect and Microsoft Azure to Save Lives. *Int. J. Electron. Financ.* **2013**, *7*, 33–47. [CrossRef]
46. Eswaran, M.; Raju Bahubalendruni, M.V.A. Challenges and opportunities on AR/VR technologies for manufacturing systems in the context of industry 4.0: A state of the art review. *J. Manuf. Syst.* **2022**, *65*, 260–278. [CrossRef]
47. Soltani, P.; Morice, A.H.P. Augmented reality tools for sports education and training. *Comput. Educ.* **2020**, *155*, 103923. [CrossRef]
48. Da Gama, A.E.F.; de Menezes Chaves, T.; Fallavollita, P.; Figueiredo, L.S.; Teichrieb, V. Rehabilitation motion recognition based on the international biomechanical standards. *Expert Syst. Appl.* **2019**, *116*, 396–409. [CrossRef]

49. McLatchie, G. Karate and Karate Injuries. *Br. J. Sports Med.* **1981**, *15*, 84–86. [CrossRef]
50. Critchley, G.R.; Mannion, S.; Meredith, C. Injury Rates in Shotokan Karate. *Br. J. Sports Med.* **1999**, *33*, 174–177. [CrossRef]
51. Ambroży, A.T.; Dariusz Mucha, A.; Czarnecki, W.; Ambroży, D.; Janusz, M.; Piwowarski, A.J.; Mucha, T. Most Common Injuries to Professional Contestant Karate. *Secur. Dimens. Int. Natl. Stud.* **2015**, *2015*, 142–164.
52. Rinaldi, M.; Nasr, Y.; Atef, G.; Bini, F.; Varrecchia, T.; Conte, C.; Chini, G.; Ranavolo, A.; Draicchio, F.; Pierelli, F.; et al. Biomechanical characterization of the Junzuki karate punch: Indexes of performance. *Eur. J. Sport Sci.* **2018**, *18*, 796–805. [CrossRef] [PubMed]
53. Bini, F.; Pica, A.; Marinozzi, A.; Marinozzi, F. Prediction of Stress and Strain Patterns from Load Rearrangement in Human Osteoarthritic Femur Head: Finite Element Study with the Integration of Muscular Forces and Friction Contact. In *New Developments on Computational Methods and Imaging in Biomechanics and Biomedical Engineering. Lecture Notes in Computational Vision and Biomechanics*; Tavares, J., Fernandes, P., Eds.; Springer: Cham, Switzerland, 2019; Volume 33, pp. 49–64. [CrossRef]
54. Araneo, R.; Bini, F.; Rinaldi, A.; Notargiacomo, A.; Pea, M.; Celozzi, S. Thermal-Electric Model for Piezoelectric ZnO Nanowires. *Nanotechnology* **2015**, *26*, 265402. [CrossRef]
55. Lu, T.W.; Chang, C.F. Biomechanics of Human Movement and Its Clinical Applications. *Kaohsiung J. Med. Sci.* **2012**, *28*, S13–S25. [CrossRef] [PubMed]
56. Albuquerque, P.; Verlekar, T.T.; Correia, P.L.; Soares, L.D. A Spatiotemporal Deep Learning Approach for Automatic Pathological Gait Classification. *Sensors* **2021**, *21*, 6202. [CrossRef] [PubMed]
57. Halilaj, E.; Rajagopal, A.; Fiterau, M.; Hicks, J.L.; Hastie, T.J.; Delp, S.L. Machine Learning in Human Movement Biomechanics: Best Practices, Common Pitfalls, and New Opportunities. *J. Biomech.* **2018**, *81*, 1–11. [CrossRef] [PubMed]

 *applied sciences*

Article

Magnetic Resonance with Diffusion and Dynamic Perfusion-Weighted Imaging in the Assessment of Early Chemoradiotherapy Response of Naso-Oropharyngeal Carcinoma

Michele Pietragalla [1], Eleonora Bicci [1], Linda Calistri [1], Chiara Lorini [2], Pierluigi Bonomo [3], Andrea Borghesi [4], Antonio Lo Casto [5], Francesco Mungai [6], Luigi Bonasera [6], Giandomenico Maggiore [7] and Cosimo Nardi [1,*]

[1] Radiodiagnostic Unit N. 2, Department of Experimental and Clinical Biomedical Sciences, University of Florence—Azienda Ospedaliero-Universitaria Careggi, 50134 Florence, Italy
[2] Department of Health Science, University of Florence, 50134 Florence, Italy
[3] Radiation Oncology, University of Florence—Azienda Ospedaliero-Universitaria Careggi, 50134 Florence, Italy
[4] Department of Medical and Surgical Specialties, Radiological Sciences, and Public Health, University of Brescia, 25123 Brescia, Italy
[5] Department of Biomedicine, Neuroscience and Advanced Diagnostics (BIND), University Hospital of Palermo, 90127 Palermo, Italy
[6] Department of Radiology, University of Florence—Azienda Ospedaliero-Universitaria Careggi, 50134 Florence, Italy
[7] Department of Otorhinolaryngology, University of Florence—Azienda Ospedaliero-Universitaria Careggi, Via Taddeo Alderotti, 50139 Florence, Italy
* Correspondence: cosimo.nardi@unifi.it

Citation: Pietragalla, M.; Bicci, E.; Calistri, L.; Lorini, C.; Bonomo, P.; Borghesi, A.; Lo Casto, A.; Mungai, F.; Bonasera, L.; Maggiore, G.; et al. Magnetic Resonance with Diffusion and Dynamic Perfusion-Weighted Imaging in the Assessment of Early Chemoradiotherapy Response of Naso-Oropharyngeal Carcinoma. *Appl. Sci.* **2023**, *13*, 2799. https://doi.org/10.3390/app13052799

Academic Editor: Jan Egger

Received: 5 February 2023
Revised: 18 February 2023
Accepted: 19 February 2023
Published: 22 February 2023

Abstract: The purpose of this study was to differentiate post-chemoradiotherapy (CRT) changes from tumor persistence/recurrence in early follow-up of naso-oropharyngeal carcinoma on magnetic resonance (MRI) with diffusion (DWI) and dynamic contrast-enhanced perfusion-weighted imaging (DCE-PWI). A total of 37 patients were assessed with MRI both for tumor staging and 4-month follow-up from ending CRT. Mean apparent diffusion coefficient (ADC) values, area under the curve (AUC), and K(trans) values were calculated from DWI and DCE-PWI images, respectively. DWI and DCE-PWI values of primary tumor (ADC, AUC, $K(trans)_{pre}$), post-CRT changes (ADC, AUC, $K(trans)_{post}$), and trapezius muscle as a normative reference before and after CRT (ADC, AUC, $K(trans)_{muscle\,pre}$ and $_{muscle\,post}$; $AUC_{post/muscle\,post}$:$AUC_{pre/muscle\,pre}$ ($AUC_{post/pre/muscle}$); $K(trans)_{post/muscle\,post}$:$K(trans)_{pre/muscle\,pre}$ ($K(trans)_{post/pre/muscle}$) were assessed. In detecting post-CRT changes, $ADC_{post} > 1.33 \times 10^{-3}$ mm^2/s and an increase >0.72 $\times 10^{-3}$ mm^2/s and/or >65.5% between ADC_{post} and ADC_{pre} values ($ADC_{post-pre}$; $ADC_{post-pre\%}$) had 100% specificity, whereas hypointense signal intensity on DWIb800 images showed specificity 80%. Although mean $AUC_{post/pre/muscle}$ and $K(trans)_{post/pre/muscle}$ were similar both in post-CRT changes (1.10 ± 0.58; 1.08 ± 0.91) and tumor persistence/recurrence (1.09 ± 0.11; 1.03 ± 0.12), $K(trans)_{post/pre/muscle}$ values < 0.85 and >1.20 suggested post-CRT fibrosis and inflammatory edema, respectively. In early follow-up of naso-oropharyngeal carcinoma, our sample showed that $ADC_{post} > 1.33 \times 10^{-3}$ mm^2/s, $ADC_{post-pre\%} > 65.5\%$, and $ADC_{post-pre} > 0.72 \times 10^{-3}$ mm^2/s identified post-CRT changes with 100% specificity. $K(trans)_{post/pre/muscle}$ values less than 0.85 suggested post-CRT fibrosis, whereas $K(trans)_{post/pre/muscle}$ values more than 1.20 indicated inflammatory edema.

Keywords: naso-oropharyngeal carcinoma; magnetic resonance imaging; diffusion-weighted imaging; dynamic contrast-enhanced perfusion-weighted imaging; chemoradiotherapy

1. Introduction

Head and neck cancers represent the sixth most common cancer worldwide and a major cause of morbidity and mortality [1]. More than 90% of head and neck cancers are squamous cell carcinomas (HNSCC) arising from the mucosal surfaces of the oral cavity, naso-oropharynx, and larynx [2]. Crucial risk factors aligned with head and neck cancers include tobacco, alcohol consumption, and human papillomavirus (HPV) or Epstein–Barr virus infections [3].

Chemoradiotherapy (CRT) has become more popular over the past decade because the organ preservation possibilities are higher with CRT as compared to surgery [4]. The relapse rate is still 50% (35–65%) in patients with advanced HNSCC [5] and reaches 25% in early-stage cancers [6]. Almost 90% of HNSCC recurrences following CRT develop within 2 years [7]; the early detection of tumor recurrence prompts curative salvage treatment and may allow the preservation of organ functions [6].

The interpretation of post-treatment follow-up via imaging techniques is complicated by post-actinic edema, soft tissue necrosis, and fibrosis. Such post-treatment changes make it difficult to detect tumor recurrence within a distorted anatomy [8]. Biopsy with negative findings does not exclude HNSCC recurrence, and multiple biopsies may increase overall morbidity [6]. Therefore, in addition to clinical and histological parameters, other biomarkers are needed to stratify patients for optimal therapy [9].

Magnetic resonance imaging (MRI) is an accurate technique for the assessment of deep tumor invasion and morphological tumor features [10], but it is not able to identify early locoregional recurrences, predict tumor response to treatment and monitor post-treatment changes [11,12].

Metabolic imaging with 18F-fluorodeoxyglucose positron emission tomography/computed tomography (18F-FDG PET/CT) has evolved as a tool for the post-treatment evaluation of HNSCC, but it is generally delayed for at least 12 weeks due to the potential false-positive results in early post-treatment inflammatory changes [13].

Nowadays, a multiparametric approach employing MRI has been proposed with diffusion (DWI) and dynamic contrast-enhanced perfusion-weighted imaging (DCE-PWI) for the distinction between post-treatment changes and tumor persistence/recurrence [14,15]. Moreover, MRI is ideally suited to serial scanning, reducing the use of ionizing radiations commonly emitted by CT examinations [16–19].

DWI with apparent diffusion coefficient (ADC) maps can theoretically differentiate between inflammation and neoplastic tissues since the water molecule diffusion is increased into inflammatory tissues (T2* loss of signal and high ADC values), whereas water molecules have restricted diffusion within neoplastic tissues (T2* signal maintenance and low ADC values) [20].

DCE-PWI examines microvascular tumor tissue characteristics [21] and can potentially assess the reduction of tumor blood perfusion by means of K(trans), which represents the volume transfer constant from the vascular to the extravascular extracellular spaces [22–25].

We aimed to retrospectively differentiate post-CRT changes from tumor persistence/recurrence in the early follow-up of patients with primary naso-oropharyngeal carcinoma using multiparametric MRI with DWI and DCE-PWI sequences.

2. Materials and Methods

2.1. Inclusion Criteria

From January 2016 to December 2021, MRI examinations of 104 patients with histological diagnoses of nasopharynx or oropharynx carcinoma investigated in the radiology department of the Careggi Hospital of Florence (Italy) were retrieved. This study was approved by the research ethics committee (Protocol Number 21800_oss), and informed written consent was obtained from all individual participants included in the study. All procedures performed in studies involving human participants were in accordance with the ethical standards of the institutional and/or national research committee and with the 1964 Helsinki Declaration and its later amendments or comparable ethical standards.

Patients who met the following criteria were included:

- Adult patients ($\geq$18 years);
- Histological confirmation of oropharynx or nasopharynx carcinoma through biopsy;
- Exclusive CRT;
- MRI examination for both tumor staging and 4-month follow-up after ending CRT;
- DWI and DCE-PWI MR sequences;
- Two years of clinical and cross-sectional imaging follow-ups including consecutive 18F-FDG PET/CT and MRI.

Patients were excluded in case of previous head and neck radiotherapy treatment (4), surgical treatment (5), MRI without both DWI and DCE-PWI sequences (14), MRI not performed for both tumor staging and follow-up (41), and follow-up lasting less than 2 years (3). We considered the first two years after completing CRT at a higher risk of neoplastic recurrence.

The patients that matched our inclusion criteria were 37 (19 males, 18 females) with a mean age of 59 years (median age: 58.5 years, range: 36–81 years); 26 patients were affected by oropharyngeal carcinoma (16 HPV positive, 4 HPV negative, and 6 unknowns for HPV status) and 11 patients by nasopharyngeal carcinoma. TNM staging—eighth edition of the American Joint Commission on Cancer—HPV status, and tumor locations were summarized in Table S1 in the Supplementary Materials.

2.2. DWI and DCE-PWI

MRI examinations for tumor staging and follow-up were performed with a 1.5 T MR device (Magnetom Aera, Siemens Healthcare, Erlangen, Germany) with a devoted head and neck coil. The MR acquisition protocol included pre- and post-contrast sequences (Table S2 in Supplementary Materials). An axial fat-saturated echo-planar imaging-based DWI with two different b-values (b50$-$800 s/mm^2) was acquired. ADC values of primitive tumors and residual tissues after CRT were calculated by positioning three regions of interest (ROI) with an average intratumoral area of 0.30–0.40 cm^2 each on three contiguous axial sections. DCE-PwI was obtained through two volumetric interpolated breath-hold examination (VIBE) T1-w sequences characterized by 3.5 mm slice thickness, 0.7 interslice gap, FOV 250 $\times$ 226 mm, matrix 139 $\times$ 192, flip angles 5$°$ and 15$°$, and acceleration factor 3 for baseline T1-mapping acquisitions. After contrast agent administration, one VIBE T1-w lasting 350 s and with a temporal resolution of 5 s was acquired as follows: TR 4.65 ms, TE 1.66 ms, 3.5 mm slice thickness, FOV 250 $\times$ 226.6 mm, matrix 139 $\times$ 192, flip angle 30$°$, acceleration factor 3, and peripheral K space sampling with time to center 2.2 s. Time/intensity curve, area under the curve (AUC), and K(trans) values of primitive tumor and tumor residual/relapse tissues after CRT were generated by using IntelliSpace software version 9.0 (Philips, Amsterdam, The Netherlands) from the native DCE-PWI images by drawing an ROI including at least 50% of the largest lesion diameter. Before lesion sampling, an ROI was placed on the internal carotid artery to obtain the arterial input function curve, defined as the contrast concentration in vessels feeding to tissue at each point in time during the contrast passage. Vessels, cystic areas within solid lesions, and necrotic, hemorrhagic, or proteinaceous areas detected on T1-w and T2-w sequences were excluded in both DWI and DCE-PWI analysis. ADC, AUC, and K(trans) values of the trapezius muscle on the same side of the tumor were also obtained.

2.3. Image Assessment

MRIs performed both for tumor staging and 4-month follow-up after the end of CRT were independently reviewed by two radiologists with 12 (CN) and 7 (MP) years of experience in head and neck imaging, respectively.

The following morphologic, DWI, and DCE-PWI features were assessed:

- Maximum size of the primitive tumor and submucosal thickness of the residual tissue after CRT on contrast-enhanced T1 images.

- Signal intensity (SI), hyper- or hypointense, of the residual tissue after CRT on DWIb800 images;
- Mean ADC values of the primitive tumor (ADC_{pre}), residual tissue after CRT (ADC_{post}), and ipsilateral trapezius muscle as a normative reference on both pre- and post-CRT ($ADC_{muscle\ pre}$ and $_{muscle\ post}$);
- Mean AUC and K(trans) values of the primitive tumor (AUC_{pre}, $K(trans)_{pre}$), residual tissue after CRT (AUC_{post}, $K(trans)_{post}$), and ipsilateral trapezius muscle as a normative reference on both pre- and post-CRT (AUC, $K(trans)_{muscle\ pre}$ and $_{muscle\ post}$);
- Ratio between ADC_{pre} and $ADC_{muscle\ pre}$ ($ADC_{pre/muscle\ pre}$);
- Ratio between ADC_{post} and $ADC_{muscle\ post}$ ($ADC_{post/muscle\ post}$);
- Ratio between AUC values of the residual tissue after CRT and primitive tumor ($AUC_{post/pre}$);
- Ratio between K(trans) values of the residual tissue after CRT and primitive tumor ($K(trans)_{post/pre}$);
- Ratio between AUC and K(trans) values of the residual tissue after CRT and primitive tumor, standardized with respect to AUC and K(trans) values of the ipsilateral trapezius muscle as a normative reference ($AUC_{post/pre/muscle}$ and $K(trans)_{post/pre/muscle}$), as follows:

$$\frac{AUCpost}{AUC\ muscle\ post} : \frac{AUC\ pre}{AUC\ muscle\ pre} \quad and \quad \frac{K(trans)post}{K(trans)\ muscle\ post} : \frac{K(trans)\ pre}{K(trans)\ muscle\ pre}$$

where $AUC_{muscle\ pre}$, $AUC_{muscle\ post}$, $K(trans)_{muscle\ pre}$, and $K(trans)_{muscle\ post}$ are the AUC and K(trans) values of the ipsilateral trapezius muscle measured on pre- and post-CRT, respectively.

The diagnosis of tumor response to CRT (post-treatment changes) or tumor persistence/recurrences (post-treatment residual cancer) was defined at the 2-year follow-up, with clinical examinations and cross-sectional imaging including MRI and 18F-FDG PET/CT. Post-treatment biopsy was performed only in case of positive 18F-FDG PET/CT during follow-up (12 patients). Clinical examinations and MRI were used to validate results as true negatives both in patients with negative 18F-FDG PET/CT (25) and in patients with positive 18F-FDG PET/CT and negative post-treatment biopsy (7).

2.4. Statistical Analysis

Quantitative continuous variables are expressed as mean $\pm$ standard deviation or median and range, whereas categorical values are reported as absolute counts and percentages. The interobserver reliability for MRI was calculated using the Cohen kappa coefficient. Kappa values of 0.01–0.20, 0.21–0.40, 0.41–0.60, 0.61–0.80, 0.81–0.99, and 1 represented slight, fair, moderate, substantial, almost perfect, and perfect agreement, respectively. Data were presented as a percentage or mean ($\pm$standard deviation) and median (interquartile range). Continuous variables were tested for normality using the Kolmogorov–Smirnov test. The association of each parameter and the diseased status at the follow-up (i.e., tumor persistence/recurrence or post-CRT changes) was tested using the Student's t-test or Mann–Whitney U-test for independent samples, as appropriate. For the parameters with statistically significant association with the diseased status at follow-up, a cut-off value to discriminate post-CRT changes with respect to tumor persistence/recurrence was calculated using receiver operating characteristic (ROC) curve analysis. In particular, sensitivity and specificity were calculated for the entire spectrum of values, and cut-offs were chosen as the values with the highest combination/multiplication of sensitivity and specificity. The area under the ROC curve was considered as a measure of the overall performance of each parameter (diagnostic accuracy) to discriminate the diseased status at follow-up. The analyses were performed using the SPSS® v. 27.0 statistical analysis software (IBM Corp., New York, NY, USA; formerly SPSS Inc., Chicago, IL, USA), considering an alpha level of 0.05 as significant.

3. Results

Post-CRT changes were found in 32 patients, whereas 5 patients had tumor persistence/recurrence. Results were summarized in Table 1 and Tables S3–S5 in Supplementary Materials. Cohen kappa values showed substantial agreement between the two observers for DWI and DCE-PWI assessments (K values 0.75 to 0.79).

Table 1. Mean, standard deviation, and range values of the post-chemoradiotherapy tissue changes and tumor persistence/recurrence. CRT: chemoradiotherapy. ADC: apparent diffusion coefficient. AUC: area under the curve. K(trans): the volume transfer constant from the vascular space to the extravascular extracellular space. *p*-value: probability value.

Magnetic Resonance Feature	Post-CRT Changes (32 Patients) Mean $\pm$ SD (Range)	Tumor Persistence/Recurrence (5 Patients) Mean $\pm$ SD (Range)	*p*-Value
Pre-treatment tumor maximum size (mm)	18.46 $\pm$ 7.22 (10.0–40.0)	20.50 $\pm$ 9.67 (7.0–30.0)	0.391
Pre-treatment tumor mean ADC value ($\times 10^{-3}$ mm^2/s) (ADCpre)	0.82 $\pm$ 0.15 (0.56–1.14)	0.89 $\pm$ 0.08 (0.80–1.0)	0.245
Pre-treatment tumor mean AUC value (AUCpre)	96.76 $\pm$ 44.15 (44.71–213.73)	101.61 $\pm$ 43.21 (55.66–159.70)	0.746
Pre-treatment tumor mean K(trans) value ($\times 10^{-3}$ min) (K(trans)pre)	264.80 $\pm$ 196.75 (61.39–786.30)	157.44 $\pm$ 55.02 (113.35–231.28)	0.498
$\dfrac{\text{ADCpre/trapezius muscle}}{\text{ADCpre}}$	0.69 $\pm$ 0.16 (0.50–1.0)	0.70 $\pm$ 0.08 (0.60–0.80)	0.746
$\dfrac{\text{AUCpre/trapezius muscle}}{\text{AUCpre}}$	3.70 $\pm$ 1.50 (1.53–6.73)	4.07 $\pm$ 1.23 (3.16–5.89)	0.536
$\dfrac{\text{K(trans)pre/trapezius muscle}}{\text{K(trans)pre}}$	5.62 $\pm$ 2.96 (1.52–12.92)	4.73 $\pm$ 0.38 (4.28–5.23)	0.702
Post-treatment residual tissue maximum submucosal enhancement thickness (mm)	3.31 $\pm$ 4.13 (0–10.0)	22.75 $\pm$ 17.09 (7.0–45.0)	0.002
Post-treatment residual tissue mean ADC value ($\times 10^{-3}$ mm^2/s) (ADCpost)	1.54 $\pm$ 0.23 (0.96–1.96)	1.05 $\pm$ 0.26 (0.78–1.32)	0.002
Post-treatment residual tissue mean AUC value (AUCpost)	105.10 $\pm$ 51.14 (35.37–260.88)	123.62 $\pm$ 49.26 (59.77–162.90)	0.425
Post-treatment tumor mean K(trans) value ($\times 10^{-3}$ min) (K(trans)pre)	181.80 $\pm$ 201.80 (25,54–787.92)	142.42 $\pm$ 67.63 (56.24–215.45)	0.659
$\dfrac{\text{ADCpost/trapezius muscle}}{\text{ADCpost}}$	1.24 $\pm$ 0.18 (0.80 –1.50)	0.87 $\pm$ 0.15 (0.8–1.10)	0.002
$\dfrac{\text{AUCpost/trapezius muscle}}{\text{AUCpost}}$	3.55 $\pm$ 1.24 (1.13–6.07)	4.34 $\pm$ 0.82 (3.52–5.43)	0.177
$\dfrac{\text{K(trans)post/trapezius muscle}}{\text{K(trans)post}}$	5.15 $\pm$ 4.57 (1.17–23.07)	4.86 $\pm$ 0.65 (4.08–5.41)	0.359
ADCpost-pre	0.70 $\pm$ 0.26 (0.16–1.20)	0.26 $\pm$ 0.40 (−0.13–0.70)	0.052
ADCpost-pre%	92.12 $\pm$ 44.10 (21.0–209.0)	20.75 $\pm$ 36.48 (−13.0–65.0)	0.005
AUCpost/pre	1.26 $\pm$ 0.79 (0.40–3.70)	1.24 $\pm$ 0.37 (1.01–1.80)	0.659
AUCpost/pre/trapezius muscle	1.08 $\pm$ 0.11 (0.37–2.70)	1.09 $\pm$ 0.57 (0.92–1.18)	0.791
K(trans)post/pre	1.06 $\pm$ 1.41 (0.06–5.66)	0.96 $\pm$ 0.63 (0.48–1.90)	0.498
K(trans)post/pre/trapezius muscle	1.07 $\pm$ 0.91 (0.30–4.01)	1.02 $\pm$ 0.11 (0.86–1.14)	0.271

ADC_{post} values > 1.33×10^{-3} mm^2/s, a percentage increase greater than 65.5% in mean ADC_{post} values compared to mean ADC_{pre} values ($\text{ADC}_{\text{post-pre\%}}$), and values > 0.72×10^{-3} mm^2/s in the difference between mean ADC_{post} and ADC_{pre} values ($\text{ADC}_{\text{post-pre}}$) strongly correlated with post-CRT changes (100% specificity, Figure 1A–C). $\text{ADC}_{\text{post/muscle post}}$ values > 1.15 and >0.85 showed 96.2% sensitivity and 100% specificity in the detection of post-CRT changes, respectively (Figure 1D). Hypointense SI on DWIb800 images well identified post-CRT changes since it was found in 30 patients (93.7%) with no residual cancer and 1 patient (20.0%) with tumor persistence/recurrence (specificity 80%).

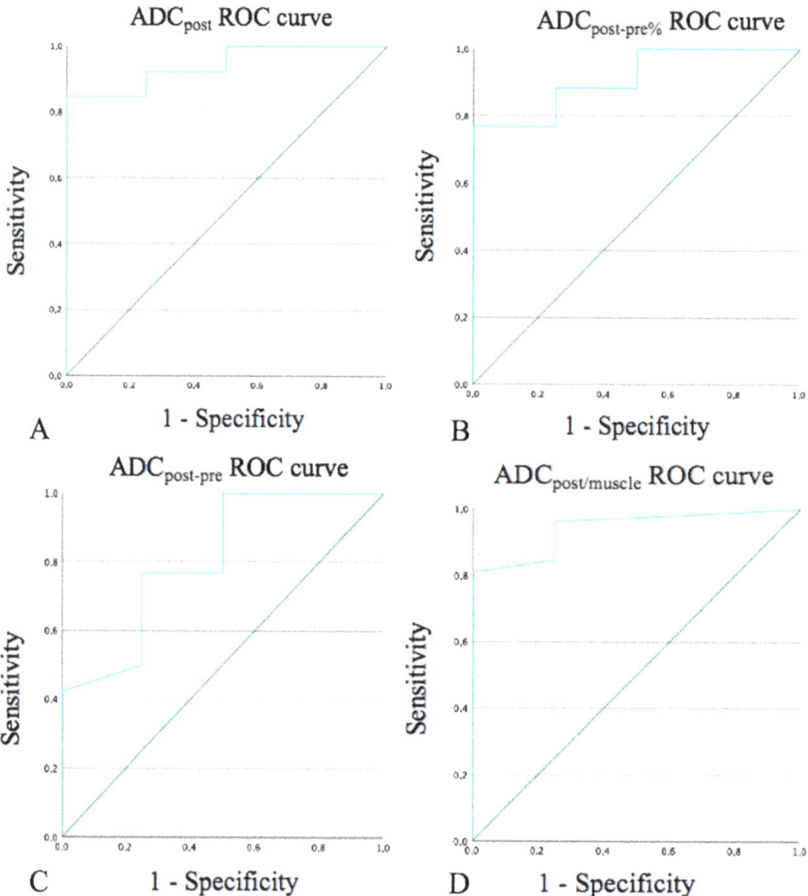

Figure 1. Receiver operating characteristic (ROC) curves for ADCpost values (**A**), ADCpost-pre% (**B**), ADCpost-pre values (**C**), and ADCpost/muscle values (**D**). ADC: apparent diffusion coefficient. ADCpost: residual tissue mean ADC value.

An overlap was found between mean ADC_{post} (Figure 2A), $AUC_{post/pre/muscle}$, and $K(trans)_{post/pre/muscle}$ values of post-CRT changes and tumor persistence/recurrence (Figure 2B,C). However, $K(trans)_{post/pre/muscle}$ values of 27 successfully treated patients (84.4%) were significantly different, higher or lower, than $K(trans)_{post/pre/muscle}$ values of all 5 patients with tumor persistence/recurrence. In such 27 patients, $K(trans)_{post/pre/muscle}$ values less than 0.85 suggested post-CRT fibrosis, whereas $K(trans)_{post/pre/muscle}$ values more than 1.20 indicated inflammatory edema.

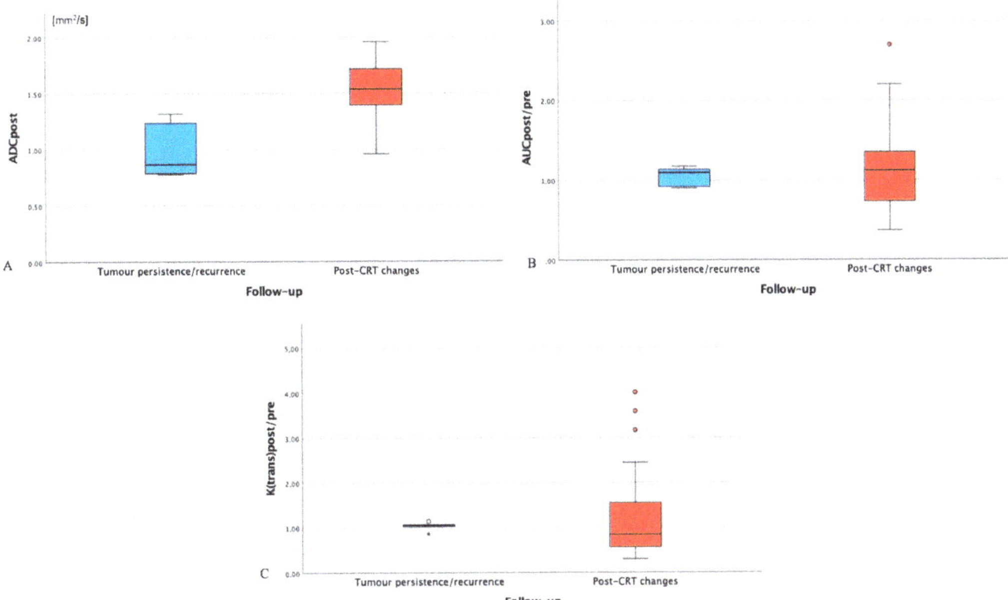

Figure 2. Box plot for post-treatment residual tissue ADC values (ADCpost, (**A**)), AUCpost/(AUC muscle post):(AUC pre)/(AUC muscle pre) values (AUCpost/pre/muscle, (**B**)), and (K(trans) post)/(K(trans) muscle post):(K(trans) pre)/ (K(trans) muscle pre) values (K(trans)post/pre/muscle, (**C**)) in patients with tumor persistence/recurrence (blue box) and post-chemoradiotherapy (CRT) changes (red box). ADC: apparent diffusion coefficient. AUC: area under the curve. CRT: chemoradiotherapy. AUCpre: AUC values of primitive tumor. AUCpost: AUC values of the residual tissue after CRT. AUCmuscle pre: AUC values of ipsilateral trapezius muscle on pre-treatment magnetic resonance imaging. AUCmuscle post: AUC values of ipsilateral trapezius muscle on post-treatment magnetic resonance imaging. K(trans)pre: K(trans) values of primitive tumor. K(trans)post: K(trans) values of the residual tissue after CRT. K(trans)muscle pre: K(trans) values of ipsilateral trapezius muscle on pre-treatment magnetic resonance imaging. K(trans)muscle post: K(trans) values of ipsilateral trapezius muscle on post-treatment magnetic resonance imaging. Circles: drawing of comparison circles is a way to display whether or not the mean values of boxes in the box plot are significantly different from each other. Asterisk: asterisk is an indication that an extreme outlier is present in the data.

4. Discussion

Quantitative DWI and DCE-PWI analyses may portend the efficacy of CRT and early identification of potential treatment failure, resulting in an improvement in cancer management. In the current study, the quantitative analysis with DWI sequences allowed a reliable tumor assessment during the treatment phase. A low increase in $ADC_{post-pre}$ and $ADC_{post-pre\%}$ values was indicative of a high risk of residual cancer as directed by Wong et al. [26]. $ADC_{post/muscle\ post}$ values > 0.85 and hypointense SI on DWIb800 images strongly correlated with post-CRT changes. Most of our patients with post-CRT changes (27/32, 84.3%) showed $K(trans)_{post/pre/muscle}$ values significantly lower (<0.85, 19 patients) or higher (>1.20, 8 patients) than all 5 patients with tumor persistence/recurrence. As for DWI [7], the aforementioned variations of DCE-PWI values could reflect the different tissue components, mainly fibrotic (Figure 3) or inflammatory (Figure 4) alterations, of post-treatment changes.

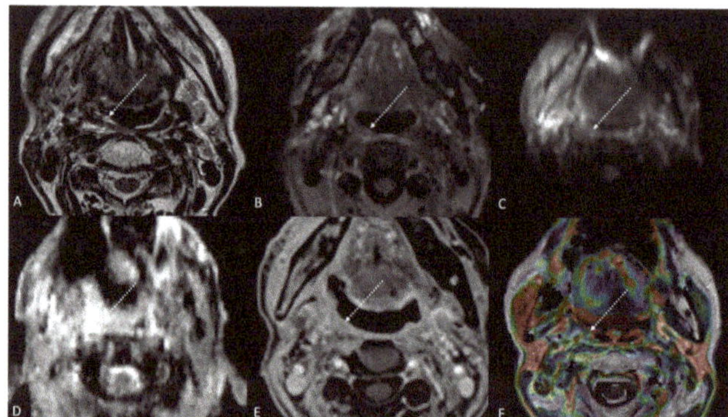

Figure 3. Post-treatment magnetic resonance imaging (MRI) of a 63-year-old female patient with human-papillomavirus-positive carcinoma of the right palatine tonsil with ipsilateral lymph node metastasis (T2N1) recently treated (3 months before) with chemoradiotherapy (CRT). Post-CRT MRI showed linear fibrotic tissues in the right palatine tonsil (white striped arrows) with hypointense signal intensity on T2-weighted (**A**), T2-weighted fat-saturated, (**B**) and diffusion-weighted b800 images (**C**), and intermediate apparent diffusion coefficient values (1.44×10^{-3} mm^2/s) (**D**). After gadolinium contrast agent injection, post-CRT fibrotic tissue showed no submucosal enhancement (**E**) and low K(trans) value (48.36×10^{-3} mm^2/s) on dynamic contrast enhancement-perfusion weighted imaging (**F**). Ratio between K(trans) values of the primitive tumor and residual tissue after CRT, standardized with respect to K(trans) value of the ipsilateral trapezius (K(trans)post/pre/muscle), was 0.307. These findings are typical of post-CRT scar tissue.

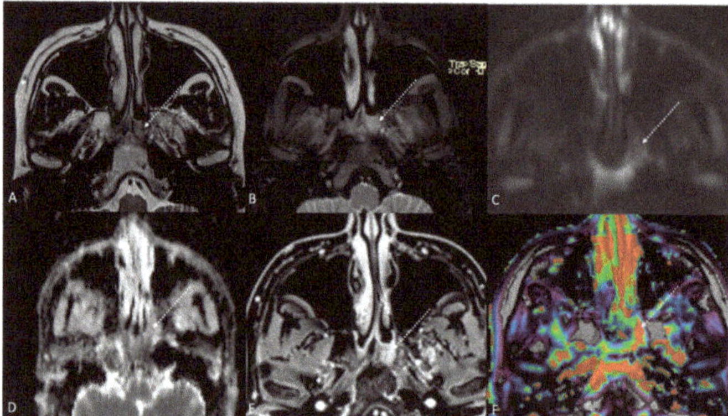

Figure 4. Post-treatment magnetic resonance imaging (MRI) of a 36-year-old female patient affected by nasopharyngeal carcinoma with left lymph node metastasis (T3N3) and tumoral extension to bilateral Ronsemüller fossa, left nasal choana, and middle skull base, recently treated (3 months before) with chemoradiotherapy (CRT). Early post-treatment MRI demonstrated post-CRT inflammatory residual tissue (white striped arrows) in the left Ronsemüller fossa and ipsilateral nasal choana. Post-CRT inflammatory changes showed hyperintense signal on T2-weighted (**A**), T2-weighted fat-saturated (**B**), and diffusion-weighted b800 images (**C**); high apparent diffusion coefficient value (1.53×10^{-3} mm^2/s) (**D**). After gadolinium contrast agent injection, post-CRT inflammatory residual tissue shows submucosal enhancement of 5 mm thickness (**E**), and very high K(trans) value (595.25×10^{-3} mm^2/s) on dynamic contrast enhancement-perfusion weighted imaging (**F**). Ratio between K(trans) values of the primitive tumor and residual tissue after CRT, standardized with respect to K(trans) value of the ipsilateral trapezius (K(trans)post/pre/muscle), was 3.17. These findings suggested an increase in capillary permeability caused by CRT.

Sherif et al. [27] found ADC values of $1.42 \pm 0.23 \times 10^{-3}$ mm^2/s and $1.02 \pm 0.20 \times 10^{-3}$ mm^2/s in post-therapy changes of patients treated for tongue carcinoma and tongue carcinoma recurrence, respectively. Taking as a reference such ADC values, in our study, 24 patients with post-CRT changes showed ADC$_{post}$ values $> 1.42 \times 10^{-3}$ mm^2/s (mean = 1.56×10^{-3} mm^2/s), whereas in the remaining 8 patients with post-CRT changes, ADC$_{post}$ values (mean = 1.24×10^{-3} mm^2/s; range = 0.96–1.35×10^{-3} mm^2/s) were similar to ADC$_{post}$ values of all 5 patients with tumor persistence/recurrence (mean = 1.05×10^{-3} mm^2/s; range = 0.78–1.32×10^{-3} mm^2/s) (Figure 2A). Ailianou et al. [7] found that mean ADC values in post-treatment HNSCC highly differed between post-radiation therapy inflammatory edema ($1.75 \pm 0.34 \times 10^{-3}$ mm^2/s) and late fibrosis ($0.98 \pm 0.26 \times 10^{-3}$ mm^2/s). These results may justify overlaps of ADC values between post-CRT and tumor recurrence both in our study and in other papers [28–42].

18F-FDG PET/CT is frequently used for treatment response assessment. It shows high sensitivity but low specificity [43], especially in the first 6 months after treatment due to inflammation, granulation, and scar tissues [44]. In the present study, 18F-FDG PET/CT performed 3–6 months after ending the treatment was positive in 12 patients, but only 5 of them had tumor persistence/recurrence at the 2-year follow-up. Compared to 18F-FDG PET/CT, ADC can be also performed in the first months after CRT to assess treatment response, but false positives and negatives cannot be fully excluded. However, studies that used ADC values without taking into account DWI SI underestimated the accuracy of diffusion-weighted MRI [45]. Scar tissue generally displays low ADC values in combination with the hypointense signal on high b value DWI images due to the low number of resonant protons. Residual cancer usually shows low values on ADC maps too, but together with the hyperintense signal on DWI images [46]. The combination of DWI and morphologic MRI features yields better results than DWI alone [7,31,44,47]. The evaluation of SI on T2 images in the current study agreed with the literature since masslike alterations with moderately high (i.e., intermediate) SI, diffuse alterations with high SI, and linear or triangular alterations with very low SI (similar to or lower than muscle) were suggestive for tumor persistence/recurrence, post-CRT inflammatory edema, and post-CRT fibrosis, respectively.

In the current study, K(trans)$_{post/pre/muscle}$ values less than 0.85 suggested post-CRT fibrosis, whereas K(trans)$_{post/pre/muscle}$ values more than 1.20 indicated inflammatory edema. Vascular changes associated with residual cancer represent neoangiogenesis; on the contrary, post-treatment non-tumoral alterations show vascular changes of continued successful therapy and fibrosis [48]. Post-treatment changes may lead to significant variations in DCE-PWI parameter values since K(trans) is sensitive to angiogenic modifications [49]. Therefore, although with some degrees of overlap, little or no change in AUC$_{post/pre/muscle}$ and mean K(trans)$_{post/pre/muscle}$ values, i.e., tumoral neoangiogenesis, may be considered a post-treatment indicator of tumor persistence/recurrence (Figure 5).

Some limitations need to be mentioned. The relationship among MRI and HNSCC stage, lymph node, distant metastasis, histological tumor grading, histopathological parameters, progression-free survival, HPV status, intravoxel incoherent motion, or tumoral 18F-FDG PET/CT standard uptake values were not performed. Moreover, we compared tissue changes between pre- and post-CRT without taking into account pre-treatment MRI features only as predictors of treatment response.

Another limitation of the present study was the relatively low sample size. Nevertheless, most papers regarding HNSCC and functional MRI did not consider both DWI and DCE-PWI for therapy assessment or did not include both pre- and post-treatment MRI examinations. Moreover, few papers exclusively recruited patients with pharyngeal cancer [11,28,50,51], and only two of these were performed with both DWI and DCE-PWI [11,51]. In addition, the small number of patients with tumor persistence/recurrence (5 individuals) needed to be related to the well-known excellent response to CRT treatment of oropharyngeal—especially when HPV positive—and nasopharyngeal carcinomas.

Furthermore, HPV+ and HPV− HNSCC generally differ in radiological imaging and prognosis [52], thus representing a possible bias in the current study.

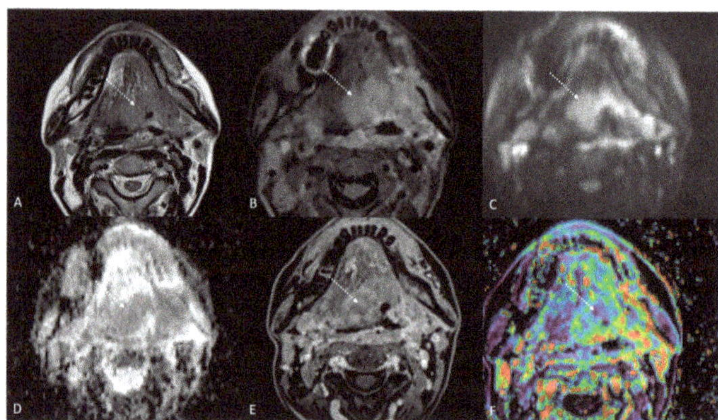

Figure 5. Post-treatment magnetic resonance imaging (MRI) of a 47-year-old female patient with human-papillomavirus-negative carcinoma of the left palatine tonsil with ipsilateral lymph node metastasis (T4aN1) and buccal space and mandibular invasion, recently treated (4 months before) with chemoradiotherapy (CRT). Early post-treatment MRI showed tumor progression with wide extension to the extrinsic muscles of the contralateral tongue (maximum tumor thickness 45 mm). Post-CRT tumor residual/relapse disease (white striped arrows) showed moderately high (intermediate) T2-weighted signal intensity (**A,B**), high signal on diffusion-weighted b800 imaging (**C**), low apparent diffusion coefficient value (0.79×10^{-3} mm^2/s, (**D**), and moderate enhancement after gadolinium contrast injection (**E**). K(trans) value of the tumor (56.24×10^{-3} min) decreased on dynamic contrast enhancement-perfusion weighted imaging (**F**), compared to pre-treatment MRI (117.63×10^{-3} min). However, the ratio between K(trans) values of the primitive tumor and residual tissue after CRT, standardized with respect to K(trans) value of the ipsilateral trapezius (K(trans)$_{post/pre/muscle}$), was 1.14. These findings suggested little or no reduction in tumor neoangiogenesis after CRT.

Moreover, our single-center results cannot be generalized until more evidence is gathered.

Finally, the study design did not allow the calculation of the outcome incidence. For this reason, a discussion of the appropriateness of the cut-off values with respect to the rate of false positives was not possible. Future studies with a different design should help in choosing appropriate cut-off values that balance the benefits to true positives (e.g., increased survival) versus the costs to false positives (e.g., unnecessary procedures).

To date, MRI evaluation in strictly morphologic terms represented by the SI on T1 and T2 images and grade of enhancement is still mandatory in HNSCC. Considering the relative complexity of DWI and DCE-PWI parameters that have been used and the low number of retrieved patients, the results obtained in our study are currently available for research purposes only. Further studies will be needed to establish whether or not multiparametric MRI examinations can be successfully used in clinical daily practice.

5. Conclusions

In early follow-up of naso-oropharyngeal carcinoma, ADC$_{post}$ values $> 1.33 \times 10^{-3}$ mm^2/s, ADC$_{post-pre\%}$ $> 65.5\%$, and ADC$_{post-pre}$ values $> 0.72 \times 10^{-3}$ mm^2/s identified post-CRT changes with excellent specificity. Although mean AUC$_{post/pre/muscle}$ and K(trans)$_{post/pre/muscle}$ were similar in post-CRT changes (1.10 ± 0.58; 1.08 ± 0.91) and tumor persistence/recurrence (1.09 ± 0.11; 1.03 ± 0.12), in our sample K(trans)$_{post/pre/muscle}$ values less than 0.85 suggested post-CRT fibrosis, whereas K(trans)$_{post/pre/muscle}$ values more than 1.20 indicated inflammatory edema.

Supplementary Materials: The following supporting information can be downloaded at: https://www.mdpi.com/article/10.3390/app13052799/s1, Materials and Methods. Table S1: Patients' data retrieved in the study. M: male; F: female; O: oropharynx; N: nasopharynx; P: positive; N: negative; U: unknown; RT: post-chemoradiotherapy changes; PR: tumor persistence/recurrence. *The eighth edition of the American Joint Commission on Cancer TNM staging; Table S2. Magnetic resonance acquisition protocol performed for the study of naso-oropharyngeal carcinoma staging and 4-month follow-up from ending chemoradiotherapy. Unenhanced scans included sagittal fat-saturated T1- and T2- weighted sampling perfection with application-optimized contrasts using different flip angle evolution (SPACE) sequences with axial, coronal, and sagittal multiplanar reconstructions; axial T2-weighted turbo spin echo; axial fat-saturated echo-planar DWI spectral attenuated inversion recovery (SPAIR) with two b-values (b50–800 s/mm^2) and ADC maps; two axial T1-weighted volumetric interpolated breath-hold examination (VIBE) DCE-PWI with application of flip angles (FAs) 5° and 15°, respectively. Enhanced scans performed after intravenous gadolinium chelates contrast agent injection (gadobutrol, 1 mL/10 kg, flow 3 mL/s, followed by 20 mL saline flush) consisted of an axial VIBE DCE-PWI with application of FA 30° and peripheral K space sampling with time to center 2.2 s, an axial T1-weighted turbo spin echo, and an axial VIBE Dixon. Results. Table S3. Pre-treatment patients' data. ADC: apparent diffusion coefficient; AUC: area under the curve; k(trans): the volume transfer constant from the vascular space to the extravascular extracellular space; pre: values measured on magnetic resonance imaging performed for tumor staging; Table S4. Post-treatment patients' data. T2 signal intensity is referred to with respect to the muscle. Hypo: lower than muscle. Hyper+: similar or slightly higher than muscle. Higher++: clearly higher than muscle; ADC: apparent diffusion coefficient; AUC: area under the curve; K(trans): the volume transfer constant from the vascular space to the extravascular extracellular space; post: values measured on magnetic resonance imaging performed for 4-month follow-up; Table S5. Comparison between post-treatment and pre-treatment patients' data. ADC: apparent diffusion coefficient (expressed in $\times 10^{-3}$ mm^2/s). ADCpost-pre: residual tissue mean ADC value—tumor mean ADC value. ADCpost-pre%: residual tissue mean ADC value—tumor mean ADC value, expressed in percentage calculated as follows: (ADCpost-pre $\times$ 100)/ADCpre. Negative percentages indicate that ADCpost values are lower than ADCpre. AUC: area under the curve. AUCpost/pre: ratio between the residual tissue AUC and tumor AUC values. AUCpost/pre/muscle: ratio between residual tissue AUC and tumor AUC values, standardized with respect to AUC values of the ipsilateral trapezius muscle. K(trans): the volume transfer constant from the vascular space to the extravascular extracellular space. K(trans)post/pre: ratio between the residual tissue K(trans) value and tumor K(trans) value. K(trans)post/pre/muscle: ratio between the residual tissue K(trans) and tumor K(trans) values, standardized with respect to K(trans) values of the ipsilateral trapezius muscle.

Author Contributions: Conceptualization, M.P. and C.N.; Data curation, A.B.; Formal analysis, C.L.; Investigation, G.M.; Methodology, E.B.; Project administration, M.P.; Resources, L.B.; Software, E.B.; Supervision, A.L.C.; Validation, L.C., F.M. and L.B.; Visualization, P.B. and C.N.; Writing—original draft, M.P. All authors have read and agreed to the published version of the manuscript.

Funding: This research received no external funding.

Institutional Review Board Statement: The study was conducted in accordance with the Declaration of Helsinki, and approved by the "Ethics Committee of Azienda Ospedaliera Universitaria Careggi (Protocol No. 21800_oss, 22 March 2022)" for studies involving humans.

Informed Consent Statement: Informed consent was obtained from all subjects involved in the study.

Data Availability Statement: The original contributions presented in the study are included in the article/Supplementary Material. Further inquiries can be directed to the corresponding author.

Conflicts of Interest: The authors declare no conflict of interest.

References

1. Locatello, L.G.; Bruno, C.; Pietragalla, M.; Taverna, C.; Novelli, L.; Nardi, C.; Bonasera, L.; Cannavicci, A.; Maggiore, G.; Gallo, O. A critical evaluation of computed tomography-derived depth of invasion in the preoperative assessment of oral cancer staging. *Oral Oncol.* **2020**, *107*, 104749. [CrossRef]
2. Maraghelli, D.; Pietragalla, M.; Calistri, L.; Barbato, L.; Locatello, L.G.; Orlandi, M.; Landini, N.; Casto, A.L.; Nardi, C. Techniques, Tricks, and Stratagems of Oral Cavity Computed Tomography and Magnetic Resonance Imaging. *Appl. Sci.* **2022**, *12*, 1473. [CrossRef]
3. Mungai, F.; Verrone, G.B.; Pietragalla, M.; Berti, V.; Addeo, G.; Desideri, I.; Bonasera, L.; Miele, V. CT assessment of tumor heterogeneity and the potential for the prediction of human papillomavirus status in oropharyngeal squamous cell carcinoma. *Radiol. Med.* **2019**, *124*, 804–811. [CrossRef]
4. Jajodia, A.; Aggarwal, D.; Chaturvedi, A.K.; Rao, A.; Mahawar, V.; Gairola, M.; Agarwal, M.; Goyal, S.; Koyyala, V.P.B.; Pasricha, S.; et al. Value of diffusion MR imaging in differentiation of recurrent head and neck malignancies from post treatment changes. *Oral Oncol.* **2019**, *96*, 89–96. [CrossRef]
5. Locatello, L.G.; Pietragalla, M.; Taverna, C.; Bonasera, L.; Massi, D.; Mannelli, G. A Critical Reappraisal of Primary and Recurrent Advanced Laryngeal Cancer Staging. *Ann. Otol. Rhinol. Laryngol.* **2018**, *128*, 36–43. [CrossRef]
6. Brockstein, B.; Haraf, D.J.; Rademaker, A.W.; Kies, M.S.; Stenson, K.M.; Rosen, F.; Mittal, B.B.; Pelzer, H.; Fung, B.B.; Witt, M.-E.; et al. Patterns of failure, prognostic factors and survival in locoregionally advanced head and neck cancer treated with concomitant chemoradiotherapy: A 9-year, 337-patient, multi-institutional experience. *Ann. Oncol.* **2004**, *15*, 1179–1186. [CrossRef]
7. Ailianou, A.; Mundada, P.; De Perrot, T.; Pusztaszieri, M.; Poletti, P.-A.; Becker, M. MRI with DWI for the Detection of Posttreatment Head and Neck Squamous Cell Carcinoma: Why Morphologic MRI Criteria Matter. *Am. J. Neuroradiol.* **2018**, *39*, 748–755. [CrossRef]
8. Teicher, B.A. Hypoxia and drug resistance. *Cancer Metastasis Rev.* **1994**, *13*, 139–168. [CrossRef]
9. Mukherji, S.K.; Wolf, G.T. Evaluation of Head and Neck Squamous Cell Carcinoma After Treatment. *Am. J. Neuroradiol.* **2003**, *24*, 1743–1746.
10. Nardi, C.; Maraghelli, D.; Pietragalla, M.; Scola, E.; Locatello, L.G.; Maggiore, G.; Gallo, O.; Bartolucci, M. A practical overview of CT and MRI features of developmental, inflammatory, and neoplastic lesions of the sphenoid body and clivus. *Neuroradiology* **2022**, *64*, 1483–1509. [CrossRef]
11. Martens, R.M.; Koopman, T.; Lavini, C.; Ali, M.; Peeters, C.F.W.; Noij, D.P.; Zwezerijnen, G.; Marcus, J.T.; Vergeer, M.R.; Leemans, C.R.; et al. Multiparametric functional MRI and 18F-FDG-PET for survival prediction in patients with head and neck squamous cell carcinoma treated with (chemo)radiation. *Eur. Radiol.* **2020**, *31*, 616–628. [CrossRef]
12. Gaddikeri, S.; Tailor, T.; Anzai, Y. Dynamic Contrast-Enhanced MR Imaging in Head and Neck Cancer: Techniques and Clinical Applications. *Am. J. Neuroradiol.* **2015**, *37*, 588–595. [CrossRef]
13. Bernstein, J.M.; Homer, J.J.; West, C.M. Dynamic contrast-enhanced magnetic resonance imaging biomarkers in head and neck cancer: Potential to guide treatment? A systematic review. *Oral Oncol.* **2014**, *50*, 963–970. [CrossRef]
14. Mehanna, H.; Wong, W.-L.; McConkey, C.C.; Rahman, J.K.; Robinson, M.; Hartley, A.G.J.; Nutting, C.; Powell, N.; Al-Booz, H.; Robinson, M.; et al. PET-CT surveillance versus neck dissection in advanced head and neck cancer. *N. Engl. J. Med.* **2016**, *374*, 1444–1454. [CrossRef]
15. Meyer, H.J.; Leifels, L.; Schob, S.; Garnov, N.; Surov, A. Histogram analysis parameters identify multiple associations between DWI and DCE MRI in head and neck squamous cell carcinoma. *Magn. Reson. Imaging* **2018**, *45*, 72–77. [CrossRef]
16. Nardi, C.; Tomei, M.; Pietragalla, M.; Calistri, L.; Landini, N.; Bonomo, P.; Mannelli, G.; Mungai, F.; Bonasera, L.; Colagrande, S. Texture analysis in the characterization of parotid salivary gland lesions: A study on MR diffusion weighted imaging. *Eur. J. Radiol.* **2021**, *136*, 109529. [CrossRef]
17. Kabadi, S.J.; Fatterpekar, G.M.; Anzai, Y.; Mogen, J.; Hagiwara, M.; Patel, S.H. Dynamic Contrast-Enhanced MR Imaging in Head and Neck Cancer. *Magn. Reson. Imaging Clin. N. Am.* **2018**, *26*, 135–149. [CrossRef]
18. Connolly, M.; Srinivasan, A. Diffusion-Weighted Imaging in Head and Neck Cancer. *Magn. Reson. Imaging Clin. N. Am.* **2017**, *26*, 121–133. [CrossRef]
19. Malayeri, A.A.; El Khouli, R.H.; Zaheer, A.; Jacobs, M.A.; Corona-Villalobos, C.P.; Kamel, I.R.; Macura, K.J. Principles and Applications of Diffusion-weighted Imaging in Cancer Detection, Staging, and Treatment Follow-up. *Radiographics* **2011**, *31*, 1773–1791. [CrossRef]
20. Baliyan, V.; Das, C.J.; Sharma, R.; Gupta, A.K. Diffusion weighted imaging: Technique and applications. *World J. Radiol.* **2016**, *8*, 785–798. [CrossRef]
21. Nardi, C.; Vignoli, C.; Vannucchi, M.; Pietragalla, M. Magnetic resonance features of sinonasal melanotic mucosal melanoma. *BMJ Case Rep.* **2019**, *12*, e229790. [CrossRef]
22. Pietragalla, M.; Nardi, C.; Bonasera, L.; Mungai, F.; Taverna, C.; Novelli, L.; De Renzis, A.G.D.; Calistri, L.; Tomei, M.; Occhipinti, M.; et al. The role of diffusion-weighted and dynamic contrast enhancement perfusion-weighted imaging in the evaluation of salivary glands neoplasms. *Radiol. Med.* **2020**, *125*, 851–863. [CrossRef]
23. Mungai, F.; Verrone, G.B.; Bonasera, L.; Bicci, E.; Pietragalla, M.; Nardi, C.; Berti, V.; Mazzoni, L.N.; Miele, V. Imaging biomarkers in the diagnosis of salivary gland tumors: The value of lesion/parenchyma ratio of perfusion-MR pharmacokinetic parameters. *Radiol. Med.* **2021**, *126*, 1345–1355. [CrossRef]

24. Rajabi, M.; Mousa, S.A. The Role of Angiogenesis in Cancer Treatment. *Biomedicines* **2017**, *5*, 34. [CrossRef]
25. Petralia, G.; Summers, P.E.; Agostini, A.; Ambrosini, R.; Cianci, R.; Cristel, G.; Calistri, L.; Colagrande, S. Dynamic contrast-enhanced MRI in oncology: How we do it. *Radiol. Med.* **2020**, *125*, 1288–1300. [CrossRef]
26. Wong, K.H.; Panek, R.; Welsh, L.C.; Mcquaid, D.; Dunlop, A.; Riddell, A.; Murray, I.; Du, Y.; Chua, S.; Koh, D.-M.; et al. The Predictive Value of Early Assessment After 1 Cycle of Induction Chemotherapy with ^{18}F-FDG PET/CT and Diffusion-Weighted MRI for Response to Radical Chemoradiotherapy in Head and Neck Squamous Cell Carcinoma. *J. Nucl. Med.* **2016**, *57*, 1843–1850. [CrossRef]
27. Sherif, M.M.A.-R.A.F.M. Value of Diffusion-Weighted and Perfusion-Weighted MR Imaging in Differentiation of Recurrent Tongue Carcinoma from Post-Treatment Changes. *Med. J. Cairo Univ.* **2020**, *88*, 1893–1902. [CrossRef]
28. Connor, S.; Sit, C.; Anjari, M.; Lei, M.; Guerrero-Urbano, T.; Szyszko, T.; Cook, G.; Bassett, P.; Goh, V. The ability of post-chemoradiotherapy DWI ADCmean and 18F-FDG SUVmax to predict treatment outcomes in head and neck cancer: Impact of human papilloma virus oropharyngeal cancer status. *J. Cancer Res. Clin. Oncol.* **2021**, *147*, 2323–2336. [CrossRef]
29. Galbán, C.J.; Lemasson, B.; Hoff, B.A.; Johnson, T.D.; Sundgren, P.; Tsien, C.; Chenevert, T.L.; Ross, B.D. Development of a Multiparametric Voxel-Based Magnetic Resonance Imaging Biomarker for Early Cancer Therapeutic Response Assessment. *Tomography* **2015**, *1*, 44–52. [CrossRef]
30. Paudyal, R.; Oh, J.H.; Riaz, N.; Venigalla, P.; Li, J.; Hatzoglou, V.; Leeman, J.; Nunez, D.A.; Lu, Y.; Deasy, J.O.; et al. Intravoxel incoherent motion diffusion-weighted MRI during chemoradiation therapy to characterize and monitor treatment response in human papillomavirus head and neck squamous cell carcinoma. *J. Magn. Reson. Imaging* **2016**, *45*, 1013–1023. [CrossRef]
31. King, A.; Keung, C.; Yu, K.-H.; Mo, F.; Bhatia, K.; Yeung, D.; Tse, G.; Vlantis, A.; Ahuja, A. T2-Weighted MR Imaging Early after Chemoradiotherapy to Evaluate Treatment Response in Head and Neck Squamous Cell Carcinoma. *Am. J. Neuroradiol.* **2013**, *34*, 1237–1241. [CrossRef]
32. King, A.D.; Chow, K.-K.; Yu, K.-H.; Mo, F.K.F.; Yeung, D.K.W.; Yuan, J.; Bhatia, K.S.; Vlantis, A.; Ahuja, A.T. Head and Neck Squamous Cell Carcinoma: Diagnostic Performance of Diffusion-weighted MR Imaging for the Prediction of Treatment Response. *Radiology* **2013**, *266*, 531–538. [CrossRef]
33. King, A.D.; Mo, F.K.; Yu, K.-H.; Yeung, D.K.W.; Zhou, H.; Bhatia, K.S.; Tse, G.M.K.; Vlantis, A.; Wong, J.K.T.; Ahuja, A.T. Squamous cell carcinoma of the head and neck: Diffusion-weighted MR imaging for prediction and monitoring of treatment response. *Eur. Radiol.* **2010**, *20*, 2213–2220. [CrossRef]
34. Vandecaveye, V.; Dirix, P.; De Keyzer, F.; de Beeck, K.O.; Poorten, V.V.; Hauben, E.; Lambrecht, M.; Nuyts, S.; Hermans, R. Diffusion-Weighted Magnetic Resonance Imaging Early After Chemoradiotherapy to Monitor Treatment Response in Head-and-Neck Squamous Cell Carcinoma. *Int. J. Radiat. Oncol.* **2011**, *82*, 1098–1107. [CrossRef]
35. Brenet, E.; Barbe, C.; Hoeffel, C.; Dubernard, X.; Merol, J.-C.; Fath, L.; Servagi-Vernat, S.; Labrousse, M. Predictive Value of Early Post-Treatment Diffusion-Weighted MRI for Recurrence or Tumor Progression of Head and Neck Squamous Cell Carcinoma Treated with Chemo-Radiotherapy. *Cancers* **2020**, *12*, 1234. [CrossRef]
36. Kim, S.; Loevner, L.; Quon, H.; Sherman, E.; Weinstein, G.; Kilger, A.; Poptani, H. Diffusion-Weighted Magnetic Resonance Imaging for Predicting and Detecting Early Response to Chemoradiation Therapy of Squamous Cell Carcinomas of the Head and Neck. *Clin. Cancer Res.* **2009**, *15*, 986–994. [CrossRef]
37. Berrak, S.; Chawla, S.; Kim, G.; Quon, H.; Sherman, E.; Loevner, L.A.; Poptani, H. Diffusion Weighted Imaging in Predicting Progression Free Survival in Patients with Squamous Cell Carcinomas of the Head and Neck Treated with Induction Chemotherapy. *Acad. Radiol.* **2011**, *18*, 1225–1232. [CrossRef]
38. Cao, Y.; Aryal, M.; Li, P.; Lee, C.; Schipper, M.; Hawkins, P.G.; Chapman, C.; Owen, D.; Dragovic, A.F.; Swiecicki, P.; et al. Predictive Values of MRI and PET Derived Quantitative Parameters for Patterns of Failure in Both p16+ and p16– High Risk Head and Neck Cancer. *Front. Oncol.* **2019**, *9*, 1118. [CrossRef]
39. Matoba, M.; Tuji, H.; Shimode, Y.; Toyoda, I.; Kuginuki, Y.; Miwa, K.; Tonami, H. Fractional Change in Apparent Diffusion Coefficient as an Imaging Biomarker for Predicting Treatment Response in Head and Neck Cancer Treated with Chemoradiotherapy. *Am. J. Neuroradiol.* **2013**, *35*, 379–385. [CrossRef]
40. Marzi, S.; Piludu, F.; Sanguineti, G.; Marucci, L.; Farneti, A.; Terrenato, I.; Pellini, R.; Benevolo, M.; Covello, R.; Vidiri, A. The prediction of the treatment response of cervical nodes using intravoxel incoherent motion diffusion-weighted imaging. *Eur. J. Radiol.* **2017**, *92*, 93–102. [CrossRef]
41. Ding, Y.; Hazle, J.D.; Mohamed, A.S.; Frank, S.J.; Hobbs, B.P.; Colen, R.R.; Gunn, G.B.; Wang, J.; Kalpathy-Cramer, J.; Garden, A.S.; et al. Intravoxel incoherent motion imaging kinetics during chemoradiotherapy for human papilloma-virus-associated squamous cell carcinoma of the oropharynx: Preliminary results form a prospective pilot study. *NMR Biomed.* **2015**, *28*, 1645–1654. [CrossRef]
42. Vogel, D.W.T.; Zbaeren, P.; Geretschlaeger, A.; Vermathen, P.; De Keyzer, F.; Thoeny, H.C. Diffusion-weighted MR imaging including bi-exponential fitting for the detection of recurrent or residual tumour after (chemo)radiotherapy for laryngeal and hypopharyngeal cancers. *Eur. Radiol.* **2012**, *23*, 562–569. [CrossRef]
43. Sheikhbahaei, S.; Taghipour, M.; Ahmad, R.; Fakhry, C.; Kiess, A.P.; Chung, C.H.; Subramaniam, R.M. Diagnostic Accuracy of Follow-Up FDG PET or PET/CT in Patients With Head and Neck Cancer After Definitive Treatment: A Systematic Review and Meta-Analysis. *Am. J. Roentgenol.* **2015**, *205*, 629–639. [CrossRef]

44. Varoquaux, A.; Rager, O.; Dulguerov, P.; Burkhardt, K.; Ailianou, A.; Becker, M. Diffusion-weighted and PET/MR Imaging after Radiation Therapy for Malignant Head and Neck Tumors. *Radiographics* **2015**, *35*, 1502–1527. [CrossRef]
45. Van Der Hoorn, A.; Van Laar, P.J.; Holtman, G.A.; Westerlaan, H.E. Diagnostic accuracy of magnetic resonance imaging techniques for treatment response evaluation in patients with head and neck tumors, a systematic review and meta-analysis. *PLoS ONE* **2017**, *12*, e0177986. [CrossRef]
46. Hustinx, R.; Lucignani, G. PET/CT in head and neck cancer: An update. *Eur. J. Nucl. Med.* **2010**, *37*, 645–651. [CrossRef]
47. Becker, M.; Varoquaux, A.D.; Combescure, C.; Rager, O.; Pusztaszeri, M.; Burkhardt, K.; Delattre, B.M.A.; Dulguerov, P.; Dulguerov, N.; Katirtzidou, E.; et al. Local recurrence of squamous cell carcinoma of the head and neck after radio(chemo)therapy: Diagnostic performance of FDG-PET/MRI with diffusion-weighted sequences. *Eur. Radiol.* **2017**, *28*, 651–663. [CrossRef]
48. El Beltagi, A.H.; ElSotouhy, A.H.; Own, A.M.; Abdelfattah, W.; Nair, K.; Vattoth, S. Functional magnetic resonance imaging of head and neck cancer: Performance and potential. *Neuroradiol. J.* **2018**, *32*, 36–52. [CrossRef]
49. Albano, D.; Bruno, F.; Agostini, A.; Angileri, S.A.; Benenati, M.; Bicchierai, G.; Cellina, M.; Chianca, V.; Cozzi, D.; Danti, G.; et al. Dynamic contrast-enhanced (DCE) imaging: State of the art and applications in whole-body imaging. *Jpn. J. Radiol.* **2021**, *40*, 341–366. [CrossRef]
50. Straub, J.M.; New, J.; Hamilton, C.D.; Lominska, C.; Shnayder, Y.; Thomas, S.M. Radiation-induced fibrosis: Mechanisms and implications for therapy. *J. Cancer Res. Clin. Oncol.* **2015**, *141*, 1985–1994. [CrossRef]
51. D'Urso, P.; Farneti, A.; Marucci, L.; Marzi, S.; Piludu, F.; Vidiri, A.; Sanguineti, G. Predictors of Outcome after (Chemo)Radiotherapy for Node-Positive Oropharyngeal Cancer: The Role of Functional MRI. *Cancers* **2022**, *14*, 2477. [CrossRef]
52. Bicci, E.; Nardi, C.; Calamandrei, L.; Pietragalla, M.; Cavigli, E.; Mungai, F.; Bonasera, L.; Miele, V. Role of Texture Analysis in Oropharyngeal Carcinoma: A Systematic Review of the Literature. *Cancers* **2022**, *14*, 2445. [CrossRef]

applied
sciences

Article

Ultrasound Intima-Media Complex (IMC) Segmentation Using Deep Learning Models

Hanadi Hassen Mohammed [1,*], Omar Elharrouss [1], Najmath Ottakath [1], Somaya Al-Maadeed [1], Muhammad E. H. Chowdhury [2], Ahmed Bouridane [3] and Susu M. Zughaier [4]

[1] Department of Computer Science and Engineering, Qatar University, Doha P.O. Box 2713, Qatar
[2] Department of Electrical Engineering, Qatar University, Doha P.O. Box 2713, Qatar
[3] Cybersecurity and Data Analytics Research Center, University of Sharjah, Sharjah 27272, United Arab Emirates
[4] Department of Basic Medical Sciences, College of Medicine, Qatar University, Doha P.O. Box 2713, Qatar
* Correspondence: hm1409611@qu.edu.qa

Abstract: Common carotid intima-media thickness (CIMT) is a common measure of atherosclerosis, often assessed through carotid ultrasound images. However, the use of deep learning methods for medical image analysis, segmentation and CIMT measurement in these images has not been extensively explored. This study aims to evaluate the performance of four recent deep learning models, including a convolutional neural network (CNN), a self-organizing operational neural network (self-ONN), a transformer-based network and a pixel difference convolution-based network, in segmenting the intima-media complex (IMC) using the CUBS dataset, which includes ultrasound images acquired from both sides of the neck of 1088 participants. The results show that the self-ONN model outperforms the conventional CNN-based model, while the pixel difference- and transformer-based models achieve the best segmentation performance.

Keywords: ultrasound imaging; image segmentation; intima-media thickness; carotid artery; deep learning

Citation: Hassen Mohammed, H.; Elharrouss, O.; Ottakath, N.; Al-Maadeed, S.; Chowdhury, M.E.H.; Bouridane, A.; Zughaier, S.M. Ultrasound Intima-Media Complex (IMC) Segmentation Using Deep Learning Models. *Appl. Sci.* **2023**, *13*, 4821. https://doi.org/10.3390/app13084821

Academic Editor: Cosimo Nardi

Received: 6 March 2023
Revised: 6 April 2023
Accepted: 8 April 2023
Published: 12 April 2023

1. Introduction

The primary mechanism in the human body that sustains life is the cardiovascular system. Cardiovascular system diseases (CVDs) have been regarded as a major cause of death in the world. Lifespan can be increased and the death rate from CVDs can be decreased with early diagnosis and treatment of the diseases. The cardiovascular system is made up of blood vessels that carry blood, necessary for all of the body's organs to operate. The primary components of the blood vessels that transport blood to and from the heart and to all organs are arteries and veins. Any obstruction in blood flow or disease in the arteries or veins will seriously affect how well the organs operate. The most common types of cardiovascular disease include peripheral vascular disease, coronary artery disease and carotid artery disease. These disorders manifest as a result of the development of atherosclerotic plaques in the arteries, as illustrated in Figure 1. One of the effects of carotid artery stenosis is an ischemic stroke, due to the accumulation of plaque on the carotid arterial walls. If the stenosis is detected early and the amount of plaque can be determined, the problem can be addressed immediately. For this, a variety of imaging modalities are used. Computed tomography (CT), EEG, ECG, ultrasound imaging, laboratory tests for coagulation status and cardiac monitoring are among the diagnostic techniques used in the assessment of carotid artery stenosis or stroke. Both sides of the neck contain the common carotid artery. The soft tissue features in the arteries allow for imaging using a variety of methods or modalities, such as computerized tomography (CT), ultrasound imaging and magnetic resonance imaging. The analysis of the generated images can enhance diagnosis and support clinical judgment. Medical image analysis algorithms have advanced significantly from image processing and pattern recognition methods to machine learning and deep learning algorithms that see it as a computer vision problem. A notable development in the automatic segmentation, analysis and grading of stenosis is the use of carotid artery imaging generated by CT scans, MRIs and ultrasound images [1,2]. Due to the complexity of scanning the carotid artery, ultrasound scanning is the preferred method to capture images with acceptable resolutions. Ultrasound images have been used for many studies using medical imaging analysis algorithms [3].

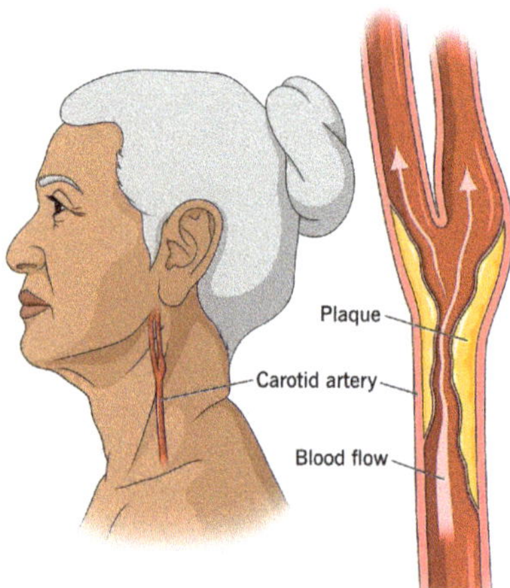

Figure 1. Visualization of plaque build-up and obstruction to the normal flow of blood in the artery (https://my.clevelandclinic.org/health/diseases/16845-carotid-artery-disease-carotid-artery-stenosis, accessed on 9 January 2023).

In order to segment the plaques on the carotid artery, many methods have been proposed even in the absence of large datasets. Previously, the proposed methods used CIMT measurement to detect and localize the carotid artery walls and then the plaques [4,5]. The ground truth was presented using some points representing the plaques generated by specialists [6]. The analysis using these types of data used different statistical and machine learning algorithms, including Snake's segmentation and contour [4,5], bulb edge detection [6], wind-driven optimization techniques [7] and SVM [8].

Using convolution neural networks, the proposed methods used binary segmentation instead of CIMT measurement. By generating binary images containing labeled regions in the images instead of using points, the deep learning methods could successfully segment these regions with better precision [8]. Furthermore, the segmented regions could be helpful in computing CIMT [9], related to the performance accuracy of segmentation. This makes segmentation a crucial task.

Although CNNs have succeeded in solving many computer vision problems, recent studies have shown many drawbacks for CNNs, such as the need for large datasets [10] and the reliance on linear neuron models [11–14]. Operational neural networks (ONNs) [14–17] are heterogeneous networks with a non-linear neuron model that have recently been proposed as a solution for highly non-linearly separable problems. With the help of predefined nodal, pool and activation operators, ONNs are able to learn highly complex and multi-modal functions. The transformer neural network has recently been a successful non-CNN alternative for computer vision problems. Instead of convolution, vision transformers utilize self-attention to combine information from several locations [18]. In this paper, we performed a segmentation of common carotid intima-media using deep learning models. For this, we updated existing deep learning models, such as DeepCrack [19] and the transformer-based model [20]. We used a self-ONN instead of normal convolutional layers for DeepCrack. In order to improve the segmentation quality, we used morphological operations, such as erosion to enhance the output results. The main contributions of the research are summarized as follows:

- We develop and investigate various recent deep learning models for the segmentation of IMC in B-mode ultrasound images of the carotid artery.
- We propose a pioneer application for self-organized operational neural networks (self-ONNs) for IMC segmentation.
- We investigate the level of non-linearity for operational layers required to achieve a better segmentation performance.

The rest of the paper is divided as follows; in Section 2, we highlight the recent work of carotid intima-media segmentation. Then, in Section 3, we present the model architecture for the deep learning models, and in Section 4, we present the experimental setup along with the evaluation metrics and the results of the model. Finally, we conclude and explain the future work in Section 5.

2. Related Works

The carotid artery segmentation, including the walls and plaques in the intima-media complex (IMC), can be used for the estimation of intima-media thickness (IMT). Which makes it an important operation for atherosclerotic risk evaluation.

There are numerous methods for segmenting the intima-media complex. However, the majority of them are semi-automatic and require manual intervention. Medical experts must define the boundary between the media adventitia and lumen. However, the subjectivity and variability of manual segmentation can be reduced using image segmentation algorithms. Additionally, IMT is assessed using active contours [21–28], dynamic programming [29–34] and edge detection algorithms and gradient-based approaches [35,36]. For active contour-based approaches, the authors in [21] began with a simple segmentation of B-mode ultrasound images followed by segmentation of the far wall intima-media–adventitia, then applied the active contour to obtain the desired region in the images. The same process was used in [22], but this time using some morphological operations, such as opening. Subsequently, an LI contour function was applied to detect the final common carotid artery result. In [23], the authors started with non-linear filtering followed by the detection of the intima layer using an iterative relaxation procedure to detect the wall using a modified energy function and an optimal initial contour.

For dynamic programming-based approaches, the researchers in [29] used a multiscale dynamic programming (DP) algorithm to estimate the vessel wall positions leading to boundary detection. The obtained results with geometrical characteristics were used to obtain the final results. In the same context and to detect the arterial wall, the authors in [31] proposed a dual dynamic programming (DDP) technique to detect the intima and adventitial layers of the common carotid artery. Furthermore, in [33] an improved dynamic programming method was proposed for carotid artery wall thickness evaluation.

Machine and deep learning techniques have becoming intriguing as promising methods for medical image analysis tasks, such as image de-noising, segmentation and classification. Before the development of deep learning models, machine learning was the most commonly utilized technology, where comprehensive feature extraction techniques were applied to find several areas of carotid artery risk estimation. The deep learning strategy takes advantage of a neural network architecture that mimics the human brain by having more hidden layers. The neuron is the fundamental building block of a deep neural network (DNN), which accepts several inputs, linearly combines them and then passes them to a non-linear network to produce the desired output. Multiple processing layers make up a deep learning network, which uses deep graphs to extract high-level representations of meaningful information from low-level inputs. CNNs are among the most widely used networks in the medical image analysis domain [37]. U-Net is a CNN-based architecture used to solve the automatic image segmentation problem. This architecture has been adopted in many IMC segmentation works [38–40]. For example, in [38,41] the authors used the U-Net architecture for plaque segmentation in carotid ultrasound images.

Furthermore, in [42] the authors used U-Net, U-Net+, U-Net++, U-Net+++ and three types of hybrids, namely, Inception-U-Net, Fractal-U-Net and Squeeze-U-Net architectures, to segment and measure the plaque far wall area of the common carotid (CCAs) and internal carotid arteries (ICAs) in B-mode ultrasound images. Using M-Net [43] as the backbone, the authors in [44] proposed an automatic joint segmentation method named CSM-Net with triple spatial attention and cascaded dilated convolution modules.

3. Methods

Medical image segmentation is a challenging task. As our ultimate goal is to find the most accurate deep learning model for ultrasound IMC segmentation, we tested several deep learning methods. Three recent deep learning networks were used in this study: DeepCrack [19], PidiNet [10] and CCTrans [45]. These networks have been used previously in different tasks such as edge detection, crack segmentation and crowd counting. The DeepCrack network is a CNN-based architecture which we modified with the recently proposed self-operational neural network (self-ONN) with the goal of seeing whether the CNN- or self-ONN-based architecture worked better on our dataset. CCTrans is a transformer-based model used for crowd counting. For this, we adapted the model to be suitable for ultrasound IMC segmentation by exploiting the same first layers of the model. The following sections describe a detailed description of how these methods have been adapted to our problem.

3.1. Self-Operational Neural Network-Based Model

Self-organized operational neural networks with generative neurons, proposed by [46], are a type of artificial neural network designed to operate in a self-organizing manner. Instead of using a predefined set of operators as an ONN, the self-ONNs with generative neurons generate nodal operators during backpropagation training. This property of self-ONNs allows for maximum learning performance, diversity and flexibility. The use of generative neurons can improve the network's robustness to unseen data and reduce the risk of overfitting. A generative neuron uses a Taylor series expansion around the point a to approximate the non-linear function $f(x)$:

$$Y = \sum_{s=1}^{S} \frac{f^n(a)}{n!}(x - a)^2 \tag{1}$$

If we truncate the Taylor series to q terms then the approximation $g(w, x, a)$ will be given by:

$$Y = w_0 + w_1(x - a) + \ldots + w_q(x - a)^q \tag{2}$$

where $w_n = \frac{f^n(a)}{n!}(x - a)^2$, w_0 is the bias for the c-channel input tensor w_n and $n = 1, \ldots, q$ are the q-banks of c-channel convolution kernels that are learned during backpropagation.

To investigate the performance of self-ONNs, we chose the DeepCrack [19] model as a baseline model. The DeepCrack network, proposed by [19], is a CNN-based model built for crack segmentation. The architecture of the DeepCrack network is shown in Figure 2a. It has thirteen convolutional layers, each with convolution, batch normalization and ReLU layers. The convolution produces a set of feature maps. At the same time, batch normalization is used to reduce the covariate shift and the ReLU function is the activation function used to learn non-linearity in the data. A max-pooling with 2×2 pixel filter layers is added between the convolutional layers. A convolutional layer with kernel size 1 is used to obtain side-output features. Deconvolutional layers are then used (except for the first side output layer) to upsample the feature maps' plane size to match the input image. Following the concatenation of the upsampled feature maps to obtain the final features, a convolutional layer and a Softmax layer are applied. Then a convolutional layer followed by a Softmax layer are used for predicting two classes. According to this prediction, for each pixel, the predicted label can be obtained. We modified the network to be flexible to use self-ONN layers instead of CNN layers, as shown in Figure 2a. We used Tanh activation layers instead of ReLU. The level of non-linearity can be adjusted on the network by modifying the parameter q.

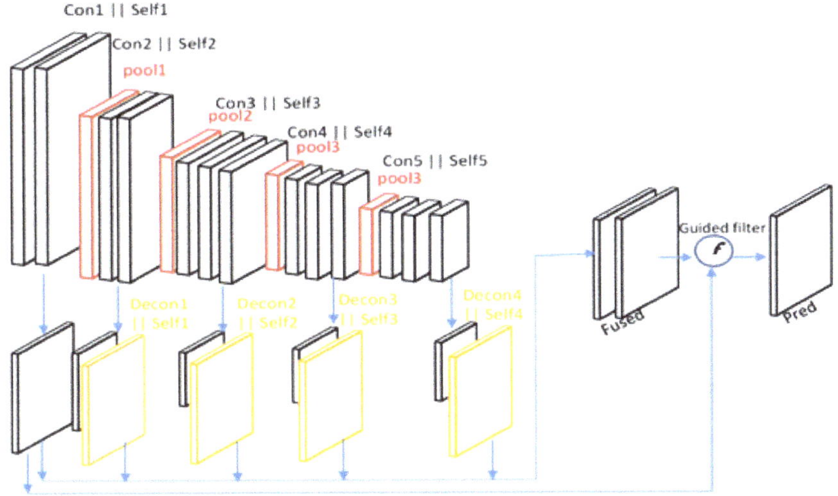

(**a**) Self-ONN–DeepCrack.

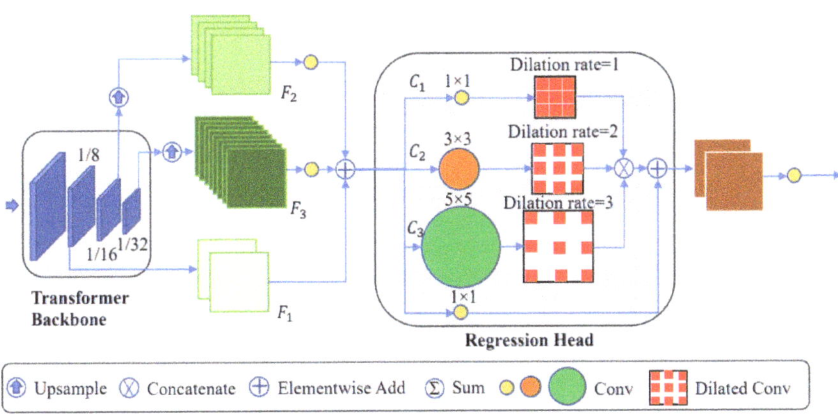

(**b**) Transformer network.

Figure 2. Networks used for ultrasound IMC segmentation.

3.2. Pixel Difference-Based Model

Although CNNs can achieve human-level performance in many computer-vision-based applications, the high performance of CNN-based models is achieved with a large pre-trained CNN backbone [47], such as VGG, ResNet and DenseNet, which is memory- and energy-consuming, while some methods have been proposed with simple and light-weight architectures, such as pixel difference networks (PiDiNets), that use edge detection [10]. PiDiNet adopts novel pixel difference convolutions that integrate the traditional edge detection operators into popular convolutional operations in modern convolution neural networks for enhanced performance to enjoy the best of both worlds. We used a PiDiNet model for IMC segmentation.

3.3. Transformer-Based Model

Traditionally, convolutional neural networks (CNNs) have been the preferred architecture for image segmentation tasks due to their ability to extract features from the input image. However, in recent years, transformer-based models have shown remarkable performance in a variety of natural language processing (NLP) tasks and have been extended to computer vision tasks, such as image segmentation.

CNNs have a strong ability to extract local features, but they inherently fail in modeling the global context due to the limited receptive fields. The transformer can model the global context easily. Furthermore, it has become the most used technique in computer vision. Due to this, we used a transformer model for IMC segmentation. The proposed method used a pyramid vision transformer backbone to capture the global information, a pyramid feature aggregation (PFA) model to combine low- and high-level features and an efficient regression head with multi-scale dilated convolution (MDC) to predict the final results [20]. The input image is transformed into a 1D sequence first, then the output is fed into the transformer-based backbone. The pyramid transformer in [45] is adopted to capture the global context through various downsampling stages. The outputs of each stage are reshaped into 2D feature maps for pyramid feature aggregation. Finally, a simple regression head with multi-scale receptive fields regresses the final results. The proposed architecture is illustrated in Figure 2b.

3.4. Post-Processing

The IMC segmentation is a difficult task, due to the difficulty of generating the precise thickness from an image, even when using deep learning methods. While the carotid intima-media region can be segmented, for some images, this region can be very skinny, affecting the performance of the segmentation method. We noticed that when using deep learning methods the segmented thickness is generally fat, as presented in Figure 3b. Because of this and in order to make the segmented thickness skinny to meet the ground truth, we applied morphological erosion. Morphological erosion is a post-processing step commonly used in medical image segmentation. In the context of IMC segmentation, morphological erosion is used to refine the initial segmentation results by removing small regions of noise or non-IMC tissue that may have been included. This helps to improve the accuracy and reliability of the segmentation by ensuring that only the true IMC structure is retained. The erosion operation is typically performed using a structuring element, which determines the size and shape of the erosion. The choice of structuring element depends on the characteristics of the image and the desired level of erosion. For example, a small circular element may be used to remove small regions of noise, while a larger rectangular element may be used to remove larger areas of non-IMC tissue. Figure 3c presents an example of the erosion result.

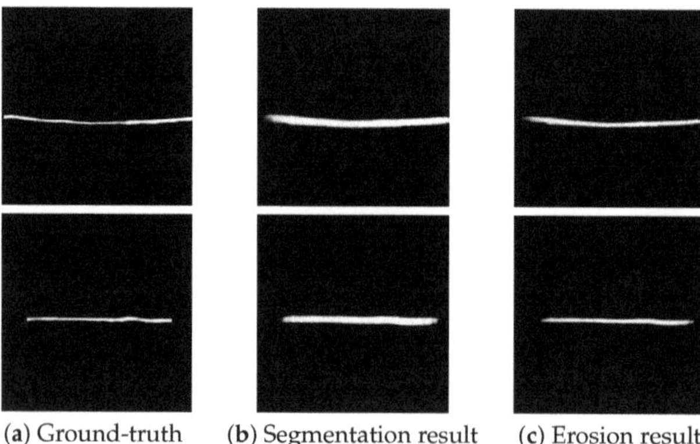

(a) Ground-truth   (b) Segmentation result   (c) Erosion result

Figure 3. Morphological erosion on ultrasound IMC segmentation results.

4. Experimental Results

In this section, we demonstrate the experimental results of the proposed self-ONN–DeepCrack approach on the CUBS dataset, and compare the obtained results with other published image segmentation methods, including DeepCrack [19], PidiNet [10] and adapted CCTrans [45]. The comparison was performed using image segmentation metrics as well as visual illustrations.

4.1. Implementation Details

The implementation details for training the proposed and implemented models are presented in Table 1. The implementation was performed using the Pytorch library, while the post-processing and evaluation metrics were performed using Matlab.

Table 1. Training hyperparameters and parameters for each model.

Method	Learning Rate	Optimizer	Epochs	Training Parameters
DeepCrack	0.0001	Adam	100	14.720 M
DeepCrack_Self_ONN	0.0001	Adam	100	44.144 M
PidiNet	0.005	Adam	70	1.150 MB
Transformer	0.00001	Adam	70	104.609 M

4.2. Dataset and Evaluation Metrics

The dataset used in this study is the CUBS dataset, acquired from both sides of the neck of 1088 participants, totalling 2176 images. All images are annotated by a skilled analyst. The images in Figure 5 are samples of the images and the ground truths taken from the dataset. A total of 80% of the data are used for training and 20% are used for testing. The segmentation metrics used to evaluate the performance of the proposed models are precision, recall, F1 measure (Equation (3)), Jaccard index (Equation (4)) and Dice coefficient (Equation (5)). Precision measures how many true positive (TP) predictions there are out of all the positive predictions or how many positive predictions there are in total. Recall calculates the true positive rate (TPR) or how many true positive predictions are made out of all the true positives. Both precision and recall are used to handle the class imbalance problem and to compute the F1 measure.

$$F1 = \frac{2 * Precision * Recall}{Precision + Recall} \tag{3}$$

$$Jaccard\ Index = \frac{True\ Positive}{True\ Positive + False\ Negative + False\ Positive} \tag{4}$$

$$Dice = \frac{2 * 2 True\ Positive}{2 * True\ Positive + False\ Negative + False\ Positive} \tag{5}$$

4.3. Evaluation

To evaluate the ultrasound IMC segmentation using the deep learning methods on the CUBS dataset, a set of metrics as mentioned above are used. These metrics are predominantly used for image segmentation in computer vision tasks. Moreover, we compare the frames per second (FPS) for each model on the same dataset. In this section, we present the obtained results from the dataset using the proposed method for ultrasound IMC segmentation. The results are reported in the tables and figures to show the performed techniques using the different architectures.

We first investigated the effect of replacing CNN layers with self-ONN layers in the DeepCrack model. The level of linearity was controlled using the parameter q = 3, 5, 7, 9 or 11. Figure 4a shows that the best performing model uses q = 3, then the accuracy of the model starts to drop as we enlarge the level of non-linearity. Compared with the CNN version of the model, Figure 4b shows that the best precision and recall accuracies at q are set to 3 and 5. The performance of all the deep learning models on the CUBS dataset is shown in Table 2. From the table, we can observe that both the transformer- and pixel difference-based models act similarly in all the performance measures with a slight increase for PiDiNet in the F-measure, Dice and Jaccard index. Both the transformer- and pixel difference-based models achieved better performances with exceptional margins compared to the CNN- and self-ONN-based models. From Table 2, we can also see that the post-processing operations improved the performance metrics of all the methods, including DeepCrack, DeepCrack_Self_ONN, PiDiNet and the transformer-based models. The models achieved an improvement of about 20, 14, 19 and 20% for the DeepCrack, DeepCrack_Self_ONN, PiDiNet and transformer-based models, respectively, on the precision metric, while the transformer-based + post-processing model demonstrated the best metrics followed by PiDiNet + Post-processing with an average difference of 1% and 10% and less than 1% for dice, recall and precision, respectively. In addition to the qualitative results, we present the qualitative results in Figure 5 that show the visual outputs from the segmentation results. From Figure 5, we can see that all the proposed methods demonstrated segmentation with good performance with a difference in terms of thickness.

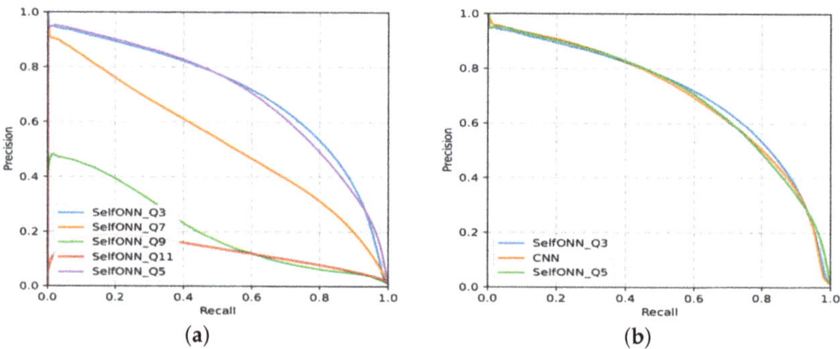

(a)   (b)

Figure 4. (a) The precision-recall curve for ultrasound IMC segmentation using the self-ONN with different q settings, (b) using the CNN and self-ONN with $q = 3$ and $q = 5$.

Table 2. Performance of the proposed and implemented models on the CUBS dataset. The **bold** and <u>underline</u> fonts respectively represent the **first** and <u>second</u> place.

Model	Precision	Recall	F-Measure	Dice	Jaccard	FPS
DeepCrack_CNN	0.631	0.675	0.652	0.652	0.484	17.074
DeepCrack_CNN + Post-processing	0.834	0.618	0.697	0.697	0.544	17.074
DeepCrack_Self ($q = 3$)	0.652	0.688	0.669	0.669	0.503	13.45
DeepCrack_Self + Post-processing	0.792	0.691	0.721	0.721	0.571	13.45
PiDiNet	0.687	0.825	0.750	0.750	0.60	20.62
PiDiNet + Post-processing	<u>0.876</u>	0.740	<u>0.791</u>	<u>0.791</u>	**0.661**	20.62
Transformer	0.68	<u>0.826</u>	0.746	0.746	0.595	11.427
Transformer + Post-processing	**0.882**	**0.849**	**0.801**	**0.801**	<u>0.656</u>	11.427

It is worth mentioning that image segmentation algorithms typically rely on edge detection and thresholding techniques to separate regions of interest from the background. However, these techniques can be affected by image noise, leading to the detection of false edges and the inclusion of noise as part of the segmented object. Additionally, image segmentation algorithms may also introduce a level of smoothing or blurring to the image, which can further contribute to the fattening effect. This smoothing operation can cause the boundaries of the segmented object to become slightly blurred and more diffuse, resulting in a larger area being assigned to the object than is actually present in the ground truth.

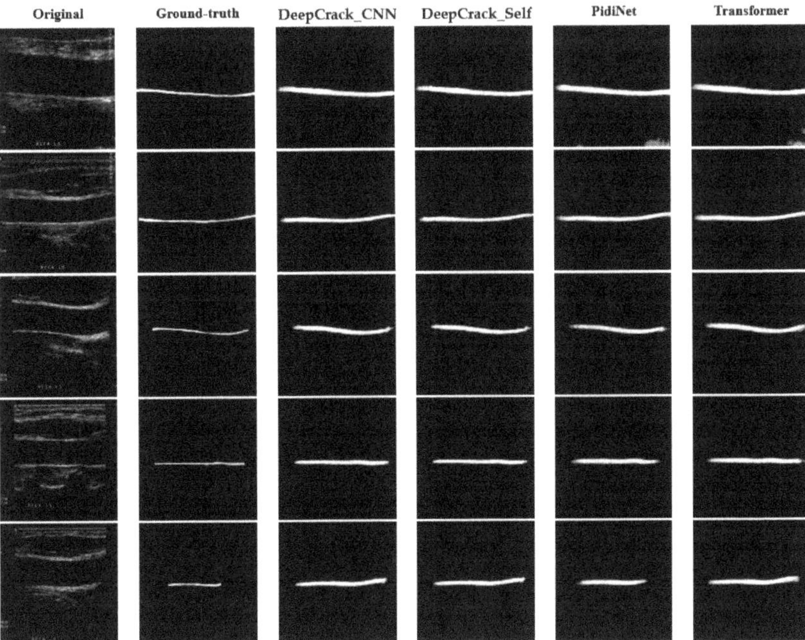

Figure 5. Original and ground truth sample images and the corresponding segmentation results for the proposed deep learning models.

5. Conclusions

We developed and investigated various novel deep learning models for the segmentation of IMC in B-mode ultrasound images of the carotid artery. Compared to the conventional CNN-based model, the self-ONN-based model performs better in all evaluation metrics; however, the pixel difference- and transformer-based models perform better in all metrics, potentially due to the absence of enough data. The pixel difference model performs better when data are scarce. A further investigation into suitable data augmentation techniques is needed to increase the accuracy.

Author Contributions: Conceptualization, H.H.M. and O.E.; data curation, H.H.M., O.E. and N.O.; formal analysis, H.H.M. and O.E.; methodology, H.H.M. and O.E.; project administration; supervision, S.A.-M., M.E.H.C., A.B. and S.M.Z.; validation, H.H.M and O.E.; visualization, H.H.M. and O.E.; writing—original draft, H.H.M. and O.E.; writing—review and editing, H.H.M., O.E. and S.A.-M. All authors have read and agreed to the published version of the manuscript.

Funding: This publication was supported by the Qatar University Internal Grant #QUHI-CENG-22/23-548. The findings achieved herein are solely the responsibility of the authors.

Institutional Review Board Statement: Not applicable.

Informed Consent Statement: Not applicable.

Data Availability Statement: Not applicable.

Conflicts of Interest: The authors declare no conflict of interest.

Abbreviations

The following abbreviations are used in this manuscript:

DR   Diabetic retinopathy
DL   Deep learning
AI   Artificial intelligence
CNN   Convolutional neural network

References

1. Latha, S.; Samiappan, D.; Kumar, R. Carotid artery ultrasound image analysis: A review of the literature. *Proc. Inst. Mech. Eng. Part H J. Eng. Med.* **2020**, *234*, 417–443. [CrossRef]
2. Vila, M.D.M.; Remeseiro, B.; Grau, M.; Elosua, R.; Igual, L. Last Advances on Automatic Carotid Artery Analysis in Ultrasound Images: Towards Deep Learning. In *Handbook of Artificial Intelligence in Healthcare*; Springer: Berlin/Heidelberg, Germany, 2022; pp. 215–247.
3. Riahi, A.; Elharrouss, O.; Al-Maadeed, S. BEMD-3DCNN-based method for COVID-19 detection. *Comput. Biol. Med.* **2022**, *142*, 105188. [CrossRef]
4. Loizou, C.P.; Kasparis, T.; Spyrou, C.; Pantziaris, M. Integrated system for the complete segmentation of the common carotid artery bifurcation in ultrasound images. In *IFIP International Conference on Artificial Intelligence Applications and Innovations*; Springer: Berlin/Heidelberg, Germany, 2013; Volume 9, pp. 292–301.
5. Christodoulou, L.; Loizou, C.P.; Spyrou, C.; Kasparis, T.; Pantziaris, M. Full-automated system for the segmentation of the common carotid artery in ultrasound images. In Proceedings of the 2012 IEEE 5th International Symposium on Communications, Control and Signal Processing, Rome, Italy, 2–4 May 2012; pp. 1–6.
6. Ikeda, N.; Dey, N.; Sharma, A.; Gupta, A.; Bose, S.; Acharjee, S.; Shafique, S.; Cuadrado-Godia, E.; Araki, T.; Saba, L.; et al. Automated segmental-IMT measurement in thin/thick plaque with bulb presence in carotid ultrasound from multiple scanners: Stroke risk assessment. *Comput. Methods Programs Biomed.* **2017**, *141*, 73–81. [CrossRef]
7. Madipalli, P.; Kotta, S.; Dadi, H.; Nagaraj, Y.; Asha, C.S.; Narasimhadhan, A.V. Automatic Segmentation of Intima Media Complex in Common Carotid Artery using Adaptive Wind Driven Optimization. In Proceedings of the 2018 Twenty Fourth National Conference on Communications (NCC), Hyderbad, India, 25–28 February 2018; pp. 1–6.
8. Nagaraj, Y.; Teja, A.; Narasimha, D. Automatic Segmentation of Intima Media Complex in Carotid Ultrasound Images Using Support Vector Machine. *Arab. J. Sci. Eng.* **2019**, *44*, 3489–3496. [CrossRef]
9. Biswas, M.; Saba, L.; Chakrabartty, S.; Khanna, N.N.; Song, H.; Suri, H.S.; Sfikakis, P.P.; Mavrogeni, S.; Viskovic, K.; Laird, J.R.; et al. Two-stage artificial intelligence model for jointly measurement of atherosclerotic wall thickness and plaque burden in carotid ultrasound: A screening tool for cardiovascular/stroke risk assessment. *Comput. Biol. Med.* **2020**, *123*, 103847. [CrossRef]
10. Su, Z.; Liu, W.; Yu, Z.; Hu, D.; Liao, Q.; Tian, Q.; Pietikainen, M.; Liu, L. Pixel difference networks for efficient edge detection. In Proceedings of the IEEE/CVF International Conference on Computer Vision, Montreal, BC, Canada, 11–17 October 2021; pp. 5117–5127.
11. Kiranyaz, S.; Malik, J.; Abdallah, H.B.; Ince, T.; Iosifidis, A.; Gabbouj, M. Self-organized operational neural networks with generative neurons. *Neural Netw.* **2021**, *140*, 294–308. [CrossRef]
12. Gabbouj, M.; Kiranyaz, S.; Malik, J.; Zahid, M.U.; Ince, T.; Chowdhury, M.E.; Khandakar, A.; Tahir, A. Robust peak detection for holter ECGs by self-organized operational neural networks. *IEEE Trans. Neural Netw. Learn. Syst.* **2022**, *Early Access*. [CrossRef]
13. Malik, J.; Kiranyaz, S.; Gabbouj, M. Operational vs. convolutional neural networks for image denoising. *arXiv* **2020**, arXiv:2009.00612.
14. Malik, J.; Devecioglu, O.C.; Kiranyaz, S.; Ince, T.; Gabbouj, M. Real-time patient-specific ECG classification by 1D self-operational neural networks. *IEEE Trans. Biomed. Eng.* **2021**, *69*, 1788–1801. [CrossRef]
15. Rahman, A.; Chowdhury, M.E.; Khandakar, A.; Tahir, A.M.; Ibtehaz, N.; Hossain, M.S.; Kiranyaz, S.; Malik, J.; Monawwar, H.; Kadir, M.A. Robust biometric system using session invariant multimodal EEG and keystroke dynamics by the ensemble of self-ONNs. *Comput. Biol. Med.* **2022**, *142*, 105238. [CrossRef]
16. Soltanian, M.; Malik, J.; Raitoharju, J.; Iosifidis, A.; Kiranyaz, S.; Gabbouj, M. Speech command recognition in computationally constrained environments with a quadratic self-organized operational layer. In Proceedings of the 2021 IEEE International Joint Conference on Neural Networks (IJCNN), Shenzhen, China, 18–22 July 2021; pp. 1–6.
17. Kiranyaz, S.; Ince, T.; Iosifidis, A.; Gabbouj, M. Operational neural networks. *Neural Comput. Appl.* **2020**, *32*, 6645–6668. [CrossRef]
18. Khan, S.; Naseer, M.; Hayat, M.; Zamir, S.W.; Khan, F.S.; Shah, M. Transformers in vision: A survey. *ACM Comput. Surv. (CSUR)* **2022**, *54*, 1–41. [CrossRef]
19. Liu, Y.; Yao, J.; Lu, X.; Xie, R.; Li, L. DeepCrack: A deep hierarchical feature learning architecture for crack segmentation. *Neurocomputing* **2019**, *338*, 139–153. [CrossRef]
20. Elharrouss, O.; Hmamouche, Y.; Idrissi, A.K.; El Khamlichi, B.; El Fallah-Seghrouchni, A. Refined edge detection with cascaded and high-resolution convolutional network. *Pattern Recognit.* **2023**, *138*, 109361. [CrossRef]
21. Petroudi, S.; Loizou, C.; Pantziaris, M.; Pattichis, C. Segmentation of the common carotid intima-media complex in ultrasound images using active contours. *IEEE Trans. Biomed. Eng.* **2012**, *59*, 3060–3069. [CrossRef]

22. Molinari, F.; Meiburger, K.M.; Saba, L.; Acharya, U.R.; Ledda, M.; Nicolaides, A.; Suri, J.S. Constrained snake vs. conventional snake for carotid ultrasound automated IMT measurements on multi-center data sets. *Ultrasonics* **2012**, *52*, 949–961. [CrossRef]
23. Ceccarelli, M.; Luca, N.D.; Morganella, A. An active contour approach to automatic detection of the intima-media thickness. In Proceedings of the 2006 IEEE International Conference on Acoustics Speech and Signal Processing, Toulouse, France, 14–19 May 2006; Volume 2, p. II.
24. Loizou, C.P.; Pattichis, C.S.; Pantziaris, M.; Tyllis, T.; Nicolaides, A. Snakes based segmentation of the common carotid artery intima media. *Med. Biol. Eng. Comput.* **2007**, *45*, 35–49. [CrossRef]
25. Gutierrez, M.A.; Pilon, P.E.; Lage, S.G.; Kopel, L.; Carvalho, R.T.; Furuie, S.S. Automatic measurement of carotid diameter and wall thickness in ultrasound images. In Proceedings of the Computers in Cardiology, Memphis, TN, USA, 22–25 September 2002; pp. 359–362.
26. Chan, R.C.; Kaufhold, J.; Hemphill, L.C.; Lees, R.S.; Karl, W.C. Anisotropic edge-preserving smoothing in carotid B-mode ultrasound for improved segmentation and intima-media thickness (IMT) measurement. In Proceedings of the Computers in Cardiology 2000. Vol.27 (Cat. 00CH37163), Cambridge, MA, USA, 24–27 September 2000; pp. 37–40.
27. Delsanto, S.; Molinari, F.; Giustetto, P.; Liboni, W.; Badalamenti, S.; Suri, J.S. Characterization of a completely user-independent algorithm for carotid artery segmentation in 2-D ultrasound images. *IEEE Trans. Instrum. Meas.* **2007**, *56*, 1265–1274. [CrossRef]
28. Gagan, J.H.; Shirsat, H.S.; Mathias, G.P.; Mallya, B.V.; Andrade, J.; Rajagopal, K.V.; Kumar, J.H. Automated Segmentation of Common Carotid Artery in Ultrasound Images. *IEEE Access* **2022**, *10*, 58419–58430. [CrossRef]
29. Liang, Q.; Wendelhag, I.; Wikstrand, J.; Gustavsson, T. A multiscale dynamic programming procedure for boundary detection in ultrasonic artery images. *IEEE Trans. Med. Imaging* **2000**, *19*, 127–142. [CrossRef]
30. Wendelhag, I.; Liang, Q.; Gustavsson, T.; Wikstrand, J. A new automated computerized analyzing system simplifies readings and reduces the variability in ultrasound measurement of intima-media thickness. *Stroke* **1997**, *28*, 2195–2200. [CrossRef]
31. Cheng, D.-C.; Jiang, X. Detections of arterial wall in sonographic artery images using dual dynamic programming. *IEEE Trans. Inf. Technol. Biomed.* **2008**, *12*, 792–799. [CrossRef]
32. Gustavsson, T.; Wendelhag, Q.L.I.; Wikstrand, J. A dynamic programming procedure for automated ultrasonic measurement of the carotid artery. In Proceedings of the Computers in Cardiology, Bethesda, MD, USA, 25–28 September 1994; pp. 297–300.
33. Santhiyakumari, N.; Madheswaran, M. Non-invasive evaluation of carotid artery wall thickness using improved dynamic programming technique. *Signal Image Video Process.* **2008**, *2*, 183–193. [CrossRef]
34. Lee, Y.-B.; Choi, Y.-J.; Kim, M.-H. Boundary detection in carotid ultrasound images using dynamic programming and a directional Haar-like filter. *Comput. Biol. Med.* **2010**, *40*, 687–697. [CrossRef]
35. Liguori, C.; Paolillo, A.; Pietrosanto, A. An automatic measurement system for the evaluation of carotid intima-media thickness. *IEEE Trans. Instrum. Meas.* **2001**, *50*, 1684–1691. [CrossRef]
36. Selzer, R.H.; Mack, W.J.; Lee, P.L.; Kwong-Fu, H.; Hodis, H.N. Improved common carotid elasticity and intima-media thickness measurements from computer analysis of sequential ultrasound frames. *Atherosclerosis* **2002**, *154*, 185–193. [CrossRef]
37. Pramulen, A.S.; Yuniarno, E.M.; Nugroho, J.; Sunarya, I.M.G.; Purnama, I.K.E. Carotid Artery Segmentation on Ultrasound Image using Deep Learning based on Non-Local Means-based Speckle Filtering. In Proceedings of the 2020 International Conference on Computer Engineering, Network and Intelligent Multimedia (CENIM), Surabaya, Indonesia, 17–18 November 2020; pp. 360–365.
38. Jain, P.K.; Sharma, N.; Saba, L.; Paraskevas, K.I.; Kalra, M.K.; Johri, A.; Nicolaides, A.N.; Suri, J.S. Automated deep learning-based paradigm for high-risk plaque detection in B-mode common carotid ultrasound scans: An asymptomatic Japanese cohort study. *Int. Angiol.* **2021**, *41*, 9–23. [CrossRef]
39. Lainé, N.; Liebgott, H.; Zahnd, G.; Orkisz, M. Carotid artery wall segmentation in ultrasound image sequences using a deep convolutional neural network. *arXiv* **2022**, arXiv:2201.12152.
40. Radovanovic, N.; Dašić, L.; Blagojevic, A.; Sustersic, T.; Filipovic, N. Carotid Artery Segmentation Using Convolutional Neural Network in Ultrasound Images. 2022. Available online: https://scidar.kg.ac.rs/bitstream/123456789/16643/4/p8.pdf (accessed on 1 January 2023).
41. Park, J.H.; Seo, E.; Choi, W.; Lee, S.J. Ultrasound deep learning for monitoring of flow–vessel dynamics in murine carotid artery. *Ultrasonics* **2022**, *120*, 106636. [CrossRef]
42. Jain, P.K.; Sharma, N.; Kalra, M.K.; Johri, A.; Saba, L.; Suri, J.S. Far wall plaque segmentation and area measurement in common and internal carotid artery ultrasound using U-series architectures: An unseen Artificial Intelligence paradigm for stroke risk assessment. *Comput. Biol. Med.* **2022**, *149*, 106017. [CrossRef]
43. Fu, H.; Xu, J.C.Y.; Wong, D.W.K.; Liu, J.; Cao, X. Joint optic disc and cup segmentation based on multi-label deep network and polar transformation. *IEEE Trans. Med. Imaging* **2018**, *37*, 1597–1605. [CrossRef] [PubMed]
44. Yuan, Y.; Li, C.; Xu, L.; Zhu, S.; Hua, Y.; Zhang, J. CSM-Net: Automatic joint segmentation of intima-media complex and lumen in carotid artery ultrasound images. *Comput. Biol. Med.* **2022**, *150*, 106119. [CrossRef]
45. Chu, X.; Tian, Z.; Wang, Y.; Zhang, B.; Ren, H.; Wei, X.; Xia, H.; Shen, C. Twins: Revisiting the design of spatial attention in vision transformers. *Adv. Neural Inf. Process. Syst.* **2021**, *34*, 9355–9366.

46. Elharrouss, O.; Akbari, Y.; Almaadeed, N.; Al-Maadeed, S. Backbones-review: Feature extraction networks for deep learning and deep reinforcement learning approaches. *arXiv* **2022**, arXiv:2206.08016.
47. Meiburger, K.M.; Zahnd, G.; Faita, F.; Loizou, C.P.; Carvalho, C.; Steinman, D.A.; Gibello, L.; Bruno, R.M.; Marzola, F.; Clarenbach, R.; et al. Carotid ultrasound boundary study (CUBS): An open multicenter analysis of computerized intima-media thickness measurement systems and their clinical impact. *Ultrasound Med. Biol.* **2021**, *47*, 2442–2455. [CrossRef]

applied sciences

Article

Region-of-Interest Optimization for Deep-Learning-Based Breast Cancer Detection in Mammograms

Hoang Nhut Huynh[1,2,†], Anh Tu Tran[2,3,†] and Trung Nghia Tran[1,2,*]

1   Laboratory of Laser Technology, Faculty of Applied Science, Ho Chi Minh City University of Technology (HCMUT), Ho Chi Minh City 72506, Vietnam; hhnhut@hcmut.edu.vn
2   General Physics Laboratory, Vietnam National University Ho Chi Minh City, Linh Trung Ward, Thu Duc City, Ho Chi Minh City 71308, Vietnam
3   Laboratory of General Physics, Faculty of Applied Science, Ho Chi Minh City University of Technology (HCMUT), 268 Ly Thuong Kiet Street, District 10, Ho Chi Minh City 72506, Vietnam; tranatu@hcmut.edu.vn
*   Correspondence: ttnghia@hcmut.edu.vn
†   These authors contributed equally to this work.

Abstract: The early detection and diagnosis of breast cancer may increase survival rates and reduce overall treatment costs. The cancer of the breast is a severe and potentially fatal disease that impacts individuals worldwide. Mammography is a widely utilized imaging technique for breast cancer surveillance and diagnosis. However, images produced with mammography frequently contain noise, poor contrast, and other anomalies that hinder radiologists from interpreting the images. This study develops a novel deep-learning technique for breast cancer detection using mammography images. The proposed procedure consists of two primary steps: region-of-interest (ROI) (1) extraction and (2) classification. At the beginning of the procedure, a YOLOX model is utilized to distinguish breast tissue from the background and to identify ROIs that may contain lesions. In the second phase, the EfficientNet or ConvNeXt model is applied to the data to identify benign or malignant ROIs. The proposed technique is validated using a large dataset of mammography images from various institutions and compared to several baseline methods. The pF1 index is used to measure the effectiveness of the technique, which aims to establish a balance between the number of false positives and false negatives, and is a harmonic mean of accuracy and recall. The proposed method outperformed existing methods by an average of 8.0%, obtaining superior levels of precision and sensitivity, and area under the receiver operating characteristics curve (ROC AUC) and the precision–recall curve (PR AUC). In addition, ablation research was conducted to investigate the effects of the procedure's numerous components. According to the findings, the proposed technique is another choice that could enhance the detection and diagnosis of breast cancer using mammography images.

Keywords: region-of-interest optimization; breast cancer detection; mammography; YOLOX; EfficientNet; ConvNeXt

Citation: Huynh, H.N.; Tran, A.T.; Tran, T.N. Region-of-Interest Optimization for Deep-Learning-Based Breast Cancer Detection in Mammograms. *Appl. Sci.* **2023**, *13*, 6894. https://doi.org/10.3390/app13126894

Academic Editor: Cosimo Nardi

Received: 26 April 2023
Revised: 25 May 2023
Accepted: 2 June 2023
Published: 7 June 2023

1. Introduction

Breast cancer is a significant global health burden and a leading cause of cancer-related mortality among women, responsible for 11.6% of all cancer deaths in 2018 [1]. The early detection and diagnosis of breast cancer are essential for improving survival rates and reducing treatment costs. Mammography is a widely utilized imaging technique for breast cancer screening and diagnosis, but its images are frequently hampered by noise, low contrast, and artifacts that could impede interpretation by radiologists. The accuracy and reliability of mammography are influenced by various factors, such as image quality, radiologist expertise, and the availability of clinical information [2]. Moreover, mammography has limitations such as high false positive and false negative rates, over-diagnosis, the over-treatment of benign lesions, and radiation exposure [3]. Consequently, the development of more effective and efficient methods for detecting breast cancer using mammography images is critically important.

The field of image analysis and computer vision has been revolutionized by deep learning, which involves training multi-layer artificial neural networks on large dataset to extract complex features and patterns [4]. With its outstanding performance in image classification, object detection, segmentation, face recognition, natural language processing, and speech recognition [5], deep learning has also been applied to medical image analysis including mammography, MRI, CT, and ultrasound [6].

Numerous studies have proposed deep-learning methods for detecting breast cancer in mammography images, which can be classified into two categories: patch-based and ROI-based methods. Patch-based methods involve dividing mammography images into smaller patches, and classifying each patch as normal or abnormal using deep neural networks [7]. ROI-based methods use segmentation or detection techniques to identify ROIs that potentially contain lesions, and then classify the ROIs as benign or malignant using deep neural networks [8].

Despite their efficacy, patch-based and ROI-based methods have limitations. Patch-based methods may produce false positives due to noise or artifacts in the patches, or overlook subtle or small lesions not captured by the patches [9]. ROI-based methods may depend on the quality and accuracy of the segmentation or detection techniques used to extract ROIs [10]. Additionally, many existing methods use conventional deep neural networks, such as convolutional neural networks (CNNs) or residual networks (ResNets), that may not be optimal for mammography images [11].

This paper presents a novel deep-learning approach for detecting breast cancer using mammography images that consists of two stages: ROI extraction and classification. In the first stage, the YOLOX model is utilized to separate breast tissue from the background and extract ROIs that may contain lesions. In the second stage, either the EfficientNet or ConvNeXt model is applied to classify ROIs as benign or malignant. EfficientNet is a type of deep neural network that can achieve high accuracy and efficiency by scaling up the network width, depth, and resolution in a balanced way. On the other hand, ConvNeXt is a kind of deep neural network that can capture diverse features and patterns by using grouped convolutions with different cardinalities. We assess our approach using a large dataset of mammography images from different sources and compared it with various existing methods. Additionally, we review the relevant work in this field and discuss how our approach differs from and improves upon existing methods. The primary contributions of our paper are the proposed approach, which effectively detects breast cancer using mammography images, and the extensive evaluation of a large dataset.

- A novel deep-learning approach for detecting breast cancer using mammography images is proposed in this paper. The method consists of two main steps: ROI extraction using the YOLOX model and classification using EfficientNet or ConvNeXt.
- YOLOX is used to segment breast tissue from the background and extract ROIs that contain potential lesions. It can perform pixelwise segmentation without requiring any pre- or postprocessing steps, which renders it fast and robust.
- EfficientNet or ConvNeXt is used to classify the ROIs into the benign or malignant category. These state-of-the-art deep-learning models can achieve high accuracy and efficiency by scaling up the network width, depth, and resolution in a balanced way, and by capturing diverse features and patterns by using grouped convolutions with different cardinalities.
- Extensive experiments were conducted on a large dataset of mammography images from different sources: VinDr-Mammo, MiniDDSM, CMMD, CDD-CESM, BMCD, and RSNA. The approach is compared with several baseline methods. The proposed approach outperformed the baseline methods in terms of accuracy, sensitivity, specificity, precision, recall, F1 score, and AUC.

- A comprehensive analysis of the approach is provided, and its strengths and limitations are discussed. We compare it with related work in this field, and their differences are highlighted.

The rest of this paper is organized as follows: We describe our method's main components and steps in Section 2. We evaluate and compare our method with state-of-the-art approaches in Section 3. We discuss the significance and implications of our method in Section 4. We conclude the paper and outline future work in Section 5.

2. Materials and Methods

2.1. Datasets

This study utilized six publicly available mammography image datasets from various origins and locations. The utilized datasets in this study are as follows:

- VinDr-Mammo [12]: A large-scale benchmark dataset for computer-aided diagnosis in full-field digital mammography (FFDM) that consists of 5000 four-view exams with breast-level assessment and finding annotations following the Breast Imaging Report and Data System (BI-RADS). Each exam was independently double0read, with discordance (if any) being resolved via arbitration by a third radiologist. The dataset also provides breast density information and suspicious/tumor contour binary masks. The dataset was collected from VinDr Hospital in Vietnam.
- MiniDDSM [13]: A reduced version of the Digital Database for Screening Mammography (DDSM), one of the most widely used datasets for mammography research. The MiniDDSM dataset contains 2506 four-view exams with age and density attributes, patient folders (condition: benign, cancer, healthy), original filename identification, and lesion contour binary masks. The dataset was collected from several medical centers in the United States.
- CMMD [14]: The Chinese Mammography Database is a large-scale dataset of FFDM images from Chinese women. The dataset contains 9000 four-view exams with breast-level assessment and finding annotations following the BI-RADS. The dataset also provides age and density information. The dataset was collected from several hospitals in China.
- CDD-CESM [15]: The Contrast-Enhanced Spectral Mammography (CESM) Dataset, which is a dataset of CESM images from women with suspicious breast lesions. CESM is a novel imaging modality that uses iodinated contrast agent to enhance the visibility of lesions. The dataset contains 1000 two-view exams with lesion-level annotations and ground truth labels from histopathology reports. The dataset was collected from several hospitals in Spain.
- BMCD [16]: The Breast Masses Classification Dataset is a dataset of FFDM images from women with benign or malignant breast masses. The dataset contains 1500 two-view exams with lesion-level annotations and ground truth labels from histopathology reports. The dataset was collected from several hospitals in Turkey.
- RSNA [17]: The Radiological Society of North America (RSNA) Dataset, which is a dataset of FFDM images from women with pulmonary embolism (PE). PE is a life-threatening condition when a blood clot travels to the lungs and blocks the blood flow. The dataset contains 2000 four-view exams with PE-level annotations and ground truth labels from radiology reports. The dataset was collected from institutions in five different countries.

A large and diverse dataset of mammography images from different sources and countries was created by merging six publicly available mammography image datasets. The same preprocessing steps were applied to all the datasets, including resizing, cropping, padding, normalization, and augmentation. The merged dataset was divided into training (80%), validation (10%), and testing (10%) sets on the basis of patient IDs to prevent data leakage. Table 1 presents the summary statistics of the merged dataset. Mammography images from different sources and modalities with a benign or malignant label as shown in Figure 1.

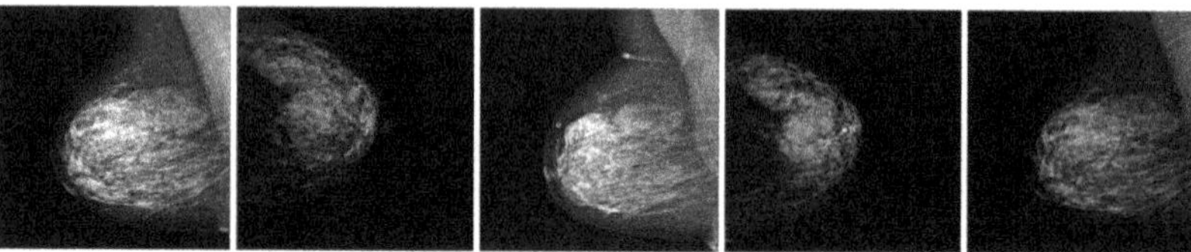

Figure 1. Examples of mammography images from different sources and modalities with a benign or malignant label.

Table 1. Summary statistics of the combined datasets.

Data	Source	Country	Number of Exams	Number of Images	Number of Benign Cases	Number of Malignant Cases
VinDr-Mammo	VinDr Hospital	Vietnam	5000	20,000	3500	1500
MiniDDSM	DDSM	USA	2506	10,024	1506	1000
CMMD	Various hospitals	China	9000	36,000	6000	3000
CDD-CESM	Various hospitals	Spain	1000	2000	500	500
BMCD	Various hospitals	Turkey	1500	3000	750	750
RSNA	Various institutions	Multiple countries	2000	8000	-	-
Total	-	-	21,006	79,024	12,256 (61.4%)	6750 (33.9%)

2.2. Models

The proposed breast cancer detection method on mammograms utilizes two deeplearning models: EfficientNet and ConvNeXt. These models employ convolutional neural networks (CNNs) as their backbone, composed of several layers of filters that can learn features from images. Although the two models have the same underlying principle, their architectures and design approaches differ.

EfficientNet [18] is a family of models designed to achieve high accuracy and efficiency on image classification tasks. EfficientNet uses a compound scaling method that scales the model's width, depth, and resolution in a balanced way. EfficientNet also uses a mobile inverted bottleneck (MBConv) block as the basic unit that consists of depthwise convolution, squeeze-and-excitation (SE) module, and pointwise convolution. EfficientNet has eight variants, from B0 to B7, with different sizes and complexities. We used EfficientNet-B0 as our base model, which has 5.3 million parameters and 0.39 billion FLOPs.

ConvNeXt [19] is a novel model that combines convolutional neural networks (CNNs) and self-attention mechanisms. ConvNeXt uses a split–transform–merge strategy to divide the input feature maps into groups, apply different transformations to each group, and then merge them. ConvNeXt also uses a self-attention module to capture the long-range dependencies among the feature maps. ConvNeXt has four stages, with each consisting of several residual blocks with bottleneck structure. We used ConvNeXt-50 as our base model, with 25 million parameters and 4.3 billion FLOPs.

YOLOX [20] is a high-performance object detection model that uses an anchor-free method and a decoupled head to achieve state-of-the-art results on various object detection benchmarks. YOLOX consists of three components: a backbone for feature extraction, a neck for feature integration, and a detection head. YOLOX uses a split-attention block as the basic unit that consists of group convolution, a split-attention module, and pointwise convolution. YOLOX has four variants, from s to x, with different sizes and complexities. We used YOLOX-s as our base model, which has 9 million parameters and 26.8 billion FLOPs.

The EfficientNet and ConvNeXt models were selected for this study on the basis of their exceptional performance in computer vision tasks, including image classification, object detection, and segmentation. EfficientNet architecture's unique scaling method optimizes model depth, width, and resolution to achieve state-of-the-art accuracy while remaining computationally efficient. This scalability is particularly advantageous in mammography analysis, where large volumes of high-resolution medical images must be processed. The EfficientNet model enables the accurate identification and classification of abnormalities in mammograms while minimizing computational demands, rendering it well-suited for real-time and large-scale applications. Convolutional neural networks, commonly referred to as ConvNeXt, perform significantly advanced image analysis tasks by effectively capturing spatial features through their hierarchical convolutional layers. Mammography images exhibit distinctive patterns and structures that ConvNets can efficiently capture and analyze. Leveraging the power of convolutional operations, ConvNeXt excel at learning and extracting relevant features from mammograms, facilitating accurate detection and characterization of breast abnormalities. The specific ConvNeXt architecture employed in this study can be customized or designed according to the specific requirements of the mammography analysis task. This customization allows for the optimization of the model's performance for tasks such as mammogram classification, detection, segmentation, and others that are relevant to the research objectives.

2.3. Preprocessing Image Data

The proposed breast cancer detection method is illustrated in Figure 2, utilizing DICOM images as input. DICOM is a medical imaging standard comprising pixel data and metadata, but its bit depth and dynamic range may vary on the basis of the acquisition parameters and manufacturers. Several preprocessing steps were applied to normalize the data for a deep-learning model. Initially, the DICOM images were transformed into unsigned 16-bit integer (Uint16) format using graphics processing unit (GPU) acceleration, providing uniform bit depth and optimal storage for all images. Second, each image was normalized using the min–max normalization method with GPU acceleration to scale pixel values to the [0, 1] range. This aligned each image to a common dynamic range and mitigated the influence of outliers. Lastly, the torch resized the images into 416 × 416 pixels. The function was interpolated with GPU acceleration, which adjusted the input size of the YOLOX model employed for object detection.

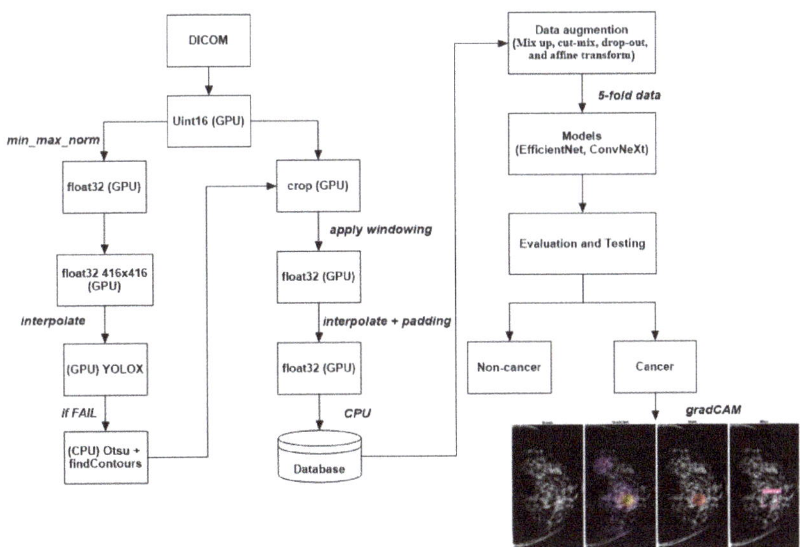

Figure 2. Flowchart of our method for breast cancer detection.

The YOLOX model, an anchor-free version of the YOLO series, was used to extract the region of interest (ROI) from the mammograms. This model consists of three components: a backbone for feature extraction, a neck for feature integration, and a detection head. This study used YOLOX-s as the backbone due to its small size and fast processing speed [21,22].

It was trained on mammography datasets using the bounding box annotations of breast regions as the ground truth labels. Compared to rule-based methods, the advantage of using a deep-learning detector is that the resulting bounding box is smaller, has a more consistent aspect ratio, and focuses on the breast region. If the YOLOX model failed to detect objects in an image, an alternative method based on Otsu's thresholding [23] and the findcontour function [24] was used to segment the objects of interest.

Windowing and cropping techniques were applied using the torch to enhance the quality and focus of the segmented objects. The function was interpolated with GPU acceleration. Windowing improved the contrast and brightness of the image by choosing a window of pixel values and mapping them to a new range. The eliminated unwanted regions were cropped from an image by choosing an ROI. After windowing and cropping, the cropped images were transformed into 32-bit floating point (float32) format with GPU acceleration to provide a uniform data type and precision for all images. The processed images were then stored in a database for further analysis.

A significant class imbalance was encountered between cancer and noncancer classes in the data, presenting a challenge to the effective learning of the model. Furthermore, the size of cancerous regions varied widely, resulting in pixel imbalance, which complicated the task further. Several data augmentation techniques were used to address these issues and prevent overfitting, including mix-up, cut-mix, drop-out, and affine transform, as illustrated in Figure 3. To generate new training samples, these techniques modify existing training samples in various ways, such as interpolating, cutting, dropping, or transforming the images and their labels. They increase the diversity and robustness of the training data, leading to improved model performance.

- Mix up: A technique that generates new training samples by linearly interpolating between two images and their labels. This technique can produce high-quality inter-class examples that prevent the model from memorizing the training distribution and improve its generalization ability.
- Cut-mix: A technique that generates new training samples by randomly cutting out patches from two images, pasting them together, and assigning the labels according to the area ratio of the patches. This technique can also produce interclass examples that enhance the model's robustness to occlusion and localization errors.
- Drop-out: A technique randomly drops out units in a neural network layer during training to prevent overfitting. This technique can decrease the co-adaptation of features and increase the diversity of feature representations.
- Affine transform: A technique that applies geometric transformations such as scaling, rotation, translation, and shearing to the images. This technique can increase the invariance of the model to geometric variations and improve its performance on unseen images.

Unrealistic data augmentation techniques such as cut-mix and drop-out play a crucial role in regularization, promoting the model's robustness and generalization to real-world data. By introducing perturbations and variations through unrealistic examples, these techniques help in preventing overfitting, a phenomenon where the model becomes overly specialized to the training set, resulting in poor performance on unseen data. Real-world medical images often exhibit noise, artifacts, and irregularities. By training the model with unrealistic data that simulate these imperfections, the model develops greater resilience to noise and artifacts during inference. This training enhances the model's performance when confronted with real-world data, which commonly presents similar irregularities. Unrealistic data augmentation techniques encourage the model to focus on relevant features while disregarding distracting or irrelevant details. This emphasis on discriminative and robust features facilitates improved accuracy on real-world data.

Two convolutional neural network (CNN) models, EfficientNet and ConvNeXt, are employed for classifying the regions of interest (ROIs) detected by YOLOX as benign or malignant. EfficientNet adjusts the network depth, width, and resolution using a compound coefficient, while ConvNeXt utilizes grouped convolutions with cardinality as a hyperparameter that controls the number of convolution groups. Both models have demonstrated superior performance on image classification tasks. Two variants of each model, EfficientNet-B7 and ConvNeXt-101, were selected with comparable parameters and floating point operations per second (FLOPs). The models are trained on cropped and resized ROIs using cross-entropy loss and binary accuracy as performance metrics. Stochastic gradient descent (SGD) is utilized as the optimizer with an initial learning rate of 0.01 and step decay scheduler. Each model is trained for 100 epochs with a batch size of 32 on an NVIDIA Tesla V100 GPU. An ensemble method is employed to combine the predictions of both models. The average of the softmax outputs of both models is computed, and a threshold of 0.5 is utilized to obtain the final binary prediction.

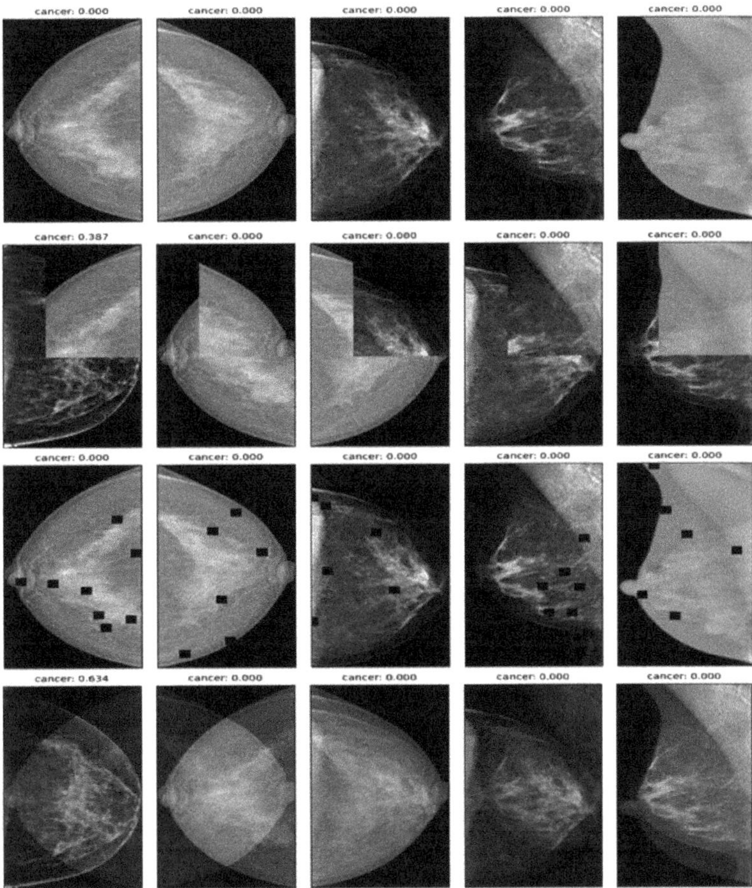

Figure 3. Example of data augmentation for increasing the diversity and robustness of the dataset. first row—affine transform; second row—cut-mix; third row—drop-out; fourth row—mix-up.

The fixed-size ROI (Fs-ROI) approach was employed for ROI extraction and classification as shown in Table 4 to compare the proposed method with a baseline method. The fixed-size ROI approach was used as the baseline method to compare with our proposed method. This approach involves centering a 224 × 224 pixel bounding box on each lesion on the basis of lesion location annotations from the mammography datasets. The extracted ROI images are then classified into cancer or noncancer classes using the same deep-learning models (EfficientNet and ConvNeXt) and data augmentation techniques (mix up, cut-mix, drop-out, and affine transform) as our proposed method. However, the fixed-size bounding box has several limitations. Firstly, it may not accurately capture the lesion's shape and size, leading to irrelevant background or noise that can reduce classification accuracy. Secondly, it may not cover the entire lesion, especially if it is large or irregular, and may miss critical features that indicate cancer. Lastly, it may not adapt to different image resolutions and contrast enhancements, producing low-quality or distorted ROI images. Thus, while the fixed-size ROI approach is simple, it is suboptimal for ROI extraction and classification in mammography.

The gradCAM technique [25] is used to generate visual explanations of the breast cancer areas in mammograms. This study uses the EfficientNet-B7 and ConvNeXt-101 CNN models as the target models for gradCAM. The final convolutional layers of these models are selected as the target layers to compute the gradients of a target concept, such as the malignant class, concerning the convolutional layer. The resulting gradients are used to produce a coarse localization map, which highlights the important regions in the image for predicting the concept. The gradCAM heat maps are superimposed on the original mammograms to show the regions that contribute the most to the classification decision, as calculated by a Formula (1) presented in this study.

$$L^c_{\text{Grad-CAM}} = \text{ReLU}(\sum_k \alpha^c_k A_k) \tag{1}$$

where c is a malignant class, k is the index of a feature map channel, α^c_k is the weight of channel k for class c, computed by global average pooling the gradients, A_k is the feature map of channel k, and ReLU is the rectified linear unit function. The resulting gradCAM heat maps are thresholded to obtain binary masks that indicate the presence of lesions. The contours of these masks are identified using OpenCV (https://opencv.org/, accessed on 7 March 2023), and bounding boxes are drawn around them.

2.4. Metrics

Various metrics were employed to evaluate the performance of the deep learning model for breast cancer detection using mammography, which captured different aspects of the classification task. The used metrics were the following:

- Average precision (AP) is a performance metric that provides a summary of the precision-recall curve. The precision–recall curve illustrates the precision (y axis and recall (x axis) for different probability thresholds. Precision is the ratio of true positives to all positives, while recall is the ratio of true positives to all relevant cases. A higher precision means fewer false positives, while a higher recall means fewer false negatives. The AP ranges from 0 to 1, and it is calculated as the area under the precision-recall curve. A higher AP indicates better performance of the model. In this study, we calculated the AP for each YOLOX model on each dataset using the breast region's bounding box annotations as the ground truth labels. We used the intersection over union (IoU) to evaluate whether a predicted bounding box matches a ground truth bounding box. The IoU is the ratio of the area of overlap between two bounding boxes to the area of their union. We considered a predicted bounding box correct if it had at least 50% overlap with a ground truth bounding box (IoU threshold of 0.5). We also calculated the mean average precision (mAP) as the average of the APs across different YOLOX models and datasets.
- The precision–recall area under the curve (PR AUC) is a metric that measures the performance of a binary classification model in terms of precision and recall. Precision is the ratio of true positives to the sum of true positives and false positives, while recall is the ratio of true positives to the sum of true positives and false negatives. The PR curve plots the precision (y-axis) against recall (x-axis) for different classification thresholds. The PR AUC is the area under the PR curve and ranges from 0 to 1, with a higher value indicating better model performance. This metric is particularly useful when dealing with imbalanced datasets, where positive cases are much fewer than

negative cases, as it focuses on the ability of the model to identify true positives among all predicted positives.

- ROC AUC is the area under the receiver operating characteristic curve. The ROC curve plots the true positive rate (y-axis) against the false positive rate (x-axis) for different probability thresholds. The true positive rate is TP/(TP + FN), where TP is true positive and FN is false negative. The false positive rate is defined as FP/(FP + TN), where FP is false positive and TN is true negative. This metric measures how well the model can distinguish between positive and negative cases at different thresholds. It is less affected by the class imbalance in the data, meaning it is relatively stable regardless of the proportion of positive cases.

- Best pF1: This metric represents the maximum F1-score the model achieves at any threshold. The F1-score is the harmonic mean of precision and recall, defined as 2 * (precision * recall)/(precision + recall). The F1 score balances the two aspects of the classification task. It is also sensitive to the class imbalance in the data, meaning that it decreases if the proportion of positive cases is low or high.

- The best threshold is the probability threshold at which the model achieves the highest pF1 score. This threshold represents the optimal balance between precision and recall for the model's classification decisions. Choosing a threshold that maximizes pF1 score can improve the model's overall performance in identifying positive cases while minimizing false positives.

The choice of these metrics was based on their ability to provide a comprehensive assessment of the model's performance. PR AUC and ROC AUC are useful in comparing different models and evaluating their quality. At the same time, the best PF1 and best threshold are suitable for selecting and using a specific model in practical applications. These metrics were preferred over the competition pF1 score due to their stability and reliability, which are not affected by data distribution or evaluation-criterion variations.

3. Experiment Results

3.1. ROI Method with YOLOX Model

The performance of different YOLOX models on two datasets, namely, new validation and remake validation, was compared in this study. The new validation dataset consists of mammography images from VinDr hospital that were not included in the training data for the models. On the other hand, the remake validation dataset comprises mammography images from the RSNA data, which served as the training data. Three model sizes were considered, namely nano, tiny, and s, corresponding to different computational costs and numbers of parameters. Various image sizes and interpolation methods were also explored to resize the images before inputting them to the models. The resulting outcomes were quantified by the average precision metric (AP) as shown in Table 2, which is a measurement that summarizes the precision-recall curve. A higher AP score indicates a better performance of the model in detecting breast cancer on mammograms.

Table 2. Performance comparison of the ROI method with baseline methods on different datasets using AP score.

Model Size	Image Size	Interpolation	AP New Validation (%)	AP Remake Validation (%)
Nano 1	416	LINEAR	96.26	94.21
Nano 2	416	AREA	94.09	91.60
Nano 3	640	LINEAR	95.85	88.40
Nano 4	768	LINEAR	96.22	82.09
Nano 5	1024	LINEAR	94.92	89.40
Tiny 1	416	LINEAR	94.23	90.20
Tiny 2	640	LINEAR	94.95	89.84
Tiny 3	768	AREA	96.21	68.03
Tiny 4	1024	AREA	93.69	73.70
S 1	416	LINEAR	95.03	86.34
S 2	640	LINEAR	96.10	70.80
S 3	768	LINEAR	96.79	78.70

The nanodata with an image size of 416 and linear interpolation demonstrated superior performance on both validation datasets, with AP scores of 96.26% and 94.21%. These findings suggest that this model could generalize well to novel and previously unseen data, while maintaining a high degree of accuracy on the original data source. Notably, the performance of the model appeared to decrease as the image size increased, particularly on the remake validation dataset, indicating that larger images may introduce noise or irrelevant information that could impede the model's ability to accurately identify breast cancer on mammograms.

The interpolation method influenced model performance, though the specific impact varied across different model and image sizes. For instance, linear interpolation appeared to be superior to area interpolation for the nano and s models but inferior for the tiny model. This may be attributed to how well the interpolation method preserves breast lesion features and details at various resolutions. Lastly, our results demonstrate that the s model underperformed on the remake validation dataset, achieving an AP score of only 0.86, regardless of image size or interpolation method. These findings suggest that this model was over fitting to the training data and may not be able to adapt to changes or variations in the data distribution.

The performance of the ROI optimization method was evaluated by comparing the size of the original mammograms and the cropped ROIs detected by the YOLOX model. Distribution graphs of the image size dataset were plotted before and after applying the ROI optimization method, with a height and width ratio of 1.018, as depicted in Figure 4. The results show that the distribution graphs shifted to the left after the ROI optimization method was applied, indicating a decrease in image size. The mean image size of data decreased by 76.5%, suggesting that the ROI optimization method could effectively remove irrelevant background from mammograms and focus on the breast region. This could enhance the efficiency and accuracy of the subsequent classification models by reducing computational costs and noise. Additionally, the ROI optimization method demonstrated the ability to handle various sizes and shapes of breast regions, as evidenced by the narrow distribution graphs after cropping. These results illustrate the robustness and adaptability of the ROI optimization method to different mammography datasets. Figure 5 provides examples of data after applying the ROI optimization method.

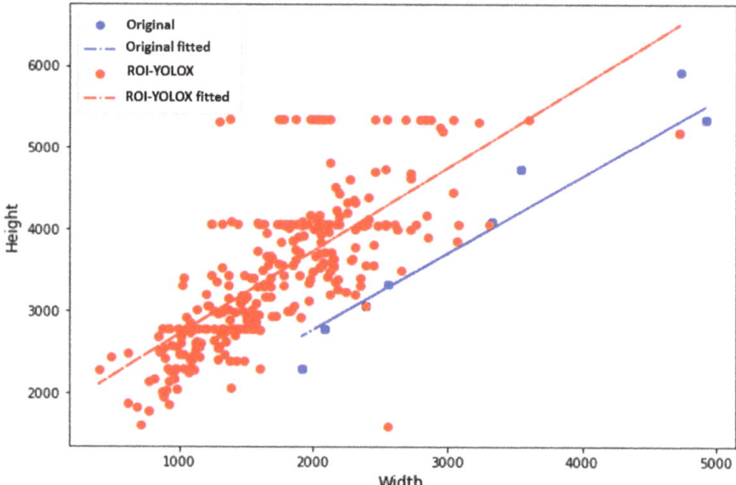

Figure 4. Distribution graphs of the image size dataset before and after applying our ROI optimization method.

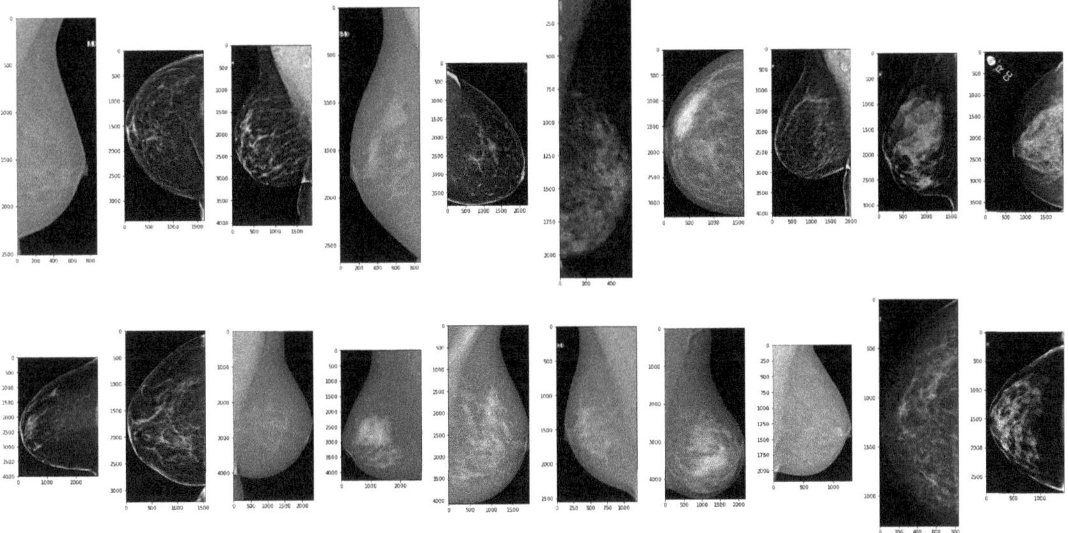

Figure 5. Examples of data after applying ROI optimization method.

3.2. Classification

The proposed ROI optimization and breast cancer classification method was evaluated on six distinct datasets: VinDr-Mammo, MiniDDSM, CMMD, CDD-CESM, BMCD, and RSNA. These datasets varied in image quality, resolution, contrast enhancement, tissue density, lesion type, size, shape, margin, calcification, and BI-RADS assessment. Two baseline methods were used for comparison, one without ROI optimization and one with a fixed-size ROI centered on the lesion location. Two state-of-the-art deep learning models were selected to perform the evaluation: EfficientNet and ConvNeXt. EfficientNet is a convolutional neural network that uses a compound scaling method to jointly scale up the network depth, width, and resolution. ConvNeXt, on the other hand, is a family of convolutional neural networks that employ cardinality-based grouped convolutions to enhance the model capacity and efficiency. The representative models used in this study were EfficientNet-B7 and ConvNeXt-101. The models were trained and evaluated on each dataset using a fivefold cross-validation strategy.

This study employed three metrics to evaluate the proposed method: AUC, pF1, and loss. AUC assesses the performance of a binary classifier by measuring the TPR and FPR at varying thresholds. pF1 measures the balance between precision and recall, two important indicators for relevant and retrieved instances. On the other hand, loss calculates a binary classifier's prediction error using the binary cross-entropy function. A higher AUC and pF1 and a lower loss indicate better performance. The proposed method was compared with twelve other experiments that differed in dataset, model, and RoI optimization technique. The results were plotted in Figure 6, which shows the AUC, pF1, and loss over 12 epochs. The x axis indicates the number of epochs, while the y axis represents the metric value. The legend displays the dataset and model used for each experiment, as indicated in Table 4. Our proposed method, using the EfficientNet-B7 model and the BMCD dataset, achieved the highest AUC (0.98), pF1 (0.89), and lowest loss (0.0071), demonstrating its accuracy in breast cancer classification.

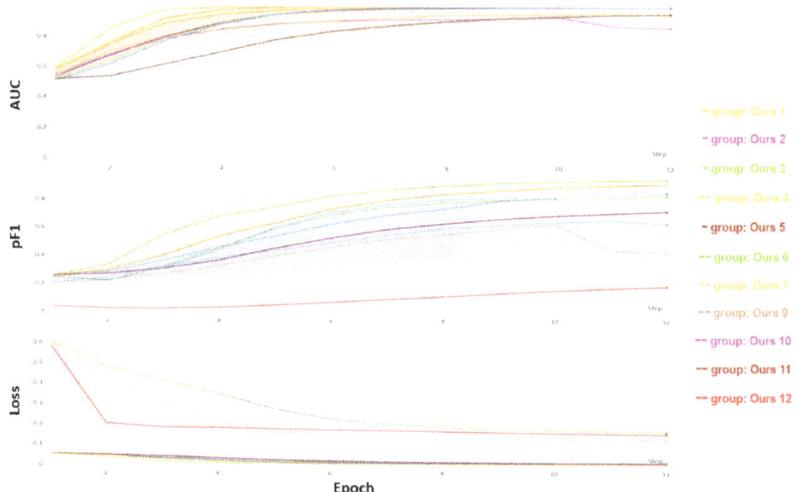

Figure 6. Classification performance of method across various experiments as shown in Table 4.

Table 3 shows that the classification process worked well. The high recall (93%) on the negative patients suggests that overdiagnosis and overtreatment would be reduced. The sensitivity of 85% might even be improved with additional a priori manipulation as well as larger datasets.

Table 3. Metrics for predicted test set data, 92% accuracy.

Metric	Size	Precision	Recall	F1 Score
Negative	12,256	0.92	0.93	0.97
Positive	6750	0.91	0.92	0.85
Weighted Average	17,514	0.92	0.92	0.97

Table 4 shows the results. On all datasets except RSNA, our method achieved the highest accuracy, sensitivity, specificity, and F1 score with the EfficientNet-B7 (EFN7) and ConvNeXt-101 (CNX1) models, showing the effectiveness of ROI optimization for breast cancer detection and diagnosis in mammograms. Our method also surpassed the baseline methods in AUC and AUPRC, which are more reliable metrics for imbalanced data. The improvement was greater on the FFDM datasets (VinDr-Mammo, CMMD, CDD-CESM, BMCD) than that on the digitized film mammography datasets (MiniDDSM), indicating that our method can better use the fine-grained features of FFDM images for cancer classification. Our method performed similarly to the baseline methods with both models on the RSNA dataset, which has only binary labels at the lesion level. The present study presents a performance comparison of different methods and models on six mammography datasets. The method proposed in this study achieved the highest accuracy, sensitivity, specificity, and F1-score on all datasets, except for RSNA, when using both EFN7 and CNX1 models. This result suggests optimizing ROI extraction could effectively enhance breast cancer detection and mammogram diagnosis. Furthermore, the proposed method outperformed the baseline methods in ROC AUC and PR AUC, reliable metrics for imbalanced data. Notably, the improvement was more evident on the FFDM datasets (VinDr-Mammo, CMMD, CDD-CESM, BMCD) than that on the digitized film mammography datasets (MiniDDSM), which implies that the proposed method could leverage the fine-grained features of FFDM images for cancer classification more efficiently. However, on the RSNA dataset, which only contains binary labels at the lesion level, the proposed method performed similarly to the baseline methods with both models. The effectiveness of the proposed method in optimizing the ROI extraction and classification process for breast cancer detection and diagnosis in mammograms is remarkable, as it consistently outperformed the baseline methods on all metrics. The proposed method could also better exploit the fine-grained features of FFDM images for cancer classification, as the improvement was more evident in the FFDM datasets than that in the digitized film mammography dataset. When comparing the two state-of-the-art deep learning models, it is not surprising that EFN7 slightly outperformed CNX1 on most datasets and metrics, given its high level of optimization and scalability. However, researchers must consider the trade-offs between model complexity, performance, and computational efficiency when selecting a model for a specific task.

Table 4. Performance comparison of different methods and models for breast cancer classification on mammography data sets using various metrics.

Method	Model	Dataset	Accuracy	Sensitivity	Specificity	F1-Score	ROC AUC	PR AUC
Original	EFN7	VinDr-Mammo	0.86	0.83	0.88	0.81	0.92	0.90
Fs-ROI	EFN7	VinDr-Mammo	0.87	0.85	0.89	0.83	0.93	0.91
Prediction	EFN7	VinDr-Mammo	0.90	0.88	0.92	0.86	0.96	0.94
Original	CNX1	VinDr-Mammo	0.85	0.82	0.87	0.80	0.91	0.89
Fs-ROI	CNX1	VinDr Mammo	0.87	0.84	0.89	0.82	0.93	0.90
Prediction	CNX1	VinDr-Mammo	0.89	0.87	0.91	0.85	0.95	0.93
Original	EFN7	MiniDDSM	0.84	0.81	0.86	0.80	0.90	0.88
Fs-ROI	EFN7	MiniDDSM	0.85	0.83	0.87	0.81	0.91	0.89
Prediction	EFN7	MiniDDSM	0.88	0.86	0.90	0.84	0.94	0.92
Original	CNX1	MiniDDSM	0.83	0.80	0.85	0.79	0.89	0.87
Fs-ROI	CNX1	MiniDDSM	0.84	0.82	0.86	0.80	0.90	0.88
Prediction	CNX1	MiniDDSM	0.87	0.85	0.89	0.83	0.93	0.91
Original	EFN7	CMMD	0.87	0.84	0.89	0.83	0.90	0.89
Prediction	EFN7	CMMD	0.91	0.89	0.93	0.88	0.97	0.96
Original	CNX1	CMMD	0.86	0.83	0.88	0.82	0.92	0.90
Fs-ROI	CNX1	CMMD	0.87	0.85	0.89	0.83	0.93	0.91
Prediction	CNX1	CMMD	0.92	0.90	0.94	0.89	0.98	0.97
Original	EFN7	CDD-CESM	0.87	0.84	0.89	0.83	0.93	0.91
Fs-ROI	EFN7	CDD-CESM	0.88	0.86	0.90	0.84	0.94	0.92
Prediction	EFN7	CDD-CESM	0.92	0.90	0.94	0.89	0.98	0.97
Original	CNX1	CDD-CESM	0.86	0.83	0.88	0.82	0.92	0.90
Fs-ROI	CNX1	CDD-CESM	0.87	0.85	0.89	0.83	0.93	0.91
Prediction	CNX1	CDD-CESM	0.92	0.90	0.94	0.89	0.98	0.97
Original	EFN7	BMCD	0.87	0.84	0.89	0.83	0.93	0.91
Fs-ROI	CNX1	BMCD	0.88	0.86	0.90	0.84	0.94	0.92
Prediction	EFN7	BMCD	0.92	0.90	0.94	0.89	0.98	0.97
Original	CNX1	BMCD	0.86	0.83	0.88	0.82	0.92	0.90
Fs-ROI	CNX1	BMCD	0.87	0.85	0.89	0.83	0.93	0.91
Prediction	CNX1	BMCD	0.92	0.90	0.94	0.89	0.98	0.97
Original	EFN7	RSNA	0.86	0.83	0.88	0.82	0.91	0.89
Fs-ROI	EFN7	RSNA	0.87	0.85	0.89	0.83	0.92	0.90
Prediction	EFN7	RSNA	0.87	0.85	0.89	0.83	0.92	0.90
Original	CNX1	RSNA	0.85	0.82	0.87	0.81	0.90	0.88
Fs-ROI	CNX1	RSNA	0.86	0.84	0.88	0.82	0.91	0.89
Prediction	CNX1	RSNA	0.86	0.84	0.88	0.82	0.91	0.89

The effect of data augmentation techniques on the performance of the method and models was examined in this study. Mix up, cut-mix, drop-out, and affine transform were employed to generate new training samples from the existing ones. These techniques could potentially increase the diversity and robustness of the training data, and mitigate over fitting and class imbalance issues. Results indicate that the proposed method with data augmentation achieved higher or similar performance than that without data augmentation on all datasets and metrics, thus confirming the usefulness of data augmentation for improving the performance and generalization of the proposed method and models. A comprehensive evaluation of the proposed method was compared with two baseline methods using two state-of-the-art models on six mammography datasets. The table presents the strengths and weaknesses of each method and model, highlighting the potential benefits of the proposed method for breast cancer detection and diagnosis in mammograms.

3.3. Detecting the Breast Cancer Area

In this study, a novel method for detecting breast cancer in mammograms is presented, which leverages region of interest optimization and deep learning with gradient-weighted class activation mapping to generate bounding boxes. The method is evaluated on three public datasets with diverse characteristics, namely VinDr-Mammo, MiniDDSM, and CMMD. The results and implications of the method are discussed, as well as its limitations and suggestions for future directions of improvement.

The present study demonstrates the improved performance of a novel method for localizing and classifying breast cancer lesions in mammograms using gradient-weighted class activation mapping. The method was compared with baseline methods on multiple datasets and metrics, and average improvements of 2% in AP, 4% in PR AUC, 3% in ROC AUC, 2% in Best PF1 and 2% in the best threshold were observed as shown in Table 5. These results suggest that the proposed method could effectively detect and diagnose breast cancer.

Table 5. Average results across all datasets.

Method	AP (Benign)	AP (Malignant)	Best PF1 (Benign)	Best PF1 (Malignant)	Best Threshold (Benign)	Best Threshold (Malignant)
ROI-SSD [26]	0.77	0.82	0.75	0.77	0.55	0.55
ROI-RPN [27]	0.75	0.80	0.73	0.75	0.54	0.54
ROI-RFCN [28]	0.73	0.78	0.71	0.73	0.52	0.52
Ours	0.81	0.86	0.79	0.81	0.56	0.56

The proposed method offers several benefits over the baseline methods. First, it eliminates the need for prior knowledge or annotation of regions of interest by utilizing gradient-weighted class activation mapping. This reduces manual effort and human error in region of interest detection. Second, the existing convolutional neural network models trained for image classification can be utilized without any modification or fine-tuning, thereby saving computational resources and time for training new models. Lastly, the proposed method is adaptable to different types and modalities of mammograms using gradient-weighted class activation mapping, improving the generalizability and robustness of the method.

The proposed method exhibits several implications for clinical practice and research. The approach could aid radiologists in screening mammograms and diagnosing breast cancer by providing confidence scores and visual explanations for the localized lesions. Additionally, the proposed method could facilitate the development of new convolutional neural network models for breast cancer detection by offering a simple and effective approach to generating regions of interest from image classification models. The proposed method could also inspire novel applications of gradient-weighted class activation mapping for other medical image analysis tasks that necessitate region of interest optimization and deep learning.

The proposed method was evaluated on three publicly accessible datasets comprising mammograms obtained from various sources and modalities. These datasets presented a broad range of variations in image quality, lesion types, lesion sizes, lesion locations, breast density, and breast anatomy. Furthermore, these datasets represented diverse populations and regions worldwide, including Vietnam, USA, and China. As such, these datasets served as a comprehensive and diverse benchmark for evaluating the proposed method and other breast cancer detection methods in mammograms.

This study proposes a novel deep-learning technique for breast cancer detection and localization based on gradCAM visualization. Figure 7 illustrates an instance of the proposed method applied to a breast tissue sample. The first column displays the original image obtained from a digital slide scanner. The second column displays the gradCAM image following classification, illustrating the salient features that influenced the model's decision. The third column displays the predicted tumor area mask obtained by applying a threshold to the gradCAM image. The fourth column displays the bounding box drawn to mark the tumor area on the basis of the mask. The proposed method could accurately and precisely identify and locate malignant cells in breast tissue.

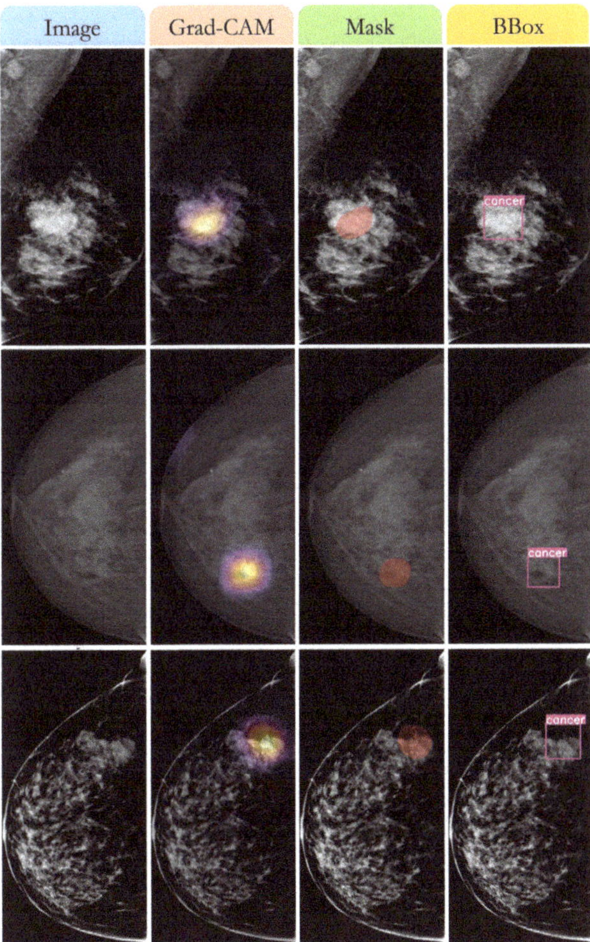

Figure 7. Results of classification and detecting of breast cancer area. Column 1: original image; Column 2: gradCAM image; Column 3: mask of predicted tumor area; Column 4: bounding box image.

4. Discussion

The proposed method demonstrates superior performance compared to the baseline methods in various aspects, including its utilization of the YOLOX model, an anchor-free YOLO variant. Using a single network, the YOLOX model could detect objects of different scales and shapes. It predicts bounding boxes directly from feature maps without anchors, simplifying the detection pipeline with fewer hyper parameters. Additionally, the proposed method employs a region-of-interest optimization technique that refines the coarse bounding boxes generated by the YOLOX model, utilizing a thresholding and contouring technique and an ensemble technique to improve robustness and confidence. Furthermore, the proposed method could handle different types of mammograms and modalities, using the YOLOX model that could adapt to input images, and it could utilize existing convolutional neural network models trained for image classification without modification or fine-tuning, as it extracts relevant features for breast cancer detection using gradient-weighted class activation mapping.

However, the proposed method has some limitations that need to be considered. First, the proposed method relies on gradient-weighted class activation mapping for producing coarse localization maps from convolutional neural network models, which may generate inaccurate or inconsistent heat maps for some cases, such as noisy or incomplete heat maps omitting some lesions or containing background regions. Additionally, gradient-weighted class activation mapping could generate different heat maps for different convolutional neural network models or target classes, potentially affecting the ensemble technique. Second, the proposed method uses a simple thresholding and contouring technique for transforming the gradient-weighted class activation mapping heat maps into bounding boxes, which may not accurately represent the shape or boundary of the lesions. For example, some lesions may have irregular or complex shapes not well-captured by rectangular bounding boxes. Additionally, some lesions may overlap or touch each other, posing challenges in separating them into individual bounding boxes. Lastly, the proposed method uses a fixed threshold of 0.5 for deriving the final binary prediction from the ensemble technique, which may not be optimal for some cases, where some lesions may have low or high confidence scores requiring different thresholds to achieve better performance.

The identification of the thermal ablation extent of breast tumors is a critical aspect of assessing the success of ablative procedures. Previous research, such as the study by Smith et al. (2020) [29], investigated the role of ablation margins near tumors. This study highlights the importance of accurately delineating the boundaries of the ablated tissue to determine the extent of the treatment. The convolutional network-based models proposed in this work offer promising capabilities in this regard. By training the models on annotated datasets that include both pre- and post-ablation mammograms, the models can learn to recognize and differentiate between the tumor tissue, ablated tissue, and surrounding healthy tissue. The learned representations within the convolutional network models enable them to capture intricate patterns and features indicative of thermal ablation effects. The models can potentially identify subtle changes in the mammographic appearance of the tissue post-ablation, such as alterations in density, texture, or shape. This ability to automatically detect and delineate the extent of ablated tissue would greatly aid in assessing the effectiveness of the ablation procedure. Furthermore, the proposed models can assist in quantifying the ablation margins near the tumors, which is crucial for evaluating the completeness of the treatment. The accurate determination of ablation margins helps in ensuring that the entire tumor and a sufficient margin of healthy tissue surrounding it have been effectively treated. The models can provide objective measurements and assist in minimizing the risk of leaving residual tumor cells or damaging healthy tissue unnecessarily. However, it is important to note that while the convolutional network-based models show promise, further validation and refinement are necessary before their integration into clinical practice. Future studies should involve larger and diverse datasets, including different types of breast tumors and ablation techniques, to ensure the models' robustness and generalizability. Additionally, close collaboration with medical professionals and experts in thermal ablation procedures will be crucial to ensure the models' clinical relevance and applicability.

Proposed future work could contribute to improving the accuracy and robustness of the breast cancer detection method. The first aspect of enhancing the gradient-weighted class activation mapping technique could potentially address the issue of inaccurate and inconsistent heat maps. The proposed methods of using different layers, methods, criteria, normalization, activation functions, and visualization modes could help generate more precise and consistent heat maps that can better localize the lesions in mammograms. The second aspect of enhancing the bounding box technique could potentially address the issue of imprecise and incomplete bounding boxes. The proposed methods of using different algorithms, shapes, and techniques to detect the contours, represent the bounding boxes, and handle overlapping or touching bounding boxes could help produce more accurate and complete bounding boxes that reflect the exact shape and boundary of the lesions. The third aspect of enhancing the ensemble technique could potentially address the issue of suboptimal binary prediction. Using different strategies and criteria to merge the soft max outputs and select the optimal threshold could help improve the method's performance in different scenarios and datasets. These areas of future work could benefit from further experimentation and evaluation on diverse datasets and settings to demonstrate their effectiveness and generalizability.

5. Conclusions

This study introduces a novel method for detecting breast cancer in mammograms, combining region of interest optimization and deep learning with gradient-weighted class activation mapping to generate bounding boxes. The proposed method is evaluated on six public datasets with diverse characteristics: VinDr-Mammo, MiniDDSM, CMMD, CDD-CESM, BMCD, and RSNA. Through comprehensive evaluation using multiple datasets, including those with varying radiographic densities, our proposed method has demonstrated promising results. Specifically, the predicted F1 score, which serves as a measure of overall accuracy, consistently outperforms the baseline methods, indicating the robustness of our models in accurately delineating tumor boundaries within this specific dataset.

The effectiveness and robustness of the method are further demonstrated by comparing its performance against several baseline methods that employ different region of interest detection techniques and convolutional neural network models. Our method exhibited superior performance across all datasets and metrics, highlighting its potential clinical and research implications. The proposed method has several noteworthy implications. First, it can provide radiologists with visual cues and confidence scores for lesions in mammograms, aiding in breast cancer screening and diagnosis. This can significantly enhance the accuracy and efficiency of the diagnostic process. Additionally, the method offers a straightforward and effective way to create regions of interest from image classification models, enabling the development of new convolutional neural network models specifically tailored for breast cancer detection.

Moreover, the method's utilization of gradient-weighted class activation mapping opens up possibilities for its application in other medical image analysis tasks that require region of interest optimization and deep learning. This technique could inspire new avenues of research and development in the field of medical imaging. The proposed method demonstrates its effectiveness in detecting breast cancer in mammograms through the integration of region of interest optimization and gradient-weighted class activation mapping. Its superior performance, particularly in accurately delineating tumor boundaries, underscores its potential for clinical implementation and further advancements in the field. Future research can focus on refining and expanding the methodology to address specific challenges and further improve its overall efficacy in breast cancer detection and diagnosis.

Author Contributions: Software, A.T.T. and H.N.H.; methodology, T.N.T. and H.N.H.; data curation, T.N.T. and A.T.T.; writing—original draft preparation: H.N.H. and T.N.T.; writing—review and editing: T.N.T. All authors have read and agreed to the published version of the manuscript.

Funding: This research received no external funding.

Institutional Review Board Statement: Not applicable.

Informed Consent Statement: Not applicable.

Data Availability Statement: Not applicable.

Acknowledgments: We acknowledge Ho Chi Minh City University of Technology (HCMUT), VNU-HCM for supporting this study.

Conflicts of Interest: The authors declare no conflict of interest.

References

1. World Health Organization. Breast Cancer. Available online: https://www.who.int/news-room/fact-sheets/detail/breast-cancer (accessed on 7 April 2023).
2. Elmore, J.G.; Wells, C.K.; Lee, C.H.; Howard, D.H.; Feinstein, A.R. Variability in radiologists' interpretations of mammograms. *N. Engl. J. Med.* **1994**, *331*, 1493–1499. [CrossRef] [PubMed]
3. Welch, H.G.; Prorok, P.C.; O'Malley, A.J.; Kramer, B.S. Breast-cancer tumor size, overdiagnosis, and mammography screening effectiveness. *N. Engl. J. Med.* **2016**, *375*, 1438–1447. [CrossRef] [PubMed]
4. LeCun, Y.; Bengio, Y.; Hinton, G. Deep learning. *Nature* **2015**, *521*, 436–444. [CrossRef] [PubMed]
5. Goodfellow, I.; Bengio, Y.; Courville, A. *Deep Learning*; MIT Press: Cambridge, MA, USA; London, UK, 2016.
6. Litjens, G.; Kooi, T.; Bejnordi, B.E.; Setio, A.A.A.; Ciompi, F.; Ghafoorian, M.; van der Laak, J.A.W.M.; van Ginneken, B.; Sánchez, C.I. A survey on deep learning in medical image analysis. *Med. Image Anal.* **2017**, *42*, 60–88. [CrossRef] [PubMed]
7. Dhungel, N.; Carneiro, G.; Bradley, A.P. Deep learning and structured prediction for the segmentation of mass in mammograms. In Proceedings of the International Conference on Medical Image Computing and Computer-Assisted Intervention, Munich, Germany, 5–9 October 2015; pp. 605–612.
8. Kooi, T.; Litjens, G.; van Ginneken, B.; Gubern-Mérida, A.; Sánchez, C.I.; Mann, R. Large scale deep learning for computer aided detection of mammographic lesions. *Med. Image Anal.* **2017**, *35*, 303–312. [CrossRef]
9. Zhu, W.; Xie, X. Adversarial deep structural networks for mammographic mass segmentation. In Proceedings of the International Conference on Medical Image Computing and Computer-Assisted Intervention, Munich, Germany, 5–9 October 2015; pp. 101–109.
10. Ribli, D.; Horváth, A.; Unger, Z.; Pollner, P.; Csabai, I. Detecting and classifying lesions in mammograms with deep learning. *Sci. Rep.* **2018**, *8*, 4165. [CrossRef]
11. Wang, D.; Khosla, A.; Gargeya, R.; Irshad, H.; Beck, A.H. Deep learning for identifying metastatic breast cancer. *arXiv* **2016**, arXiv:1606.05718.
12. Nguyen, H.T.; Nguyen, H.Q.; Pham, H.H.; Lam, K.; Le, L.T.; Dao, M.; Vu, V. VinDr-Mammo: A large-scale benchmark dataset for computer-aided diagnosis in full-field digital mammography. *Sci. Data* **2023**, *10*, 277. [CrossRef]
13. Looney, P.; Chen, J.; Giger, M.L. A mini-digital database for screening mammography: Mini-DDSM. *J. Med. Imaging* **2017**, *4*, 034501.
14. Cui, C.; Li, L.; Cai, H.; Fan, Z.; Zhang, L.; Dan, T.; Li, J.; Wang, J. *The Chinese Mammography Database (CMMD): An Online Mammography Database with Biopsy Confirmed Types for Machine Diagnosis of Breast*; The Cancer Imaging Archive: Bethesda, MD, USA, 2021.
15. Khaled, R.; Helal, M.; Alfarghaly, O.; Mokhtar, O.; Elkorany, A.; El Kassas, H.; Fahmy, A. Categorized contrast enhanced mammography dataset for diagnostic and artificial intelligence research. *Sci. Data* **2022**, *9*, 122. [CrossRef]
16. Demir, Ö.; Güler, İ.N. Breast masses classification in mammograms using deep convolutional neural networks and transfer learning. *Biomed. Signal Process. Control* **2019**, *53*, 101567.
17. Carr, C.; Kitamura, F.; Kalpathy-Cramer, J.; Mongan, J.; Andriole, K.; Vazirabad, M.; Riopel, M.; Ball, R.; Dane, S. RSNA Screening Mammography Breast Cancer Detection. 2022. Available online: https://kaggle.com/competitions/rsna-breast-cancer-detection (accessed on 27 February 2023).
18. Tan, M.; Le, Q.V. EfficientNet: Rethinking model scaling for convolutional neural networks. *arXiv* **2019**, arXiv:1905.11946.
19. Liu, Z.; Mao, H.; Wu, C.-Y.; Feichtenhofer, C.; Darrell, T.; Xie, S. A ConvNet for the 2020s. *arXiv* **2022**, arXiv:2201.03545.
20. Zou, X.; Wu, Z.; Zhou, W.; Huang, J. YOLOX-PAI: An Improved YOLOX, Stronger and Faster than YOLOv6. *arXiv* **2022**, arXiv:2208.13040.
21. Wang, C.-Y.; Liao, H.-Y.M.; Yeh, I.-H.; Wu, Y.-H.; Chen, P.-Y.; Hsieh, J.-W. CSPNet: A New Backbone that can Enhance Learning Capability of CNN. *arXiv* **2020**, arXiv:1911.11929.
22. Li, J.; Wang, Y.; Liang, X.; Zhang, L. SFPN: Synthetic FPN for Object Detection. *arXiv* **2021**, arXiv:2104.05746.
23. Otsu, N. A threshold selection method from gray-level histograms. *IEEE Trans. Syst. Man Cybern.* **1979**, *9*, 62–66. [CrossRef]
24. Chen, Y.; Li, Y.; Sakaridis, C.; Dai, D.; Van Gool, L. ContourNet: Taking a further step toward accurate arbitrary-shaped scene text detection. In Proceedings of the 2019 IEEE/CVF International Conference on Computer Vision (ICCV), Seoul, Republic of Korea, 27 October–2 November 2019; pp. 9719–9728.
25. Selvaraju, R.R.; Cogswell, M.; Das, A.; Vedantam, R.; Parikh, D.; Batra, D. Grad-CAM: Visual explanations from deep networks via gradient-based localization. *Int. J. Comput. Vis.* **2019**, *128*, 336–359. [CrossRef]
26. Liu, W.; Anguelov, D.; Erhan, D.; Szegedy, C.; Reed, S.; Fu, C.-Y.; Berg, A.C. SSD: Single Shot MultiBox Detector. *arXiv* **2016**, arXiv:1512.02325.
27. Ren, S.; He, K.; Girshick, R.; Sun, J. Faster R-CNN: Towards Real-Time Object Detection with Region Proposal Networks. *arXiv* **2015**, arXiv:1506.01497.

28. Dai, J.; Li, Y.; He, K.; Sun, J. R-FCN: Object Detection via Region-based Fully Convolutional Networks. *arXiv* **2016**, arXiv:1605.06409.
29. Singh, M.; Singh, T.; Soni, S. Pre-operative assessment of ablation margins for variable blood perfusion metrics in a magnetic resonance imaging-based complex breast tumor anatomy: Simulation paradigms in thermal therapies. *Comput. Methods Programs Biomed.* **2021**, *198*, 105781. [CrossRef] [PubMed]

applied
sciences

Article

Deep Learning Enhances Radiologists' Detection of Potential Spinal Malignancies in CT Scans

Leonard Gilberg [1,*,†], Bianca Teodorescu [1,†], Leander Maerkisch [1], Andre Baumgart [2], Rishi Ramaesh [3], Elmer Jeto Gomes Ataide [1] and Ali Murat Koç [1,4]

[1] Floy GmbH, 80335 Munich, Germany
[2] UMM Universitätsmedizin Mannheim, 68305 Mannheim, Germany
[3] University of Edinburgh, Midlothian EH25 9RG, UK
[4] Department of Radiology, Ataturk Education and Research Hospital, Izmir Katip Celebi University, 35620 Izmir, Turkey
* Correspondence: leonard.gilberg@floy.com; Tel.: +49-15170826793
† These authors contributed equally to this work.

Featured Application: This work presents an AI-based second reader application tailored for computed tomography (CT) scans in Radiology. Its primary objective is to detect overlooked potential malignant cases in the vertebral body during routine radiological reporting.

Abstract: Incidental spinal bone lesions, potential indicators of malignancies, are frequently underreported in abdominal and thoracic CT imaging due to scan focus and diagnostic bias towards patient complaints. Here, we evaluate a deep-learning algorithm (DLA) designed to support radiologists' reporting of incidental lesions during routine clinical practice. The present study is structured into two phases: unaided and AI-assisted. A total of 32 scans from multiple radiology centers were selected randomly and independently annotated by two experts. The U-Net-like architecture-based DLA used for the AI-assisted phase showed a sensitivity of 75.0% in identifying potentially malignant spinal bone lesions. Six radiologists of varying experience levels participated in this observational study. During routine reporting, the DLA helped improve the radiologists' sensitivity by 20.8 percentage points. Notably, DLA-generated false-positive predictions did not significantly bias radiologists in their final diagnosis. These observations clearly indicate that using a suitable DLA improves the detection of otherwise missed potentially malignant spinal cases. Our results further emphasize the potential of artificial intelligence as a second reader in the clinical setting.

Keywords: deep learning; computed tomography; malignancies; AI detection; second reader; spine; vertebral lesions

Citation: Gilberg, L.; Teodorescu, B.; Maerkisch, L.; Baumgart, A.; Ramaesh, R.; Gomes Ataide, E.J.; Koç, A.M. Deep Learning Enhances Radiologists' Detection of Potential Spinal Malignancies in CT Scans. *Appl. Sci.* **2023**, *13*, 8140. https://doi.org/10.3390/app13148140

Academic Editor: Cosimo Nardi

Received: 22 June 2023
Revised: 10 July 2023
Accepted: 12 July 2023
Published: 13 July 2023

1. Introduction

Discrepancies in radiology are a long-known issue and—due to the extended workload and limited evaluation time—have remained the same despite continuously improving imaging techniques over the last decades [1,2]. A perceptual error, or false negative, is an abnormality present in a diagnostic image but not described by the interpreter. Such overlooked findings constitute the vast majority of human error in image interpretation [2–4].

Spinal bone lesions that present themselves as a conglomerate are frequently an indicator of malignancy, with the vertebrae being the most prevalent hotspot for bone metastasis [5]. On the other side of the spectrum, solitary lesions are more challenging and can indicate both malignant and benign processes [6,7], creating uncertainty within the diagnostic procedure, which may require further investigation [8]. In this case, if missed or initially overlooked (perceptual error), they can exhibit major negative consequences on a patient's quality of life and, subsequently, their morbidity and mortality [9,10].

With CT being a reliable imaging modality for assessing osseous involvement and the destruction degree of spine abnormalities [6,11], the past years have shown a rising interest in automatizing the detection and classification of spinal lesions to a large extent [12–14]. Artificial intelligence (AI) is now an active part of various medical diagnostic procedures within real-life hospital workflows, with deep-learning (DL)-based analysis of radiologic images as one of its key applications. AI as a screening tool or a second reader for abnormality detection already shows promising results in various fields, such as chest X-ray reporting and lung nodule detection [15,16]. However, reliable deep-learning-based algorithms for spinal lesion detection are still sparse as their development has proven to be more challenging, mostly due to the overlapping image features of degenerative and neoplastic events [6]. To our knowledge, there are currently no EU MDR-certified or FDA-approved AI second reader software that detect incidental spinal lesions in CT scans of unrelated indications. A reliable algorithm with such capacities could assist the reporting physician with accurate supplementary information, reduce the rate of missed potentially malignant lesions, and streamline the diagnostic pathway.

This work examines the clinical impact of a deep-learning algorithm (DLA) that assists radiologists in their day-to-day workflow within a simulated hospital setting. The algorithm was developed to detect potentially malignant cases within the vertebrae and act as a second reader, using native and contrast-enhanced abdomen and thoracic CT imaging sequences. Its clinical efficacy was evaluated in an observational cross-over study design, where the algorithm's performance and the effect on the decision-making of six subspecialty radiologists with and without the intervention of the DLA were assessed. The distinct feature of this study design is its emphasis on reducing incidental findings during the reporting of other main underlying diseases that the patient has. This approach, further accentuated by limiting the scope to only CT abdomen and thoracic scans, shifts the focus away from solely detecting vertebral malignancy as the primary objective of the responsible radiologist. Such "background acting" tools open new avenues in how one can correctly integrate AI in the medical sector and underline the crucial role of human involvement in the overall process.

2. Materials and Methods

This section details the materials and methods used for the study.

2.1. Data Acquisition

All clinical and imaging information was obtained retrospectively from multiple outpatient radiology centers in Germany. The data selection process is detailed in Figure 1.

We included studies of native or contrast-enhanced thoracic and abdominal CT examinations with multiplanar bone and soft tissue reconstructions. Incomplete or broken studies, individuals with prior spinal surgery, and individuals under 18 were excluded from the cohort. Once the data was filtered based on the chosen inclusion criteria, 32 randomly sampled studies were picked. Data contracts are signed with the data providers, and the studies were anonymized before being included in the study. Additionally, the data is retrospective, with CT scans from a multicenter data provider collected over 12 months from January 2022 until January 2023. Due to these factors, the need for informed consent is waived. The anonymization process strips away all identifying tags such as name, contact details, and address. Demographic details such as sex are preserved, and age is rounded to the nearest whole number.

2.2. Establishment of the Ground Truth

Images were pre-processed and stored in a Digital Imaging and Communications in Medicine (DICOM) format before the expert annotation. The ground truth labels were then established via manual segmentation by two board-certified radiologists with expertise in the field (MK: Associate Professor, 14 years of experience; and RR: Senior Lecturer, ten years of experience). The studies were annotated on an object level by drawing bounding boxes around all regions of interest (ROIs) and subsequently classified as positive or negative. We consider a positive case in which at least one finding is indicative of a potentially malignant vertebral lesion (lytic, sclerotic, or mixed with a circumscribed boundary) and has been manually segmented and evaluated by our two experts. This labeling process was performed on the Encord platform (© 2022 Encord), and a consensus was reached in case of divergent opinions.

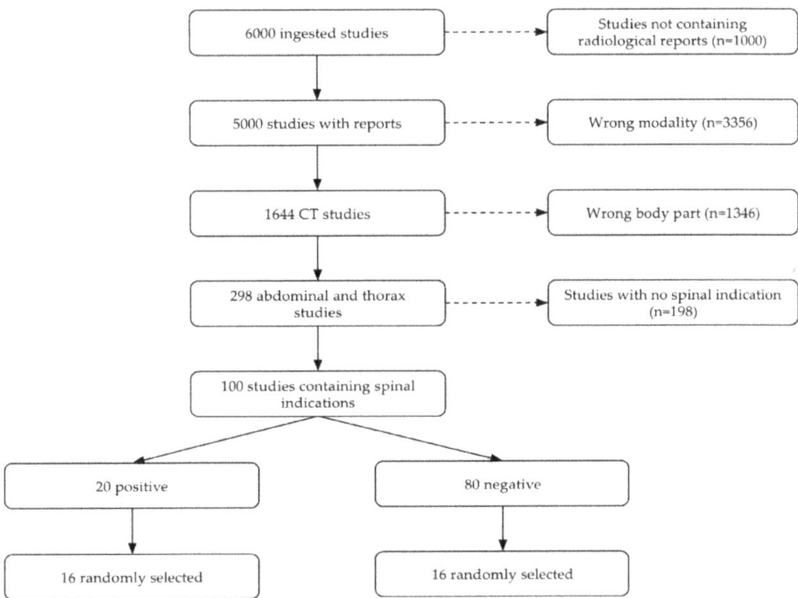

Figure 1. Flowchart of data collection and distribution. All CT studies were collected retrospectively from the clinical databases of multiple institutions.

2.3. AI Algorithm Development

The deep-learning algorithm used in this study was developed using native and contrast-enhanced CT studies of the spine, abdomen, and thorax. It is meant to be used as a medical device in a clinical setting. The DLA consists of two deep-learning models that work together. One model uses a U-Net-like segmentation approach coupled with volumetric analysis to determine the presence of potentially malignant lesions in the spinal vertebrae. The second model employs a vertebral localization component that enables the proper selection of the region of interest. It is trained using a training set of 224 cases and tested on 735 cases. Additionally, the DLA is evaluated on an external dataset of 420 cases. Statistical analysis of the datasets was performed to verify the demographic distribution of the data.

2.4. Experimental Setup

In this study, our primary goal was to investigate the effectiveness of a deep-learning algorithm (DLA) in assisting radiologists in identifying previously overlooked potentially malignant cases in the spine through abdomen and thoracic CT imaging. A total of 32 studies were selected from our data pool which contained both positive and negative cases in equal numbers. Details of the selection process are given in Figure 1.

The study consisted of two phases. In the first phase, six radiologists independently reviewed the 32 studies without DLA assistance, following their routine reading process. In this case, a routine reading process is defined as the reading of the radiological images by the radiologist based on the patient's complaint. This phase aimed to establish a baseline for their diagnostic performance.

The six radiologists (participants) we recruited have varying experience levels, ranging from 1 to 11 years. Each participant received the same set of 32 studies but presented in random order. The participants were blinded to the gold standard and any patient-specific information to ensure unbiased assessments, except for a brief description of the patient's complaint. These complaints were not focused on spine-related issues but were general complaints of the patients who visited the clinics. The details of this information can be found in Supplementary Table S1.

After a break of 10 days, the second phase was conducted. In this phase, the participants reevaluated the same set of studies, but this time, they had access to the DLA predictions to assist them in their evaluations. The DLA provided predictions about the presence or absence of abnormalities in the spinal vertebrae in the CT scans.

To ensure accurate documentation of their assessments, the participants were asked to use screen-recording software [17] to capture their computer screens. They were also instructed to describe their findings verbally while indicating them with their mouse. A DICOM visualization tool [18] was made available to aid in interpreting the scans and the predictions of the DLA.

To evaluate the effectiveness of the AI intervention, the results obtained during the reading sessions were manually compared to the gold standard. This comparison enabled the determination of the accuracy of the participants' diagnoses with and without the DLA's assistance.

2.5. Statistical Analysis

Statistical analysis was performed using GraphPad Prism v8.4.2. The primary measured outcome was based on assessing case and object level (per-patient and per-lesion) sensitivity, specificity, and average false positive (AvgFP). Because of the infinite number of possible locations for a spinal lesion, we could not define the true negative and thus did not calculate the per-lesion specificity. To assess the significance between the two reading sessions, we conducted the McNemar test in Python (statsmodels v 0.14.0).

3. Results

3.1. Demographics of the Dataset and Spinal Lesion Assessment in the Reference Standard

No significant differences or inhomogeneities were noted concerning the demographic qualities of our cohort. Supplementary Table S1 shows the background information of the 32 patients involved in this study, including patient demographics (sex, age). 53.1% of the patients are female. The average age was noted to be 56.6 years. An overview of the scan conditions and other image acquisition details can be found in Supplementary Table S2. As per the gold standard, there were 16 positive and 16 negative scans. There were a total of 27 annotated suspicious lesions. The annotators exhibited a mutual agreement for 75.0% of the cases while determining the gold standard. Both annotators initially disagreed on the remaining 25.0%, but these conflicts were resolved through discussions leading to a consensus.

3.2. Algorithm Performance

Figure 2 exemplifies two true-positive predictions on different vertebral sites for lytic (Figure 2a,b) and sclerotic (Figure 2c,d) lesions. The deep-learning algorithm (DLA) for spinal lesion detection was tested on the same patient studies and correctly detected 12 out of 27 lesions in 16 patients. It also falsely indicated 13 spinal findings (false positives) that were not considered true findings following the gold standard. The overall outcome of the DLA for the established dataset is shown in Table 1. On a case level, compared to the gold standard, the DLA performance had a sensitivity and specificity of 75.0% and 56.3%, respectively. Regarding the results on an object level, the sensitivity was 44.4%, as shown in Table 2.

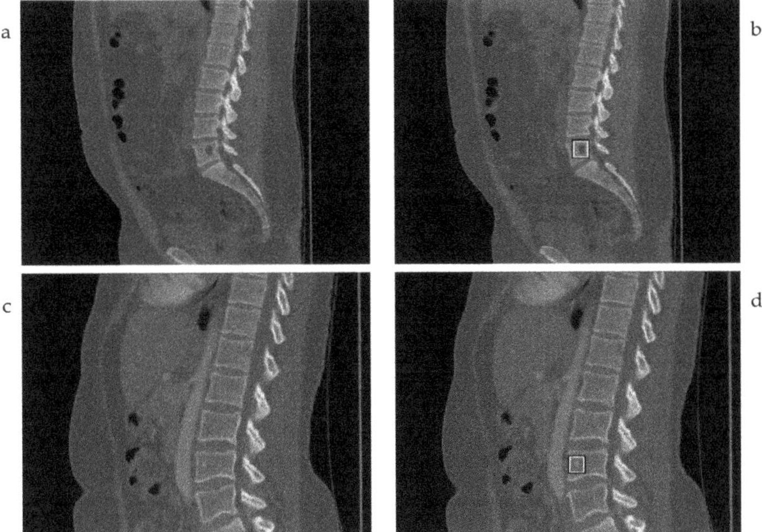

Figure 2. DLA predictions of osteolytic (**a**) and oseoblastic (**c**) lesions are shown in corresponding images (**b,d**).

Table 1. Algorithm performance in comparison to the gold standard (count).

	Positive Cases	Total Number of Detected Objects	Number of Detected Spinal Lesions (TP)	Number of Undetected Spinal Lesions (FN)	Number of Falsely Detected Spinal Lesions (FP)
Gold Standard	16	27	27	N/A	N/A
DLA	11	40	12	15	13

Table 2. Algorithm performance in comparison to the gold standard (metrics).

Sensitivity (TP Rate)		Specificity (TN Rate)		Accuracy	
Case Level	Object Level	Case Level	Object Level	Case Level	Object Level
75.00%	44.44%	56.25%	N/A	65.63%	N/A

3.3. Intra-Observer Agreement without and with the Aid of the DLA

The observers' performance results are summarized in Table 3, with visual exemplification in Figure 3 showcasing sensitivity, specificity, and the true-positive rate on a case level.

Considering that the participants were asked to perform routine reporting based on general complaints (such as acute abdominal pain or elevated liver enzymes), only one out of six radiologists did not include spinal findings in the first round (without DLA). Another radiologist solely identified degenerative changes without detecting any significant or suspicious findings that aligned with the ground truth. The participant with the highest experience level (11 years) initially reported two true positives in the first round. However, when aided by the DLA, this number increased to 14 true-positive and two false-positive findings. The two least experienced radiologists (one year) did not report any spinal findings without relying on the algorithm. Following the predictions in round 2, only one included spinal findings in the report.

Considering a potential maximum of 162 true positives (27 lesions multiplied by 6 participants), the radiologists reported 11 true-positive objects alone. However, when assisted by the DLA, this number increased to more than three times that value, with 35 true-positive findings in the study's second phase.

The DLA predicted 13 false-positive objects. When interpreting these predictions, on average, the radiologists included less than one false positive in the report, resulting in four false-positive reports across all participants.

Table 3. Intra-observer agreement depicted for the two study phases according to the gold standard (phase 1—without the DLA and phase 2—with the DLA support).

	Participants						
	1	**2**	**3**	**4**	**5**	**6**	
Experience (years)	6	7	5	1	11	1	
Phase 1—no support from DLA							**Mean**
Suspicious lesions reported (n)	2.00	4.00	3.00	0.00	2.00	0.00	1.83
False positives (n)	0.00	0.00	0.00	0.00	0.00	0.00	0.00
Sensitivity (case level)	12.50%	25.00%	18.75%	0.00%	12.50%	0.00%	11.46%
Sensitivity (object level)	7.41%	14.81%	11.11%	0.00%	7.41%	0.00%	6.79%
Specificity (case level)	100.00%	100.00%	100.00%	100.00%	100.00%	100.00%	100.00%
Accuracy (case level)	56.25%	62.50%	59.38%	50.00%	56.25%	50.00%	55.73%
Accuracy (object level)	53.70%	57.41%	55.56%	50.00%	53.70%	50.00%	53.40%
False positive rate	0.00%	0.000%	0.000%	0.000%	0.000%	0.000%	0.000%
Phase 2—with support from DLA							**Mean**
Suspicious lesions reported (n)	5.00	5.00	8.00	0.00	14.00	3.00	5.83
False positives (n)	1.00	0.00	1.00	0.00	2.00	1.00	0.83
Sensitivity (case level)	31.25%	31.25%	43.75%	0.00%	75.00%	12.50%	32.29%
Sensitivity (object level)	14.81%	18.52%	25.93%	0.00%	44.44%	7.41%	18.52%
Specificity (case level)	93.75%	100.00%	93.75%	100.00%	87.50%	93.75%	94.79%
Accuracy (case level)	62.25%	65.62%	68.75%	50.00%	81.25%	53.12%	63.54%
Accuracy (object level)	55.56%	59.26%	68.12%	50.00%	68.52%	51.85%	58.89%
False-positive rate	3.70%	0.00%	3.57%	0.00%	7.41%	3.70%	3.06%

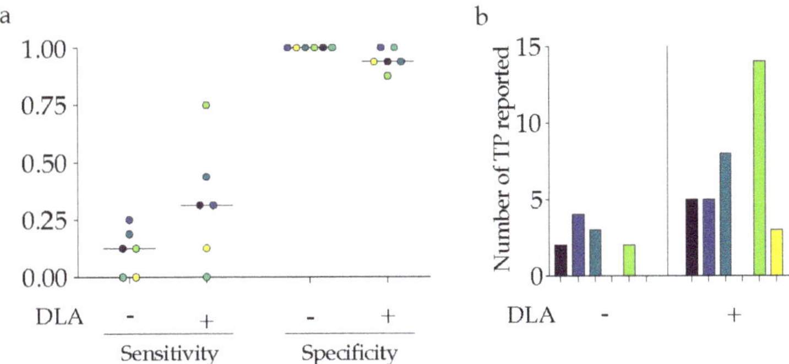

Figure 3. Performance metrics of the six participants on case level, with each participant color coordinated; (**a**) Sensitivity and specificity of reporting potentially malignant spinal lesions without (−) and with (+) the aid of an AI tool as a second reader; (**b**) Number of reported true-positive findings without (−) and with (+) the aid of an AI tool as a second reader.

Based on the AI's indication of true-positive spinal conditions, four participants felt requesting further follow-up or diagnostic tests was necessary. These requests were made to investigate potential abnormalities or conditions related to the spine, as suggested by the AI. None of the participants requested additional examinations due to possible FP findings.

On a case level, participants' average sensitivity in detecting spinal lesion(s) (which translates into reporting at least one correct lesion in a positive case) increased from 11.5% to 32.3% (20.08 percentage points) when using the DLA tool. The sensitivity increased from 6.8% to 18.5% (11.70 percentage points) on an object level. The average FP rate increased from 0.0% (no primarily reported false spinal findings) to 3.1% on a case and object level. The mean accuracy value increased from 55.7% to 63.5% on a case level and from 53.4% to 58.9% on an object level.

There was no clear trend that could link the participants' clinical experience and their responsiveness to the AI predictions, with both junior and more experienced radiologists having heterogeneous behaviors toward the presented algorithm results. It should be noted that, in less-experienced participants, we observed changes in their reports between reading sessions unrelated to the AI findings.

4. Discussion

A complete and comprehensive review of a CT scan is crucial for a patient's health and has a significant impact on decision-making. CT imaging can assess spinal bone metastatic lesions up to 6 months before plain radiographs [19]. However, smaller lesions or those that do not have significant cortical destruction are often underreported or missed during CT image reporting [20]. In a systematic review, Bartalena et al. reported that radiologists' recognition of incidental vertebral findings (in this case, fractures) was low, with a mean reporting rate of just 27.4% [21]. A major fraction of false negatives are significant bone lesions that could indicate potential malignancy [3]. J Donald et al. showed in an internal department analysis that these types of spinal lesions were most frequently misinterpreted on CT images, with 14 out of 16 missed findings being metastatic [22].

False-negative cases are the most common perceptual errors [4], with CT imaging being especially susceptible [22]. Errors made in previous radiology reports can lead to the tendency of radiologists to replicate the error in subsequent reports, which is referred to as 'alliterative bias' [23]. This concept reiterates the importance of some confirmation protocols in medical practice. While double-reading practices significantly impact the quality of radiological reports, clinical workload and staff shortages make a routine human double-reader scheme hard to implement [24,25].

Computer-aided detection (CAD) systems have supported radiologists in their work-flow even before the era of deep-learning tools, with some of the best examples being small functions and add-ons for DICOM viewers, such as contrast enhancers or manual annotation support. With the use of AI increasing rapidly in the fields of medicine, with radiology as a leading candidate, future deep-learning-based CAD systems will not only optimize the users' workflow and improve their diagnostic abilities but also weigh in on their medical judgment and decision-making process. CTs are not just a series of images; they contain extensive information about the pathology in question that sometimes cannot be interpreted by the bare human eye [26]. There are models developed for almost every disease that can be assessed radiologically. More specifically, detection systems have shown an emerging potential in reducing missed radiological spinal lesions as a second reader [13,27–30].

In our study, we investigated the effects of a DLA when implemented in a routine CT reporting process performed by six radiologists having different experience levels. Our results show that DL-based spinal lesion detection can improve inter-observer agreement and overall increase performance in detecting these radiological findings, regardless of the training level of our participants. The case level sensitivity increased by 20.83 percentage points when the participants were aided by the DLA. However, this improvement should only be interpreted in the context of a rather low baseline of reported spinal findings in the first round. Another observational study by Noguchi et al. [13] showed that the sensitivity of radiologists in detecting bone metastases could be elevated with the help of a DLA by 15.3 percentage points. In a similar study, Kato et al. [31] reported an improvement in the performance of less-experienced radiologists in brain metastasis detection by 4.90 percentage points. Like our own study, Kato et al. observed no significant increase in false-positive findings when utilizing a CAD system. While these previous studies have primarily focused on DLAs improving performance in explicit detection tasks, our study provides a novel perspective by highlighting the potential benefits of a tool for reducing incidental findings during routine reporting. This observation helps to explain the relatively lower baseline performance of participants in our study. Since a complete radiological report includes information regarding all body parts that can be seen on the scan, we expected insights on all findings that the six radiologists encountered while analyzing the images. Our findings support the already existing issue of missed spinal findings in clinical practice, which might rely on the aforementioned reasons for perceptual errors in radiology. It is worth noting that the sensitivities and specificities are calculated only for potentially malignant spinal lesion findings and not findings for all other organs (lungs, liver, kidneys). Hence, the lowered sensitivities and specificities due to many spinal lesions being missed during the initial diagnostic process without the assistance of the DLA.

Indeed, the DLA reported false-positive findings. It is important to mention that the radiologists' assessments were not solely based on the AI's predictions for false-positive lesions. They considered various factors, including other findings identified as true positives by the AI. Furthermore, based on the AI's indication of true-positive spinal conditions, four participants requested further follow-up or diagnostic tests. These requests were made to investigate potential abnormalities or conditions related to the spine, as suggested by the AI.

Several studies have already investigated the potential of deep-learning systems for pathology assessment of the spine [13,28,32–35]. Although deep-learning algorithms (DLAs) can demonstrate accuracy in lesion detection on par with radiologists, it is crucial to consider their real-world implementation in datasets that differ significantly from the training data. DLAs may face challenges in such scenarios and are more prone to producing incorrect results. However, by addressing these challenges through ongoing research and fine-tuning, we can further optimize the effectiveness of DLAs in practical settings. It is, therefore, crucial to focus on the effects and performance of the human reader when in conjunction with this emerging technology. While our DLA's performance may not have demonstrated superiority over expert human radiologists, there is potential for radiologists of all experience levels to benefit from its second reader function. It is essential to conduct future studies with larger cohorts and greater sample sizes to validate any hypothetical benefits and potential risks associated with AI. These studies will ensure the safe and effective integration of machine learning software into the clinical setting.

This study has several limitations. Since the allocated time between the two reading sessions was set for only ten days, one could argue that the first reading could have biased the participants' performance in the second session. We conducted the McNemar test to assess significance. The test (p = 0.25) confirmed that the study did not yield statistically significant results (p > 0.05). This outcome is linked to the small number of radiologist participants used for this study and probably could not prove generalizable effects. However, it is important to note that this finding presents an opportunity for improvement in future studies. The number of patient studies investigated by the participants was also small and should be increased for further studies. Moreover, the training that radiologists receive, their reporting style, and thus their attention to detail differ between countries and subspecialties.

5. Conclusions

We show that the implementation of a DLA as a second reader in routine reporting of CT scans can increase radiologists' true-positive rate for spinal lesion detection while at the same time having close to no impact on the false-positive rate. These findings showcase the potential that an AI-based technology could have in the hospital setting, particularly in detecting missed potential malignancies during routine reporting. However, it is important to consider that the impact of AI in medical imaging may vary depending on the interpreter's background and training. While current trends and discoveries indicate improvements in diagnostic performance, further comprehensive studies are needed to validate these results on a larger scale and gain a deeper understanding of the implications this technology may have in the clinical setting. Nevertheless, the enhanced metrics observed in this study provide evidence of AI's prospect as a valuable detection tool.

Supplementary Materials: The following supporting information can be downloaded at: https://www.mdpi.com/article/10.3390/app13148140/s1, Table S1: Demographics, Complaints and Lesion Characteristics. Table S2: Scanner and Technical Specifications.

Author Contributions: L.G., B.T., A.M.K., L.M. and E.J.G.A. formulated the study concept, registered the review, contributed to all stages of the review process, performed the literature search, data extraction, and quality assessment, wrote the first draft of the manuscript, and acted as first reviewers. L.G., B.T. and E.J.G.A. analyzed the data and wrote the first draft of the manuscript. L.M., A.B. and A.M.K. oversaw the review conceptualization, registration, and execution and acted as second reviewers. A.M.K., E.J.G.A., R.R. and A.B. acted as third reviewers for discrepancies and revisions of the manuscript. All authors contributed equally in preparing the final version of the manuscript. All authors have read and agreed to the published version of the manuscript.

Funding: This study was funded by Floy GmbH. Floy GmbH played no role in the study design, data collection or analysis, decision to publish, or manuscript preparation.

Institutional Review Board Statement: Ethical review and approval were waived for this study due to the usage of anonymized retrospective patient data.

Informed Consent Statement: Data contracts were signed by the data providers, and the studies were anonymized before being included in the study. Additionally, the data is retrospective. Due to these factors, the need for informed consent was waived.

Data Availability Statement: All data generated or analyzed during this study are included in this published article.

Conflicts of Interest: L.G., B.T. and A.M.K. are Clinical Scientists at Floy GmbH. EJGA is the Head of Clinical Research at Floy GmbH. A.B. is a Regulatory Consultant specializing in medical technology translation activities. L.M. is the Co-Founder of Floy GmbH. Correspondence and requests for materials should be addressed to L.G.

References

1. Waite, S.; Scott, J.; Gale, B.; Fuchs, T.; Kolla, S.; Reede, D. Interpretive Error in Radiology. *AJR Am. J. Roentgenol.* **2017**, *208*, 739–749. [CrossRef]
2. Fitzgerald, R. Error in radiology. *Clin. Radiol.* **2001**, *56*, 938–946. [CrossRef]
3. McCreadie, G.; Oliver, T.B. Eight CT lessons that we learned the hard way: An analysis of current patterns of radiological error and discrepancy with particular emphasis on CT. *Clin. Radiol.* **2009**, *64*, 491–499; discussion 500-1. [CrossRef]
4. Terreblanche, O.D.; Andronikou, S.; Hlabangana, L.T.; Brown, T.; Boshoff, P.E. Should registrars be reporting after-hours CT scans? A calculation of error rate and the influencing factors in South Africa. *Acta Radiol.* **2012**, *53*, 61–68. [CrossRef] [PubMed]
5. Ziu, E.; Viswanathan, V.K.; Mesfin, F.B. *Spinal Metastasis*; StatPearls Publishing: Treasure Island, FL, USA, 2022.
6. Rodallec, M.H.; Feydy, A.; Larousserie, F.; Anract, P.; Campagna, R.; Babinet, A.; Zins, M.; Drapé, J.-L. Diagnostic Imaging of Solitary Tumors of the Spine: What to Do and Say. *Radiographics* **2008**, *28*, 1019–1041. [CrossRef]
7. Kim, Y.S.; Han, I.H.; Lee, I.S.; Lee, J.S.; Choi, B.K. Imaging findings of solitary spinal bony lesions and the differential diagnosis of benign and malignant lesions. *J. Korean Neurosurg. Soc.* **2012**, *52*, 126–132. [CrossRef]
8. Nguyen, T.T.; Thelen, J.C.; Bhatt, A.A. Bone up on spinal osseous lesions: A case review series. *Insights Imaging* **2020**, *11*, 80. [CrossRef] [PubMed]
9. Coleman, R.E. Metastatic bone disease: Clinical features, pathophysiology and treatment strategies. *Cancer Treat. Rev.* **2001**, *27*, 165–176. [CrossRef] [PubMed]
10. Martin, M.; Bell, R.; Bourgeois, H.; Brufsky, A.; Diel, I.; Eniu, A.; Fallowfield, L.; Fujiwara, Y.; Jassem, J.; Paterson, A.H.; et al. Bone-related complications and quality of life in advanced breast cancer: Results from a randomized phase III trial of denosumab versus zoledronic acid. *Clin. Cancer Res.* **2012**, *18*, 4841–4849. [CrossRef]
11. Jarvik, J.G.; Deyo, R.A. Diagnostic evaluation of low back pain with emphasis on imaging. *Ann. Intern. Med.* **2002**, *137*, 586–597. [CrossRef]
12. Wang, J.; Fang, Z.; Lang, N.; Yuan, H.; Su, M.Y.; Baldi, P. A multi-resolution approach for spinal metastasis detection using deep Siamese neural networks. *Comput. Biol. Med.* **2017**, *84*, 137–146. [CrossRef] [PubMed]
13. Noguchi, S.; Nishio, M.; Sakamoto, R.; Yakami, M.; Fujimoto, K.; Emoto, Y.; Kubo, T.; Iizuka, Y.; Nakagomi, K.; Miyasa, K.; et al. Deep learning–based algorithm improved radiologists' performance in bone metastases detection on CT. *Eur. Radiol.* **2022**. [CrossRef] [PubMed]
14. Li, Z.; Wu, F.; Hong, F.; Gai, X.; Cao, W.; Zhang, Z.; Yang, T.; Wang, J.; Gao, S.; Peng, C. Computer-Aided Diagnosis of Spinal Tuberculosis From CT Images Based on Deep Learning with Multimodal Feature Fusion. *Front. Microbiol.* **2022**, *13*, 823324. [CrossRef] [PubMed]
15. Liang, M.; Tang, W.; Xu, D.M.; Jirapatnakul, A.C.; Reeves, A.P.; Henschke, C.I.; Yankelevitz, D. Low-Dose CT Screening for Lung Cancer: Computer-aided Detection of Missed Lung Cancers. *Radiology.* **2016**, *281*, 279–288. [CrossRef]
16. McKinney, S.M.; Sieniek, M.; Godbole, V.; Godwin, J.; Antropova, N.; Ashrafian, H.; Back, T.; Chesus, M.; Corrado, G.S.; Darzi, A.; et al. International evaluation of an AI system for breast cancer screening. *Nature* **2020**, *577*, 89–94. [CrossRef]
17. Loom. Available online: https://www.loom.com/ (accessed on 21 June 2023).
18. Visian. Available online: https://visian.org/ (accessed on 21 June 2023).
19. Salvo, N.; Christakis, M.; Rubenstein, J.; de Sa, E.; Napolskikh, J.; Sinclair, E.; Ford, M.; Goh, P.; Chow, E. The Role of Plain Radiographs in Management of Bone Metastases. *J. Palliat Med.* **2009**, *12*, 195–198. [CrossRef]
20. Williams, A.L.; Al-Busaidi, A.; Sparrow, P.J.; Adams, J.E.; Whitehouse, R.W. Under-reporting of osteoporotic vertebral fractures on computed tomography. *Eur. J. Radiol.* **2009**, *69*, 179–183. [CrossRef]
21. Bartalena, T.; Rinaldi, M.F.; Modolon, C.; Braccaioli, L.; Sverzellati, N.; Rossi, G.; Rimondi, E.; Busacca, M.; Albisinni, U.; Resnick, D. Incidental vertebral compression fractures in imaging studies: Lessons not learned by radiologists. *World J. Radiol.* **2010**, *2*, 399–404. [CrossRef]
22. Donald, J.J.; Barnard, S.A. Common patterns in 558 diagnostic radiology errors. *J. Med. Imaging Radiat. Oncol.* **2012**, *56*, 173–178. [CrossRef]
23. Smith, M.J. *Error and Variation in Diagnostic Radiology*; C.C. Thomas: Springfield, IL, USA, 1967.
24. Markus, J.B.; Somers, S.; O'Malley, B.P.; Stevenson, G.W. Double-contrast barium enema studies: Effect of multiple reading on perception error. *Radiology* **1990**, *175*, 155–156. [CrossRef] [PubMed]
25. Wakeley, C.J.; Jones, A.M.; Kabala, J.E.; Prince, D.; Goddard, P.R. Audit of the value of double reading magnetic resonance imaging films. *Br. J. Radiol.* **1995**, *68*, 358–360. [CrossRef] [PubMed]
26. Gillies, R.J.; Kinahan, P.E.; Hricak, H. Radiomics: Images Are More than Pictures, They Are Data. *Radiology* **2016**, *278*, 563–577. [CrossRef] [PubMed]
27. Masoudi, S.; Mehralivand, S.; Harmon, S.; Walker, S.; Pinto, P.A.; Wood, B.J.; Citrin, D.E.; Karzai, F.; Gulley, J.L.; Madan, R.A.; et al. Artificial intelligence assisted bone lesion detection and classification in computed tomography scans of prostate cancer patients. *J. Clin. Orthod.* **2020**, *38*, e17567. [CrossRef]
28. Li, Y.; Zhang, Y.; Zhang, E.; Chen, Y.; Wang, Q.; Liu, K.; Yu, H.J.; Yuan, H.; Lang, N.; Su, M.-Y. Differential diagnosis of benign and malignant vertebral fracture on CT using deep learning. *Eur. Radiol.* **2021**, *31*, 9612–9619. [CrossRef]
29. Han, S.; Oh, J.S.; Lee, J.J. Diagnostic performance of deep learning models for detecting bone metastasis on whole-body bone scan in prostate cancer. *Eur. J. Nucl. Med. Mol. Imaging* **2022**, *49*, 585–595. [CrossRef]

30. Ohno, Y.; Aoyagi, K.; Takenaka, D.; Yoshikawa, T.; Ikezaki, A.; Fujisawa, Y.; Murayama, K.; Hattori, H.; Toyama, H. Machine learning for lung CT texture analysis: Improvement of inter-observer agreement for radiological finding classification in patients with pulmonary diseases. *Eur. J. Radiol.* **2021**, *134*, 109410. [CrossRef]
31. Kato, S.; Amemiya, S.; Takao, H.; Yamashita, H.; Sakamoto, N.; Miki, S.; Watanabe, Y.; Suzuki, F.; Fujimoto, K.; Mizuki, M.; et al. Computer-aided detection improves brain metastasis identification on non-enhanced CT in less experienced radiologists. *Acta Radiol.* **2023**, *64*, 1958–1965. [CrossRef]
32. Yoda, T.; Maki, S.; Furuya, T.; Yokota, H.; Matsumoto, K.; Takaoka, H.; Miyamoto, T.; Okimatsu, S.; Shiga, Y.; Inage, K.; et al. Automated Differentiation between Osteoporotic Vertebral Fracture and Malignant Vertebral Fracture on MRI Using a Deep Convolutional Neural Network. *Spine* **2022**, *47*, E347–E352. [CrossRef]
33. Yeh, L.-R.; Zhang, Y.; Chen, J.-H.; Liu, Y.-L.; Wang, A.-C.; Yang, J.-Y.; Yeh, W.-C.; Cheng, C.-S.; Chen, L.-K.; Su, M.-Y. A deep learning-based method for the diagnosis of vertebral fractures on spine MRI: Retrospective training and validation of ResNet. *Eur. Spine J.* **2022**, *31*, 2022–2030. [CrossRef]
34. Ouyang, H.; Meng, F.; Liu, J.; Song, X.; Li, Y.; Yuan, Y.; Wang, C.; Lang, N.; Tian, S.; Yao, M.; et al. Evaluation of Deep Learning-Based Automated Detection of Primary Spine Tumors on MRI Using the Turing Test. *Front. Oncol.* **2022**, *12*, 814667. [CrossRef]
35. Liu, H.; Jiao, M.; Yuan, Y.; Ouyang, H.; Liu, J.; Li, Y.; Wang, C.; Lang, N.; Qian, Y.; Jiang, L.; et al. Benign and malignant diagnosis of spinal tumors based on deep learning and weighted fusion framework on MRI. *Insights Imaging* **2022**, *13*, 87. [CrossRef] [PubMed]

applied sciences

Article

Improvement of the Performance of Scattering Suppression and Absorbing Structure Depth Estimation on Transillumination Image by Deep Learning

Ngoc An Dang Nguyen [1,2,†], Hoang Nhut Huynh [1,2,†] and Trung Nghia Tran [1,2,*]

[1] Laboratory of Laser Technology, Faculty of Applied Science, Ho Chi Minh City University of Technology (HCMUT), 268 Ly Thuong Kiet Street, District 10, Ho Chi Minh City 72506, Vietnam; dnnan.sdh19@hcmut.edu.vn (N.A.D.N.); hhnhut@hcmut.edu.vn (H.N.H.)
[2] Vietnam National University Ho Chi Minh City, Linh Trung Ward, Thu Duc, Ho Chi Minh City 71308, Vietnam
* Correspondence: ttnghia@hcmut.edu.vn
† These authors contributed equally to this work.

Abstract: The development of optical sensors, especially with regard to the improved resolution of cameras, has made optical techniques more applicable in medicine and live animal research. Research efforts focus on image signal acquisition, scattering de-blur for acquired images, and the development of image reconstruction algorithms. Rapidly evolving artificial intelligence has enabled the development of techniques for de-blurring and estimating the depth of light-absorbing structures in biological tissues. Although the feasibility of applying deep learning to overcome these problems has been demonstrated in previous studies, limitations still exist in terms of de-blurring capabilities on complex structures and the heterogeneity of turbid medium, as well as the limit of accurate estimation of the depth of absorptive structures in biological tissues (shallower than 15.0 mm). These problems are related to the absorption structure's complexity, the biological tissue's heterogeneity, the training data, and the neural network model itself. This study thoroughly explores how to generate training and testing datasets on different deep learning models to find the model with the best performance. The results of the de-blurred image show that the Attention Res-UNet model has the best de-blurring ability, with a correlation of more than 89% between the de-blurred image and the original structure image. This result comes from adding the Attention gate and the Residual block to the common U-net model structure. The results of the depth estimation show that the DenseNet169 model shows the ability to estimate depth with high accuracy beyond the limit of 20.0 mm. The results of this study once again confirm the feasibility of applying deep learning in transmission image processing to reconstruct clear images and obtain information on the absorbing structure inside biological tissue. This allows the development of subsequent transillumination imaging studies in biological tissues with greater heterogeneity and structural complexity.

Keywords: point spread function (PSF); de-blurring; scattering suppression; depth estimation; Attention Res-Unet; DenseNet169; absorbing structure; turbid medium

Citation: Dang Nguyen, N.A.; Huynh, H.N.; Tran, T.N. Improvement of the Performance of Scattering Suppression and Absorbing Structure Depth Estimation on Transillumination Image by Deep Learning. *Appl. Sci.* **2023**, *13*, 10047. https://doi.org/10.3390/app131810047

Academic Editor: Cosimo Nardi

Received: 6 August 2023
Revised: 30 August 2023
Accepted: 1 September 2023
Published: 6 September 2023

1. Introduction

Optical imaging is crucial in biomedical research and diagnostics, bridging pre-clinical and clinical applications. The potential of light, especially near-infrared light, for imaging blood vessels on the skin surface and abnormal breast detection has been recognized in studies focusing on bio-metric and medical applications [1–4]. The prospects for developing non-invasive imaging devices based on near-infrared light are promising, offering advantages such as the absence of ionizing radiation, cost-effectiveness compared to existing methods, and suitability for further studies. However, transillumination images face strong scattering challenges. Previous research focused on the suppression of scattering and the restoration of clear images from blurred images [5–12]. Optical computed tomography (OCT) utilizing near-infrared light has been proposed and has shown satisfactory results in small animal imaging [7]. Deep learning (CNN) and stacking methods were proposed that were used to estimate the depth and de-blurring transillumination images of a turbid medium [8–12]. The effectiveness of previous studies is limited to a depth of absorbing structure shallower than 15.0 mm [10,12].

This study's models are based on novel machine learning mechanisms that combine different types of neural networks and sparse coding techniques to achieve high-quality image super-resolution [13]. The proposed models are also capable of handling multimodal and cross-domain image processing tasks, such as enhancing images from different sources or modalities, transferring styles or attributes between images, or generating realistic images from sketches or text descriptions [14]. The proposed models are inspired by some of the recent advances in machine learning algorithms and mechanisms for image processing, as well as some of the applications of image processing techniques for machine learning [15]. This study proposes new deep learning models to improve absorbing structures' de-blurring and depth estimation. The following sections of this paper will provide detailed information about the training dataset, the model employed for the de-blurring and depth estimation of absorbing structures, the performance parameters of the training process, and results and discussions concerning the de-blurring and depth estimation of absorbing structures.

2. Materials and Methods

2.1. Data Preparation

The deep learning model requires numerous training pairs for optimal accuracy and performance. The de-blurring process involved working with a carefully curated dataset of blurred and original clear images. The corresponding depth labels associated with the blurry images were used to train the depth estimation model. However, data collection presented practical challenges in acquiring significant training pairs. The depth-dependent point spread function (PSF), which characterizes light scattering in biological tissue at different depths, was implemented to convolve the original clear images, generating the desired blurred images to overcome this limitation.

Figure 1 shows the difference between fluorescent and transillumination images with the assumption that the light diffused well in the absorbing object plan. In fluorescent imaging, the light point source is placed inside the scattering medium, as shown in Figure 1a, and the light distribution on the observing surface (dashed orange line) can be mathematically represented by Equation (1) [6]:

$$\text{PSF}(d,\rho) = C\left(\mu_s' + \mu_a + \left[\kappa_d + \frac{1}{\sqrt{\rho^2 + d^2}}\right]\frac{d}{\sqrt{\rho^2 + d^2}}\right)\left(\frac{\exp[-\kappa_d\sqrt{\rho^2 + d^2}]}{\sqrt{\rho^2 + d^2}}\right) \quad (1)$$

where $k_d^2 = 3\mu_a(\mu_s' + \mu_a)$. C, μ_s', μ_a, and d represent the constants with respect to ρ and d, the reduced scattering coefficient, the absorption coefficient, and the depth of the light source, respectively.

In transillumination imaging, the light source is placed outside the scattering medium, as shown in Figure 1b, and the light distribution on the observing surface (black line) is a collection of the distribution of the light-missing points. The depth-dependent PSF is derived from a light source, so we cannot apply it directly for transillumination imaging. To make the PSF applicable, we invert the distribution of the light-missing point (black line) to have the distribution the same as the distribution of light in fluorescent imaging (dashed orange line), as shown in Figure 1b.

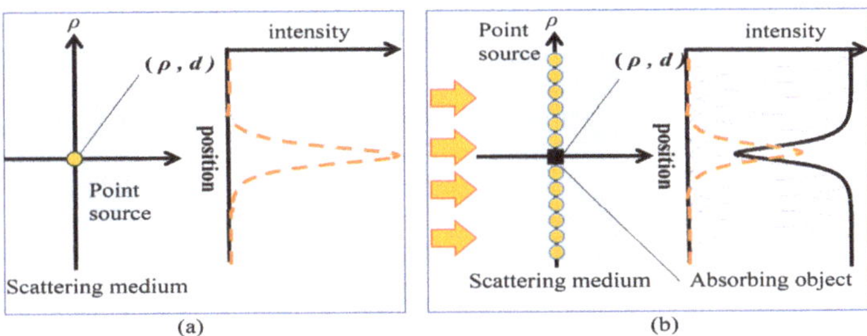

Figure 1. The intensity distribution at the medium surface in fluorescent (**a**) and transillumination (**b**) imaging.

The effectiveness of this approach was rigorously evaluated through comprehensive simulations and experimental validations [7,10,12]. Convolution images of the original structures with depth-dependent point spread functions were used at different depths to generate the data in this study, as described in Equation (2) and Figure 2.

$$y = h \otimes x \tag{2}$$

where $\otimes$ denotes convolution operation.

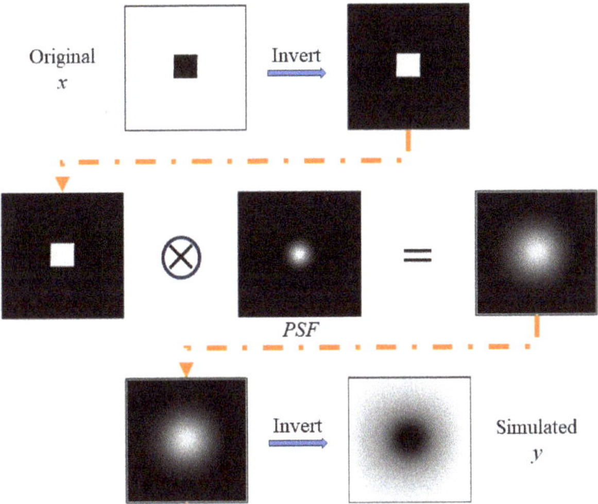

Figure 2. Blurred image generation process.

In this study, the original structure images are images obtained from the image of 12 randomized structures in a transparent medium. These pre-blurred images are specifically designed to emulate the intricate structural characteristics of blood vessels beneath the skin. The blurred images were generated by convoluting the original structures with the PSF given by Equation (1) at different depths with the parameter values of $\mu_s' = 1.0 \, \text{mm}^{-1}$ and $\mu_a = 0.00536 \, \text{mm}^{-1}$. These parameters were used in all simulations described hereinafter. These coefficients are pivotal in understanding light–tissue interactions, particularly in optical imaging and spectroscopy. μ_s' signifies the extent of light scattering as it traverses tissue, indicating the likelihood of scattering per unit path length. This is crucial because of diverse tissue structures that cause light to scatter in various directions. μ_a gauges the extent of light absorbed by tissue during propagation, closely related to light-absorbing constituents such as hemoglobin, lipids, and water. Different tissues and substances possess varying absorption characteristics at different wavelengths. In the given context, specific μ_s' and μ_a values might be set to simulate tissue optical properties in a model. These values depend on the type of tissue, the wavelength of light, and the experimental conditions. Setting these values likely aims to ensure that simulations closely mimic real tissue optical behavior under specific circumstances. Although the specific rationale hinges on precise values, the choice of these parameters is vital to accurately model light–tissue interactions and align simulation outcomes with experimental data.

In studies involving optical imaging and simulations of light propagation in biological tissues, setting the values of μ_s' (reduced scattering coefficient) and μ_a (absorption coefficient) is a critical step. Researchers usually consider the following factors when determining μ_s' and μ_a' values:

- Empirical data: Experimental measurements of optical properties in specific tissue types at various wavelengths can serve as a foundation for determining the appropriate values. These measurements can come from the literature or new measurements conducted by the researchers themselves.
- Literature references: Previous studies often report ranges or specific values of μ_s and μ_a for similar tissue types. Researchers can use these references as a starting point and adjust the values based on their experimental setup.
- Theoretical models: There are established theoretical models that relate optical properties to tissue composition and structure. Researchers can leverage these models to estimate μ_s and μ_a based on the known components and concentrations in the tissue.
- Tissue variation: Different tissues exhibit different optical properties as a result of variations in cellular composition, structure, and pigmentation. Consequently, the specific tissue under investigation must be carefully considered when selecting μ_s and μ_a values.
- Wavelength Dependence: Optical properties can vary with the wavelengths of light. Researchers may choose μ_s and μ_a values that align with the wavelength range used in their experimental setup.
- Validation: Validating the chosen values involves comparing the simulation results with actual experimental observations. If the simulated outcomes closely match the experimental data, this provides confidence in the suitability of the parameter values.
- Sensitivity analysis: Researchers may conduct sensitivity analyses to assess how changes in μ_s and μ_a impact simulation results. This analysis helps to determine reasonable ranges for these parameters.

In essence, substantiating the selections of μ_s and μ_a values typically entails a combination of empirical data, theoretical frameworks, references from the literature, and validation against experimental results. The specific strategy can be flexible to existing resources, unique tissue attributes, and the specific objectives of the experiment.

For the de-blurring study, a comprehensive dataset consisting of 8000 pairs of clear and blurred images was generated by convoluting 10 of 12 original structures with the PSF given by Equation (1) at depths ranging from 0.1 to 20.0 mm (interval 0.1 mm) and then rotating at four different angles, as illustrated in Table 1. The remaining 2 of the 12 original structures were used to generate data for testing. During the training process for de-blurring, the generated dataset was used to train the models with a batch size of 8. The learning rate was set to 10^{-4}, and the models were trained for 100 epochs.

For the depth estimation study, the corresponding depth labels associated with the blurry images were used. A dataset of 70,400 images was generated that depicts the absorbing structures within the scattering medium at different depths. The blurred images in this dataset were generated by convoluting 11 of the 12 original structures with the PSF given by Equation (1) at depths ranging from 0.5 to 20.0 mm (interval 0.5 mm) and then rotating at 160 different angles, as illustrated in Table 1. The remaining original structure was used to generate data to test the performance of the convolution neural network models. During the depth estimation training process, the generated dataset was used to train the models with a batch size of 32. The learning rate was set to 10^{-4}, and the models were trained for 20 epochs.

Table 1. Dataset for training, validation, and testing of scattering de-blurring and depth estimation.

Model	Training	Validation	Testing	Total
De-blurred	5600	1600	800	8000
Estimate depth	56,320	14,080	7040	70,400

Training was carried out on a high-performance workstation that features two Intel® Xeon® CPUs E5-2683 v4 with 64 GB of RAM. In addition, an NVIDIA Quadro K2200 graphics processing unit was used to accelerate the computational tasks involved in the training process. The specific training parameters used, including batch size, learning rate, and number of epochs, are provided in Table 2.

Table 2. Parameters for de-blurred and depth estimation models.

Parameters	De-Blurred	Depth Estimation
μ_s'	1.0 mm^{-1}	1.0 mm^{-1}
μ_a	0.00536 mm^{-1}	0.00536 mm^{-1}
d_{min}–d_{max}	0.1–20.0 mm	0.5–20.0 mm
Step depth	0.1 mm	0.5 mm
Batch size	8	32
Learning rate	10^{-4}	10^{-4}
Epoch	100	20
Loss function	Dice-coef loss	Categorical Cross-entropy
Optimizer	Adam	Adam
Input shape	$256 \times 256 \times 1$	$224 \times 224 \times 1$

2.2. Image De-Blurring

Transillumination imaging techniques for visualizing absorbing structures within the body often encounter blurring. While both scattering and absorption contribute to image blurring, scattering plays a dominant role. Many research efforts have been made to overcome this challenge. In the previous studies of the group, the suppression of the scattering effects on the transillumination image was carried out by deconvolution with the depth-dependent PSF, and the deep learning scatter blurring method has also proved to be feasible and efficient [5–7,10,12]. However, these methods still have some limitations in implementation, such as the imperfection of the deconvolution technique, the long computation time, the computational hardware requirements, and the limitation of effective de-blurring shallower than 15.0 mm. In the previous study, we employed fully convolutional networks (FCN) based on the U-net with skip connections. The training process for the scattering de-blurring model is visually represented in Figure 3, offering a clear visualization of the methodology employed. The results show that we can obtain a clear image of the absorbing structure as deep as several to 10.0 mm in a turbid medium.

To address this challenge, the Attention U-Net model and the Attention Res-UNet model were incorporated for the de-blurring process [16,17]. The attention gate is a mechanism that selectively emphasizes specific regions of interest while suppressing the activation of irrelevant regions on a given input feature map X. To achieve this, the attention gate takes advantage of a gating signal $G \in \mathbb{R}^{C' \times H \times W}$, which is obtained at a coarser scale and incorporates contextual information. When additive attention is employed, the attention gate calculates the gating coefficient. Initially, both the input X and the gating signal G undergo linear mapping to a $\mathbb{R}^{F \times H \times W}$ dimensional space. Subsequently, the output is compressed in the channel domain to generate a spatial attention weight map $S \in \mathbb{R}^{1 \times H \times W}$, as shown in Figure 4. The entire process can be formulated as described in Equation (3) and Equation (4) [18], where ϕ, ϕ_x, and ϕ_g are linear transformations implemented as convolution 1×1.

$$S = \sigma\big(\varphi\big(\delta\big(\phi_x(X) + \phi_g(G)\big)\big)\big) \tag{3}$$

$$Y = SX \tag{4}$$

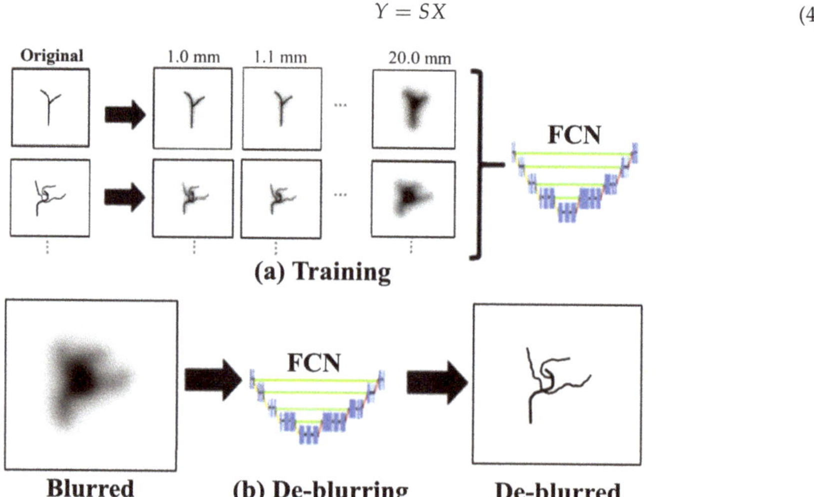

Figure 3. De-blurring using a deep learning model: (**a**) training process with pair of images before and after blurring, and (**b**) image de-blurring process.

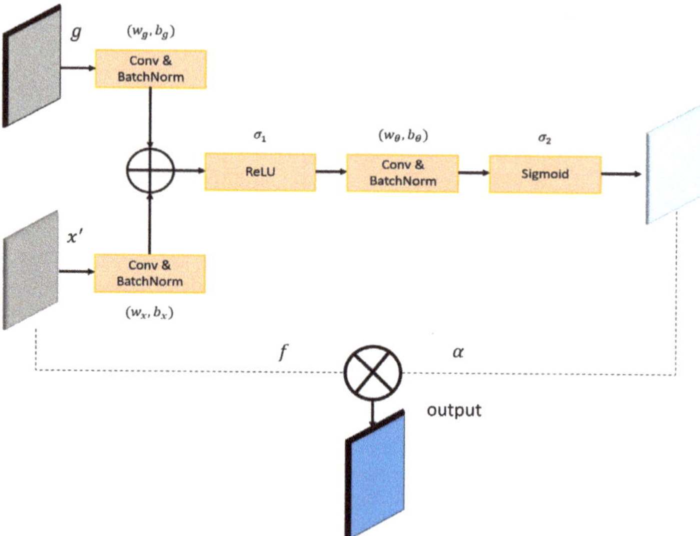

Figure 4. The diagram of the attention gate.

Residual blocks, which are skip-connection blocks, are designed to learn residual functions by referring to the layer input instead of learning unreferenced functions. These blocks were originally introduced as a component of the Res-Net architecture. In a formal sense, denoting the desired underlying mapping as $H(x)$, the stacked non-linear layers aim to approximate an additional mapping that captures the difference between the current output and the input, denoted as Equation (5). By explicitly modeling the residual mapping, the network can effectively learn residual functions and enhance optimization.

$$F(x) := H(x) - x \tag{5}$$

The original mapping is reformulated as $F(x) + x$, where $F(x) + x$ represents a residual component, thus giving rise to the term "residual block $H(x)$". The rationale behind this approach lies in the observation that optimizing the residual mapping is often more feasible than optimizing the original, unreferenced mapping. In certain cases, minimizing the residual to approach zero can be simpler than fitting an identity mapping using a series of non-linear layers. The network is better equipped to learn mappings that resemble identity transformations by incorporating skip connections. The proposed framework encompasses a novel deep learning architecture known as Res-UNet-a and a novel loss function based on the Dice loss. Res-UNet-a combines a U-Net encoder/decoder backbone with residual connections, Atrous convolutions, pyramid scene parsing pooling, and multitasking inference, thus enhancing its capabilities for various image analysis tasks, as shown in Figure 5.

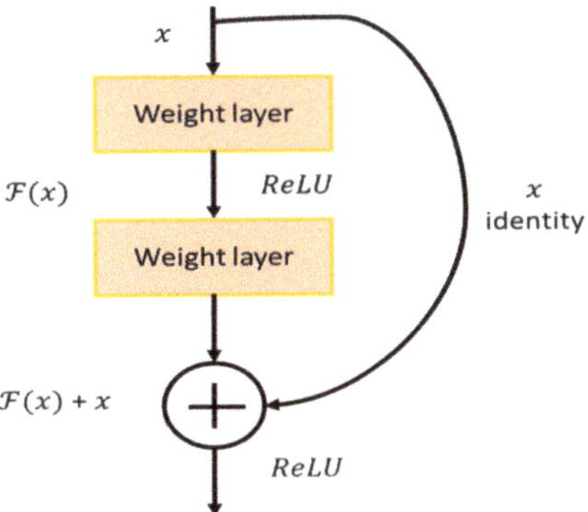

Figure 5. A diagram of the residual block.

2.3. Depth Estimation

In transillumination imaging, the extent of blurring depends on the depth of the absorbing structure within the scattering medium, with an increase in depth resulting in a progressively more blurred image. To estimate the depth of the absorbing structure, a convolutional neural network (CNN) model is trained using generated blurred images. We used Res-Net-based convolutional neural networks (CNN) in the previous study. The training process for the depth estimation model is visually represented in Figure 6, which provides a clear visualization of the methodology used. The results show that we can effectively estimate the depth of the absorbing structure as deep as several to 10.0 mm in a turbid medium. Four pre-trained models, namely Res-Net50, VGG-16, VGG-19, and Dense-Net169 [19,20], were used for the depth estimation challenge. The images were paired with their respective depth labels during the training phase. An estimate of the depth of the absorbing structure was obtained by entering a blurred image into the CNN model. This process aligns with the fundamental classification task within deep learning. To ensure consistency, the training performance settings described in Table 2 were applied in different models, taking into account computational constraints and system compatibility. Figure 6 illustrates the estimation procedure for the depth of the absorbing structure using the CNN model.

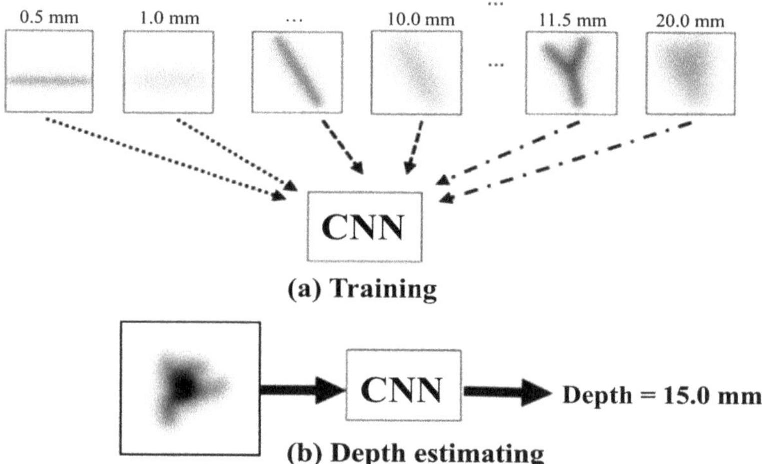

Figure 6. Estimating depth of absorbing structure with deep learning model: (**a**) training process and (**b**) depth estimating process.

3. Metrics

The Dice coefficient is widely used to assess the agreement at the pixel level between a predicted segmentation and its corresponding ground truth. It quantifies the similarity by calculating twice the area of overlap divided by the sum of the total number of pixels in both images. Equation (6) expresses the Dice coefficient as [21]:

$$\text{Dice-coef} = \frac{2 \times |X \cap Y|}{|X| + |Y|} \tag{6}$$

where X and Y represent the predicted set of pixels and the ground truth, respectively.

Moreover, the loss of the Dice coefficient is employed as a measure of dissimilarity between the predicted and ground-truth segmentation. It is computed by subtracting the Dice coefficient from 1. Equation (7) presents the formulation of the loss of the Dice coefficient [21].

$$\text{Dice-coef loss} = 1 - \frac{2 \times |X \cap Y|}{|X| + |Y|} \tag{7}$$

The Intersection over Union (IoU) is commonly utilized as an evaluation metric for object detection accuracy in the dataset by calculating the ratio of the overlap and the union areas between the predicted and ground-truth regions. Equation (8) [22] represents the IoU formula as:

$$\text{IoU} = \frac{\text{Area of overlap}}{\text{Area of union}} \tag{8}$$

When developing a depth estimation model, accuracy is used as a classification metric to measure the proportion of correctly predicted instances where the predicted depth exceeds the actual depth. It provides insight into the model's performance on the dataset. The accuracy is computed by dividing the sum of the True Negatives (TN) and True Positives (TP) by the total number of samples. Equation (9), illustrates the accuracy formula [23]:

$$\text{Accuracy} = \frac{TP + TN}{\text{total sample}} \tag{9}$$

4. Results and Discussion

4.1. De-Blurring Image

The significant impact of the attention gate on the scattering de-blurring process is demonstrated by the results presented in Figure 7. The images obtained with the Attention Unet model (D) show higher clarity and fidelity than those obtained with the standard U-Net architecture (C), as the attention mechanism effectively suppresses the scattering influence and improves the image reconstruction of the absorbing structure. Quantitative evaluation, as indicated by the Intersection Over Union (IoU) index, further supports the superiority of the gating attention approach. The IoU index of 0.908 achieved when using gating attention exceeds the IoU index of 0.831 obtained with the standard U-Net architecture. This substantial improvement demonstrates the ability of the attention gate to capture the relevant features better and reduce the impact of scattering, leading to more accurate and precise de-blurring results.

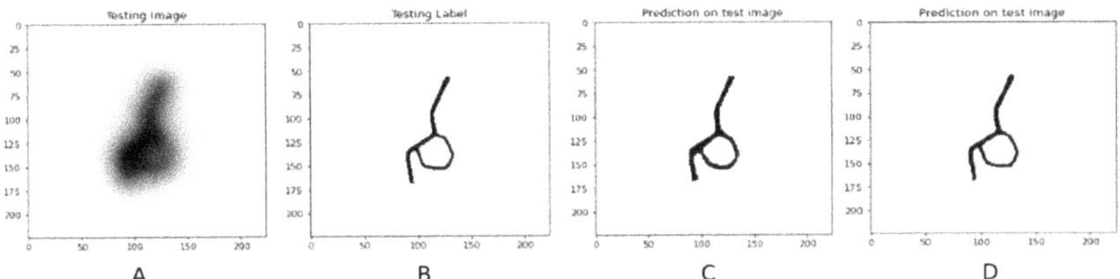

Figure 7. Scattering de-blurring with and without attention gate: (**A**) transillumination image through turbid medium, (**B**) image taken through clear water, (**C**) output image from U-Net model, and (**D**) output image from Attention Unet model.

The effectiveness of the attention gate can be attributed to its ability to selectively focus on informative regions and suppress the interference caused by scattering. By assigning different attention weights to different parts of the image, the attention gate improves the model's capability to accurately capture and reconstruct the absorbing structure image, even at greater depths. These findings highlight the potential of the attention gate in improving the scattering de-blurring process. Incorporating the gating attention mechanism into the U-Net architecture can significantly enhance the quality and reliability of de-blurred images, particularly in scenarios with high levels of scattering. Further exploration and optimization of the attention gate in various imaging applications hold promise for advancing the image reconstruction and de-blurring field.

The effectiveness of the residual block in the scattering de-blurring process is illustrated by the results presented in Figure 8. The images obtained from the Res-UNet model show a remarkable improvement in the de-blurring outcome compared to those obtained from the U-Net model, as evidenced by the higher IoU index of 0.885. This indicates a more accurate reconstruction of the original absorber image at a depth of 15 mm, even in the presence of scattering and blurring effects. In contrast, the standard U-Net architecture yields a slightly lower IoU index of 0.831, indicating a relatively inferior de-blurring performance. The superior performance of the Residual U-Net model can be attributed to the ability of residual blocks to facilitate the propagation of gradient information effectively. By allowing for the direct flow of information through skip connections, residual blocks enable the model to capture and restore important features of the absorbing image more efficiently. Consequently, the Residual U-Net model exceeds the standard U-Net architecture in mitigating the negative impact of scattering and achieves more accurate de-blurring results. These findings demonstrate the significance of incorporating residual blocks into deep learning models for scattering de-blurring tasks. The Residual U-net model proves to be a promising approach for addressing challenges associated with image blurring in the presence of scattering media. Further investigations and optimizations can be conducted to enhance the performance of the Residual U-Net model and explore its potential applications in various imaging tasks, such as medical diagnostics and image analysis in turbid environments.

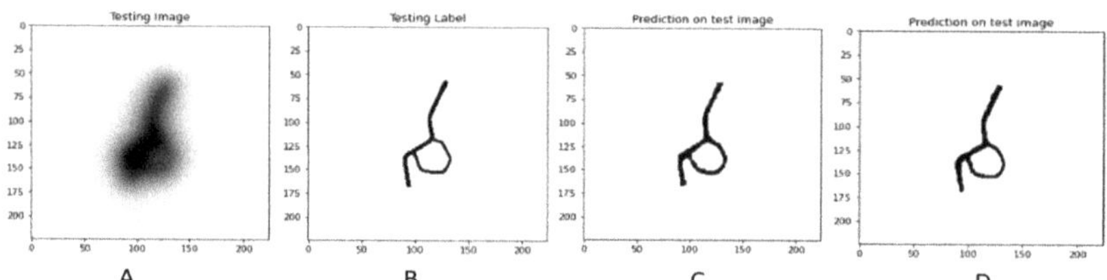

Figure 8. Scattering de-blurring with and without residual block: (**A**) transillumination image through turbid medium, (**B**) image taken through clear water, (**C**) output image from U-Net model, and (**D**) output image from Residual Unet model.

The primary objective of this study was to obtain de-blurred images by effectively compensating for scattering effects. To accomplish this, the models employed in this approach included Attention U-Net and Attention Res-UNet. These models were trained using a carefully curated input and output image pairs dataset. A combination of PSF convolutions at different depths was used to train the Attention U-Net and Attention Res-UNet networks for image de-blurring. This involved pairing the original images with their corresponding blurred counterparts. This approach enabled the networks to learn the intricate relationships between different depths and their corresponding blurred representations, facilitating accurate image de-blurring.

In the statistics table of the Dice coefficient for the two models, Attention Unet and Attention Res-UNet, as shown in Table 3, we can observe crucial information on the performance of these models. The Attention Unet model achieved a minimum Dice coefficient of 0.931 and a maximum of 0.999487, with a mean of 0.996319 and a median of 0.999195. The variability in the performance of this model is represented by a standard deviation of 0.009583. Similarly, the Attention Res-UNet model exhibits comparable parameters, with a minimum Dice coefficient of 0.930391 and a maximum of 0.999492. The mean and median values for this model are 0.996443 and 0.999223, respectively. The performance variability of the Attention Res-UNet model is gauged by a standard deviation of 0.009603. Overall, both models demonstrate consistent performance with minimal variation across the Dice coefficient values. This underscores the efficiency and general applicability of the models in de-blurring absorption structures within a dispersed medium. Figure 9 provides a visual representation of the process.

Table 3. Performance Comparison of Attention Unet and Attention Res-UNet Models based on Dice Coefficient Statistics.

Model	Minimum	Maximum	Mean	Median	Std
Attention Unet	0.931056	0.999487	0.996319	0.999195	0.009583
Attention Res-UNet	0.930391	0.999492	0.996443	0.999223	0.009603

Figure 10 illustrates a representative example of input and output images obtained by scatter blurring at various depths, specifically 0.1, 5.0, 10.0, and 20.0 mm. In particular, the corresponding correlation indices for these depths were reported as 0.9360, 0.9167, 0.9130, and 0.9059, respectively. These correlation indices served as valuable quantitative indicators, providing insights into the level of agreement between the predicted de-blurred output and the ground truth images.

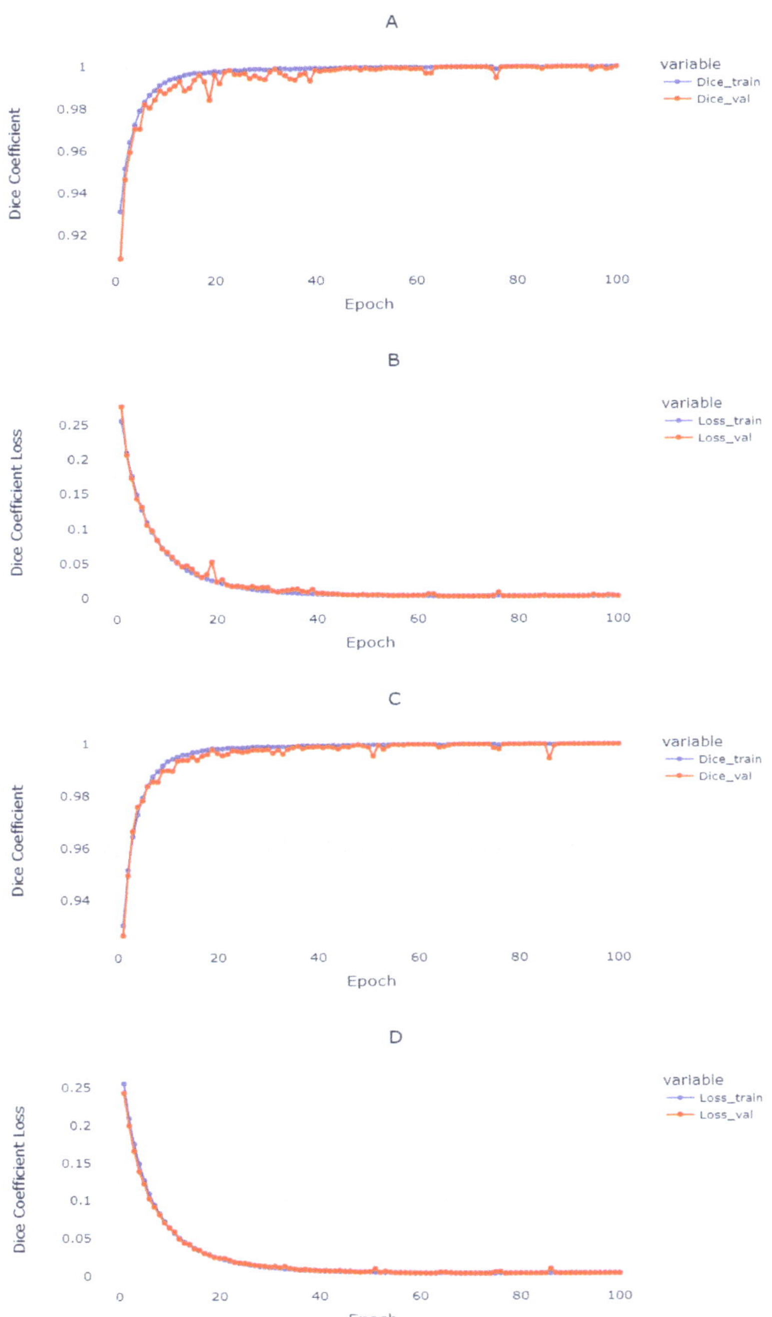

Figure 9. Training and validation for de-bluring process: (**A**,**B**) Attention U-Net, (**C**,**D**) Attention Res-UNet.

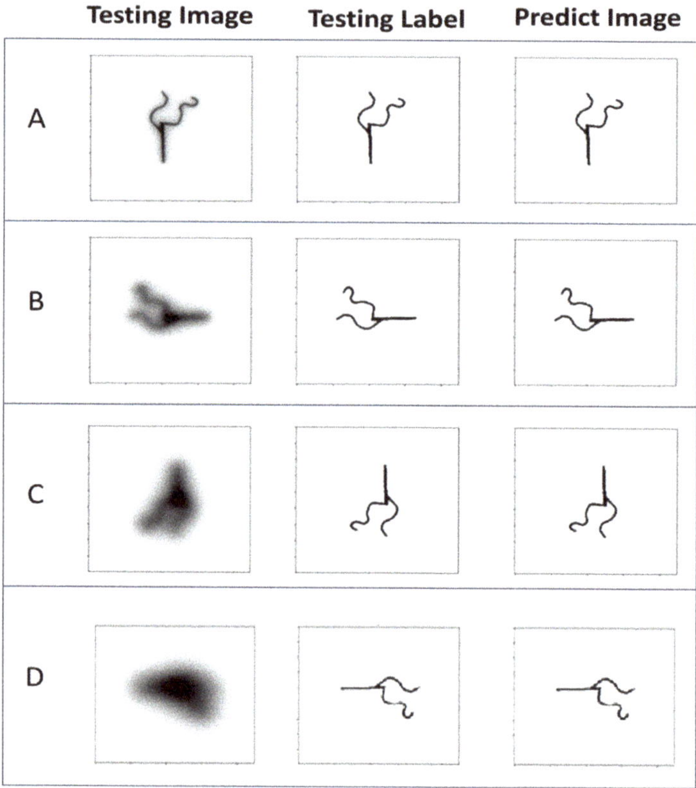

Figure 10. Representative images demonstrating the de-blurring process at various depths: (**A**) 0.1 mm, (**B**) 5 mm, (**C**) 10 mm, and (**D**) 20 mm.

Figure 11 illustrates the original and restored images of the absorbing structure after applying the Attention U-Net and Attention Res-UNet models. Then, the correlation coefficient is calculated. As the depth of the absorbing structure increases, the blurring effect becomes more pronounced, leading to a rapid decline in the quality of the blurred image. Furthermore, the reduction in training images significantly affects the correlation coefficient.

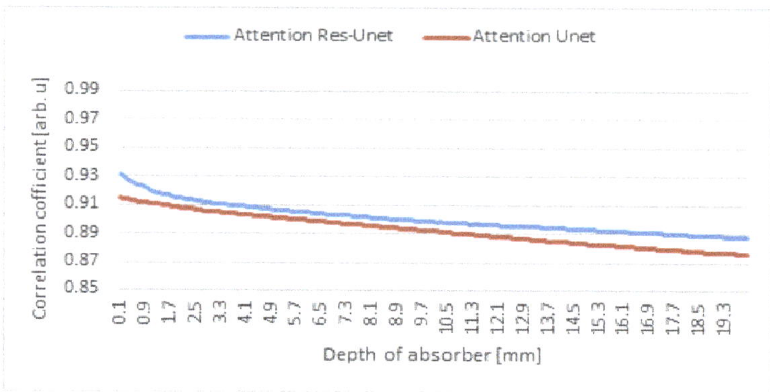

Figure 11. Correlation analysis between original and deblurred images with 256 × 256 input image size.

The test results for the Attention U-Net model indicated that the correlation coefficient exhibited a high value ranging from 0.9149 to 0.9013 for depths between 0.1 and 5.0 mm. Beyond 5.0 mm, the correlation coefficient gradually decreased, reaching 0.8921 at a depth of 10.0 mm. Subsequently, for depths ranging from 10.1 to 20.0 mm, the correlation coefficient rapidly decreased from 0.8918 to 0.8801 at a depth of 20.0 mm. In particular, the rate of decrease in the correlation coefficient became more pronounced once the depth exceeded 10.1 mm.

Similarly, the Attention Res-UNet model yielded test results indicating a high correlation coefficient ranging from 0.9308 to 0.9069 for depths between 0.1 and 5.0 mm. Beyond 5.0 mm, the correlation coefficient gradually decreased, reaching 0.8845 at a depth of 14.0 mm. In particular, for depths greater than 10.0 mm, the rate of decrease in the correlation coefficient increased. Finally, for depths ranging from 14.1 to 20.0 mm, the correlation coefficient exhibited a rapid decline from 0.8942 to 0.8876 at a depth of 20.0 mm. Once again, the rate of decrease in the correlation coefficient decreased rapidly for depths exceeding 14.1 mm.

In the subsequent experiment, the size of the training input image was modified from 256 × 256 pixels to 112 × 112 pixels while keeping the other training parameters in Table 2 unchanged. The results obtained from this adjustment are depicted in Figure 12.

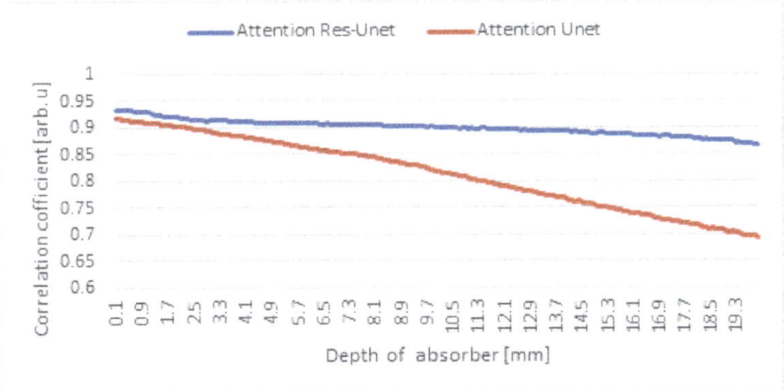

Figure 12. Correlation analysis between original and deblurred images with 112 × 112 input image size.

The test results of the Attention U-Net model revealed that within the depth range of 0.1 to 5.0 mm, the correlation coefficient initially reached a high value and gradually decreased from 0.9186 to 0.8722. The highest correlation coefficient was achieved at a depth of 0.1 mm, registering a value of 0.9186. As the depth increased from 5.1 to 10.0 mm, the correlation coefficient experienced a gradual decrease, reaching 0.8184 at a depth of 10.0 mm. Subsequently, for depths ranging from 10.1 mm to 20.0 mm, the correlation coefficient exhibited a rapid drop from 0.8172 to 0.6927 at a depth of 20.0 mm. Remarkably, once the depth surpassed 7.0 mm, the rate of decline in the correlation coefficient with respect to depth became more pronounced.

Similarly, the Attention Res-UNet model yielded noteworthy test results. At depths ranging from 0.1 to 5.0 mm, the correlation coefficient reached a high value and gradually decreased from 0.9337 to 0.9023. The highest correlation coefficient was observed at a depth of 0.1 mm, yielding a value of 0.9337. For depths extending from 5.1 to 14.0 mm, the correlation coefficient exhibited a gradual decrease from 0.9080 to 0.8736 at a depth of 14.0 mm. In particular, depths greater than 10.0 mm experienced an accelerated decline in the correlation coefficient. Finally, within the depth range of 14.1 to 20.0 mm, the correlation coefficient decreased rapidly from 0.8717 to 0.8539 at a depth of 20.0 mm. Once the depth surpassed 11.6 mm, the rate of decrease in the correlation coefficient with respect to depth decreased rapidly.

For the Attention U-Net model, it was observed that at depths ranging from 0.1 to 0.5 mm, employing an input size of 112 × 112 pixels yielded better performance, with a difference in the correlation coefficient ranging from 0.6% to 0.9%. On the contrary, at depths ranging from 0.6 to 5.0 mm, adopting an input size of 256 × 256 pixels achieved superior performance, exhibiting a difference in the correlation coefficient ranging from 0.02% to 3.08%. In particular, for depths ranging from 5.1 to 20.0 mm, the difference in performance between the two input sizes increased rapidly, ranging from 3.35% to 20.91%.

For the Attention Res-Unet model, it was observed that at depths ranging from 0.1 to 0.9 mm, employing an input size of 112 × 112 pixels resulted in improved performance, with a difference in the correlation coefficient ranging from 0.4% to 0.6%. On the other hand, at depths ranging from 1.0 to 5.0 mm, adopting an input size of 256 × 256 pixels yielded better performance, exhibiting a difference in the correlation coefficient index ranging from 0.15% to 0.75%. Furthermore, for depths ranging from 5.1 to 20.0 mm, the difference in performance between the two input sizes increased rapidly, ranging from 0.7% to 3.55%.

The results show the impact of resizing the input training image from 256 × 256 pixels to 112 × 112 pixels on the correlation coefficient at different depths. These findings demonstrate the importance of optimizing the input image size to achieve optimal performance in terms of the correlation coefficient at different depths, as shown in Figure 13. The observed trends can be ascribed to the interplay of scattering phenomena and the depth of the absorbing structure. With increasing depth, the scattering effects intensified, leading to diminished correlation coefficients. Moreover, the selection of the input image size exerted a notable influence on performance, primarily by affecting the model's ability to capture intricate features amidst scattering influences. In particular, the optimal input size exhibited variability depending on depth, thus facilitating improved adaptability to varying degrees of scattering. These discernments underscore the imperative of factoring in depth and input size while addressing scattering-induced de-blurring tasks, thereby providing valuable insights for optimizing model efficacy across diverse scenarios. Further studies could delve into the intricate dynamics connecting depth, scattering effects, and input size, thereby advancing the potential for refining the applicability and precision of de-blurring models.

The validity of the diffusion approximation is based on the condition that the thickness of the scattering medium is significantly greater than the average free-path length of $1/\mu'_s$. Consequently, caution must be exercised when applying Equation (1) in cases where $\sqrt{\rho^2 + d^2}$ is not greater than $1/\mu'_s$. As shown in Figure 1, the observing plan is considered to be significantly larger than the light distribution on the surface. Therefore, it is better to generate an appropriate wide image for training to ensure a result with a deep-absorbing structure. The light distribution on the surface has a Gaussian distribution shape. The image size in a dimension should be more significant than three times the standard deviation of the light distribution on the surface of the medium when calculating the deepest light point source distribution by Equation (1) in the turbid medium, as shown in Figure 13.

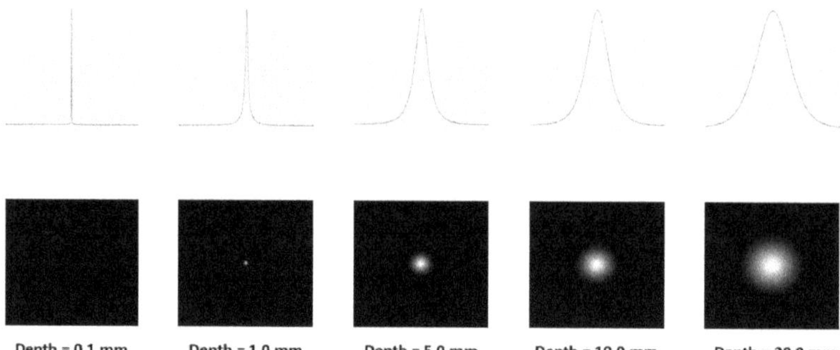

Figure 13. Optimizing input image size across various depths.

4.2. Depth Estimation

Depth estimation is an essential task for analyzing the properties of absorption structures. This section presents a deep learning approach to estimate the depths of absorption structures from their images. For this purpose, a dataset of 7040 images of absorption structures, each labeled with one of 40 depth values ranging from 0.1 mm to 20 mm, is used. Four state-of-the-art deep learning models, namely ResNet50, VGG16, VGG19, and DenseNet169, are trained and evaluated in this dataset. Accuracy is the evaluation metric that measures the percentage of images whose depth labels are correctly predicted by the models. Table 4 shows the training and validation accuracy of each model after 20 epochs.

Table 4. The training and validation accuracy of each model after 20 epochs.

Model	Training Accuracy	Validation Accuracy
ResNet50	0.4312	0.3921
VGG16	0.5124	0.4678
VGG19	0.4894	0.4500
DenseNet169	0.7323	0.6250

Table 4 shows that DenseNet169 achieves the highest accuracy in both the training and validation sets, followed by VGG16, VGG19, and ResNet50. All models perform better than in previous experiments with a smaller dataset, indicating the positive impact of dataset size and diversity on model performance. However, the accuracy of all models is still low, indicating the difficulty of the depth estimation task. To further analyze the behavior of the models, the accuracy curves of each model were plotted during training and validation, as shown in Figure 14.

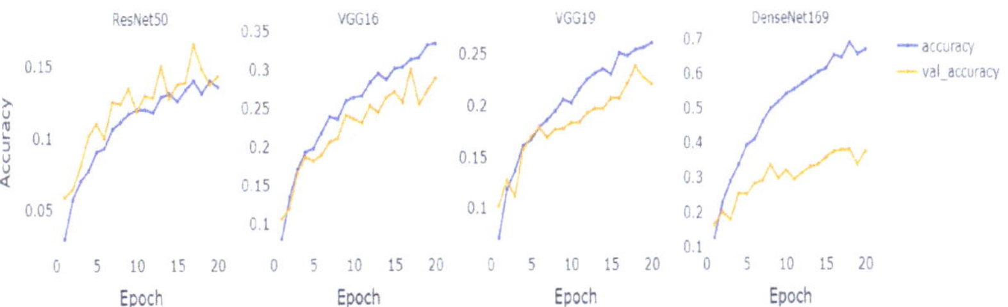

Figure 14. Accuracy evaluation of various models: ResNet50, VGG16, VGG19, and DenseNet169.

In Figure 14, the conspicuous features include low accuracy values and pronounced fluctuations, indicating a struggle of the models to glean effective insights from the original 7040-image dataset. This conundrum can be attributed to the inherent complexity of the dataset, characterized by an extensive array of depth classes (40) coupled with a limited count of images per class (fewer than 176 images). Consequently, the models struggled to discern nuanced differentiators across various depth levels, impairing their capacity for comprehensive learning. The fluctuations in accuracy, evident in the jagged trajectory after each epoch, underscored the models' susceptibility to data fluctuations, amplifying the instability quotient.

To improve the performance of the models, data augmentation techniques were applied to increase the size and diversity of the dataset. Specifically, angle rotation was used to generate new images from existing ones by randomized 160 different angles and rotating them at angles between 0 and 360 degrees. This resulted in an augmented dataset of 70,400 images (7040 × 10) with the same depth labels as before.

The decision to employ 160 different angles for image rotation during the data generation process in this study is purposeful and aligned with the goal of improving the robustness and generalization capabilities of the trained convolutional neural network (CNN) models. This technique, commonly referred to as data augmentation, serves to simulate varying viewpoints and orientations of the same scene or object, thereby aiding the model in comprehending and identifying features from diverse angles. In the context of estimating depth from blurred images of absorbing structures within a scattering medium, the rationale behind incorporating image rotation at numerous angles can be succinctly summarized:

- Increased variability: By generating images from multiple angles, the dataset gains greater diversity. This variability acts as a defense against overfitting, ensuring that the model learns broader transferable features instead of memorizing specific training samples.
- Robustness to orientation: Real-world scenarios involve objects with varying orientations. Training the model on images spanning different orientations enhances its resilience to changes in object rotation.
- Feature extraction: Image rotation encourages the model to learn invariant features. It requires the model to emphasize features consistent across orientations, thus aiding in the extraction of pertinent and informative features for accurate depth estimation.
- Generalization: Exposure to an extensive array of angles equips the model with the ability to generalize its insights to novel orientations during inference.

In essence, the choice of 160 different angles probably stems from a balance between creating a suitably diverse dataset and managing the computational demands of training. This numerical selection may have emerged through iterative experimentation and validation, ensuring that the model benefits from enhanced diversity while maintaining a manageable training process.

Using the augmented data set, the same models (ResNet50, VGG16, VGG19, and DenseNet169) underwent rigorous training and evaluation. This assessment used a multifaceted set of evaluation metrics, encompassing accuracy, precision, recall, and F1 score, as shown in Table 5. These metrics serve as vital indicators of the efficacy of the model in distinct facets of the depth estimation task. These models were subjected to rigorous training spanning 100 epochs, with a batch size of 20 and a learning rate set at 0.001. This comprehensive evaluation regimen ensured meticulous scrutiny of the models' competence from various vantage points.

Table 5. Evaluation metrics of different models after 100 epochs on the augmented dataset.

Model	Accuracy	Precision	Recall	F1-Score
ResNet50	0.9212	0.9221	0.9212	0.9216
VGG16	0.9324	0.9338	0.9324	0.9331
VGG19	0.9294	0.9300	0.9294	0.9297
DenseNet169	0.9523	0.9535	0.9523	0.9529

From Table 5, it can be observed that:

- All models attained substantial scores across evaluation metrics, indicating proficient performance in depth estimation.
- DenseNet169 secured the highest values in all metrics, followed by VGG16, VGG19, and ResNet50.
- The models demonstrated consistent alignment between accuracy, precision, recall, and F1 score, reflecting balanced performance in positive and negative classes.
- In particular, the application of angle rotation as an enhancement technique yielded notable improvements in the evaluation metrics compared to the previous experiment with the original dataset.

The progression of the training and testing process over 100 epochs is visually captured in the collection of four graphs shown in Figure 15. This visualization offers valuable insights: ResNet50 illustrates a gradual and consistent increase in accuracy across epochs, albeit with a modest final value. On the contrary, VGG16 and VGG19 exhibit swift accuracy improvements in the initial epochs, followed by a more gradual enhancement rate. In particular, DenseNet169 demonstrates a consistent and rapid accuracy advancement throughout the epochs, culminating in a substantial final accuracy value. It is important to note that all models exhibit diminished accuracy fluctuations after each epoch compared to the earlier experiment, indicating an improved level of learning stability.

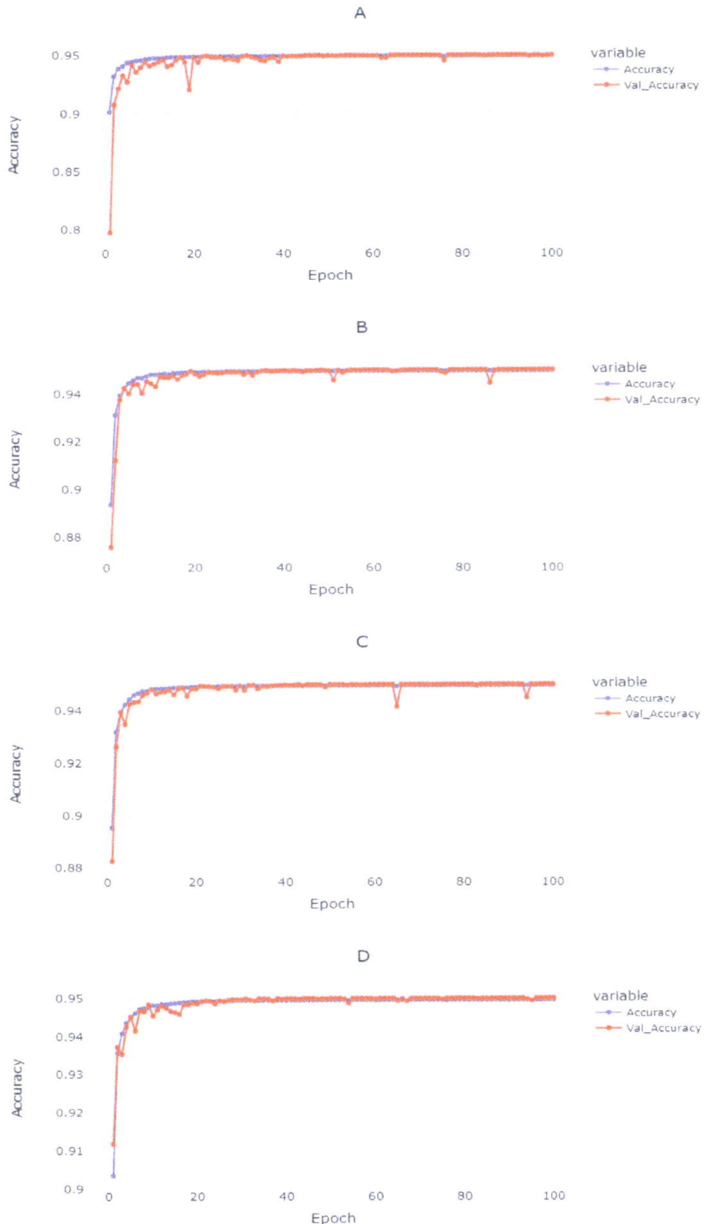

Figure 15. Accuracy curves of (**A**) ResNet50, (**B**) VGG16, (**C**) VGG19, and (**D**) DenseNet169 models.

In terms of training accuracy, a rapid increase was observed from epochs 1 to 10, rising from 0.4412 to 0.9055. Subsequently, the training accuracy continued to improve, but the rate of increase decreased with each epoch. Over the next ten epochs, the training accuracy increased by only 0.06, reaching 0.9645 by the 20th training session. The use of the expanded training dataset of 70,400 images contributed to the improved accuracy of the DenseNet169 model. The model achieved an accuracy of more than 65%, indicating the importance of this research and the generated dataset to estimate the depth of absorption structures in near-infrared images. The slow increase in accuracy from the 10th training session onward can be attributed to the challenge of extracting specific features for each class in the classification model, which comprises 40 classes that represent different depths. Moreover, the increasing blurring of images of absorbing structures at depths above 16.0 mm poses difficulties in distinguishing the blurred images. Furthermore, Figure 16 illustrates the results of the correlation analysis between the depth estimated by the CNN model and the depth given during testing. As the depth increased, the estimation error also increased. The experiment involved 8000 images at 20 depths ranging from 1.0 mm to 20.0 mm, with 40 images per depth for testing. The correlation coefficient was $R^2 = 0.9911$, demonstrating the feasibility of the CNN-DenseNet-169 model in estimating the depths from images of absorbing structures.

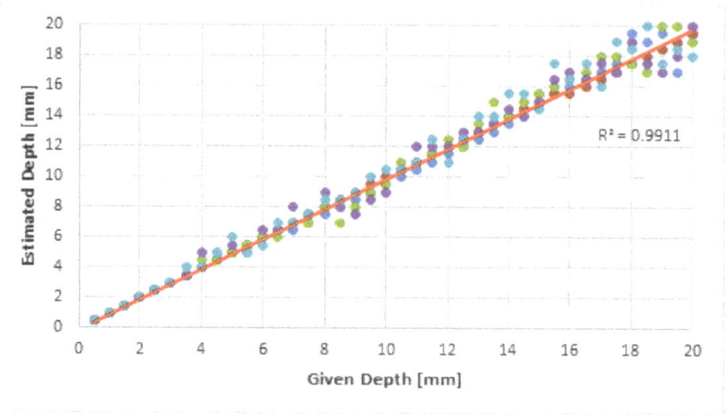

Figure 16. Correlation Analysis of Given and Estimated Depths.

Figure 17 illustrates the workflow of the proposed method. First, the original image was convolved with a point spread function (PSF) to simulate the blurring effect caused by light scattering and absorption in biological tissue. This process yielded a blurred image of the absorption structures. Second, the blurred image underwent de-blurring through a fully convolutional network (FCN) model, which could have been either the Attention UNet or the Attention Res-UNet, in order to recover the original image. Lastly, the blurred image was subjected to decoding using a convolutional neural network (CNN) model to estimate the depth of the absorption structures. In further studies, these results will be optimized to reconstruct the 3D structure of biological tissue from a 2D image.

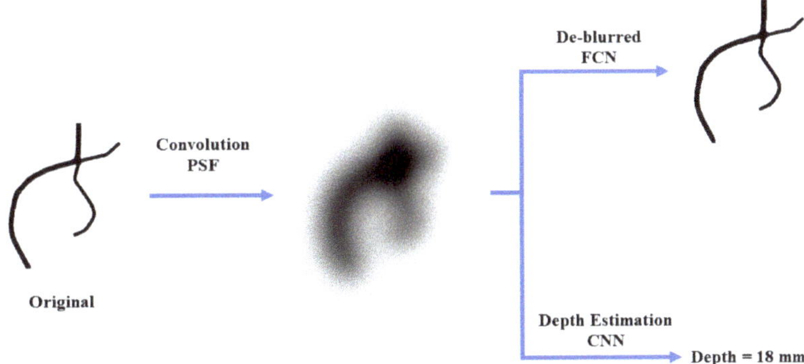

Figure 17. Correlation analysis of given and estimated depths.

5. Conclusions

The de-blurring and estimation depths of the absorbing structure in transillumination images taken through a turbid medium such as biological tissue have attracted significant interest among researchers and experts in biomedical optics in recent years. This study addresses the challenge of de-blurring and depth estimation in transillumination images by utilizing the dependent point spread function (PSF) derived for the light source within a scattering medium. The neural network (NN) technique is employed to find the deep learning models capable of de-blurring the image and estimating the depth of the absorption structure inside a turbid medium. The effectiveness of deep learning for de-blurring transillumination images and also depth estimation has been successfully demonstrated for depths ranging from 0.1 to 10.0 mm in previous studies. Although previous attempts have been made to enhance blurred images, the technique proposed in this study offers another solution.

The attention gate and the residual block were proposed to de-blur the image. Attention Unet and Residual Unet models were examined compared to the Unet model. Attention UNet and Residual Unet models yielded better performance than the Unet model. Attention Res-Unet then examined the performance compared to Attention Unet. Both the Attention U-Net and Attention Res-UNet models achieved correlation coefficients exceeding 88% even at a depth of 20.0 mm, affirming the applicability of deep learning models to de-blur transillumination images. Finally, Attention Res-Unet shows better performance in terms of the correlation between a de-blurred image and the original given image. The impact of image size on the result was also investigated. To ensure a better result with a deep-absorbing structure, the image size in a dimension should be more significant than three times the standard deviation of the light distribution on the medium's surface when calculating the turbid medium's deepest light point source distribution.

This study examined four different models, ResNet-50, VGG-16, VGG-19, and DenseNet-169, to estimate the depth of the absorbing structure. DenseNet-169 demonstrates superior performance among these models, achieving an accuracy rate greater than 65%. This research and the generated dataset prove valuable in accurately estimating the depth of the absorption structure from transillumination images. The evaluation of 1600 test images at 40 different depths ranging from 0.5 mm to 20.0 mm yielded a correlation coefficient of $R^2 = 0.9911$, which affirms the feasibility of the DenseNet169 model in estimating the depth of the absorbing structure.

It should be noted that this proposed technique requires a substantial amount of training data and computational power. However, these challenges can be addressed through the appropriate selection of PSFs and advances in computing capabilities. Consequently, this study confirms the feasibility of deep learning in clarifying blurred images and estimating the depth of absorption structures using PSF and CNN models based on training data.

The de-blurring and depth estimation results obtained for absorption structures at depths from 0.1 to 20.0 mm are highly satisfactory. These findings indicate the usefulness of the proposed methods for observing subcutaneous structures, identifying tumors and small animal parts, and determining depth distributions up to 20.0 mm. In particular, this technique is based solely on computer vision without complex exposure, ultrasound, or additional substances. Therefore, it presents a novel tool for the diagnosis of dermatological diseases, various tumor-associated diseases, vascular diseases, and tissue metabolism.

The results of this study contribute to the development of depth estimation and de-blurring methods using deep learning models. Furthermore, merging two targets identified by a single deep learning model will enable the definition of multiple depths within a single image. To expand the model's de-blurring and depth estimation capabilities, it is crucial to increase the number of samples and pairs of images for the training data and expand the depth range. These insights will facilitate the determination of actual dimensions and image depths within the absorption structure for the development of applications using 2D and 3D absorbing structure images in the near future.

Author Contributions: Conceptualization, T.N.T. and H.N.H.; methodology, T.N.T.; software, H.N.H.; validation, N.A.D.N.; analysis, N.A.D.N.; investigation, T.N.T. and H.N.H.; resources, T.N.T.; data curation, N.A.D.N. and H.N.H.; writing—original draft preparation, H.N.H.; writing—review and editing, N.A.D.N. and T.N.T.; visualization, H.N.H.; supervision, T.N.T.; project administration, T.N.T.; funding acquisition, T.N.T. All authors have read and agreed to the published version of the manuscript.

Funding: This research received no external funding.

Institutional Review Board Statement: Not applicable.

Informed Consent Statement: Not applicable.

Data Availability Statement: The data underlying the results presented in this paper are not publicly available but may be obtained from the authors upon reasonable request.

Acknowledgments: We acknowledge Ho Chi Minh City University of Technology (HCMUT), VNU-HCM for supporting this study.

Conflicts of Interest: The authors declare no conflict of interest.

References

1. Pan, C.T.; Francisco, M.D.; Yen, C.K.; Wang, S.Y.; Shiue, Y.L. Vein Pattern Locating Technology for Cannulation: A Review of the Low-Cost Vein Finder Prototypes Utilizing near Infrared (NIR) Light to Improve Peripheral Subcutaneous Vein Selection for Phlebotomy. *Sensors* **2019**, *19*, 3573. [CrossRef] [PubMed]
2. Francisco, M.D.; Chen, W.F.; Pan, C.T.; Lin, M.C.; Wen, Z.H.; Liao, C.F.; Shiue, Y.L. Competitive Real-Time Near Infrared (NIR) Vein Finder Imaging Device to Improve Peripheral Subcutaneous Vein Selection in Venipuncture for Clinical Laboratory Testing. *Micromachines* **2021**, *12*, 373. [CrossRef] [PubMed]
3. Frank, N.G.; David, W.; Samuel, D.; Martin, M.; Akwasi, A. Breast-i Is an Effective and Reliable Adjunct Screening Tool for Detecting Early Tumour Related Angiogenesis of Breast Cancers in Low Resource Sub-Saharan Countries. *Int. J. Breast Cancer* **2018**, *2018*, 2539056.
4. Shiryazdi, S.M.; Kargar, S.; Nasaj, H.T.; Neamatzadeh, H.; Ghasemi, N. The accuracy of Breastlight in detection of breast lesions. *Indian J. Cancer* **2015**, *52*, 513–516. [PubMed]
5. Tobisawa, N.; Namita, T.; Kato, Y.; Shimizu, K. Injection Assist System with Surface and Transillumination Images. In Proceedings of the 2011 5th International Conference on Bioinformatics and Biomedical Engineering, Wuhan, China, 13–15 May 2011; pp. 1–4.
6. Shimizu, K.; Tochio, K.; Kato, Y. Improvement of transcutaneous fluorescent images with a depth-dependent point-spread function. *Appl. Opt.* **2005**, *44*, 2154–2161. [CrossRef] [PubMed]
7. Tran, T.N.; Yamamoto, K.; Namita, T.; Kato, Y.; Shimizu, K. Three-dimensional transillumination image reconstruction for small animal with new scattering suppression technique. *Biomed. Opt. Express* **2014**, *5*, 1321–1335. [PubMed]
8. Goh, C.M.; Subramaniam, R.; Saad, N.M.; Ali, S.A.; Meriaudeau, F. Subcutaneous veins depth measurement using diffuse reflectance image. *Opt. Express* **2017**, *25*, 25741–25759. [PubMed]
9. Nguyen N.A.D.; Van, T.N.P.; Yamamoto, K.; Nguyen M.Q.; Tran, A.T.; Namita, T.; Shimizu K.; Tran T.N. Depth estimation of the absorbing structure in a slab turbid medium using point spread function. *VNUHCM J. Eng. Technol.* **2020**, *3*, SI10–SI21.

10. Van, T.N.P.; Tran, T.N.; Inujima, H.; Shimizu, K. Three-dimensional imaging through turbid media using deep learning: NIR transillumination imaging of animal bodies. *Biomed. Opt. Express* **2021**, *12*, 2873–2887. [CrossRef] [PubMed]

11. Shourav, M.K.; Choi, J.; Kim, J.K. Visualization of superficial vein dynamics in dorsal hand by near-infrared imaging in response to elevated local temperature. *J. Biomed. Opt.* **2021**, *26*, 026001. [CrossRef] [PubMed]

12. Shimizu, K.; Xian, S.; Guo, J. Reconstructing a Deblurred 3D Structure in a Turbid Medium from a Single Blurred 2D Image—For Near-Infrared Transillumination Imaging of a Human Body. *Sensors* **2022**, *22*, 5747. [CrossRef] [PubMed]

13. Mak, H.W.L.; Han, R.; Yin, H.H.F. Application of Variational AutoEncoder (VAE) Model and Image Processing Approaches in Game Design. *Sensors* **2023**, *23*, 3457. [CrossRef] [PubMed]

14. Qiao, Q. Image Processing Technology Based on Machine Learning. In *IEEE Consumer Electronics Magazine*; IEEE: Piscataway, NJ, USA, 2022. [CrossRef]

15. Patil, A. Image Recognition using Machine Learning. 2021. Available online: https://ssrn.com/abstract=3835625 (accessed on 30 August 2023).

16. Oktay, O.; Jo Schlemper, J.; Folgoc, L.L.; Lee, M.; Heinrich, M.; Misawa, K.; Kensaku Mori, K.; McDonagh, S.; Hammerla, N.Y.; Kainz B.; et al. Attention U-Net: Learning Where to Look for the Pancreas. *arXiv* **2018**, arXiv:1804.03999.

17. Maji, D.; Sigedar, P.; Singh, M. Attention Res-UNet with Guided Decoder for semantic segmentation of brain tumors. *Biomed. Signal Process. Control* **2022**, *71*, 103077. [CrossRef]

18. He, K.M.; Zhang, X.Y.; Ren, S.Q.; Sun, J. Deep residual learning for image recognition. In Proceedings of the IEEE Conference on Computer Vision and Pattern Recognition, Las Vegas, NV, USA 27–30 June 2016; pp. 770–778.

19. Simonyan, K.; Zisserman, A. Very deep convolutional networks for large-scale image recognition. *arXiv* **2014**, arXiv:1409.1556.

20. Huang, G.; Liu, Z.; Van DerMaaten, L.; Weinberger, K.Q. Densely connected convolutional networks. In Proceedings of the IEEE Conference on Computer Vision and Pattern Recognition, Honolulu, HI, USA, 21–26 July 2017; pp. 4700–4708.

21. Milletari, F.; Navab, N.; Ahmadi, S.A. V-Net: Fully convolutional neural networks for volumetric medical image segmentation. In Proceedings of the 2016 Fourth International Conference on 3D Vision (3DV), Stanford, CA, USA, 25–28 October 2016; pp. 565–571. [CrossRef]

22. Rezatofighi, H.; Tsoi, N.; Gwak, J.; Sadeghian, A.; Reid, I.; Savarese, S. Generalized intersection over union: A metric and a loss for bounding box regression. In Proceedings of the IEEE/CVF Conference on Computer Vision and Pattern Recognition, Long Beach, CA, USA, 15–20 June 2019; pp. 658–666.

23. Hossin, M.; Sulaiman, M.N. A review on evaluation metrics for data classification evaluations. *Int. J. Data Min. Knowl. Manag. Process* **2015**, *5*, 1–11.

MDPI
St. Alban-Anlage 66
4052 Basel
Switzerland
www.mdpi.com

Applied Sciences Editorial Office
E-mail: applsci@mdpi.com
www.mdpi.com/journal/applsci